AF753970

SOLAR FLARE MAGNETOHYDRODYNAMICS

THE FLUID MECHANICS OF ASTROPHYSICS AND GEOPHYSICS

A series of books edited by PAUL H. ROBERTS, *University of Newcastle-upon-Tyne, U.K.*

International Standard Serial Number (ISSN): 0260-4353

Volume 1 SOLAR FLARE MAGNETOHYDRODYNAMICS
Edited by E. R. Priest
University of St Andrews, U.K.

This book is part of a series. The publisher will accept continuation orders which may be cancelled at any time and which provide for automatic billing and shipping of each title in the series upon publication. Please write for details.

SOLAR FLARE MAGNETOHYDRODYNAMICS

Edited by

E. R. PRIEST

University of St Andrews

GORDON AND BREACH SCIENCE PUBLISHERS

New York London Paris

Copyright © 1981 by Gordon and Breach, Science Publishers, Inc.

Gordon and Breach, Science Publishers, Inc.
One Park Avenue
New York NY10016

Gordon and Breach Science Publishers Ltd.
42 William IV Street
London WC2N 4DE

Gordon & Breach
7-9 rue Emile Dubois
Paris 75014

Library of Congress Cataloging in Publication Data
Main entry under title:

Solar flare magnetohydrodynamics.

(The Fluid mechanics of astrophysics and geophysics; v. 1)
Bibliography: p.
Includes indexes.
1. Solar flares. 2. Magnetohydrodynamics. I. Priest, Eric Ronald, 1943- II. Series: Fluid mechanics of astrophysics and geophysics; v. 1.
QB526.F6S59 523.7'5 80-67417
ISBN 0-677-05530-7

Library of Congress catalog card number 80-67417. ISBN 0 677 05530 7. ISSN: 0260-4353.
All rights reserved. No part of this book may be reproduced or utilized in any form or by any means, electronic or mechanical, including photocopying, recording, or by any information or retrieval system, without prior permission in writing from the publishers. Printed in Great Britain.

Contents

Introduction to the Series . . . ix

Preface . . . xi

1. INTRODUCTION – E. R. PRIEST

1.1 Basic Description of a Flare . . . 2
1.2 Recent Advances in Understanding . . . 7
1.3 An Overview of the Flare Phenomenon . . . 12
1.4 Earlier Theories for the Solar Flare . . . 14
1.5 The Magnetohydrodynamic Equations . . . 25
1.6 Summary of the Book . . . 29
References . . . 45

2. FLARE OBSERVATIONS – Z. ŠVESTKA

2.1 Introduction . . . 47
2.2 Flare Observations in the Hα Line . . . 55
2.3 Chromospheric Flare Spectra . . . 64
2.4 Thermal Flares in Soft X-Rays . . . 73
2.5 Impulsive Hard X-Ray and Microwave Bursts . . . 81
2.6 Energy Transfer through the Transition Layer . . . 88
2.7 Metric Radio Bursts . . . 95
2.8 Flare Ejecta . . . 102
2.9 Production and Effects of High-Energy Particles in Flares . . . 112
References . . . 127

3. CURRENT SHEETS – E. R. PRIEST

3.1 Introduction . . . 139
3.2 Means of Formation . . . 144
3.3 Behaviour of Current Sheets . . . 172
3.4 Role of Current Sheets in Flare Models . . . 193
3.5 The Emerging (or Evolving) Flux Model . . . 196
3.6 Conclusion . . . 212
References . . . 213

4. SIMPLE-LOOP FLARES: MAGNETIC INSTABILITIES – G. VAN HOVEN

4.1 Introduction . . . 217
4.2 Physical Conditions . . . 218
4.3 Equilibrium Magnetic Fields . . . 223
4.4 Infinite–Conductivity Stability . . . 229
4.5 Resistive Magnetic Instability . . . 248
4.6 Attempts at a Loop-Flare Model . . . 271
References . . . 273

5. SIMPLE-LOOP FLARES: THERMAL EVOLUTION – I. J. D. CRAIG

5.1 Introduction . . . 277
5.2 Observational Background . . . 278
5.3 The Hydrodynamic Equations . . . 284
5.4 Quasi-static Models of the Solar Atmosphere . . . 292
5.5 Analytic Conduction Models . . . 305
5.6 Hydrodynamic Models of a Flaring Loop . . . 317
5.7 Summary: Theory and Observation Revisited . . . 332
References . . . 334

6. TWO-RIBBON FLARES: MAGNETOSTATIC EQUILIBRIA – J. BIRN AND K. SCHINDLER

6.1 Introduction . . . 337
6.2 Basic Equations and Assumptions of the Quasi-Static Theory . . . 338
6.3 Existence of Equilibrium Solutions in Two Dimensions . . . 342
6.4 Special Solutions . . . 346
6.5 Stability . . . 362
6.6 Slow Passive Evolution . . . 372
6.7 Discussion . . . 376
References . . . 377

7. TWO-RIBBON FLARES: (POST)-FLARE LOOPS – G. W. PNEUMAN

7.1 Introduction . . . 379
7.2 Observations . . . 380
7.3 Loop Prominences as a Reconnection Process . . . 392
7.4 Magnetic Field Dynamics . . . 405
7.5 Mass Supply . . . 413
7.6 Heating and Cooling of the Loops . . . 418
7.7 Summary and Conclusions . . . 424
References . . . 426

8. PARTICLE ACCELERATION IN SOLAR FLARES – J. HEYVAERTS

8.1 Observational Evidence for Accelerated Particles . . . 429
8.2 Energetic Particles at 1 au . . . 431
8.3 Hard X-ray Emission in Flares . . . 433
8.4 Implications of X-Ray Observations: How Many Electrons? . . . 441
8.5 Microwave Impulsive Bursts . . . 442
8.6 Gamma-Ray Emission . . . 446
8.7 Elemental and Isotopic Abundances of Solar Cosmic Rays . . . 448
8.8 Fast Particle Impact on the Chromosphere and Photosphere from EUV and Hα Bursts . . . 454
8.9 Directed Flow of Fast Particles to the Chromosphere or Hot Fireball Observations . . . 455
8.10 Directed Flow of Fast Particles to the Chromosphere or Hot Fireball: Theory . . . 457
8.11 First-Phase Acceleration Process: An Overview . . . 466
8.12 Runaways Through Turbulence . . . 470
8.13 Current-Driven Micro-Instabilities: Weak Field ($B \approx 0$) . . . 473
8.14 Current-Driven Micro-Instabilities: Magnetized Medium . . . 480

8.15 Production of a Langmuir Wave Spectrum Suitable for Acceleration . 482
8.16 Strong-Turbulence Effects 485
8.17 Proposed Microturbulent Acceleration Models 493
8.18 Double-Layers 496
8.19 Approaching Microturbulent Conditions: a Difficulty 506
8.20 Tearing Modes and Fast Particles 509
8.21 Second-Phase Acceleration: an Overview 517
8.22 Second-Phase Fermi Acceleration is Compatible with Observational Data 518
8.23 Fermi Mechanism and Resonant MHD Processes 523
8.24 First-Order Fermi Mechanism near Shocks 532
8.25 Nuclei Preheating. 537
8.26 Acceleration in and near Shock Fronts 544
References 551

Index 556

Introduction to the Series

Anyone who surveys the shelves of an average science library will soon convince himself that fluid mechanics has taken more than its share of the rapid growth of scientific publication following World War II. And within fluid mechanics, publications of astrophysical or geophysical interest have enjoyed increasing prominence. There are a number of obvious reasons for this, ranging from the academic to the practical. Prominent is the long term aim of understanding and partially controlling the Earth's atmosphere, with implications that range from the environmental one of minimizing the spread of harmful industrial wastes to the humanitarian one of maximizing agricultural yields. No less pressing in these days of developing ocean resources is the need to understand sea movements, how they might diffuse radioactive wastes placed on the ocean floor, how they can be harnessed to help quench man's thirst for energy, how they affect oceanic life and the harvest man takes from the sea. Seismology and vulcanology are of scarcely less practical significance to many populous areas, and they owe their existence to crustal stresses set up by the motion of an underlying mantle. Beneath the mantle lies the Earth's core in whose internal motions lies the cause of the Earth's magnetism.

Like geomagnetism, most topics in astrophysical fluid mechanics are hard to model in the laboratory and are in any case principally of academic interest. Exceptions are the interaction of the solar wind with the Earth's external magnetic field, magnetic storms and their ionospheric consequences on radio reception. The number of fluid mechanical topics of academic interest in astrophysics are legion, and include the study of stellar winds in general, the motion of gas clouds and H II regions, the pulsation of stars, their rotation, their magnetism, and their formation by gravitational collapse, the propagation of density waves in galaxies, the maintenance of galactic magnetic fields, pulsar magnetospheres, supernovae explosions, the motion of planetary atmospheres, and so forth. Some of these subjects have an old and respected history but many are young. Sometimes the impetus has come from discoveries made during space exploration or from orbiting observatories, but often it can be understood as part of a natural progression; for example, as the classic problem of the structure and evolution of a static spherically symmetric star becomes increasingly well solved, what more natural than to examine the 'exceptional' departures: the pulsation of a star, its distortion by rotation, tides or magnetism?

A thread common to many of these topics is the fluid mechanics of rotating, stratified (and perhaps electrically-conducting) fluids. Not surprisingly, concepts useful in large-scale atmospheric dynamics and oceanography find parallels in solar, planetary and cosmical physics. A scientist in one of these disciplines may well find himself reading, comprehending, and perhaps being helped by a paper in another. It was with this idea in mind that the publishers encouraged me in 1977 to broaden the scope of the Journal *Geophysical Fluid Dynamics* to include papers of astrophysical interest. The response has been so favourable that they have urged me to start a series of books to complement the Journal. This is the first in that series, entitled *The Fluid Mechanics of Astrophysics and Geophysics*. Further volumes are planned at irregular intervals.

The present book, which deals with a topic of great significance in solar physics, should appear at a propitious moment, soon after results from the Solar Maximum Mission begin to be known. Although it may be reasonably expected that these will overturn many long cherished ideas, a clear statement of the current 'state of the art' such as this book provides will, it is believed, serve as a secure foundation for future developments. Scientists in other specialities will also, it is hoped, be stimulated by the unusual physical phenomena elucidated herein.

P. H. ROBERTS

Preface

This book is aimed at first-year postgraduate students, with each chapter presenting a simple description of previous work and also some more recent unpublished calculations. The chapters are meant to be self-contained, so that, for instance, each includes its summary of relevant observations and its list of references.

Tremendous advances in the *observations* of solar flares have taken place over the past ten years, both from high-resolution ground-based instruments and space telescopes such as those on board the Skylab and Solar Maximum Mission satellites. This has also been an exciting time for theorists: enormous strides have been taken towards understanding and modelling the basic flare process and so it seems a good time to summarise the present position. At last one feels the flare is basically being understood, although there are still many detailed aspects to explain. The fundamental energy conversion process has been described and several plausible theories concerning the basic instability responsible for the two major types of flare have been proposed.

A flare can be exceedingly complex and incorporate a multitude of secondary phenomena. Its study demands the cooperation of many types of expertise, including observation, magnetohydrodynamics, plasma physics, atomic physics and radiation theory. We are here concerned mainly with the *magnetohydrodynamics* of the flare process, but this is, to a certain extent, complemented in the book by an account of flare observations from Z. Švestka (Chapter 2) and by a summary of particle acceleration mechanisms from J. Heyvaerts (Chapter 8). The first chapter provides an introduction to the flare problem, together with a summary (§1.6.2) of the contents of each chapter, while Chapter 3 describes the release of magnetic energy in current sheets. Flares may be divided into two types — *compact* (or *simple-loop*) *flares* and *two-ribbon flares*. The magnetic and thermal aspects of the former are described by G. Van Hoven (Chapter 4) and by I. Craig (Chapter 5), respectively. Then the two-ribbon event is analysed by J. Birn and K. Schindler, who discuss the cause of the initial magnetic eruption (Chapter 6), and by G. W. Pneuman, who models "post"-flare loops (Chapter 7).

Acknowledgements are to be found at the end of some chapters and I would in addition personally like to thank, for their cheerful encouragement, my colleagues at St. Andrews (B. Roberts, A. Hood, I. Rae, P. Cargill, M. Wragg, T. Tur, A. Milne). Also, I am very grateful to the authors of each chapter for their enthusiasm and efficiency and for agreeing to use a standard set of equations in MKS units. I myself grew up with CGS units and then suffered the trauma of changing to MKS in response to a plea by the Royal Society some years ago!

E. R. PRIEST
University of St Andrews

CHAPTER 1

Introduction

E. R. PRIEST

Applied Mathematics Department, The University, St. Andrews, Scotland

A great solar flare is a truly beautiful event that captivates the onlooker and fills him with a sense of awe. He may witness, over the space of an hour, the release of an enormous quantity of energy — the largest in the solar system — and its transformation into an incredibly complex variety of forms. Afterwards, he may wonder: how did it all begin? What gave rise to this violent instability? How was the magnetic energy converted into heat, kinetic energy and particle energy? Such questions have for many years tantalized the theorist, but recently great strides have been taken towards answering them.

The purpose of this introduction is to lay a foundation for the chapters that follow. We shall briefly outline the two main types of solar flare, namely the *simple-loop* (or *compact*) *flare* and the *two-ribbon flare*, together with the flare theories that have been proposed. Then there follows a summary of the equations of magnetohydrodynamics, which will form the basis of our study and which describe much of the interaction between the Sun's plasma atmosphere and its magnetic field. Chapter 2 gives a more detailed account of solar flare observations, especially those that have come recently from space instruments and high-resolution ground-based telescopes. Chapter 3 discusses the behaviour of current sheets, which may be important in the instability and in the energy conversion process. Chapters 4 and 5 are concerned with a simple-loop flare, including the magnetic instability that may give rise to it and the thermal response of loop plasma to a heat input. This is followed in Chapters 6 and 7 by an analysis of two-ribbon flare theory, both the preflare evolution through a series of equilibria and the production of so-called "post"-flare loops. Finally, a review of particle acceleration mechanisms is presented in Chapter 8.

Further background material may be found in several complementary texts. For example, the excellent book by Švestka (1976) gives a comprehensive description of flare observations. Also, the magnetohydrodynamic behaviour of general cosmical magnetic fields is analysed in the magnificent tome by Parker (1979), while the theory of solar magnetic fields in particular is outlined by Priest (1981).

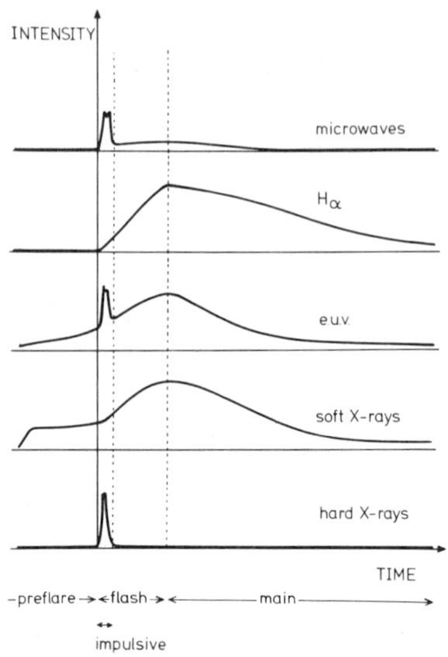

FIGURE 1.1 A schematic profile of the flare intensity in several wavelengths. There is great variation in the duration and complexity of the various phases, but for a large event the preflare phase lasts typically 10 minutes, the impulse phase 1 minute, the flash phase 5 minutes and the main phase 1 hour. An example of the intensity at different X-ray energies is given in Figure 2.8.

1.1 BASIC DESCRIPTION OF A FLARE

A solar flare may be naively defined as a rapid brightening in Hα, but it can simultaneously have manifestations right across the electromagnetic spectrum (§2.1.2) and may eject high energy particles and blobs of plasma out into the solar wind. The Hα flare (§2.2 and §2.3) is formed in the chromosphere and has two basic stages. During the *flash phase*, which lasts typically 3 minutes (but sometimes an hour), the intensity and area of the emission rapidly increase in value. Then, in the *main phase*, the intensity slowly declines over about an hour or occasionally as much as a day.

However, above the cool Hα flare there lies a high-temperature coronal region, which may be heated during the flare to tens of millions of degrees (§2.4); it exhibits variations co-incident with the flash and main phases but also shows two more distinct phases. As seen in Figure 1.1, the soft X-ray emission (<10 keV) perhaps possesses a *pre-flare phase* for minutes (or possibly longer) before flare onset, because of an enhanced thermal emission from the coronal plasma. Also, for 10–100 seconds at the start of the flare, an *impulsive phase* (§2.5) is sometimes present, as indicated by the appearance of a microwave burst (§8.5) and a hard X-ray burst (>30 keV) caused by highly accelerated electrons (§2.5, §8.3). The hard X-rays exhibit a complex spikey structure, with smallest real time-scales of about 2 seconds in moderate events and 10 seconds in large ones; it is these which give the time-scale for electron acceleration during the impulsive phase. After the impulsive phase some of the largest events show a distinct second hard X-ray component due to a second phase of particle acceleration (§§8.21–8.26). For other events, the impulsive phase may take place after the Hα intensity has started increasing or it may be absent. In the latter case, where there is little particle acceleration, the events are known as *thermal flares*; they tend to occur in less complex regions and have a slow rise to flare maximum.

Flares have been classified in a multitude of different ways (§2.1.3), but here we wish to be rather simple-minded and suggest just two main types of flare (§2.1.4 and §2.4.2), which appear to require quite different physical mechanisms.

i) *Simple-Loop Flare* (or *compact* flare). Most flares and subflares are of this type. It is a small flare, in which essentially a single magnetic loop or flux tube brightens in X-rays and remains apparently unchanged in shape and position throughout the event (Figure 1.2). The loop may have a structure consisting of several (or a continuum of) loops and may cause a simple brightening in Hα at the feet of the loop. The flare can occur within a large-scale unipolar region or near a simple sunspot, where little excess magnetic energy is stored. (Occasionally, it may be accompanied by a surge as a stream of plasma is squirted upwards.) Simple-flare loops vary considerably in size and, at the lower end of the spectrum, one may even regard *X-ray bright points* as being part of the family.

Theoretical models for the response of plasma to a sudden extra heat source in a rigid magnetic loop are described in Chapter 5. However, it is not known at present whether the magnetic field of such a loop-flare changes so as to allow a release of magnetic energy or whether it does not. In the former case, a magnetic instability could be responsible for the flare; some possibilities are discussed in §4.4 and §4.5. In the latter case, the energy source cannot be magnetic and some non-magnetic process, such as thermal nonequilibrium (§5.4.6), must be the cause.

ii) *Two-Ribbon Flare*. All major events are of this type (Figures 1.3 and 1.4). It is much larger and more dramatic than a compact flare and takes place near a solar prominence, which shows up in Hα pictures as a dark ribbon called a *filament*. When the filament is located in part of the quiet Sun, such as a remnant active region, the flare tends to be slow, long-lived and not very energetic, presumably because the magnetic field is relatively weak near such a *quiescent filament*. However, *active-region filaments* are located near the intense and sometimes complex fields of sunspots and they are associated with the most violent and energetic flares (§2.2.4).

During the flash phase of a two-ribbon flare, two ribbons of Hα emission form, one on each side of the filament (or filament channel) and, throughout

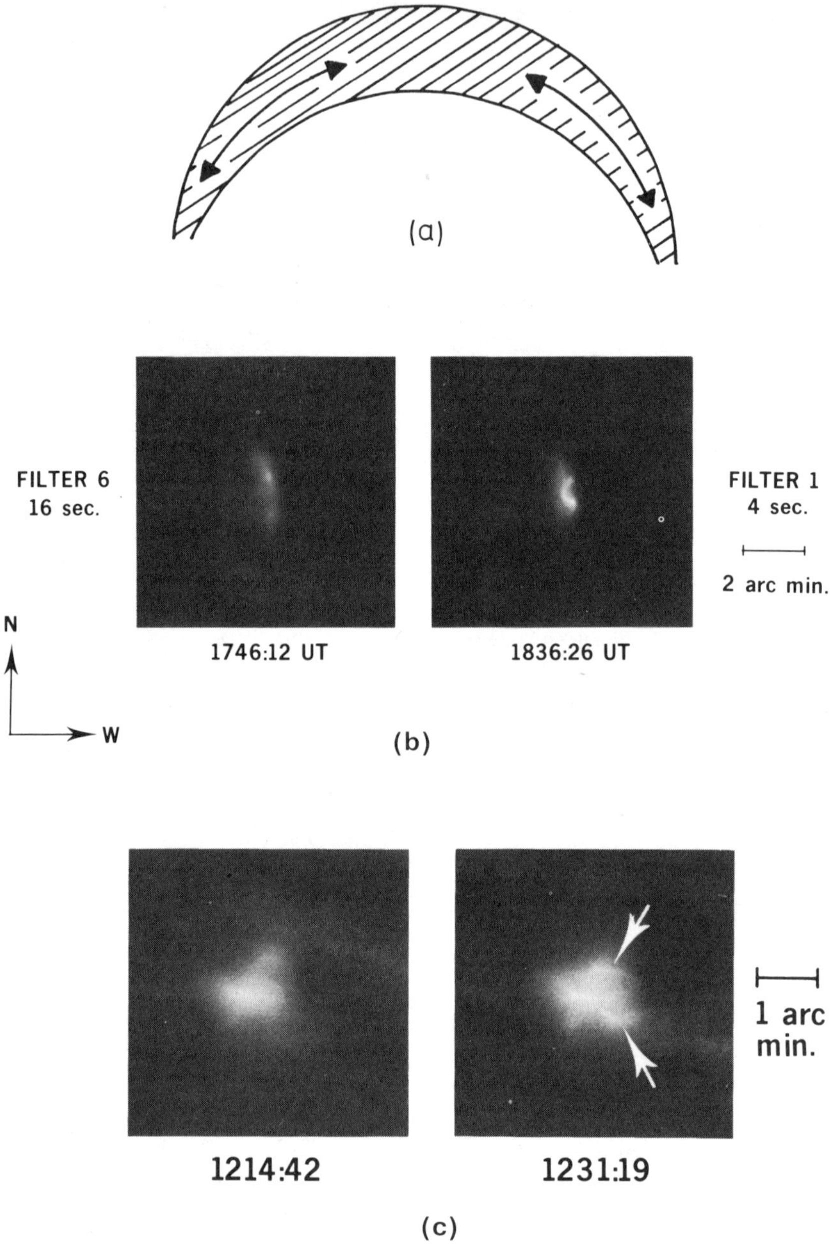

FIGURE 1.2 The simple-loop flare. (a) Schematically, a single loop brightens but maintains its shape and position, while motions are present along the loop. (b) and (c) Examples from the ASE soft X-ray telescope on Skylab on August 29 and June 22, respectively. (Courtesy D. Webb, American Science and Engineering.)

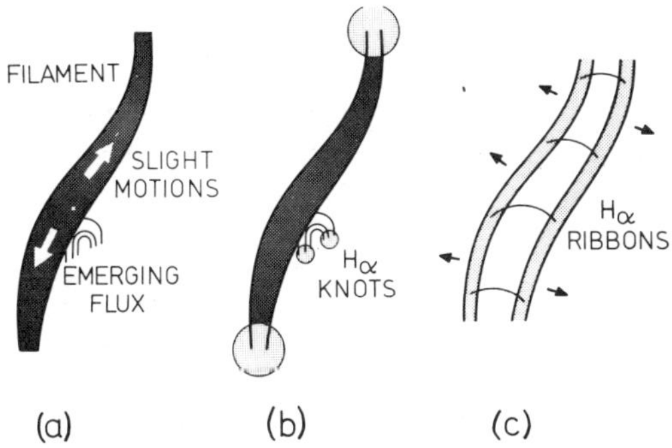

FIGURE 1.3 A schematized two-ribbon flare as observed on the disc. (a) Preflare phase: new flux is sometimes observed to emerge close to the filament, which slowly rises and along which slight motions are observed. (b) Impulsive phase: intense knots of Hα emission are formed, both near the emerging flux and the ends of the filament. (c) Flash and main phase: the filament erupts rapidly, while Hα ribbons develop on both sides of the filament channel and move away from it; loops that resemble a potential field structure join the two ribbons.

the main phase, the ribbons move apart at 2–10 km s^{-1}. Frequently, they are seen to be connected by a rising arcade of so-called "*post*"-*flare loops* (Figure 1.3). Occasionally, the filament remains intact, though slightly disturbed, but usually it rises and disappears completely. Such an eruption of the filament begins slowly in the preflare phase, typically 10 minutes (but up to an hour) before flare onset, and continues at the flash phase with a much more rapid acceleration than before (§2.8.1). The magnetic fields near filaments vary greatly in strength, depending on whether the filament snakes its way right through the centre of an active region or bounds the edge of it or is located far from sunspots. In the latter case, the eruption of a quiescent filament from a weak magnetic field region may only give rise to X-ray emission with no Hα brightening, but it is probable that as far as the magnetic instability is concerned it still belongs to the same family.

The disturbance of the solar atmosphere by a large two-ribbon flare may be manifested in many ways. Except when the flare occurs well within a closed field region, the hard X-ray emission (§2.5, §8.3) is accompanied by type III radio bursts (§2.7.3), which indicate the presence in the corona of streams of *high energy electrons* (§8.1), accelerated up to about a third the speed of light. Type II radio emissions (§2.7.2) and a Moreton wave both suggest the outwards propagation from the flare site at more than 1000 km/s of a fast *magnetohydrodynamic shock wave*. A *coronal transient* (§2.8.2) indicates the propagation out into the solar wind of plasma contained in a huge magnetic loop system; such a *mass ejection* moves faster than the ambient solar wind and a shock wave is formed at its leading edge. Also, long-lasting type IV bursts (§2.7.1) show the presence of an *energetic plasma cloud* above the flare site. On the limb one may observe the "post"-flare loops as a *loop-prominence* system,

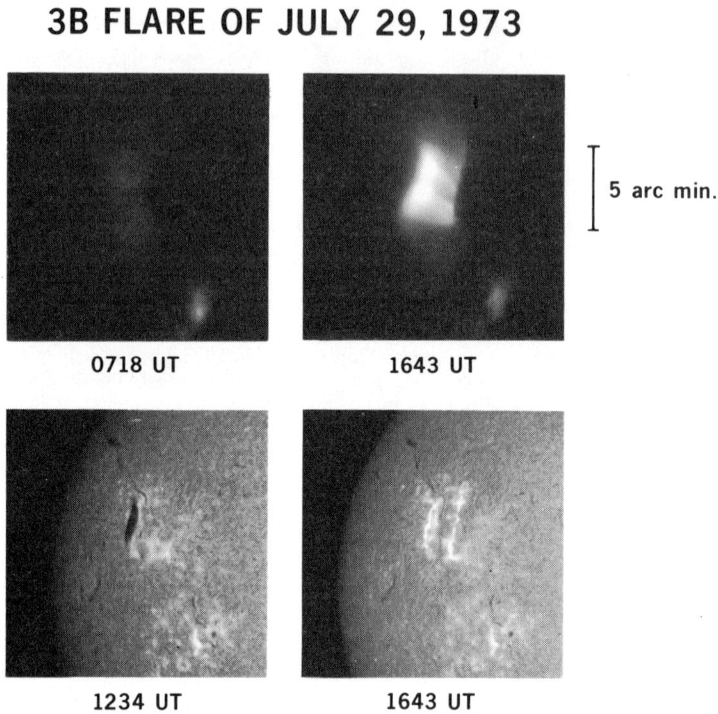

FIGURE 1.4 Hα and X-ray pictures before and after the start of a two-ribbon flare. Beforehand there is an Hα filament and a faint X-ray cloud. Afterwards, one sees the Hα flare ribbons joined by an arcade of X-ray loops, which have a temperature of 10^7 K and are brightest at their tops. As the flare proceeds, the ribbons move apart and the arcade of loops rises. (Courtesy of A. Krieger, American Science and Engineering.)

which roughly outlines the loops of a potential magnetic field; down their legs drains an immense amount of material — far exceeding, in fact, the amount in the surrounding corona!

The energy released in a flare (§2.1.5) varies from, say, 10^{22} Joules in a subflare to 3×10^{25} Joules (3×10^{32} ergs) in the largest of events. The division between various types of energy in the latter case has been estimated roughly as follows:

Electromagnetic radiation up to soft X-rays	10^{25} J
Interplanetary blast wave	10^{25} J
Fast electrons (hard X-rays)	5×10^{24} J
Subrelativistic nuclei	2×10^{24} J
Relativistic nuclei	3×10^{24} J
Total energy output	3×10^{25} J ($= 3 \times 10^{32}$ ergs)

It should be noted that the hard X-ray contribution is uncertain but may possibly be as much as 2×10^{25} J (§8.4.2).

There are several properties of the flare which are of particular importance when it comes to constructing a theoretical model. For example, active regions which are complex and rapidly evolving are the most likely flare producers; by contrast, simple bipolar regions show little non-thermal activity (§2.2.1). Also there are no large-scale changes in the photosphere beneath a flare; variations in the magnetic flux through sunspots during a flare are usually no different from normal evolutionary changes. In any case, one would not expect the changes to be very great, since the flare energy is only a small fraction of the energy (10^{27} J) of a large sunspot group. Other interesting properties of some flares are those of "homology" and "sympathy". In the first case, a flare may repeatedly occur in the same place and with very similar characteristics. In the second, a flare may be triggered by the occurrence of another flare, even though the two are widely separated.

1.2 RECENT ADVANCES IN UNDERSTANDING

Our knowledge of the solar flare has increased greatly over the past decade through two complementary sources, namely high-resolution (up to 1 arcsec) ground-based observations of the cool Hα part of the flare and space observations (up to 2–5 arcsec) of the high temperature EUV and X-ray emitting part. This has led to a much more detailed and broader description of the flare phenomenon. The main results are:

i) the flare appears to originate in the corona, with the Hα emission being just a secondary response: the magnetic energy is probably released in the corona and then carried down to the chromosphere by fast particles and thermal conduction, together with mass motions and radiation (§2.6.3, §5.5, §5.6);

ii) the transition-region density and coronal density increase dramatically during the flare (§2.4.1) but reach their maximum values after the temperature;

iii) flares often occur after the emergence of new flux (§2.2.4);

iv) the orientations of fibrils sometimes suggest a reconnection during the flare;

v) the term "post"-flare, applied to the arcade of loops that joins the two ribbons, is a misnomer, since the energy release *continues* during this main phase and indeed can last much longer than the optical emission (§7.1);

vi) the temperature minimum may be depressed in location (by two scale-heights) during a flare and the value of the temperature there may be increased (§2.3.5 and §2.6.3); it may be caused by a transfer of energy from the corona in the form of EUV photons or 10–20 MeV protons, or, alternatively, the energy may be released locally (Sturrock, 1980);

vii) the presence of nuclear reactions (when protons and α-particles are accelerated to more than about 30 MeV per nucleon) has been confirmed by detection of the 2.23 MeV gamma-ray line (§2.9.4, §8.6).

1.2.1 Ground-based observations

Very high quality Hα photographs of flares have been produced at the Big Bear Solar Observatory, and these have enabled Zirin (1974) to give a detailed description of the development of the Hα flare. He stresses that the majority of flares arise after the emergence of new magnetic flux from below the photosphere, as sometimes evidenced by the appearance of small satellite sunspots close to large sunspots of the opposite polarity (Rust, 1968; Martres *et al.*, 1968). Flares are especially likely when the satellite is of "following" polarity and appears just in front of the main spot with "preceding" polarity. The most flare-productive active regions have steep field gradients (i.e., high electric currents) and also polarity inversions which result from the arrival of these satellites.

During the course of the flare, changes in fibril direction may sometimes be seen which suggest a reconnection of the magnetic field. An example which seems to indicate reconnection and also emerging flux is shown in Figure 1.5.

Zirin also finds that great flares may begin at several points and then expand over a large area, giving the impression that the initial flare energy does not simply spread out but stimulates an extra energy release over a large region. Sometimes the initial brightening occurs simultaneously at points situated at opposite ends of a line of force. There is a tendency for slowly-rising flares to be thermal (no impulse phase) and fast-rising ones to produce non-thermal particles; also *knots* or *kernels* of Hα emission appear (often near satellite spots) during the impulsive phase simultaneously with the hard X-ray and microwave bursts.

Magnetic field measurements (§2.2.1 and §2.2.4) have been important in aiding our understanding of the flare process and in supporting the inferences from Hα structure. Flares tend to occur near the "$H_{\parallel} = 0$ line" or "polarity inversion line", along which the longitudinal magnetic field component changes sign and the field is largely horizontal. At Sacramento Peak Observatory, Rust (1972) demonstrated that, except possibly in the largest flares, flare-related photospheric magnetic field changes are confined to satellite spot regions; they show an increase in strength throughout the preflare phase (half-an-hour) followed by a decrease during the flare itself. Moreover, it is those satellites which are evolving that tend to give rise to

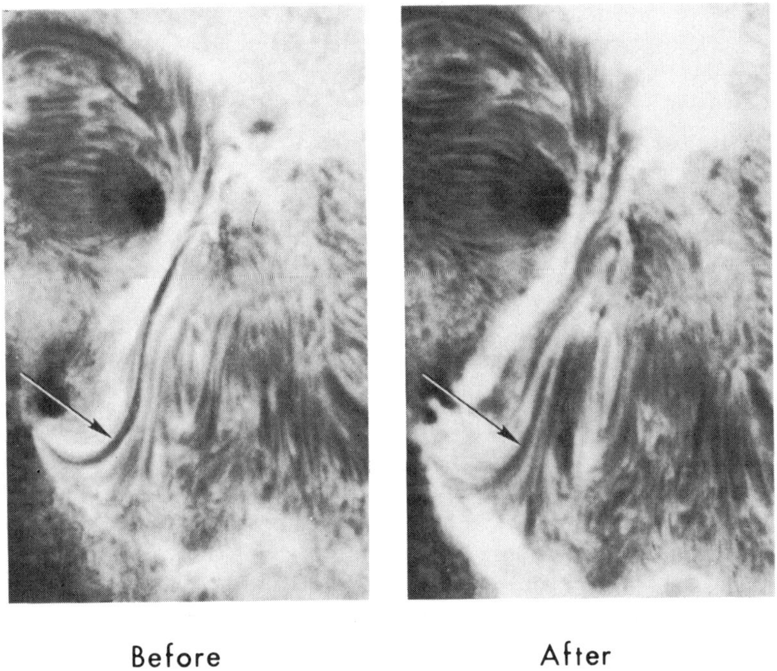

FIGURE 1.5 An example of an Hα structure which is suggestive of reconnection during a flare. Arrows point to the lower end of the filament and show how it changed from being part of the filament *before* the flare to being part of an arch filament system (which indicates emerging flux) *after* the flare. (Courtesy of D. Rust, American Science and Engineering.)

flares; but there is no obvious feature of the field morphology which indicates whether the resulting flare will be small or large. When observing a two-ribbon flare, one's attention may tend to be focussed on the filament and the ribbons, but it is the local emergence of new flux which seems to initiate the activation or eruption of the filament (Rust, 1976) and the brightest flare knots appear near the points of emerging flux (Rust and Bridges, 1975).

August of 1972 presented the opportunity to study several great flares which were produced by one unusually complex sunspot group. The group possessed inverted polarity and extremely high field gradients with a continual invasion of one polarity into the other and a distortion of the polarity inversion line. Zirin and Tanaka (1973) have described its evolution. They see, in one impulsive flare, small (1″), short-duration (5 s) emission knots at the foot-points of magnetic flux loops; the knots occur at the same time as a hard X-ray burst and may be caused by the dumping of high energy (>50 keV) electrons in the low chromosphere. In another two-ribbon flare most of the Hα emission is produced in small kernels and there is a substantial velocity shear (6 km/s) across the polarity inversion line. Rust and Bar (1973) discuss the loop prominences which join the two ribbons during the main phase of the same flare; the loops are situated progressively higher in the atmosphere and, as time progresses, they become aligned more normal to the filament channel (indicative of a potential field). Rust and Bar suggest that particle acceleration continues through the main phase and that the fast particles bombard the chromosphere at the bright outer edges of the flare ribbons, whereas the dense, loop prominence material falls on to the inner edges. Tanaka and Nakagawa (1973) demonstrate further that the observed proper motions of the sunspots before the flare could store enough energy in a force-free configuration to supply the flare.

Photospheric velocity measurements (§2.2.3) before and during flares have not been well-studied in the past. For the 7th August (1972) flare, Rust (1973) observes upward motions of 1–2 km/s in the half-hour before the flare, and downward motions during and after the flare of as much as 5 km/s. The upward motions are said by Harvey (1974) to be a fairly general preflare property; maybe they are linked with emerging flux. Harvey also stresses that flares occur when there are shearing motions, in particular when the velocity and magnetic inversion lines cross (Martres *et al.*, 1971), whereas Levine and Nakagawa (1974) find flare activity to be located where the rate of strain is a local maximum and the vertical magnetic field or its gradient is zero.

1.2.2 Space observations from Skylab

Perhaps the most striking of the recent advances in solar flare observations has concerned the structure and time-evolution of the *high energy flare* plasma (§§2.4–2.6). The EUV and X-ray emitting parts have been studied in great detail with instruments of the Apollo Telescope Mount on board Skylab, as summarised in Sturrock (1980) and §§2.4–2.6, where detailed references may be found. Most of these observations, however, have been of small events. They indicate the appearance of a high temperature (up to 3×10^7 K) *kernel* or *flare core* during the impulsive phase and the start of the flash phase. It may be situated at some point (often the top) of a loop structure, which presumably outlines a magnetic flux tube in the upper chromosphere or lower corona; also, the loop may brighten gradually 10 minutes before the flare. The loop is typically 4–13 Mm high (1 Mm = 1000 km) and the kernel has a plasma density of at least 10^{17} m^{-3}, or possibly as much as 3×10^{18} m^{-3}; its dimension may be as small as 4 Mm or even 1.5 Mm and its temperature as large as 2×10^7 K. As the flare progresses, the EUV and X-ray emission may spread along the loop through the kernel and jump to other nearby loops; in one example, an arcade of loops brightened sequentially as a disturbance appeared to propagate along at about 200 km/s.

Studies have been made of the association between filament activation and the preflare soft X-ray enhancement, which starts an average of 2 minutes before the Hα emission. Preflare filament activation is accompanied by a slight soft X-ray enhancement, with X-ray increases synchronized to filament expansion. The preflare soft X-ray increase is localized near an erupting filament. Most filament disappearances outside active regions exhibit X-ray brightening but no Hα emission.

Although most of the Skylab observations were of subflares, one two-ribbon event, which was covered with good time resolution throughout its life, is of particular interest (Figure 1.4). The preflare X-ray increase is seen to come from a linear structure overlying a filament channel and possibly associated with the activation of a filament. At flare onset a high-temperature flare-kernel is located at the intersection of the linear structure with a loop. During the early phase of the flare, the filament erupts, being most distorted near the loop, which then becomes the dominant part of the flare. As the main phase proceeds, consecutive X-ray loops straddling the filament channel ($H_{\parallel}=0$ line) are seen to increase in height and become located more nearly perpendicular to it. Evidence is presented for continued heating during the main phase. During the main phase, the two Hα ribbons are found to be connected by an arcade of "post"-flare loops, which rises at a speed of initially 10–20 km/s and later less than 1 km/s. The lowest loops show up as a cool loop-prominence system and above them one finds much hotter soft X-ray loops.

Some of the main conclusions of the Skylab Workshop on Solar Flares (Sturrock, 1980, pages 5–13) are as follows:

i) The flare of 5 September, 1973, appeared to start in a cool EUV loop bridging the polarity inversion line at an angle, and then it progressed to higher altitudes where it became a higher-temperature X-ray loop with a more nearly potential field orientation.

ii) Coronal transients are preceded by a broad halo, called a *forerunner*, which contains 10–20% of the mass.

iii) The observed cooling-time for flaring X-ray loops is longer than the classical conduction-time; this may be due to either continued energy release or the inhibition of classical conduction by geometrical effects or plasma instabilities (§5.3).

iv) The acceleration of ions to high energies probably occurs in the corona at such a rapid rate that the ionization states are not changed from normal coronal values. (A summary of the observational effects of fast-particle acceleration is given in §§8.1–8.10.)

v) Streams of electrons flowing from the corona to the chromosphere may drive a reverse current, which can give significant heating (§8.10.1).

vi) It is uncertain whether one can discriminate between thermal and non-thermal electron spectra, but hard X-ray emission does appear to come from altitudes sometimes as high as 20,000 km (§2.5.1, §8.3.3).

vii) Sprays consist of helical strands confined in expanding (magnetic) loops and may well represent the eruption of active-region (or plage) filaments.

viii) The estimated pressure gradient along a surge is sufficient to drive surge material outward, and the characteristics of a surge may be reproduced by a model which has impulsive heating low in the chromosphere.

ix) Just after the initial flare energy release, thermal conduction down from the corona is the dominant energy loss mechanism, but then the chromosphere responds by evaporating mass upwards until radiative cooling takes over (§5.5, §5.6).

x) The energy release may take place only in the impulsive phase of a compact flare, but it continues through the main phase of a two-ribbon flare.

xi) The flare of 5 September, 1973, was found to have a mechanical energy output of at least 2×10^{24} J (2×10^{31} erg) and a radiative energy output of only 4×10^{22} J (4×10^{29} erg).

xii) In a final chapter, Sturrock divides the requirements of a successful flare model into two types: primary ones comprise energy storage, energy release, particle acceleration, mass ejection and heating of the temperature-minimum region; secondary requirements include X-ray and EUV radiation, UV and visible radiation, radio emission and internal mass flow. He considers several flare models and concludes that they are becoming more complex and that it is unclear whether or not several models will be required to explain several flare types.

1.3 AN OVERVIEW OF THE FLARE PHENOMENON

The theory of solar flares is a vast subject, incorporating many different areas of speciality. Each area is important for understanding the whole event, but the danger is of concentrating one's attention on only one such aspect and losing an overview of the total phenomenon. The aim of the present book is certainly not to survey all of flare theory but rather to focus on some of the key problems associated with the magnetic field itself. The whole subject may be divided as follows.

i) *Preflare build-up* This includes an analysis of how enough magnetic energy is stored in a largely force-free field prior to its release as a flare. For simple-loop flares, the ends of a magnetic loop may be twisted by photospheric motions, whereas for two-ribbon flares the whole magnetic arcade in which a filament is suspended may become sheared. One also needs to understand various other preflare processes such as photospheric velocity patterns, sunspot motions, magnetic flux emergence, slow coronal heating, slow filament rise. This is all well-summarised in Chapter 2 of the Skylab Solar Flare Workshop Proceedings (Sturrock, 1980).

ii) *Instability* The preflare coronal magnetic configuration is believed to evolve passively through a series of equilibria in response to photospheric motions until, at some critical point, it becomes unstable and the flare commences. In the case of the simple-loop flare the loop begins to heat up dramatically, whereas for the two-ribbon event the slowly-rising filament suddenly begins to ascend much more quickly. One needs to analyse the instabilities that are responsible for this behaviour and to demonstrate that they can be triggered. A linear analysis is not sufficient, since it must be demonstrated that the preflare configuration is *metastable* (Sweet, 1969): in other words, nonlinear effects promote the instability, which is then said to be *explosive* (Sturrock, 1966).

iii) *Energy release* During the main part of the flare, magnetic energy is converted into heat, bulk kinetic energy and fast particle energy (Chapter 3). This may be accomplished by reconnection at a single current sheet or by resistive instabilities at many such sheets throughout the configuration. One needs to show that the conversion can take place fast enough in the magnetic configuration appropriate to each type of flare.

iv) *Bulk motions* In simple-loop flares, the release of heat causes motions along the loop, which must be explained in detail. For two-ribbon flares, the resulting motions may be extremely complex. As well as the eruption of the filament itself, surges may appear as streams of plasma are ejected; plasma ahead of the rising filament becomes compressed and shock waves and coronal transients may be generated; matter falls down the ends of the filament and motions are set up in the "post"-flare loops that form beneath the filament. All these various types of mass ejection are described graphically in the Proceedings of the Skylab Workshop on Solar Flares (Chapter 7 of Sturrock, 1980).

v) *Particle acceleration* High energy particles are accelerated, especially during the impulsive phase of the flare. In the simple-loop type they spiral down along field lines and bombard the lower atmosphere. In the two-ribbon flare the particles may also escape along open field lines during the main phase, and acceleration may continue at both the reconnection site and the shock waves. It is crucial to determine how the acceleration of so many particles occurs and what are the detailed plasma processes involved in the propagation of a beam of particles from the acceleration site. A comprehensive discussion of the observations is given in (§§8.1–8.10) and is followed by a summary of theoretical considerations for first-phase (§§8.11–8.20) and second-phase (§§8.21–8.26) acceleration.

vi) *Radiation* Radiation over an immense range of the electromagnetic spectrum is produced during a solar flare and it is important to determine how each type is formed. Radio waves may arise from the acceleration of particles in many ways, such as their spiralling in the magnetic field (§8.5). Hard X-rays and small knots of Hα emission may be generated by the dumping of high energy particles in the chromosphere (§§8.3, 8.9, 8.10). Also, the addition of heat to coronal plasma enhances the soft X-ray emission, and, after being conducted down to the chromosphere, it may produce the Hα ribbons. A review of this field has been given by Brown and Smith (1980).

1.4 EARLIER THEORIES FOR THE SOLAR FLARE

A multitude of flare mechanisms has been proposed in the past; many have not progressed far beyond a qualitative description, while some others have failed on new observational or theoretical grounds. Two basic magnetic configurations have been proposed: the flux loop and the current sheet. The current-sheet model has received the most attention, though the flux-loop model for simple-loop flares is presently undergoing a revival, due perhaps to the observation of such loops in X-ray pictures. In this section, we discuss briefly a few of the flux-loop and current-sheet theories; some of the former may involve current sheets in a secondary manner. It may be noted that the more recent theories mentioned in (§1.6.1) and detailed in other parts of this book have in many respects developed out of the earlier ideas.

The solar flare problem (for either the simple-loop flare or the two-ribbon flare) may be dealt with in five stages of explanation:

i) to show that a particular stable configuration has enough stored magnetic energy to supply the flare;

ii) to find a means of converting this energy into heat, bulk kinetic energy and fast particle energy sufficiently rapidly;

iii) to demonstrate how the release is triggered;

iv) to explain the preflare phase, the impulsive phase, the flash phase and the main phase of a flare;

v) to describe the many secondary effects of the energy release (Hα, X-ray and radio emission).

A comprehensive theory should cover all these stages, both qualitatively and quantitatively, but it does not yet exist; the present book summarizes recent advances in key areas of the problem.

1.4.1 The storage of magnetic energy and its release

The magnetic energy contained inside a cube of side L is

$$W = L^3 \frac{B^2}{2\mu}$$

and, in principle, values of L and B characteristic of active regions give a magnetic energy that is easily adequate to account for a flare. The typical dimensions of a large Hα disc-flare is 3–70 Mm (1 Mm = 1000 km), whereas the height of a limb-flare may be as much as 20 Mm and the magnetic field strength in an active region is probably at least 0.01–0.05 T (100–500 G). The 3×10^{25} J (3×10^{32} erg) of energy characteristic of a large flare would be released if, for instance, the whole of a 0.05 T (500 Gauss) magnetic field in a cube of side 30 Mm (or of a 0.01 tesla field in a cube of side 90 Mm) were annihilated; it also corresponds to the energy released if the field in a cube of side 40 Mm (= 40,000 km) falls from 0.05 tesla to 0.04 tesla.

In order to solve problem (I) above, one needs to show that the particular configuration under consideration can store energies of this size in a stable manner. A magnetic arcade appears to be able to do so (Chapter 6); for example, an arcade of field strength 0.05 T, radius 20 Mm (= 20,000 km) and length 100 Mm contains an energy of 6×10^{25} J. However, *single flux loops appear in practice to be able to store only enough excess energy for a sub-flare*; for example, the effect of twisting a loop of radius 5 Mm, length 100 Mm and field strength 0.05 T (500 G) through an angle 2π is to store only an extra 7×10^{23} J.

If the magnetic energy for a flare is stored above the photosphere and the normal photospheric magnetic field component does not change during the flare, the preflare magnetic field cannot be current-free (or potential). The extra energy in excess of potential can then be provided only by force-free or current-sheet configurations, since the plasma pressure gradients in the ambient chromosphere and corona above an active region are too small by up to four orders of magnitude to balance the Lorentz forces. For a loop, energy may be stored in excess of potential by twisting the ends and creating a force-free field. The loop then becomes kink unstable when a field line makes more than one or two revolutions about the axis (Raadu, 1972; Hood and Priest, 1979; §4.4.8). Sufficient energy for a flare may also be stored in a more general sheared force-free field (Tanaka and Nakagawa, 1973). On the other hand, for a current-sheet configuration, the energy stored actually inside the sheet is not usually sufficient for a flare; in this case, the excess energy in the surrounding field, which is distorted by the presence of the current sheet, may be sufficient, provided preflare reconnection is inhibited (Priest and Raadu, 1975; Tur and Priest, 1976).

In the solar atmosphere, the Spitzer value of the electrical conductivity,

$$\sigma = 10^{-3}\ T^{3/2}\ \text{mho m}^{-1},$$

is so large that, for typical scale-lengths L and plasma speeds V, the magnetic Reynolds number

$$R_m \equiv LV\mu\sigma$$

is much larger than unity. For instance, with $T = 10^4$ K, $V = 20$ km s^{-1} and $L = 20$ Mm ($= 20{,}000$ km), we find $R_m = 5 \times 10^8$. The result is that the magnetic field lines are 'frozen' to the plasma. Furthermore, the corresponding time-scale

$$\tau_D \equiv L^2\mu\sigma$$

for the conversion of magnetic energy into heat is very large. With the above values for L and T, we have $\tau_D = 5 \times 10^{11}$ s, much longer than the typical duration of 10^3 s for a solar flare. This means that, for length-scales characteristic of an active region as a whole, magnetic energy is released throughout the whole volume by ohmic dissipation at an extremely slow rate. Even if turbulent values are adopted for the electrical conductivity, this conclusion is unaltered. In order to explain the rapid time-scales characteristic of solar flares one therefore needs much smaller length-scales — typically as small as a kilometre (or larger than this if the conductivity is turbulent). However, the price that is paid for considering current filaments, current sheets, current sheaths or shock waves with such small widths is that the energy release now occurs only within the current concentrations and not throughout the whole volume. These intense currents may well be the result of an instability and it has been suggested that either there is basically one such concentration or that there are many of them distributed through the flaring volume (§8.19.2).

Another property of the overall solar atmosphere is that the electric current densities are relatively small; thus a field strength of amount $B = 5 \times 10^{-2}$ T (500 G), say, which varies over the length L, corresponds by Ampère's Law to a current density

$$j \equiv B/(\mu L) = 2 \times 10^{-3}\ \text{A m}^{-2}.$$

(For one particular sunspot group, Moreton and Severny (1968) deduce a maximum current density which is about four times this value.) It should be noted that for loops as a whole, the current densities are usually much too small to excite micro-instabilities and so create plasma turbulence (§8.19).

1.4.2 Flux-loop models

One of the earliest flux tube models was due to Gold and Hoyle (1960). It considered a pair of flux tubes that are oppositely directed but twisted in the same sense and pressed together. The resulting reconnection at the current sheet that separates the two tubes means that the model is not a pure flux-tube one. It is only qualitative and its main disadvantage is that the configuration does not agree with observations.

Alfvén and Carlqvist (1967) and Carlqvist (1969) considered a magnetic flux-tube which undergoes some magnetohydrodynamic instability, resulting in a local increase in the current density j (§8.18.1). They suggest that a micro-instability occurs when the current density is so large that the electron conduction speed u $(\equiv j/(ne))$ exceeds a critical value, namely,

$$u > (kT_i/m_e)^{1/2},$$

where n is the plasma number density and T_i the ion temperature. The qualitative effect of the instability is to cause a local density rarefaction, so that the electrical conductivity approaches zero and the current-flow is interrupted. Particles are accelerated to high energies while the flux-tube untwists and all its excess energy is dumped at the site of the instability in about 100 seconds. Alfvén and Carlqvist argued by analogy with a laboratory experiment and they were perhaps pushing the analogy too far, since the plasma conditions in their experiment and the solar atmosphere are by no means identical. Smith and Priest (1972) have pointed out that the relevant microinstability is the ion acoustic instability, which takes place when

$$u > (kT_e/m_e)^{1/2}$$

for equal electron and ion temperatures ($T_e = T_i$). The effect of the instability is to produce plasma turbulence, which heats the electrons and lowers the threshold to

$$u > (kT_e/m_i)^{1/2}$$

for $T_e \gg T_i$. (This suggests the existence of a metastable situation, which is believed necessary for the occurrence of a flare (§1.3 above).) The onset of turbulence means that the electrical conductivity has a turbulent value; this has the effect of impeding the acceleration of electrons and causing the energy to be dumped much more slowly. Recently, however, Carlqvist (1972, 1979) has taken a more quantitative look at his model and considered the formation of space-charge regions called double layers (§8.18). It appears that the nonlinear development of the ion acoustic instability may, in theory, produce either turbulence or double-layers; evidently, more work is required to determine under what conditions in practice double-layers may develop in the solar atmosphere rather than pure turbulence. In any case, it is hard to see how the current density in a loop that is large enough to store the necessary energy can exceed the above extremely high threshold for instability (§8.19.1), and, furthermore, a single loop configuration cannot allow the escape of high energy electrons to give type III radio emission.

A qualitative development of the Alfvén–Carlqvist model has been given by Spicer (1977a, b) for the simple-loop flare. It relies heavily on laboratory plasma physical results, which is valid provided of course the crucial differences between the behaviour of a laboratory and solar plasma are borne in mind (§1.4.4). Spicer has promising ideas such as invoking the resistive kink instability, but they require more detailed analysis in the solar context. In §4.6 comments can be found on his work as well as that of Colgate (1978), who also considers energy dissipation in a single flux-tube.

An interesting idea for explaining a flare has been put forward by Pneuman (1967), though it has not been further-developed. He considers the injection of energy into the magnetic structure of an active region by means of Alfvén waves travelling up along field lines from the photosphere. The energy in small-scale motions builds up in time as a resonance until it exceeds the mean-field energy and a disruption occurs. The large wave amplitudes required seem rather unlikely, but a relevant observation, which needs to be verified, is that flares occur about 20 minutes after chromospheric oscillations over a large area have become synchronized (Tanaka, 1974).

A similar "Alfvén-wave theory" of solar flares has been put forward by Piddington (1973, 1974); it is still of a largely qualitative nature but deserves serious attention. He discounts the need to store the flare energy, but instead suggests that the flare energy is provided by Alfvén waves which propagate up from the photosphere during the course of the flare; however, there is no evidence for such waves. The basic magnetic configuration for this model is shown in Figure 1.6. Several typical field-lines are shown, all assumed to possess helical twist. B_1 is a low-lying flux strand, along which Alfvén waves propagate from below and produce two-ribbons of Hα emission. B_2 is a similar strand which projects into the lower corona, where more energy is released. B_3 is grossly deformed by the presence of a large-amplitude Alfvén wave and, finally, B_4 has been blown open during the flare. The Alfvén waves do not by themselves dissipate easily, but the way they couple nonlinearly to slow magneto-acoustic waves, which dissipate very efficiently, has been calculated by Uchida and Kaburaki (1974). Although such ideas are probably relevant for general heating above an active region (Wentzel, 1974), it is not clear to what extent they are applicable to the more violent flare heating.

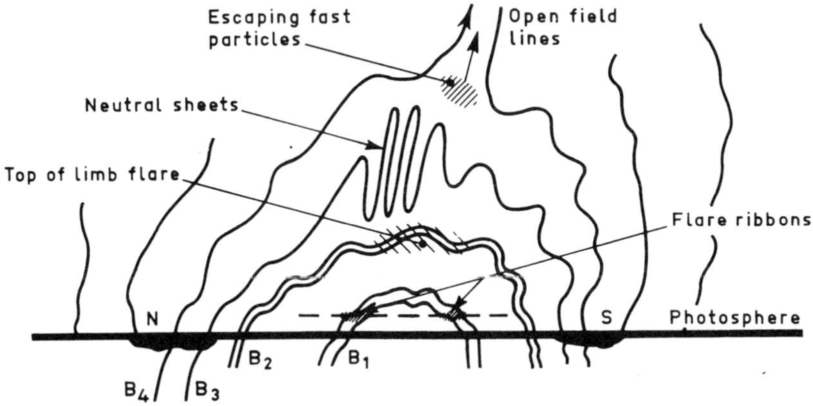

FIGURE 1.6 The magnetic configuration for the Alfvén wave theory of solar flares. (From Piddington, 1974.)

1.4.3 Current-sheet theories

As pointed out in §1.4.1, throughout most of the solar atmosphere, the magnetic field is nearly "frozen" to the plasma and the time-scale τ_d for converting magnetic energy into heat by ohmic dissipation is very long. However, if one considers shock waves or current sheets, where the scale-length can be as little as metres or kilometres, then the diffusion time τ_d is much reduced and the current density considerably higher.

Current sheets have thus been suggested as the site for flare energy release, because they can convert magnetic energy very rapidly into both heat and directed plasma energy. They also contain strong electric fields which have the potential to accelerate fast particles. Fairly sophisticated treatments have now been given of the quasi-steady energy release in a current sheet. Despite much earlier debate, it now seems established that Petschek's mechanism (Petschek, 1964) can provide energy release fast enough for a flare (Vasyliunas, 1975; Priest and Soward, 1976; §3.3.4). The rate of release is governed mainly by the speed with which external conditions make the plasma approach the current sheet, but there does exist a maximum value for this speed; it depends weakly on the electrical conductivity and is typically a tenth or a hundredth of the ambient Alfvén speed (Roberts and Priest, 1975; Soward and Priest, 1977). Much of this work on magnetic reconnection has aimed to explain only the point (II) mentioned at the beginning of §1.4, but it has led to a great advance in our understanding of the process. It is only relatively recently, in, for example, Sturrock's model or the emerging flux model (§3.5), that an attempt has been made to incorporate reconnection in an overall flare model that accounts for the observations. A review of the behaviour of current sheets has been given by Priest (1976) and is expanded in Chapter 3. In it a description of the processes of magnetic field annihilation and reconnection, as well as the formation of a sheet and the tearing mode instability is given, together with a critical review of the main current-sheet flare models.

There have been many flare theories based on current sheets, but the main ones are due to Syrovatsky, Sweet and Sturrock, which we outline below. Following earlier work by Giovanelli (1946) on the behaviour of the magnetic field near a neutral point, where the field vanishes, Dungey (1953) suggested that an X-type neutral point may be unstable and collapse into a current sheet. This idea was taken up by Syrovatsky (1966, 1969, 1970, 1972), who considers a potential magnetic field containing a neutral point. He supposes that at some time the sources of the field move together faster than the sound speed so that a current sheet forms about the neutral point (Figure 1.7). Under the assumption of no magnetic field reconnection, he gives a simple analysis of the process of sheet formation and finds that in certain regions of the current sheet, the current density j increases while the plasma number density n decreases. This, of course, implies that the electron conduction speed $u = j/(ne)$ rises considerably in value. If the original source-displacement is large enough, u exceeds the threshold value $(kT_e/m_e)^{1/2}$ for the onset of ion sound turbulence (§§8.13, 8.14). In Syrovatsky's model (§3.2.1) the resulting plasma heating produces the preflare increase in soft X-ray emission. The impulsive phase then occurs when either the electron conduction speed achieves a relativistic value or the tearing mode instability takes place, so that a strong electric field accelerates particles to high energies. The fast particles produce hard X-ray, optical and radio emission as a secondary effect. For the main phase, the hot plasma evolves and cools, with perhaps a continuation of turbulent heating. Two criticisms have been levelled at the details of Syrovatsky's mechanism (§3.2.1). First of all, Smith and Priest (1972) have shown that u is unlikely to become relativistic, since the ion acoustic turbulence leads to a diffusion of plasma across field lines. But, more fundamentally, Anzer (1973) has studied the formation of the current sheet more carefully and claims that neither an extreme density rarefaction nor a strong field compression occurs, so that relativistic values of u are not attained. Even though the model may not work for a flare, the time-dependent formation of a current sheet is most important and Syrovatsky's analysis has helped our understanding greatly. Moreover, his ideas may be relevant to general coronal heating, since the response of a general active-region field to footpoint motions appears to be to create current sheets and so to give a localised magnetic dissipation (Parker, 1972; Syrovatsky, 1978), as described in §3.2.2.

Sweet (1958) analysed a very similar problem to Syrovatsky. He considered the magnetic field shown in Figure 1.8a due to a pair of bipolar regions. When the regions approach one another the region around the neutral point collapses and forms into a current sheet (Figure 1.8b). The subsequent

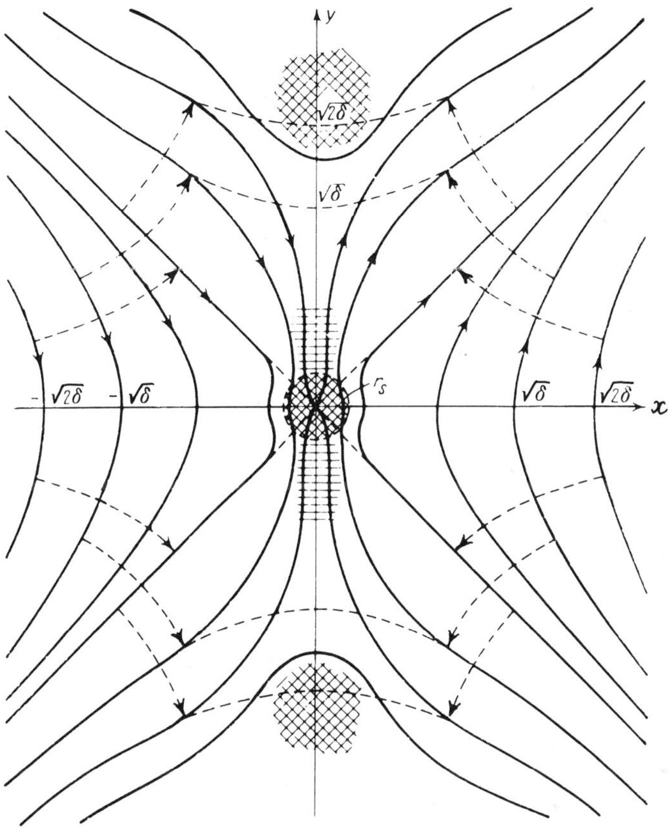

FIGURE 1.7 The formation of a current sheet from a configuration initially containing an X-type neutral point (N) in Syrovatsky's flare model. The sources of the field have moved a distance δ and cause the field-lines at large distances ($r > r_s$) to move as indicated by the arrowed, dashed lines. Close to N the field lines are greatly distorted so as to form a current sheet, since no reconnection is allowed. Two regions (cross-hatched) of compressed plasma are formed on the y-axis and one near the neutral point, while plasma in the rest of the sheet is rarefied, as shown by the horizontal shading. (From Syrovatsky, 1966.)

dissipation of energy gives the flare. An analytic solution involving complex variable theory was obtained for the current-sheet configuration by Priest and Raadu (1975), who worked out the quantitative details of the current-sheet length and energy storage. The main inadequacies of the model are that in practice magnetic reconnection would set in to produce a highly dynamic process and, further, that the configuration is not typical of flaring regions. These have been overcome by the *emerging flux model* of Heyvaerts, Priest and Rust (1977), which is described in §3.5 and may be regarded as a sophisticated development of Sweet's work; it incorporates reconnection and was proposed in an attempt to explain the observations as much as possible.

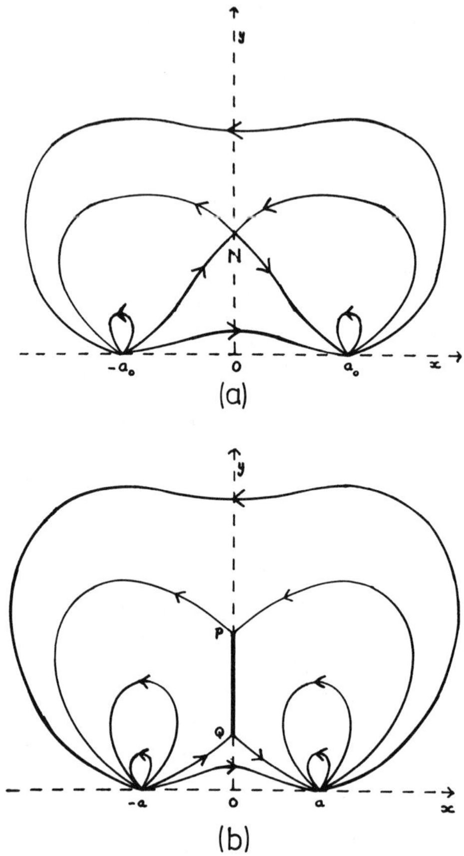

FIGURE 1.8 Sweet's flare model. (a) The current-free magnetic configuration due to two line dipoles of equal moment, which are situated at $x = \pm\alpha_0$ on the x-axis. An X-type neutral point, where the magnetic field vanishes, is located at N. (b) When the dipoles approach one another to $x = \pm\alpha$ and no reconnection is allowed, the resulting equilibrium field contains a current sheet PQ.

The basic feature of Sturrock's model (1966, 1968, 1972, 1974), as originally put forward, is the existence of a helmet-streamer-like magnetic configuration, shown on the left of Figure 1.9. The beauty of the configuration is that it can explain both the two Hα ribbons and the escape of high-energy particles in a natural way. The impulsive phase of the flare is due to a tearing mode instability which generates a strong electric field in the open current sheet. This leads to fast electrons, some of which escape along the current sheet and produce type III radio bursts, while others travel down to the chromosphere and give a hard X-ray burst. During the flash phase, reconnection continues due to Petschek's mechanism (with a turbulent electrical conductivity) and two ribbons of emission are formed when heat and energetic particles reach the

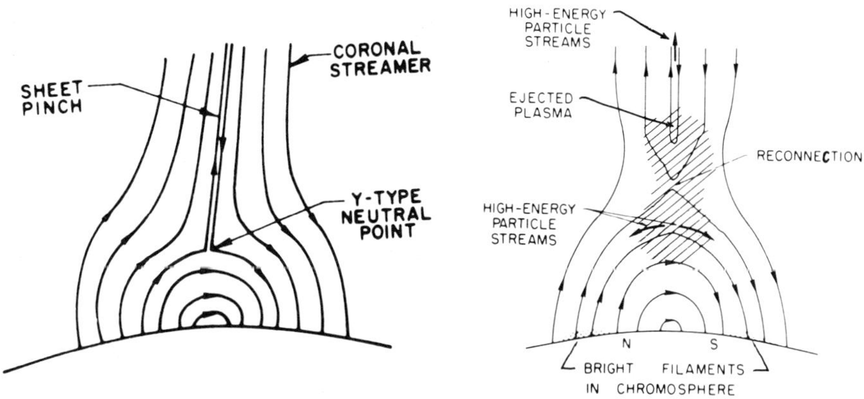

FIGURE 1.9 Sturrock's solar flare model. (From Sturrock, 1968.)

chromosphere. At the main phase, an outwards drift of the ribbons takes place, because the reconnection process continues with field-lines which are successively further out from the inversion line. Various secondary flare effects, such as loop prominences, type II bursts, and second-phase particle acceleration, are explained in a qualitative way. In Sturrock's original model, the open current sheet was presumed to be high in the corona, with the field lines dragged out by the solar wind. But realistic field strengths are then too low to provide enough energy, so Sturrock subsequently proposed a means of forming such a configuration much lower down inside an active region (Barnes and Sturrock, 1972; Sturrock, 1974). He considers an initially potential field such as a simple flux tube, whose footpoints move and so generate a force-free field. When the energy in the force-free field exceeds that of an open field with the same photospheric boundary conditions, the force-free field is, he suggests, metastable and erupts into the open configuration. This is a most interesting idea, which one would like to see demonstrated more quantitatively, but there are a few theoretical reservations (Priest, 1976) such as the effect of boundary conditions at large distances and a doubt as to whether the force-free configuration can indeed erupt into the open one without reconnection. Further work on this problem is described in §§6.3–6.5.

Sturrock's qualitative idea for the flare configuration shown in Figure 1.9 is now accepted by many people to explain the main phase of a two-ribbon flare and is incorporated in, for example, the emerging flux model; however, the open current sheet is believed to be located much lower down in the atmosphere than a coronal streamer so that the magnetic field is much stronger. A quantitative analysis of the formation of "post"-flare loops has been presented by Kopp and Pneuman (1976), as described in Chapter 7. At present, one of the main emphases of flare theory is to try and explain the early phase of the flare when a closed magnetic arcade erupts out to give the open field (Chapter 6).

1.4.4 Basic themes

There are many themes which run through solar flare theory. One of them is the development of microscopic instabilities within macroscopic ones. Thus it is thought that some relatively *slow magnetohydrodynamic evolution* or instability takes place, until critical conditions are attained for the onset of small-scale or *micro-instabilities* (§§8.13, 8.14). Several ways of reaching the critical point have been suggested. Alfvén and Carlqvist originally relied on a magnetohydrodynamic instability to increase the electric current density j, while the subsequent decrease in density n was a result of their micro-instability (§8.18.1). Piddington's magnetohydrodynamic stage is the propagation of Alfvén waves and their coupling to other waves which steepen into shocks. Syrovatsky, on the other hand, suggests that j increases and n decreases during the formation of a current sheet; the electric field for fast-particle acceleration in his model results when the electron conduction speed approaches the speed of light. For Sturrock's theory, the initial evolution is of a force-free field, which erupts into a current sheet, and it is a tearing instability which gives the electric field for particle acceleration. Finally, in the emerging flux model, the magnetohydrodynamic evolution is that of newly-emerging magnetic regions (and, for the two-ribbon flare, of an over-lying force-free field). The value of the electron conduction speed inside the reconnecting current sheet between new and ambient flux increases with height (both because of a rise in j and in a fall in n) up to a critical level, and the impulsive electric field results from a rapid expansion of the sheet at turbulence onset.

Another theme is the presence of current sheets to allow a drastic reduction in the magnetic dissipation time-scale. Also, the onset of *plasma turbulence*, ion acoustic or otherwise, is commonly evoked (§8.16). Maybe this turbulence starts in small current sheets and is spread outwards by beams of particles (§8.11). It is clear that kinetic plasma theory is of crucial importance here in providing the details of the small-scale processes and in determining the micro-instabilities and transport properties of the plasma (see Chapter 8). One warning, however, is that *results obtained in the laboratory should not be carried over to the Sun uncritically*: for example, the energy balance in solar plasma is most important and produces different effects from those found commonly in the laboratory; also, the electric current is distributed throughout the plasma rather than flowing in discrete tubes, which means that circuit theory cannot be applied naively; finally, it should be noted that in the laboratory one often has

$$\mathbf{j}=\sigma\mathbf{E}$$

for Ohm's Law, so that electric currents are driven by electric fields, but in many solar applications Ohm's Law becomes approximately

$$\mathbf{E}+\mathbf{v}\times\mathbf{B}=\mathbf{O}$$

and the current is then determined as a secondary variable from the magnetic field by

$$\mathbf{j}=\text{curl}\,\mathbf{B}/\mu.$$

1.5 THE MAGNETOHYDRODYNAMIC EQUATIONS

The aim of magnetohydrodynamics is to describe the interaction between a magnetic field and a plasma, treated as a continuous medium. Most of the matter in the universe is in a plasma state and it behaves quite differently from the other three forms, namely solid, liquid and gas. In the solar atmosphere the relationship between the plasma and the magnetic field is an intimate one — so much so that most of the detailed solar structure we see owes its existence to the magnetic field!

The magnetic field influences a plasma in many ways. It *exerts a force* $\mathbf{j}\times\mathbf{B}$, which is perpendicular to the magnetic field and may be described as the sum of a *magnetic tension* and a *magnetic pressure*. This has the effect of structuring the solar atmosphere in a direction normal to the field, which is in addition to the natural structuring along the field caused by gravity. The magnetic field also *stores energy* of amount $B^2/(2\mu)$ per unit volume and may *heat the plasma* through ohmic dissipation j^2/σ per unit volume. Finally, the field tends to *insulate the plasma* thermally in a direction normal to the field, since heat conduction tends to be much less efficient in that direction than along the field.

The fundamental equations which we shall adopt throughout most of this book for the magnetic induction $\mathbf{B}$, plasma velocity $\mathbf{v}$, plasma density ρ and plasma temperature T are, in MKS units,

the *induction equation*

$$\frac{\partial\mathbf{B}}{\partial t}=\boldsymbol{\nabla}\times(\mathbf{v}\times\mathbf{B})-\boldsymbol{\nabla}\times(\eta\boldsymbol{\nabla}\times\mathbf{B}) \tag{1.1}$$

the *equation of motion*

$$\rho\frac{\mathrm{D}v}{\mathrm{D}t}=-\boldsymbol{\nabla}p+\mathbf{j}\times\mathbf{B}+\rho\mathbf{g}, \tag{1.2}$$

the *mass continuity equation*

$$\frac{D\rho}{Dt}+\rho\boldsymbol{\nabla}.\mathbf{v}=0, \tag{1.3}$$

and the *energy equation*

$$\frac{\rho^{\gamma}}{\gamma-1}\frac{D}{Dt}\left(\frac{p}{\rho^{\gamma}}\right)=\boldsymbol{\nabla}.(\boldsymbol{\kappa}\boldsymbol{\nabla}T)+j^2/\sigma-L_{\text{rad}}+H_w, \tag{1.4}$$

where the electric current density **j** and plasma pressure p follow from *Ampère's Law*

$$\mathbf{j}=\boldsymbol{\nabla}\times\mathbf{B}/\mu, \tag{1.5}$$

and the *perfect gas law*

$$p=(R/\tilde{\mu})\rho T=nk_BT. \tag{1.6}$$

Also, the electric field follows from *Ohm's Law* as

$$\mathbf{E}=-\mathbf{v}\times\mathbf{B}+\mathbf{j}/\sigma, \tag{1.7}$$

and the magnetic field is restricted by

$$\text{div}\,\mathbf{B}=0. \tag{1.8}$$

In these equations

$$\eta=(\mu\sigma)^{-1} \tag{1.9}$$

is the magnetic diffusivity, μ is the magnetic permeability (roughly the vacuum value μ_0) and σ is the electrical conductivity;

$$\frac{D}{Dt}\equiv\frac{\partial}{\partial t}+\mathbf{v}.\boldsymbol{\nabla} \tag{1.10}$$

is the material time derivative;

$$\rho=\tilde{\mu}m_pn \tag{1.11}$$

is the plasma density, written in terms of the proton mass m_p, the total particle density n and the mean atomic weight $\tilde{\mu}$, which is 0.5 for a fully-ionized hydrogen plasma and approximately 0.6 for much of the solar atmosphere; γ is the ratio of specific heats, equal to $\frac{5}{3}$ for fully-ionized hydrogen; $\boldsymbol{\kappa}$ is the thermal conduction tensor; L_{rad} is the optically thin radiative loss, which may be written in terms of the electron density n_e for a fully-ionized hydrogen plasma as

$$L_{\text{rad}} = n_e^2 Q(T), \tag{1.12}$$

where the function $Q(T)$ has been tabulated by many authors (e.g. Cox, and Tucker, 1969; McWhirter *et al.*, 1975; Raymond, Cox and Smith, 1976) and possesses a maximum at around 10^5 K; H_w represents the heating due to small-scale effects such as waves — it is poorly known and is often assumed to be uniform or proportional to density; R is the gas constant and k_B is Boltzmann's constant.

A discussion of the equations can be found in, for example, Priest (1981). Several general comments may be made here about their significance. They have been written in MKS units, so that B is measured in teslas (1 T $=10^4$ Gauss), v is in metres per sec (1 m s$^{-1}=100$ cm s^{-1}), ρ is in kilograms per cubic metre (1 kg m$^{-3}=10^{-3}$ g cm^{-3}), T is in degrees Kelvin, j is in Amps per square metre (1 A m$^{-2}=3\times10^5$ statamp cm^{-2}), p is in Newtons per square metre (1 N m$^{-2}=10$ dyne cm^{-2}).

The induction equation gives magnetic field changes at a point in time due to advection of the magnetic field with the plasma (the first term on the right) and diffusion of the magnetic field through the plasma (the second term on the right). Except within extreme current concentrations, diffusion is usually negligible and so the magnetic field is frozen to the plasma. A typical coronal value for the magnetic diffusivity is $\eta=10^9\ T^{-3/2}$ m^2 s^{-1}.

The equation of motion assumes the plasma is acted on by a plasma pressure gradient $-\nabla p$, a Lorentz force $\mathbf{j}\times\mathbf{B}$, and a gravitational force $\rho\mathbf{g}$ acting towards the solar centre. The magnetic force may be written, after using (1.5) and a vector identity, as

$$\mathbf{j}\times\mathbf{B} = -\nabla(B^2/(2\mu)) + (\mathbf{B}\cdot\nabla)\mathbf{B}/\mu. \tag{1.13}$$

The first term on the right represents the magnetic pressure force and it acts from regions of high magnetic field strength towards those of low field strength. The second term is the magnetic tension force; it is non-zero when the field lines are curved and acts towards the centre of curvature.

The energy equation gives the time rate of change of the entropy due to various sources and sinks on the right-hand-side of (1.4), namely heat conduction, ohmic heating, radiative loss and small-scale heating (possibly waves). For many purposes, heat conduction across the field may be neglected and the conduction term reduces to

$$\nabla.(\kappa_{\|}\nabla_{\|}T)=\frac{d}{ds}\left(\kappa_{\|}\frac{dT}{ds}\right)-\frac{\kappa_{\|}}{B}\frac{dB}{ds}\frac{dT}{ds} \tag{1.14}$$

$$=\frac{1}{A}\frac{d}{ds}\left(\kappa_{\|}\frac{dT}{ds}A\right), \tag{1.15}$$

where s is measured along the magnetic field and $A(s)$ is the cross-sectional area of a flux tube, which is related to the magnetic field strength by

$$\frac{d}{ds}(BA)=0.$$

A typical coronal value for the coefficient of heat conduction along the field is

$$\kappa_{\|}=9\times10^{-12}\ \mathrm{T}^{5/2}\ \mathrm{W\ m}^{-1}. \tag{1.16}$$

In some cases the full equation (1.4) is used, while at other times it is approximated: when a change is so rapid (adiabatic) that no heat is lost or gained by a plasma element we have

$$\frac{p}{\rho^{\gamma}}=\text{constant} \tag{1.17}$$

following the motion; when thermal conduction dominates or one wants to neglect changes in temperature for simplicity the energy equation reduces to

$$T=\text{constant}. \tag{1.18}$$

In adopting the above equations we have made the following assumptions, some of which may be dropped for a more sophisticated treatment: the plasma is treated as a continuum, which is valid provided the length-scale for magnetic variations greatly exceeds the typical internal plasma lengths; the plasma is regarded as a single fluid and its properties are isotropic, except for the thermal conductivity; the simple form of Ohm's Law is applicable; relativistic and rotational effects are negligible; the plasma is in thermodynamic equilibrium and is optically thin.

Typical plasma speeds are the *sound speed*

$$C_s=\left(\frac{\gamma p}{\rho}\right)^{1/2} \tag{1.19}$$

and the *Alfvén speed*

$$V_A = \frac{B}{(\mu\rho)^{1/2}}.$$

The dimensionless number

$$\beta = \frac{p2\mu}{B^2} \approx 3.5 \times 10^{-29} \frac{nT}{B^2}$$

gives the ratio of the plasma pressure to magnetic pressure and is a measure of the relative sizes of the two forces in the equation of motion (1.2). Within active regions β is generally small, and has a typical coronal value of 10^{-2} or 10^{-4}. The *magnetic Reynolds number*,

$$R_m = \frac{lV}{\eta} \approx 10^{-9}\ lVT^{3/2},$$

gives the ratio of the advective to the diffusive term in (1.1); for length-scales characteristic of active regions as a whole it is extremely large (say 10^8).

1.6 SUMMARY OF THE BOOK

1.6.1 Possible solar flare mechanisms

The solar flare owes its existence to the magnetic field and the aim of the present book is to show how that magnetic field behaves during the flare and how it controls the flare plasma. We shall concentrate on the fundamentals of the flare process and leave aside many aspects of the secondary effects such as radiation mechanisms, mass motions and interplanetary consequences, which have been well-discussed elsewhere. Our treatment will be mainly from a magnetohydrodynamic viewpoint, save for the plasma particle processes described in Chapter 8.

Great progress has been made on many basic questions of flare theory, as described in various parts of the book. Why does the magnetic configuration become unstable for a simple-loop flare (Chapter 4) or for a two-ribbon flare (Chapter 6)? How is the magnetic energy converted into heat and plasma kinetic energy (Chapter 3)? What is the response of the plasma to a heat input in a simple loop (Chapter 5)? How does the plasma react to the closing down of open magnetic field lines during a two-ribbon flare (Chapter 7)? What are the main mechanisms for accelerating particles (Chapter 8)?

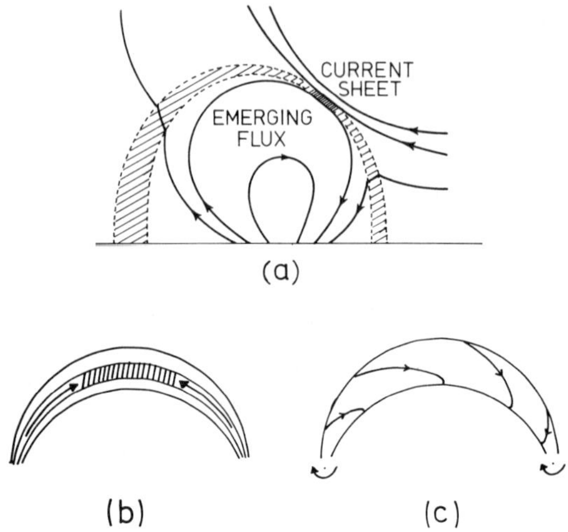

FIGURE 1.10 Possible mechanisms for a simple-loop flare: (a) emerging or evolving flux; (b) thermal nonequilibrium, during which a cool loop core heats up; (c) kink or resistive kink instability.

Several new models have been proposed for solar flares in the light of recent observations and advances in theory. Indeed, it is likely that a number of different mechanisms are operating to produce flares of different types. Once a mechanism has been worked out in detail, the two main tests are whether it agrees with observations (particularly in Hα and in X-rays) and whether it is internally consistent so that it works in theory. In the past, many models have either been put forward only in qualitative terms or have failed one or other of these tests. The present situation is that there appear to be several possible explanations for a flare, and only future work and observations can distinguish between them.

A *simple-loop flare* may be caused by one of the following mechanisms (Figure 1.10). It may result when a new magnetic flux tube emerges from below the photosphere and interacts with the overlying magnetic field as in the emerging flux model (§3.5); triggering takes place when the current sheet separating the two flux systems reaches a critical height. Alternatively, it may be an example of thermal nonequilibrium, when a cool loop ceases to be in thermal equilibrium and heats up (§5.4.6); the flare is, in this case triggered when the loop length, pressure or heating reach critical values. A third possibility is that the loop becomes kink unstable when it is twisted too much (§4.4). Finally, it has been suggested that the tearing mode or resistive kink instability is responsible (§3.2.4C, §3.3.3, §4.5), but a method of triggering the

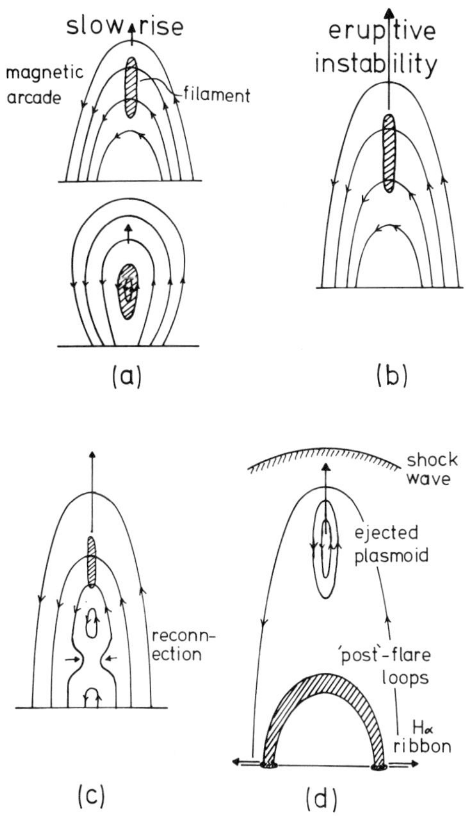

FIGURE 1.11 The overall mechanism for a two-ribbon flare, as seen in a section through the magnetic arcade. (a) Slow preflare rise of a filament, possibly due to thermal nonequilibrium or the initial stages of an MHD instability. The surrounding field may just be sheared or it may contain a magnetic island so that the filament lies along a flux tube (as in the lower figure). (b) Eruptive instability of a magnetic arcade supporting the filament. (c) Field lines below the filament are stretched out until reconnection can start. (d) As reconnection proceeds, "post"-flare loops rise and Hα ribbons move apart.

release of stored energy for this mechanism has not yet been demonstrated and so perhaps it just contributes to the general heating of the corona rather than producing a loop flare.

A *two-ribbon flare* is generally believed to be caused by the instability of the magnetic configuration in which a filament is situated (Figure 1.11). The filament and magnetic field erupt outwards and then it is thought that, during the main phase, the magnetic field reconnects back down to give rise to the "post"-flare loops (Chapter 7). The original eruptive instability has several possible explanations (Figure 1.12). It may be triggered by emerging flux,

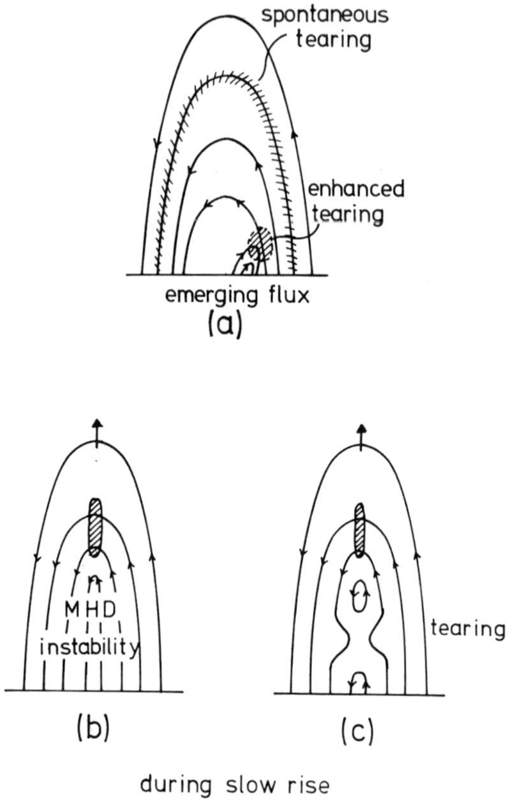

FIGURE 1.12 Possible mechanisms for the onset of the fast, eruptive instability in a two-ribbon flare. (a) Tearing mode instability, either enhanced by emerging flux or occurring spontaneously. (b) MHD instability when the filament reaches a critical height. (c) The onset of reconnection early in the flare may drive the eruption rather than be a consequence of it.

which locally enhances the resistivity and so lowers the growth-time for resistive instabilities (§3.5). Alternatively, the tearing mode may occur spontaneously throughout the sheared magnetic arcade, but again this may just produce a general heating rather than a flare. Another possibility is that there are two stages to the instability, namely a slow rise of the filament while, for instance, it loses thermal equilibrium (§5.4.6), followed by the flare instability proper: this may be caused either by the magnetohydrodynamic instability of the overall configuration (§6.5.3) or it may be due to tearing-mode-induced reconnection in the magnetic field that is being stretched out below the rising filament.

1.6.2 Summary of chapter contents

Flare observations Chapter 2 by Švestka gives a comprehensive description of solar flare observations, especially those that have been made in the five years since his book on Solar Flares was completed. It begins with a discussion of Hα observations (§2.2 and §2.3), including the relation of the flare to the magnetic field, to the velocity field and to various preflare phenomena such as emerging flux and filament activation. Next, the manifestations in soft and hard X-rays are summarised (§2.4 and §2.5), together with those from the transition region joining the cool and hot parts of the flare (§2.6). Finally, the remaining products of a large flare are described, namely radio bursts, mass ejecta and high energy particles (§§2.7–2.9). Some of the points made by Švestka are as follows:

1) it is important to continue to make (the relatively cheap) Hα observations along with the more recent space observations, since the Hα resolution can be better than 1″ and a great body of knowledge about the Sun's sophisticated behaviour in Hα has been built up (§2.1.2);

2) energy must be supplied throughout a two-ribbon flare, but need only be released at the beginning of a compact flare (§2.1.4);

3) quasi-thermal flares appear to be possible in a purely bipolar region, although there may be some small-scale complexity (§2.2.1);

4) small bipolar regions that flare always lie close to a supergranule boundary and so may be reconnecting with the network field (§2.2.1);

5) a combination of newly-emerging flux and sheared fields is most favourable for flare occurrence (§2.2.4);

6) the electron density in flares is typically 10–100 times bigger than normal (§2.3.2);

7) the geometrical thickness of the chromospheric flare is astonishingly small ($\sim$10–100 km), which probably means a filamentary shell structure (§2.3.3);

8) white-light emission coincides in time with the impulsive phase and so is produced by impinging energetic particles, but it is unclear whether the particles are electrons or protons and whether the light comes from the photosphere or chromosphere (§2.3.5);

9) compact flares have a small volume ($\sim 10^{20}$–10^{21} m^3), a low height (usually $<10^4$ km) and a high coronal density (10^{17}–10^{18} m^{-3}); they last only tens of minutes and are either uniformly bright in X-rays along their length or are brighter near their feet (§2.4.2);

10) two-ribbon flares have their maximum temperature at the summit of the arcade of "post"-flare loops; the cool Hα loops are located below the X-ray

loops and their feet are located on the inner edge of the Hα ribbons; the X-ray loops may exceed 100,000 km in altitude; the density is constant along the length of an X-ray loop or is possibly enhanced at the summit; the density and temperature are typically 10^{17} m^{-3} and 2.5×10^7 K early in the flare and decrease to 10^{16} m^{-3} and 5×10^6 K after a few hours (§2.4.2);

11) two-ribbon flares can give rise to coronal transients, type II and IV radio bursts, and high energy particles in space (with second-stage acceleration), whereas compact flares cannot (§2.4.2);

12) most hard X-ray bursts are asymmetric, with a smaller rise-time than decay-time, but some short ones may be symmetric; most of the hard X-rays are produced at low altitude, but occasionally some may arise from an altitude of 10,000 km or higher (§2.5.1); hard X-ray and microwave bursts may show some quasi-periodicity, with typical periods between 10 and 100 secs, though sometimes longer (§2.5.3);

13) the EUV flare-emission from the transition layer has an impulsive component and a thermal component, similar to the X-ray emission (§2.6); various line ratios have been used to estimate the electron density, with results ranging from 10^{16} m^{-3} to 10^{19} m^{-3} (§2.6.2);

14) moving type IV radio bursts are associated with erupting filaments and often appear to come from a cloud of plasma containing a magnetic field that is oppositely directed from the ambient field (§2.7.1); type III bursts occur mostly in the peripheral parts of active regions (§2.7.3);

15) all mass-ejection coronal transients probably occur with the disruption of filaments and the faster ones are associated with two-ribbon flares; broad regions of slightly enhanced density, called *forerunners*, move out ahead of transients (§2.8.2); erupting filaments may generate a slow-mode wave (§2.8.3);

16) all proton flares are of two-ribbon type occurring in complex active regions with high magnetic field gradients, but they tend to avoid the sunspot maximum when the active regions may be too active to allow conditions for such big events to be built up (§2.9.1); low-energy proton streams may be accelerated at interplanetary shock waves.

Current sheets Chapter 3 discusses the behaviour of current sheets and their role in solar flare models. There are several ways in which such sheets may form. The region near an X-type neutral point may collapse into a current sheet (§3.2.1) over a few Alfvén transit-times. The most detailed analysis of this instability is by Syrovatsky, who finds that plasma is expelled rapidly from the neutral-point region; the resulting increase in the electron conduction speed ($j/(ne)$) forms the basis for his dynamic dissipation theory. Current sheets may also develop when a magnetostatic equilibrium no longer exists, a situation known as *non-equilibrium* (§3.2.2); Parker and Syrovatsky have demonstrated, for instance, that when the photospheric footpoints move slowly the coronal field cannot always adjust to form a new force-free equilibrium. When topologically distinct flux systems are pushed together, a current sheet can grow at the boundary between them (§3.2.3); this process has been simulated for the curved sheet between unequal dipoles in two dimensions and for the annular sheet between equal dipoles in three dimensions. The final means of current sheet formation is by an instability, which may be ideal, thermal or resistive in nature (§3.2.4).

The behaviour of a current sheet once it has formed is outlined in §3.3. The magnetic field strength tends to decrease in time as the field is annihilated and the magnetic energy is converted ohmically into heat (§3.3.1). If the sheet is curved, it can be unstable to the interchange mode, which creates small-scale ripples (§3.3.2). Reconnection may either grow spontaneously (as a result of the tearing or rippling mode instabilities) or it may be driven from outside by an external disturbance or it may be driven locally by a resistivity enhancement (§3.3.3). The nonlinear growth of reconnection depends crucially on the assumed boundary conditions (§3.3.3); in some cases a phase of steady nonlinear reconnection may be produced, known as Petschek's mechanism (§3.3.4), which is now on a sound mathematical basis as the only model for fast reconnection. A recent analysis of the central diffusion region shows that it may be an order of magnitude thicker than previously thought and that there is a β-limitation, with no steady solution when β is too small. Laboratory experiments and numerical experiments are summarized in §§3.3.5 and 3.3.6.

The role of current sheets in flare models is outlined in §3.4 and, in particular, the emerging flux model is described in §3.5, where it is pointed out that the mechanism also works when neighbouring flux systems are evolving by horizontal motions rather than vertical ones.

Simple-loop flares Chapters 4 and 5 deal with simple-loop (or compact) flares, for which a single flux tube flares and often remains in one position. Van Hoven (Chapter 4) surveys the magnetic instabilities which may produce the flare, including both ideal and resistive types. Then Craig (Chapter 5) assumes some sudden heat deposition in a loop and describes the dynamic response of the loop plasma.

Simple-loop flares: Magnetic instabilities Chapter 4 begins by describing the physical conditions in an active-region coronal loop before it flares. The electron number density lies between 10^{15} and 10^{16} m^{-3}, while the temperature ranges from less than 10^5 K to more than 10^6 K (§4.2.1). Loop lengths lie between 10^4 and 10^5 km, whereas their widths are typically a factor of ten smaller. At least some loops have an annular structure, with a cool central core having a temperature of 10^5 K or less; the density variation across the loop is not well-known and may be roughly constant (§4.2.2). These EUV results imply a pressure minimum on the axis, whereas X-ray observations in the 2–3 $\times 10^6$ K range suggest that loops possess a pressure that is up to a factor of ten larger than the external coronal value. The core of a loop may possibly contain a mixture of hot and cool material and it may be surrounded by a high pressure sheath. The coronal magnetic field in loops is not at all well-known (§4.2.3). The ratio β (of plasma to magnetic pressure) is at most 10^{-2}–10^{-1} in preflare loops, which maintain their form through a ten-fold flaring increase in both density and temperature. Equating the magnetic pressure to the observed flare pressure gives a surprisingly large magnetic field strength of 0.02 T (200 G) at 50,000 km altitude.

The preflare equilibrium for a coronal loop is generally modelled by a cylindrically symmetric magnetic field, which is either force-free or magnetostatic (with a pressure gradient balancing the Lorentz force). Several such equilibrium fields are described in §4.3. Force-free fields include the linear (constant-α) and uniform-twist fields, but more general nonlinear fields are also possible. Magnetostatic fields have been considered by Chiuderi *et al.* (a modified constant-α field) and by Hood and Priest (a uniform-twist field and a variable-twist field).

The ideal MHD stability of coronal loops is treated in some detail in §4.4 using the energy principle of Bernstein *et al.* Newcomb has considered an infinitely long cylindrical flux tube, for which the minimizing perturbation is incompressible and the problem is reduced to the solution of an Euler–Lagrange equation (4.22) for the radial perturbation; the equation is singular both at $r=0$ and at all layers where $\mathbf{k}\cdot\mathbf{B}$ vanishes. If the radial pressure gradient at the singular layers is larger than a negative value which depends on the field structure (Equation (4.27)), the equilibrium is stable to localized interchanges (Suydam's criterion). Global perturbations can be more dangerous than local ones in promoting instability. For example, the linear force-free field of infinite length is unstable to a kink perturbation with a long enough axial wavelength (Voslamber and Callebaut); so too is any cylindrical force-free flux tube of infinite length (Anzer). The question therefore arises: what makes solar coronal loops stable for long periods of time? In an attempt to answer this, two extra effects have been included in the stability analysis, namely radial pressure gradients and photospheric line-tying. The stabilizing effect of a positive pressure gradient in a modified constant-α field has been demonstrated by Van Hoven *et al.*, but the main reason for coronal loop stability is likely to be line-tying. It has been modelled for force-free fields by Raadu, and Hood and Priest, who find that a uniform-twist field is stable when the twist is less than 2π but has become kink unstable by the time the twist exceeds 3.3π; results are also given for the effect of line-tying on magnetostatic fields.

Ideal instabilities possess very fast growth-rates, of order the Alfvén travel-time, but they may saturate nonlinearly or lead to the production of current sheets. Furthermore, when the shear $(rB_z/B_\theta)'$ is large they are dominated by resistive instabilities, which involve magnetic reconnection and are discussed in §4.5. They occur in extremely narrow diffusion regions, which may be treated as planar and form around the singular layers where $\mathbf{k}\cdot\mathbf{B}$ vanishes (§4.5.1). Van Hoven describes tearing mode instability in a linear force-free field, in both planar and cylindrical geometries. The fastest growing modes have a growth-time

$$\tau \sim (\tau_d \tau_A)^{1/2}$$

intermediate between the resistive diffusion time $\tau_d = \mu\sigma a^2$ and the Alfvén travel-time $\tau_A = a/v_A$ for an overall length-scale of a. The diffusion region has a width of only εa, where

$$\varepsilon \sim (\tau_A/\tau_d)^{1/4},$$

so that the diffusion time $(\varepsilon^2\tau_r)$ for the region is of order the instability time τ (§4.5.2). Furthermore, the wavelength along the singular layer is large ($ka \ll 1$). The mechanical energetics of the instability may be understood as a rough balance between the input of magnetic energy to the diffusion layer and the output of plasma kinetic energy. At high magnetic Reynolds numbers, the nonlinear development shows typically a balance between the transverse magnetic field energy and the kinetic energy, each saturating at about 15% of the initial energy density in 3 or 4 linear growth-times. Finally, in §4.6 some phenomenological models of loop flares are outlined and criticised.

Simple-loop flares: Thermal evolution Chapter 5 by Craig describes theories for the thermal evolution of simple-loop flares, the object being to model the thermal response of plasma in a rigid magnetic loop to a sudden extra heat input. §5.2 summarises the observations, such as those which show that the density in a flaring loop reaches a maximum value after the temperature. Difficulties in interpreting the differential emission measure are mentioned. The basic equations are those of continuity, momentum and energy for one-dimensional, time-dependent motion along a rigid loop (§5.3.1). The conductive time-scale τ_{cond} for typical flare parameters ($T = 10^7$ K, $n_e = 10^{16}$–10^{17} m^{-3}, length-scale $\sim 10^{7.5}$ m) is 10^2–10^3 sec, while the radiative time-scale is 4×10^3–4×10^4 sec. This implies that conduction predominates in the initial flare cooling, but radiation becomes comparable in the later stages. However, the dynamical time-scale for mass motions is also of order 10^2 sec and so dynamical effects are likely to be of great importance.

Craig discusses quasi-static models of the solar atmosphere in §5.4.

Shmeleva and Syrovatsky, for example, assume a balance between radiation and conduction from a hot overlying source and suppose that either the pressure or the density are uniform. Instead, Brown suggests that the flash phase of the chromospheric flare results from a balance between radiation and the collisional heating of non-thermal electrons. Several energy-balance models for static loops have recently been set up (§5.4.4), but it appears that thermally-isolated loops are thermally unstable (§5.4.5); also the distribution of loops with different temperatures in a given volume is important, since the emission measure is unlikely to be associated with a single loop (§5.5.6). A new possible non-magnetic cause for a loop-flare, namely thermal non-equilibrium in cool loop-cores, is outlined in §5.4.6.

A number of analytic time-dependent models for different stages in the flare process have been proposed (§5.5). In the initial heating phase, the density may be constant and conduction the dominant energy transfer mechanism away from the heat source (Zeldovich and Raiser). Then during the flare decay, the hot plasma is at roughly uniform pressure. Initially, it cools by expansion and conduction (Craig and McClymont) and later a balance is achieved between conduction and evaporation (Antiochos and Sturrock), until the mass flows die down and a static equilibrium is approached.

Numerical experiments on the hydrodynamic behaviour of a flaring loop are described in §5.6. Kostyuk and Pikelner model the response of a loop atmosphere to steady, chromospheric heating by a beam of electrons: the atmosphere above about 5×10^4 K expands upwards to form a hot coronal region at 2×10^7 K; later, the plasma falls back down, compressing and heating the underlying chromosphere in the process. Craig describes in detail the formation of a (uniform-pressure) loop-atmosphere from an initially uniform state in response to a transient heat impulse at the loop summit. The loop is assumed to remain thermally isolated and has rigid boundaries at its feet. Initially, the plasma near the loop summit is heated at constant density, but, when the temperature exceeds 10^5 K, the plasma becomes radiatively unstable and heats up rapidly to about 2×10^6 K, when conduction quashes the instability. A pressure wave is then driven down towards the ends of the loop and behind it a radiative condensation forms. Subsequently, material is evaporated dynamically from the chromosphere to the corona and eventually a steady-state atmosphere is established. Various time-scales are evident in the behaviour of the summit temperature: the initial rise-time is determined by the imposed heating time-scale of 100 sec and the later overall decay on 1,000 sec is comparable with the radiative and conductive times; however, superimposed on this behaviour, one finds an oscillation at 200–300 sec (the hydrodynamic time-scale). Flare heating calculations of McClymont due to an impulsive heat injection at the loop summit are also described and have similar features: coronal material expands from the injection site and generates an evaporative upflow when it reaches the loop feet. The main conclusions of these numerical experiments are that natural hydrodynamic oscillations are superimposed on the conductive and radiative relaxation to a steady state and that dynamic effects are crucial in the loop evolution as mass is exchanged between the chromosphere and corona. Also the numerical difficulty in treating the narrow transition layer adequately is stressed.

Two-ribbon flares Chapters 6 and 7 deal with two-ribbon flares. Birn and Schindler (Chapter 6) consider a preflare magnetic arcade containing a filament, in an attempt to explain why it becomes unstable and erupts at the beginning of the flare. Then Pneuman (Chapter 7) models the evolution of so-called "post"-flare loops, which are most prominent in the later stages of the event.

Two-ribbon flares: Magnetostatic equilibria Chapter 6 discusses general results for the passive evolution of a two-dimensional magnetic configuration (such as an arcade) through a sequence of magnetostatic equilibria. The equation for a balance between a pressure gradient and a $\mathbf{j}\times\mathbf{B}$ force may be written in the form (§6.3)

$$-\nabla^2 A = \frac{\partial}{\partial A}\{\mu p(A, y) + \tfrac{1}{2}B_z^2(A)\}, \qquad (1.20)$$

where the y-axis is taken perpendicular to the solar surface and the z-axis is along the filament axis; A is the z-component of the magnetic vector potential. In general, this equation is extremely difficult to solve, since the plasma pressure p couples it to an energy equation and one would like to impose the positions of foot-points on the photospheric boundary so as to study the magnetic evolution as the foot-points slowly move. So far, a simpler problem has been considered in which the right-hand side of (1.20) is regarded as a prescribed function of A and y which changes linearly with some parameter λ:

$$-\nabla^2 A = \lambda f(A, y). \qquad (1.21)$$

Under certain circumstances there exists at least one solution of (1.21) when λ is smaller than a critical value λ^* but no solutions when $\lambda > \lambda^*$. Thus Birn *et al.* suggested that the critical point $\lambda = \lambda^*$ (where the solution curve bends back) represents the onset of a magnetic eruption, since there is no neighbouring equilibrium if λ can be forced to exceed λ^*. However, it is not yet known whether this feature is still in general present if the foot-point displacement is prescribed rather than the function $\lambda f(A, y)$.

Several particular sequences of force-free equilibria have been discovered (§6.4). The topic was initiated mainly by the paper of Barnes and Sturrock, which in turn arose from a previous paper by Sturrock and Woodbury (1967). Barnes and Sturrock give a numerical solution for a single twisted sunspot field. When the twist exceeds about π, the force-free field has an energy in excess of an open field energy and so they suggest that it erupts outwards into the open configuration: however, the boundary conditions are not identical for the two configurations and no stability analysis was performed. Low and Nakagawa computed numerically some arcade solutions for two prescribed forms of $B_z(A)$. Low found a pair of analytic solutions with $B_z = \sqrt{\lambda e^{-A}}$, which have helical field lines and $\lambda^* = 4$; one solution has the axis of the helix below the photosphere and the other has it above, so that the projection in the x-y plane has closed field lines. Priest and Milne extended Low's solutions to a family of pairs of cylindrically symmetric solutions; they also obtained two sequences of solutions with the same foot-point displacement! Jockers calculated numerically some pairs of solutions for a given $f(A, y)$ and found that a maximum value λ^* does exist, but the upper solutions do not always contain closed field lines in the x-y plane. Birn *et al.* found pairs of analytic solutions with $B_z = \sqrt{\lambda e^{-A}}$. They are not cylindrically symmetric but have a singularity at the origin and neither solution possesses closed field lines. The maximum value of λ is $\lambda^* \approx 0.44$; the footpoint displacements in the x-z plane lie on a straight line, whose slope increases monotonically as one follows λ^*! Priest and Milne generalised the solutions of Birn *et al.*, and found similar behaviour. Heyvaerts *et al.*, found numerically *three* topologically different solutions in a finite region, with Dirichlet conditions on the photospheric boundary and Neumann conditions on the open boundary. The need in future is to extend these numerical calculations to the case when the footpoint displacement is imposed and, ultimately, it would be interesting to couple the magnetostatics with the energetics.

The MHD stability of arcade structures is discussed in §6.5. The general result is established that any equilibrium situated on the lower branch is stable against two-dimensional disturbances. This suggests that the critical point $\lambda = \lambda^*$ may be reached by a quasi-static evolution along the lower equilibrium branch. Furthermore, Birn and Schindler establish that Low's lower-branch solutions are stable to locally three-dimensional disturbances. Hood and Priest have incorporated the stabilizing effect of photospheric line-tying in a stability analysis of some particular magnetic arcade configurations. They were unable to demonstrate instability for simple sheared arcades, but a large flux tube contained within an arcade was shown to be kink unstable if the length, twist or filament height become too great.

In §6.6 a general method is established for calculating the electric field and plasma velocity that are implied by a slow passive evolution of a magnetic configuration through a series of equilibria. The pressure and parallel velocity component $v_{\|}$ can be prescribed only at one end of field lines which intersect the photospheric boundary. Furthermore, in general, a change of boundary conditions for the Euler potentials (α and β), the pressure and the electrostatic potential Φ leads to a situation in which continuous equilibrium conditions are violated, at least locally. The result is that MHD waves or thin current sheets may be produced.

Two-ribbon flares: (post)-flare loops Chapter 7 by Pneuman discusses the (post)-flare-loop phenomenon in two-ribbon flares. These loop systems rise upward into the corona, at first rapidly (10–20 km s^{-1}) and later much more slowly (0–1 km s^{-1}), and they may endure for 10–20 hours. The term "post"-flare is misleading since it implies that the loops occur after the flare and are a by-product of it, whereas in fact they begin near the onset of the flare and may well produce the chromospheric flare. Pneuman suggests that the flare loops are formed by magnetic reconnection as the configuration closes back down again after being blown open by the initial eruption.

The observations are summarised in §7.2. The loops develop from a small bright mound after a filament eruption and they consist of a nested set of loops with different temperatures. The cool Hα loops are located below EUV loops; these are in turn situated below hot soft X-ray loops, which may reach an altitude of 100,000 km or more and outlive the optical flare by many hours. the Hα loops are rooted on the inside edges of the Hα ribbons, whereas the X-ray loops join the middle and outer portions of the ribbons. In the loop legs one sees strong downflows at the free-fall speed; they deliver to the chromosphere a mass of typically 10^{12-13} kg, which is comparable with the total mass of the corona. At the summit of the loop system, the temperature may reach 10^7 K; the density is typically 10^{16}–10^{17} m^{-3} (10^{10}–10^{11} cm^{-3}) and the magnetic field strength may be 0.01 tesla (100 Gauss). In X-rays the loop feet are cooler by a few million degrees but the density appears to be roughly uniform along the loop. Both the temperature and density decay slowly with time, but this should not be interpreted as the cooling of a single loop; rather, it represents a fall in the rate of energy supply to the newly-forming loops.

Equations for an isothermal flow along moving magnetic field lines are derived in §7.3 and these are used to derive the pressure and flow speed along the field in terms of a prescribed magnetic configuration and field-line velocity. In particular, the magnetic field is taken to be a dipole superimposed on a uniform field in such a way that the neutral point is located at a radial distance which increases with time. The initial flow and pressure are taken from an isothermal solar wind solution and the base pressure is assumed to remain constant. The resulting flow along a single flux tube is found to be much higher than the solar wind flow, with speeds in excess of 100 km s^{-1} in the low corona. When the tube closes, a shock propagates downwards and leaves behind it a stationary state; the shock is modelled as an ordinary gas-dynamic type.

Since the above analysis is purely kinematic, Pneuman proceeds to consider some of the magnetohydrodynamic aspects in §7.4, such as an order of magnitude model for the rising filament at the start of the flare (§7.4.1); this shows that, if the flux tube width is proportional to distance, the driving force can continue to exceed the restoring forces of gravity and tension as the prominence erupts. It is also demonstrated that the forces in the open-field configuration are such as to make the reconnection start at the coronal base (§7.4.2).

For the flare of 29th July, 1973, the total mass in the loop-prominence system is estimated to be 7.5×10^{13} kg, which is larger than previous estimates because of the inclusion of the mass contained in the X-ray loops (§7.5). The closing down of field lines can enhance the mass flow by about a factor of ten over the solar wind value, but this still leads to a total mass of only 3.2×10^{12} kg for the above event. Pneuman therefore suggests that the additional mass may be supplied by chromospheric evaporation.

The cooling-time for a single loop (by radiation or conduction) is typically less than half-an-hour, which is very much shorter than the lifetime of the loop-prominence system (10–20 hours) and so energy is probably being fed into the loops throughout the event by magnetic reconnection (§7.6). An upper limit for the magnetic energy released in an open configuration that is closing down gives reasonable agreement with observation, especially late in the event.

Particle acceleration in solar flares Finally, Chapter 8 by Heyvaerts gives a comprehensive survey of mechanisms for accelerating particles. It is split into three sections: §§8.1–8.10 deals with the implications of the observations, while §§8.11–8.20 summarizes processes for first-phase acceleration and §§8.21–8.26 treats the second phase.

Observational evidence for particle acceleration is reviewed in §8.1, where it is suggested that the number of particles involved is likely to be 10^{33}–10^{34}. Proton spectra show a power-law shape from 20 MeV to 80 MeV with a spectral index between 2.5 and 3.7, while the electrons between 22 and 45 keV have an index between 3 and 4.5 (§8.2.1). The first phase accelerates electrons to energies of about 100 keV and then the second phase, which is present in some large events, accelerates both electrons and protons to much higher energies (§8.2.2). The hard X-ray radiation is mainly due to bremsstrahlung of electrons on the ambient plasma ions and it seems to come from a compact, dense source, with a dimension of less than 10,000 km and a density in excess of 10^{16} cm^{-3} (§8.3.3); in a few cases, the brightest part of the X-ray burst can be explained as the adiabatic compression and expansion of an extremely hot (10^8 K) plasma (§8.3.5). Assuming a thick-target bremsstrahlung process with non-thermal electrons, the number of electrons required to account for the observed X-rays is immense (10^{36}–10^{38}), and so this process may well overestimate the number of electrons (§8.4.2), especially since the number required to explain the EUV and Hα impulsive emission is a factor of 100 lower (§8.8.1). Impulsive microwave radiation is probably produced in the corona by gyrosynchrotron radiation (§8.5.1), but it is necessary to know the geometry of the source before one can distinguish between thermal and non-thermal models (§8.5.3). Also, various γ-ray lines indicate the presence of high energy nuclei accelerated in the flare region (§8.6). In some events, the ratio of He^3 to He^4 is increased by a factor of 10^3, which may be caused by either a nuclear reaction effect or a preferential plasma mechanism (§8.7.1); in addition, some flares are Fe-rich and may imply that the second-phase cosmic rays are accelerated in a rather cool plasma (§8.7.2). Recent evidence has been presented for significant heating deep in the atmosphere at photospheric levels (§8.8). In §8.9 and §8.10, Heyvaerts discusses the question as to whether fast electrons simply stream down to the chromosphere or form a compact, extremely hot (10^8 K) plasma in the corona (a fireball). Observations cannot yet decide (§8.9); on theoretical grounds, the electrons in the source cannot all be accelerated in the same direction by a DC electric field, but, once they leave the acceleration region, they may be shielded by a return current (§8.10.1) and there are various effects for converting the electron energy efficiently into heat (§8.10.2). A fireball is likely to be trapped between two ion-acoustic turbulent walls, which expand downwards at the ion-sound speed and reach the chromosphere after typically 3–10 seconds; ahead of the conduction-fronts, electron beams propagate such that the return-current has its maximum stable value (§8.10.3).

For the first phase of particle acceleration the electric fields may be DC (§§8.12–8.14) or turbulent (§§8.15–8.17). *A DC field* may produce a beam of particles which interacts with the surrounding plasma by either an ion-acoustic instability of the return-current or by an oscillating-two-stream-instability; the resulting anomalous dissipation (due to low-frequency turbulence) increases the local magnetic energy conversion, which accelerates more particles and spreads the dissipation further (§8.11.1). For this model one needs to detail the trigger mechanism, the absence of effective heat removal, the self-sustaining of turbulent conditions and the dissipation spreading (§8.11.2). The possibility of partial runaway through low-frequency turbulence is demonstrated in §8.12, because the collision-frequency can scale as v^{-3}, and so one may introduce the concept of a turbulent Dreicer field; Heyvaerts also stresses that the persistence of acceleration depends on boundary conditions (such as the behaviour of E, j and n) just as much as on the internal microphysics (§8.12.2). The properties of current-driven microinstabilities are summarised in §8.13 and §8.14, both in a weak magnetic field and a strong field. In most cases when the field is weak, the ion-acoustic instability appears to be the relevant one (§8.13.2) and mechanisms responsible for its saturation are given in §8.13.3; when the field is strong, the electrostatic ion-cyclotron instability is of interest (§8.14.2). *Stochastic acceleration* of selected electrons by high-frequency turbulence is at present considered less viable than DC acceleration (§8.11.2), because of the difficulty of generating a broad spectrum of Langmuir waves (§8.15); it is apparently impossible to generate high-frequency turbulence from low-frequency turbulence (§8.15.2). Models of strong turbulence are outlined in §8.16; for example, the modulation instability tends to trap plasma waves in low-density cavities, which may collapse, so lowering their phase velocities and allowing an efficient coupling to particles. Another possible particle-accelerator is the *double-layer* (§8.18), but it is unlikely to be relevant for the solar flare, because its formation is rather obscure (§8.18.5) and requires a high current-threshold similar to that of the Buneman instability (§8.18.3); the easiest example of a self-consistent solution of the Vlasov equations able to sustain a large potential drop is the BGK mode (§8.18.2). In §8.19.1 Heyvaerts points out that the current densities required for microturbulence onset are extremely high, so that the currents must be highly concentrated rather than pervading the whole volume: such current concentrations may exist initially as shocks or current layers (possibly reconnecting) or they may be caused by a global (ideal or resistive) MHD instability; one possible concentration mechanism is a superheating instability known as the Joule mode, but the resulting current densities are unlikely to be high enough (§8.19.2). The *tearing-mode instability* or any localised reconnection process produces bulk heating, which may explain the acceleration of 100 keV electrons, provided that the plasma β is low enough ($<10^{-6}$), but little is known about the details (§8.20.1); cylindrical ($m=1$) tearing modes have a large growth-rate even when the magnetic islands are wide and so they are most likely to be able to produce fields in excess of the Dreicer value (§8.20.2).

Second-phase acceleration may be caused by *collisionless shock waves* or by the *Fermi mechanism* (§8.21), which has been better studied so far. The first-order Fermi mechanism accelerates particles by head-on collisions of moving magnetic mirrors. The second-order mechanism treats the relative velocity of particle and mirror as random; this is much more relevant in practice and gives an energy gain proportional to the square of the relative velocity (§8.22.1). The Fermi mechanism is consistent with observations provided it takes place in the high, tenuous corona, where the density is less than 10^{15} m^{-3}, and the γ-ray emission is consistent with a thick-target bombardment of the atmosphere (§8.22.2). Various features of the Fermi mechanism are described in §8.23, including the necessity for scattering to make the energy gain irreversible (§8.23.2) and the generalisation to various interactions between particles and fast magnetosonic waves or Alfvén waves (§8.23.3). Near a shock front, particles may bounce between magnetic irregularities in the upstream and downstream sides and so suffer a systematic first-order Fermi acceleration

(§8.24.1), which leads to power-law spectra (§8.24.2). An interesting explanation for selective preheating of He^3 (§8.25.2) involves the electrostatic ion-cyclotron instability in a low-β plasma ($\leqslant 10^{-3}$). Finally, the possibility of acceleration in and near shock fronts is discussed in §8.26, in particular by the modified two-stream and lower-hybrid drift instabilities, which have long wavelengths parallel to the magnetic field (§8.26.4).

References

Alfvén, H. and Carlqvist, P., *Solar Phys.* **1**, 220 (1967).
Anzer, U., *Solar Phys.* **30**, 459.
Barnes, C. W. and Sturrock, P. A., *Astrophys. J.* **174**, 659 (1972).
Brown, J. C. and Smith, D. F., *Rep. Prog. Phys.* **43**, 125 (1980).
Carlqvist, P. *Solar Phys.* **7**, 377 (1969).
Carlqvist, P., *Cosmic Electrodynamics*, **3**, 377 (1972).
Carlqvist, P., *Wave Instabilities in Space Plasmas*, Palmadesso, P. J. and Papadopoulos, K. (eds), 83, (1979).
Colgate, S. A., *Astrophys. J.* **221**, 1068 (1978).
Cox, D. P. and Tucker, W. H., *Astrophys. J.* **157**, 1157 (1969).
Dungey, J. W., *Phil. Mag.* **44**, 725 (1953).
Giovanelli, R. G., *Nature*, **158**, 81 (1946).
Gold, T. and Hoyle, F., *Mon. Not. Roy. Astron. Soc.* **120**, 89 (1960).
Harvey, J. W., in *Flare-related Magnetic Field Dynamics*, Nakagawa, Y. and Rust, D. M. (eds), NCAR, Boulder, Colo. (1974).
Heyvaerts, J., Priest, E. R. and Rust, D. M., *Astrophys. J.* **216**, 123 (1977).
Hood, A. W. and Priest, E. R., *Solar Phys.* **64**, 303 (1979).
Kopp, R. A. and Pneuman, G. W., *Solar Phys.* **50**, 85 (1976).
Levine, R. H. and Nakagawa, Y., *Astrophys. J.* **190**, 703 (1974).
Martres, M. J., Michard, R., Soru-Iscovici, I. and Tsap, T. *I.A.U. Symp.* **35**, 318 (1968).
Martres, M. J., Soru-Escaut, I. and Rayrole, J. *I.A.U. Symp.* **43**, 435 (1971).
McWhirter, R. W. P., Thonemann, P. C. and Wilson, R., *Astron. & Astroph.* **40**, 63 (1975).
Moreton, G. E. and Severny, A. B. *Solar Phys.* **3**, 282 (1968).
Parker, E. N., *Astrophys. J.* **174**, 499 (1972).
Parker, E. N., *Cosmical Magnetic Fields*, Oxford University Press (1979).
Petschek, H. E., in Hess, W. N. (ed) *AAS-NASA Symp. on the Physics of Solar Flares*, NASA Sp. 50, 425 (1964).
Piddington, J. H., *Solar Phys.* **31**, 229 (1973).
Piddington, J. H., *Solar Phys.* **38**, 465 (1974).
Pneuman, G. W., *Solar Phys.* **2**, 462 (1967).
Priest, E. R., *Solar Phys.* **47**, 41 (1976).
Priest, E. R., *Solar Magnetohydrodynamics*, D. Reidel, Dordrecht, Holland (1981).
Priest, E. R. and Raadu, M. A., *Solar Phys.* **43**, 177 (1975).
Priest, E. R. and Soward, A. M., *I.A.U. Symp.* **71**, 353 (1976).
Raadu, M. A., *Solar Phys.* **22**, 425 (1972).
Raymond, J. C., Cox, D. P. and Smith, B. W., *Astrophys. J.* **204**, 290 (1976).
Roberts, B. and Priest, E. R., *J. Plasma Phys.* **14**, 417 (1975).
Rust, D. M., *I.A.U. Symp.* **35**, 77 (1968).
Rust, D. M., *Solar Phys.* **25**, 141 (1972).
Rust, D. M., *Solar Phys.* **33**, 205 (1973).
Rust, D. M., *Phil. Trans. Roy. Soc., London*, **A281**, 427 (1976).
Rust, D. M. and Bar, V., *Solar Phys.* **33**, 445 (1973).
Rust, D. M. and Bridges, C. A., *Solar Phys.* **43**, 129 (1975).
Smith, D. F. and Priest, E. R., *Astrophys. J.* **176**, 487 (1972).

Soward, A. M. and Priest, E. R., *Phil. Trans. Roy. Soc., London*, **284**, 369 (1977).
Spicer, D. S., *Solar Phys.* **53**, 249 (1977a).
Spicer, D. S., *Solar Phys.* **53**, 305 (1977b).
Sturrock, P. A., *Nature*, **211**, 695 (1966).
Sturrock, P. A., *I.A.U. Symp.* **35**, 471 (1968).
Sturrock, P. A. *Solar Phys.* **23**, 438 (1972).
Sturrock, P. A., in *Flare-related Magnetic Field Dynamics*, Nakagawa, Y. and Rust, D. M. (eds), N.C.A.R., Boulder, Colo. (1974).
Sturrock, P. A., *Proceedings of the Skylab Workshop on Solar Flares*, Colo. Ass. Univ. Press (1980).
Sturrock, P. A. and Woodbury, E. T., in *Enrico Fermi Course* 39, 160 (1967).
Švestka, Z., *Solar Flares*, D. Reidel, Dordrecht, Holland (1976).
Sweet, P. A., *I.A.U. Symp.* **6**, 123 (1958).
Sweet, P. A., *Ann. Rev. Astron. Astroph.* **7**, 149 (1969).
Syrovatsky, S. I., *Soviet Astron.* **10**, 270 (1966).
Syrovatsky, S. I., in *Solar Flares and Space Research*, de Jager, C. (ed), 346 (1969).
Syrovatsky, S. I., in *Solar Terrestrial Physics*, E. R. Dyer (ed), 119 (1970).
Syrovatsky, S. I., *Comm. on Astrophys. and Space Sci.* **4**, 65 (1972).
Syrovatsky, S. I., *Solar Phys.* **58**, 89 (1978).
Tanaka, K., *I.A.U. Symp.* **56**, 153 (1974).
Tanaka, K. and Nakagawa, Y., *Solar Phys.* **33**, 187 (1973).
Tur, T. and Priest, E. R., *Solar Phys.* **48**, 89 (1976).
Uchida, Y. and Kaburaki, O., *Solar Phys.* **35**, 451 (1974).
Vasyliunas, V. M., *Rev. Geophys. and Space Phys.* **13**, 303 (1975).
Wentzel, D. G. *Solar Phys.* **39**, 129 (1974).
Zirin, H., *Vistas in Astron.* **16**, 1 (1974).
Zirin, H. and Tanaka, K., *Solar Phys.* **32**, 173 (1973).

CHAPTER 2

Flare Observations

ZDENĚK ŠVESTKA

Space Research Laboratory, Beneluxlaan 21, 3527 HS Utrecht, The Netherlands Centre for Astrophysics and Space Science, C-011, University of California, San Diego, La Jolla, Calif. 92093, USA

2.1 INTRODUCTION

2.1.1 General characteristics of flares

The solar flare represents a sudden release of energy in the atmospheric layers of the Sun. In tens of seconds or a few minutes the temperature rises throughout the solar atmosphere, reaching its peak, often well above 10^7 K, in the solar corona. There the temperature can be enhanced by an order of magnitude. The lower we go in the atmosphere, the smaller is the temperature enhancement, but it reaches as deep as the temperature minimum, where still a temperature rise of some two percent can be recognized.

Whereas the ultimate site of the energy release is not yet definitely known, the temperature enhancement is clearly accomplished in coronal loops and arches rooted in moderately strong magnetic fields in active regions. These flaring loops and arches have generally also enhanced density in comparison with their surroundings. Therefore, the flaring region is greatly structures and the resulting flare emission may come from threads and filaments, whose thickness is below the resolving power of the observing instruments.

At least in some flares (and possibly in all of them) this thermal emission is accompanied by an apparently non-thermal component, which, when present, always appears at the onset of or during the rising part (the flash phase) of the thermal flare. This component is very impulsive, consisting of elementary peaks with lifetimes of the order of a second. Whereas the lifetime of a thermal flare can last for hours, this impulsive component rarely survives more than a few minutes; in smaller flares it lasts only tens of seconds or less and in many flares it is either not present, or its intensity and/or duration is below the resolution of our instruments.

If one tries to interpret this impulsive component in thermal terms, the temperature needed exceeds 10^9 K! Therefore, a non-thermal process, impulsively accelerating electrons and nuclei to hundreds of keV (and to still higher energies in some extreme events) seems to be a more likely interpretation of the process involved. This impulsive, apparently non-thermal, process seems to be accomplished with greatly variable intensity in the various thermal-flare loops, depending on the gradient and complexity of the magnetic field involved. In many loops (and for some flares perhaps in all of them) it can be missing. In those loops where the most intense impulsive acceleration process is accomplished, the flare effects also penetrate to the greatest depth in the atmosphere — in strong events exciting emission even in the uppermost layers of the photosphere.

2.1.2 Flare emissions

Solar flares emit electromagnetic radiation within a very broad range of wavelengths: in extreme cases from 0.002 Å (2×10^{-11} cm, 6.1 MeV, the highest γ-ray line so far observed) up to more than 10 km (10^6 cm, 30 kHz, radio emission produced by flare-emitted electrons passing near the Earth). This radiation may be continuous emission in some parts of the spectrum, line emission in others, or a combination of both.

The different kinds of radiation come, of course, from different parts of the flare, and may originate in different phases of the flare development. The thermal flare plasma in the *solar corona* is visible in soft X-rays and its emission, depending on temperature, can extend up to energies of about 10 keV. Both the bremsstrahlung continuum and lines of heavily stripped ions contribute to this radiation. The highest-temperature and lowest-wavelength lines, prominent during the flash phase and maximum of flares, reach an optimum intensity at $T \simeq 6 \times 10^7$ K and are found in the line complex of the helium-like iron ion at 1.86 Å. Similar parts of the flare can be seen also on radio microwaves. The *flare transition layer*, with T decreasing from 10^6 K through 10^4 K, is seen best in EUV lines (300–1500 Å) such as, e.g., Si XII (2×10^6 K), Mg XII (10^6 K), Ne VII (5×10^5 K), O VI (3×10^5 K), C III (9×10^4 K), He II (2×10^4 K), and C II (3×10^4 K). The upper *chromosphere* is best described by the Lyman series of hydrogen, the middle chromosphere by the Balmer series, and the bottom chromospheric layers by the wings of the H and K lines of Ca II. Up to now, far the most observations of chromospheric flares have been carried out in the Hα line of the hydrogen Balmer series.

The impulsive, apparently non-thermal flare component, occurring during the flare flash phase, gives rise to additional kinds of radiation on both ends of the flare spectrum. The accelerated electrons produce hard X-rays through bremsstrahlung and microwave radio emission through gyrosynchrotron radiation in the strong ambient magnetic fields. The hard X-rays, mixed with the thermal soft X-ray component below 10 keV, extend commonly to tens of keV, in rare events to hundreds of keV, and in extreme cases above a few MeV.

Microwave impulsive bursts can be seen throughout the decametric and centimetric radio spectrum down to millimeter wavelengths. Many-a-time, associated with this impulsive microwave burst is a metric radio emission, called a type III radio burst, produced by streams of accelerated electrons moving with speeds up to 0.6 c out of the Sun through the corona and interplanetary space. The radio emission produced by these electron streams can extend down to ~ 30 kHz.

In contrast to the electrons, accelerated protons do not produce any observable kind of radiation, unless they are energetic enough to produce γ-ray emission or to penetrate deep into the atmosphere and excite emission in the low layers which usually do not participate in the flare radiation. One needs protons in excess of 20 or 30 MeV to be able to do that and, therefore, the γ-ray and white-light emissions are associated only with the most outstanding flare events. The strongest two lines of the γ-ray spectrum are the 2.23 MeV line, produced when neutrons are captured by protons in the solar atmosphere, and the 511 keV line produced by positrons. Both the neutrons and positrons must be first created in the flare process. The highest γ-ray line so far detected in flares is the 6.14 MeV line due to ^{16}O nuclei de-excitation.

Energetic flares also eject plasma into the upper corona and interplanetary space and this ejection creates a shock wave that propagates with supersonic speed in front of the plasma cloud. As this shock travels through the corona, it excites plasma oscillations which give rise to a drifting radio emission called a type II radio burst. A part of the accelerated electrons is trapped behind the shock in the magnetic field above the active region and produces (through gyro-synchrotron radiation) metric radio continuum, known as a radio type IV burst. This continuum emission can be either stationary, from electrons trapped in closed fields above the active region, or moving, from electrons in the rising plasma cloud. The plasma cloud itself can be seen high in the corona by means of white-light coronagraphs aboard spacecraft.

Space probes and satellites also record energetic electrons and nuclei coming from flares. Non-relativistic electrons are recorded commonly after flares associated with the type III radio bursts, while relativistic electrons and accelerated protons and heavier nuclei are usually detected after important two-ribbon flares associated with type II and type IV radio bursts.

These various kinds of the flare emission also represent the different ways in which the observed data about flares can be obtained. The space research, and the Skylab mission in particular, have yielded completely new data about flares, extending the range of observations through the EUV and X-rays to the γ-rays. Only since Skylab do we have real three-dimensional information about the flare process.

In spite of that, however, the ground-based data, and the Hα observations in particular, are not obsolete. There are three important reasons for that: (1) The

TABLE 2.I.
The importance classification of flares

Area at center of solar disk		
in millionths of hemisphere	in square degrees*	Importance
< 100	< 2.06	*S*
100—250	2.06—5.15	1
250—600	5.15—12.4	2
600—1200	12.4—24.7	3
> 1200	> 24.7	4

* 1 square degree $= 1.476 \times 10^8$ km^2 of solar surface.

spatial resolution of flare observations in Hα can be better than 1″, which is by one order of magnitude better than most observations from space. (2) The Hα observations can run all the time, whereas a sophisticated solar satellite is not a frequent occurrence in space. (3) The Hα observations have been carried out for several decades, so that we know a great variety of solar phenomena seen in this light, and we know how these phenomena behave in Hα in different phases of solar activity and how they behaved in different solar cycles. Therefore, we should always try to relate the new space observations to the Hα data, because only then can we be pretty sure that we know, in the general context of solar activity, what we have actually observed.

2.1.3 Flare classification

According to the appearance that they have in the light of the Hα line, all flares are classified in four groups by their 'importance', 1, 2, 3, and 4. Their area at the time of maximum brightness is the primary criterion for this important classification.

A transient brightening is called a flare if its area in the chromosphere exceeds 3×10^8 km^2. Smaller events are called subflares and their importance is denoted by an *S* (Table 2.I). The maximum brightness of the flare in the Hα line is added as a suffix to the importance class: f = faint, n = normal, b = brilliant. The lower limit in size for subflares is obviously far below the resolving power of solar patrol telescopes. It is known, for instance, that some 10% of coronal X-ray bright points (Golub *et al.*, 1975), which can be identified with ephemeral active regions (Harvey and Martin, 1973; Harvey *et al.*, 1975), produce flare-like brightenings that are not reported (and almost never noticed) by the Hα observers. Thus all lists of subflares are incomplete and also greatly inhomogeneous, because different solar observatories set different lower limits to their reports. The most complete lists of flares can be found in *Solar and Geophysical Data* (published by NOAA Boulder).

TABLE 2.II.
NOAA X-ray flare classification

Class C1 to C9	Peak flux 1 to 9×10^{-6} watts m^{-2}
M1 to M9	1 to 9×10^{-5} watts m^{-2}
X1 to X9+	1 to in excess of 9×10^{-4} watts m^{-2}

The duration of flares in Hα light varies from a few minutes for reported subflares up to more than 7 hours for the most powerful flares of importance 4. However, the life-time of the flare in Hα light may be quite different from the flare duration at other wavelengths. Thermal X-ray and gradual microwave bursts have durations similar to the Hα flare. Impulsive bursts on the other hand, in the hard X-ray, EUV and millimeter radio regions last for only tens of seconds in small flares, and for a few minutes in more important event. Type III radio bursts at metric wavelengths have similar durations. On the other hand, long-lived continuum bursts at metric wavelengths can last for many hours after disappearance of the chromospheric flare in Hα light.

Thus, clearly the Hα emission is only one aspect of the complex flare phenomenon and it does not necessarily reflect the flare behaviour at other wavelengths. Efforts therefore have been made to establish other flare classifications which would take other aspects of the flare phenomenon into account. A widely used one is the NOAA X-ray flare classification based on the flare emission in the 1–8 Å band as measured by the SOLRAD satellites (Baker, 1970; Simon and McIntosh, 1972). It is summarized in Table 2.II, and an example is presented in Figure 2.1. Thus, for example, a flare may be of *Hα importance* 2n (having Hα area of 10 square degrees) and *X-ray class* M8 (with 1–8 Å peak flux of 8×10^{-5} watts m^{-2}*).

Another useful classification system is the proposal made by Dodson and Hedeman (1971). They introduce a Comprehensive Flare Index (CFI) = A + B + C + D + E, where A = importance of ionizing radiation as indicated by time-associated SID, on the scale 1–3 (it might be reasonable to relate it instead to the X-ray class); B = importance of Hα flare, on the scale 1–3 (3 and 4 being classified as 3, subflare as 0); C = log of 10 cm radio flux, in units of 10^{-22} Wm^{-2} Hz^{-1}; D = effects in dynamic radio spectrum: Type II = 1, continuum = 2, Type IV = 3; E = log of 200 MHz flux in the same units as C.

The values of CFI then range from 0 for subflares without any significant ionizing or radio emission to 15 to 17 for flares that were the most outstanding in all aspects of electromagnetic flare radiation. Flares with CFI > 10 can be considered as verified major flares.

During the years 1956–1969, 20 flares had CFI > 15. All but two of them were associated with strong proton events (polar cap absorption) and 4 of

* It should be noted that MKS units have been adopted for uniformity with the rest of the book.

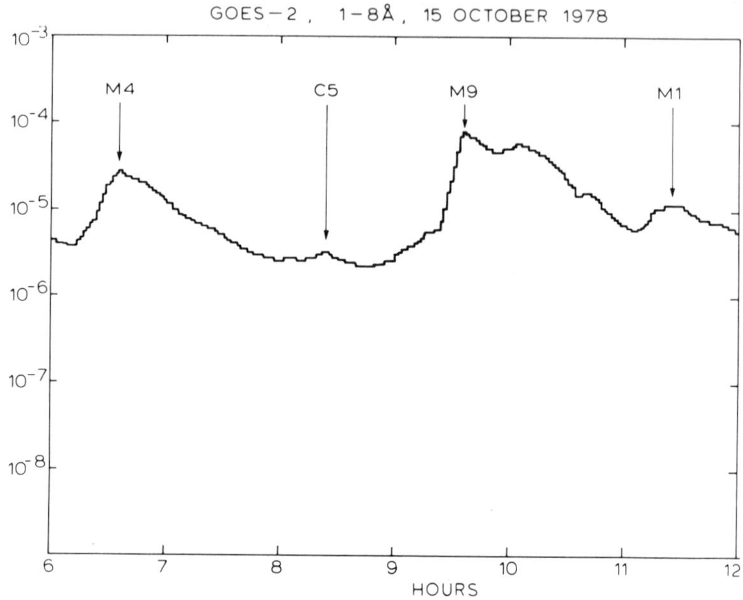

FIGURE 2.1 The X-ray flux (Wm^{-2}) in the 1–8 Å band as a function of time, showing examples of the X-ray classification for flares.

them were cosmic-ray flares. Also all but two were of Hα importance 3. Thus the most important flares are usually also important in the Hα light. But the opposite is not true. Out of a total of 164 flares of importance 3 during 1955–1969, 18 had CFI > 15, but 86 of them had CFI < 10. Out of the 210 flares that had CFI $\geqq 10$, and hence can be considered as major events, 78 were of importance 3, 106 of importance 2, 23 of importance 1, and 3 were subflares.

2.1.4 Two-ribbon and compact flares

One should be aware of the fact that the generally used term 'solar flare', with the general characteristics described above, can comprise a set of phenomena of widely different basic natures, which any of the classifications do not distinguish. There may be flares with, or without, the impulsive phase, i.e. flares with, or without, a particle acceleration involved. There may be flares, all with impulsive phase, in which the pre-accelerated particles are, or are not, accelerated to much higher energies (probably in shocks propagating from the flare source). We still do not know whether these are real qualitative differences or whether they only reflect a scaling in the intensity of the acceleration processes involved. However, as the Skylab Flare Workshop demonstrated (Sturrock, 1979), there appear definitely to be at least two basically different kinds of flare:

A *two-ribbon flare*, which we also call a *dynamic flare*, and a *compact flare* (or a *simple-loop flare*), which we might call a *static flare*.

A two-ribbon flare starts with a process which rearranges, for quite some time before the flare onset, the magnetic field. When a dark filament is seen overlying the $H_{\|}=0$ line in that area, this rearrangement is demonstrated by an activation and ultimately disruption of the filament. Only when the filament begins to disrupt does the two-ribbon flare itself begin to appear, as a set of newly formed loops crossing the $H_{\|}=0$ line at the places where the filament disappeared. The roots of these loops form two rows of bright patches in the chromosphere, paralleled to the $H_{\|}=0$ line, which look in Hα like two brilliant ribbons and give to these flares their name.

As time proceeds, new, progressively higher, loops are formed, their foot-points (i.e. the two chromospheric ribbons) move continuously farther from the $H_{\|}=0$ line, and this process can last for hours, in extreme cases for tens of hours. Energy is supplied to the flare not only at its onset, but at any time when a new loop is formed — the atmosphere is then newly heated and we have, as a matter of fact, another 'elementary flare'. The cooling time of these individual loops is much shorter than the total lifetime of the flare, which at any time is composed of tens of new loops formed at progressively higher altitudes.

In contrast to that, a compact (or simple-loop) flare is seen as an enhancement of a pre-existing arch, loop, or system of loops, which sometimes could be seen before, and the flaring loops do not show any striking motions, expansion, or shape variations. The lifetime of the flare is comparable with the expected cooling time, so that there is no obvious need to add energy to the flare at any other time than the onset of the phenomenon.

Both these basically different types of flares show a wide spectrum of events which greatly differ in their intensity and details. The brightest two-ribbon flares are the most powerful and important activity events on the Sun. The acceleration of nuclei in their impulsive phase may exceed 30 MeV/nucleon, thus producing γ-ray lines and photospheric emission, and particles ejected from them can be further accelerated to hundreds of MeV in a shock which propagates from the flare into space in front of a huge plasma cloud. At the other end of the event-spectrum for two-ribbon flares are the disparition brusques of quiescent filaments in old remnants of active regions, which only heat the corona, without any temperature enhancement in the lower atmospheric layers and without any known production of accelerated particles. In the category of compact flares, we see brightened arches that comprise a whole fully developed active region, as well as short-lived brightenings of tiny X-ray points not detectable at all in the chromosphere.

There may be still other varieties of flare. We do not understand yet the relation of flares to the various kinds of mass ejecta (surges, sprays, eruptive prominences, coronal transients). In particular, activity events on the limb often show characteristics that one can hardly reconcile with what we believe to know about flares from their observations on the disk.

2.1.5 Flare theory

The total amount of energy involved in the flare process varies within huge limits: from some 10^{19} J (10^{26} erg) in the smallest flaring phenomena on the Sun, through $\sim 10^{21}$–10^{23} J (10^{28}–10^{30} erg) in subflares and 10^{23}–10^{24} J (10^{30}–10^{31} erg) in an average flare, up to more than 10^{25} J (10^{32} erg) in the most important flare events.

This energy can be deposited quite differently in flares of different types. In a compact flare without any impulsive phase essentially all the energy is going into heating and turbulent motions. In compact flares that accelerate electrons, i.e. those with microwave and type III bursts, far the most energy appears to go into the electron stream. Whereas in two-ribbon flares the main bulk of energy goes into the coronal transient and, in the most energetic events, into the shock wave.

We do not know enough yet about this distribution of energy into the different kinds of flare emission. Besides, complete observations of all the emissions are never available. Therefore, we can only estimate the energies involved, being aware of large error bars in all these estimates.

The reader is referred to Canfield *et al.* (1979b) and Webb *et al.* (1979) who used the opportunity of fairly complete observations of the subflare of 5 September, 1973 (during the Skylab mission) to estimate the various kinds of energy released in this rather typical flaring event. The total radiated energy could be estimated to be 4×10^{29} erg, whereas the mechanical energy output possibly exceeded 10^{30} erg.

2.1.6 Arrangement of the chapter

I intend to describe in this Chapter the results of flare observations in the different kinds of flare emission, starting with the flare observations in Hα, and following approximately the same scheme I used in my recent book *Solar Flares* (Švestka, 1976). I will refer frequently to this book, in order to save space here, and also to avoid an excessive number of individual references. Complete reference will be made to all relevant papers that appeared after my book manuscript was finished (i.e. after 1974/1975), but I may refer to older papers through the book: e.g. (SVE 159 REF) means: 'see Švestka (1976), p. 159 and references therein'.

Since the present chapter should concentrate on the *observations* of flares, and is followed by several other chapters of a theoretical nature, I intend to give an interpretation of the observed data only in those cases when this is necessary for reader's understanding of the observations. In other cases, the interpretation may be found in the chapters that follow.

In spite of the fact that the 'two-ribbon' and 'compact' flares probably differ a lot in their physical nature, we often cannot distinguish between them in the following text, simply because there have not yet been enough separate observations of these two classes of flare.

2.2 FLARE OBSERVATIONS IN THE Hα LIGHT

When the Sun is observed in the Hα line, active regions appear as bright chromospheric plages surrounding and overlying the sunspots which characterize the active region in the photosphere. A flare appears either as a brightening of a part of existing plage (the most common case) or as the formation of new bright areas in places where the plage did not exist before. In regions close to the peak of their activity, brightness variations in the plage occur almost continually. The characteristic feature of actual flare brightenings is a very fast increase in brightness followed by a much slower decay.

Regular flare observations are carried out with a pass-band of 0.25 to 0.5 Å in the center of the Hα line. Thus we observe the most intense and optically thick portion of the flare in the core of the Hα line. This yields only limited information about the structure of the flare and about the associated moving phenomena which, due to the Doppler shift, emit radiation at the wings of the line. More information can be obtained if the pass-band is moved and observations carried out at various distances from the line center in the wings of the line.

To some extent, similar measurements can be made visually by means of a spectrohelioscope with a moving line-shifter. Figure 2.2 shows an example of a flare development curve measured with a spectrohelioscope. The effective widths of the red and blue wings of the flare Hα emission line are plotted here against time. It shows the typical very fast rise to the maximum and short duration of the maximum phase. This part of the flare development is often called the 'flash phase'. This appears to be the most important time interval of the flare development. The impulsive burst, if present, sets in and peaks during this phase; in compact flares all the energy is released, and in two-ribbon flares the most energetic release of energy is accomplished during these initial few minutes of the flare development.

One should not interchange the term 'flash phase' with the term 'explosive phase', often used in the literature. The flash phase means a rapid increase of the Hα line width and intensity in the brightest region of the flare, whereas the

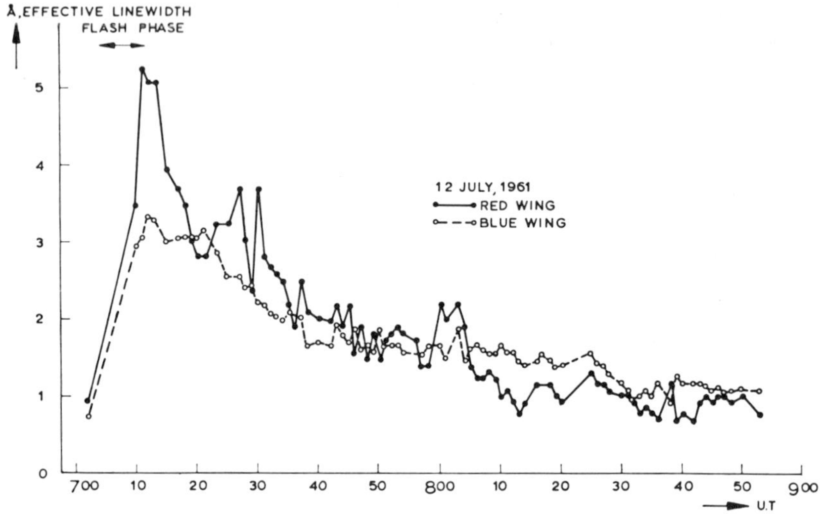

FIGURE 2.2 The time variation of the Hα effective half-width in the flare of July 12, 1961, as measured by the spectroheliοscope at Ondřejov. The effective half-width is the distance (in Å) between the center of the line and the point at the Hα line wing where the flare emission intensity decreases to $\sim 10\%$ of the adjacent continuum. Note the very steep rise during the flash phase, the pronounce red asymmetry at the maximum, and the occurrence of several secondary flashes, also characterized by the red asymmetry.

explosive phase indicates an unusually fast increase of flare area. Whilst all 'typical' flares show a flash phase, an occurrence of the explosive phase is a relatively rare phenomenon.

Individual flare knots may develop differently and the curve in Figure 2.2 is in fact only an envelope obtained by always following the brightest point in the flare region. This is also the reason for some of the secondary brightenings, in the curve: they often represent maxima of secondary flare knots.

2.2.1 Flares in relation to magnetic fields

A necessary condition for the occurrence of a flare is enhanced magnetic field in the solar atmosphere. X-ray photographs on Skylab have shown that flare-like brightenings (flaring bright points) occur as soon as tiny bipolar magnetic structures emerge on the Sun (Golub *et al.*, 1974). Petrasso *et al.* (1975) have observed an X-ray flare which occurred in a bipolar active region only 1 day old. (See SVE Figures 10, 11.) Also Hα flares have been observed many-a-time in newly born active regions prior to the appearance of the first recognizable sunspots. All this indicates that flares begin to occur as soon as a bipolar magnetic configuration is formed in the solar atmosphere.

However, the frequency and importance of flares generally increase with the rate of change of the magnetic field and with its increasing complexity (Giovanelli, 1939; Kleczek, 1953; Bell and Glaser, 1959; Künzel, 1960). The problem of whether flares can occur in a simple bipolar magnetic configuration or whether they necessarily need a magnetically complex structure is of essential importance for their theoretical interpretation. There is no evidence of a deviation from a bipolar configuration for many flaring regions, in particular if the flares in them do not show any non-thermal impulsive components. The fact that the occurrence of non-thermal acceleration processes (impulsive bursts) greatly increases in magnetically complex active regions (Švestka and Simon, 1969; Krüger, 1969; Švestka, 1971a; de Jager, 1975) supports the conclusion that the occurrence of quasi-thermal flares in simple bipolar fields is real. Nevertheless, one has to expect that complexities of the magnetic field may be so small that they are below the resolving power of our magnetographs. Thus certainly at least some 'bipolar' fields are actually complex.

It is also necessary to say that 'bipolar' need not resemble a simple dipole field. The bipolar fields can be twisted or sheared. It has been shown, e.g. by Parker (1975), how twisted magnetic flux tubes, while rising up through the atmosphere as a consequence of their buoyancy, are subject to rapid dissipation manifested by a release of energy in the corona. (Theories (such as Spicer, 1977) for flares occurring as a consequence of highly twisted loops or arches are outlined in §1.6.1 and Chapter 4.) On the other hand, Marsh (1978) has shown that all ephemeral active regions that produce Hα flares are very close to the boundaries of supergranular network, which may indicate the presence of a complex magnetic field even in these small flaring features. This seems to be supported by Moore *et al.* (1977), who have identified some of the flaring X-ray bright points with EUV macrospicules, and suggest that both the X-ray flares, occurring in apparently bipolar ephemeral regions, and the macrospicules are produced through magnetic field reconnection between the bipolar elements and the unipolar network field at the boundaries of supergranular cells.

By far the most flares occur in active regions with developed sunspots. Flares which appear in an active region without spots are exceptional events. According to a study made by Dodson and Hedeman (1970), in the years 1956–1968 only 7% of all flares of importance >2 occurred in plages with only small or no spots. These 'spotless' flares usually appear during the late, decaying phase of an active region which had been highly developed some time before, follow the disruption of a dark filament and belong to the category of the two-ribbon flares.

However, the flares, even when concentrated in sunspot groups, avoid big spots. The Hα flare patches form outside spots and even when they expand to cover a large area they prefer the spot-free space in the active region. Quite

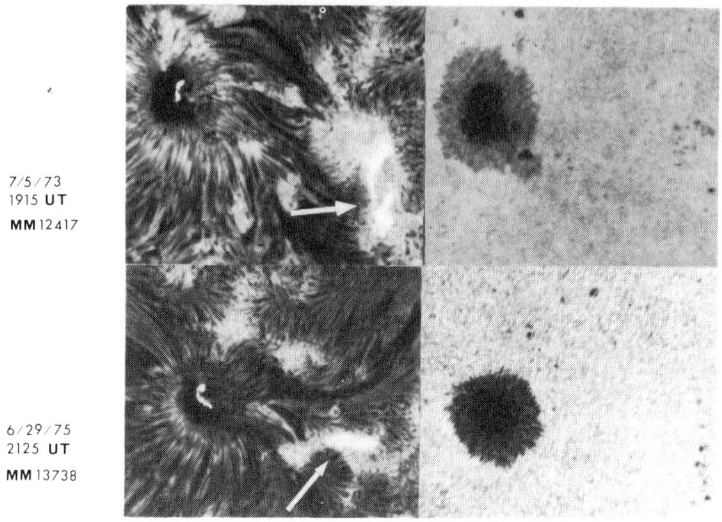

FIGURE 2.3 Examples of umbral flares. (After Tang, 1978.) While one flare ribbon is in the plage (arrow), the other ribbon is located in the sunspot umbra.

often, flare expansion is stopped at the boundary of a large spot and the $H\alpha$ emission increases at this space (the brightest flare parts are often found at the edges of sunspot penumbrae (Švestka *et al.*, 1961). If a flare penetrates onto a spot, which does happen with large flares and with exceptional flares of smaller size, the flare, as a rule, is very productive in X-rays and radio microwave radiation (Dodson and Hedeman, 1960; Martres and Pick, 1962; Malville and Smith, 1963; Hagen and Neidig, 1971). Generally, the hardness of the spectrum of an impulsive microwave burst increases with flare closeness to the spot umbra (Neidig, 1977).

However, there is a special and very rare class of 'umbral flares' when one of the ribbons of a two-ribbon flare is formed, since its beginning, inside a sunspot umbra (Figure 2.3). Only four such events have been observed in detail so far (Tang, 1978), all being subflares, and all occurred in a simple magnetic configuration with a large dominant p-spot and tiny or non-existing f-spots.

The flare usually starts with two or more bright points or areas. When there are two of them, they appear on opposite sides of the zero line ($H_{\parallel}=0$) of the longitudinal magnetic field (Martres *et al.*, 1966; Moreton and Severny, 1962). Compact flares usually consist of small, separated bright areas, which, for the most part, are brightenings of parts of the chromospheric plage. Using $H\alpha$ observations Dizer (1969) has shown that subflares very often occur at places that were brighter than the surroundings prior to the subflare onset and he calls these bright pre-subflare patches the 'live points'. During the CINOF period Švestka *et al.* (1974; also see de Jager, 1975) confirmed Dizer's observation for two subflares they studied in detail. In addition they showed that the five knots which formed the two subflares were distributed along the border of one supergranule cell, at, or very close to, the places where three supergranules met.

In two-ribbon flares the first Hα brightening often coincides with the places where the transverse field bifurcates or becomes greatly disordered (Zvereva and Severny, 1970), even to the extent that the transverse field lines near the $H_{\|} = 0$ line run nearly parallel to it, instead of crossing the line in the direction of the gradient of $H_{\|}$. Observations carried out during the strong flare activity in August 1972 confirmed the sheared structure of the magnetic field near the zero line in regions where the big flares occurred (Zirin and Tanaka, 1973; Tanaka and Nakagawa, 1973). The penumbral fibrils, which probably trace the transverse field, ran almost parallel to the zero line; also the initial loops connecting the two bright flare ribbons in the large white-light flare of August 7 crossed the zero line at a sharp angle. As time proceeded, however, the new loops, formed at larger altitudes, made progressively greater angles to the zero line. Thus, either the low magnetic loops were more sheared than the higher ones, or the shear relaxed with time.

When all components of the magnetic field are measured, there is also the possibility to compute the electric currents. Having done this, Severny and his co-workers (SVE 28 REF) have found a coincidence between the bright knots which appear during the initial phase of flares and areas with an increased vertical current density. Discrepancies between computed current-free configurations and the observed flare-loop orientation (Rust and Bar, 1973) give another evidence that strong currents indeed play a significant role in the initial phase of flares, as Severny's photospheric observations have indicated.

High resolution (in particular off-band) Hα photographs show intense short-lived brightenings in small parts of the flare area which coincide in time with the impulsive microwave and hard X-ray bursts (de Jager, 1967; Vorpahl and Zirin, 1970; Zirin *et al.*, 1971; Vorpahl, 1972; Zirin, 1978a). They have been called *flare kernels*, and are probably produced by streams of energetic electrons penetrating from the acceleration site at higher altitude into the chromosphere.

In some active regions, flares show a striking tendency to recur in nearly identical form. Ellison *et al.* (1960) have proposed for such events the term 'homologous flares' (SVE Figure 13). A very detailed study of this homology was carried out for three active regions by Dodson and Hedeman (1949) and for the interesting sunspot group of September 1963 by Zirin and Werner (1967). Homologous radio bursts have been studied by Fokker (1967).

The strength of magnetic field does not seem to be the most decisive characteristic of the flare locations. More important seems to be the gradient of the magnetic field. Gopasyuk *et al.* (1962) have shown that the energies of particles accelerated in flares increase with the increasing magnetic field gradient. Extraordinarily high horizontal gradients in the longitudinal field also were observed in association with the large 1972 August flares (Rust, 1973a) and 1966 July 7, cosmic-ray flare (Severny, 1969).

2.2.2 Flare occurrence

Flares are most frequent in magnetically complex groups (γ or δ types of the Mt. Wilson classification). As we mentioned before, the complex groups also produce the most important flares, and flares from them are the most productive in impulsive X-rays and radio bursts.

The flare occurrence rate roughly follows the sunspot solar cycle; nevertheless, there are some bursts of activity which deviate from this simple rule. In solar cycle 19, the most flare-abundant year was 1960, three years after the maximum, and a burst of flare activity occurred even in 1963, one year before the solar minimum. Similarly, the famous flare activity in August 1972 (Lincoln and Leighton, 1972; Coffey, 1973) occurred 3.7 yr after the maximum of the last solar cycle.

The very large 'catastrophic' events which produce a cosmic-ray increase at the Earth, or are characterized by CFI $\geqslant 15$, are very rare phenomena. There are many years when no such event occurs at all, and even in the very active years the annual number of such flares is only 2 to 3 on average. The number of major flares, characterized by CFI $\geqslant 10$ or by an associated polar cap absorption (PCA) event, was about 20 to 30 yr^{-1} in the years of the maximum of cycle 19, and about 10 to 15 during the maximum of the last cycle No. 20. On the other hand, the total number of all flares reported during that period was many thousands, and thousands of other events were classified as subflares. This clearly shows that, whilst solar flares are very common on the Sun, a really large flare is a fairly rare phenomenon (SVE 17 REF).

2.2.3 Flares in relation to velocity fields

Severny (1960) and Martres *et al.* (1971) have found that flares tend to occur in places where the line-of-sight component of the photospheric velocity $V_{\parallel}$ is zero; according to them, crossings of the $H_{\parallel}=0$ and $V_{\parallel}=0$ lines are the most likely positions of the initial flare brightenings, and more recent studies tend to confirm this rule (Harvey and Harvey, 1976; Martres and Soru-Escaut, 1977).

Yoshimura *et al.* (1971) observed curved absorption lines under a subflare region, which indicated large-scale mass motions in the photosphere, decaying during the flare development. However, the subflare tentatively associated with these motions was so insignificant that the association is open to some doubt.

More convincing is an observation associated with the great flare of 1972, August 7, reported by Rust (1973) and Zirin and Tanaka (1973). They observed unusual blue shifts in the photospheric lines close to the $H_{\parallel}=0$ line below the flare site prior to the flare, but the same region became red-shifted after the flare onset. However, the interpretation of these velocity changes is ambiguous. Zirin and Tanaka speak about a shearing (i.e. lateral) motion with velocity of 4 to 5 km s^{-1}, whereas Rust interprets the same observations as upward and subsequently downward motions in a sunspot penumbra.

Harvey and Harvey (1976) observed in the photosphere significant Doppler shifts towards the observer prior to several flares at large central meridian distances. This confirms that either shearing takes place at flare locations or a strong upward motion of the photosphere triggers some flares.

2.2.4 Flare-associated changes in the magnetic field

As we have seen, many observations demonstrate a strong coupling between the magnetic field and the flare phenomenon. Also, most theoreticians believe that flares derive their energy from the magnetic field. Van Tend (1979), e.g., defines the flare as a sudden conversion of magnetic energy into other forms of energy in the solar atmosphere. Therefore, the obvious question arises as to whether there appear any observable changes in the magnetic field strength and configuration during the occurrence of a large flare on the Sun.

Many studies of this problem were made in the past, with conflicting results (SVE 29 REF; Rust, 1976; the reader is referred to these papers for a more detailed discussion), and only the following facts have been well established.

A. Relation to newly emerging flux The Meudon group (Martres *et al.*, 1968a, b; Ribes, 1969; Michard, 1971; Martres and Soru-Escaut, 1977) has divided each complex pattern of longitudinal field into a number of what they call 'evolving magnetic features', each containing one polarity. They have shown that in many cases flares involve at least two adjacent features of this kind, and these evolve differently from each other in the period of flare occurrence.

In more recent years Rust (1972) and his co-workers (Rust and Roy, 1974; Rust *et al.*, 1975; Rust and Bridges, 1975) have confirmed that flares occur close to evolving magnetic features. The likelihood of occurrence of a flare in a large bipolar region increases if, in the vicinity of the zero line, there is rapid development of small, reversed-polarity magnetic features (Rust, 1973b). As a matter of fact this only confirms, by using greatly improved resolving power, what we already mentioned earlier: that magnetically complex active regions are more capable of producing flares.

Almost all the flares which Rust and his co-workers studied apparently were triggered by emergence of new magnetic flux through the photosphere. This also agrees with Vorpahl's (1973) Hα observation that most flares are triggered when and where emerging flux regions appear. These can be recognized in the Hα line as a bright plage crossed by a dark arch-filament system (after Bruzek, 1967 and Zirin, 1972).

Emergence of new flux in an active region is certainly not a sufficient condition for producing flares, since many such variations are also observed without any associated flarings. According to Rust (1973b) even reversed polarity in the emerging flux need not necessarily produce a flare. Neidig *et al.* (1978) suggest, from detailed observation of McMath region 14943 in Heyvaerts *et al.* (1977) have presented a plausible model of a flare originating through reconnection of a newly emerging flux with old pre-existing flux in the solar atmosphere. Rust *et al.* (1975), Canfield and Fisher (1976), and Levine (1978) presented observations of flares which seem to be consistent with this model.

B. Shear relaxation If a flare takes its energy from the magnetic field in the solar atmosphere, the magnetic field configuration in the atmospheric layers must get greatly simplified during the flare process. One such expected simplification is the relaxation of shears that often are observed near the $H_{\parallel}=0$ line prior to flares (Zvereva and Severny, 1970; Zirin and Tanaka, 1973; Tanaka, 1976; Neidig *et al.*, 1978); this means a removal of strong preflare currents flowing through the atmosphere and restoration of a current-free post-flare situation.

As we mentioned earlier, Zirin and Tanaka (1973) believe they have seen such a relaxation process in the big flares of August 1972.

In a plasma where the magnetic energy density exceeds the plasma pressure by many orders of magnitude, the currents flow parallel to the lines of force; i.e. we have a force-free field. This approximation has been applied to Zirin and Tanaka's observations by Tanaka and Nakagawa (1973). In the current-free (potential) configuration the field lines (and the visible loops marking them) should be perpendicular ($\gamma=\pi/2$) to the $H_{\parallel}=0$ line, representing the minimum energy state. On the other hand they may be sheared in a force-free configuration (so that $\gamma<\pi/2$) and the surplus energy stored in the non-potential field is then proportional to $\sin^{-1}\gamma$, i.e. to the amount of shearing. Thus the stored energy can be computed from the observations made by Zirin and Tanaka (1973) for the period covered by their data.

According to Tanaka and Nakagawa's computations the stored surplus energy increased from $\sim 3\times10^{30}$ erg on July 31, through 8×10^{31} erg on August 2, to values exceeding 10^{32} erg on all the following days. During the flares of August 4 and 7 the shear became smaller, thus indicating relaxation towards a potential magnetic field configuration. Similar results have been reported by Rust *et al.* (1975), Tanaka (1976, 1978), and Neidig *et al.* (1978) for other flares. Roy (1972) found a surprisingly good fit of a current-free approximation with the shape of a loop-prominence system, which also indicates that the (assumed) preflare currents had to be released during the flare phenomenon.

Brueckner *et al.* (1976) report from EUV observations that low-lying loops, often showing a chaotic pattern in newly emerged regions, relax after flare activity and expand into a bipolar configuration. This tends to support Neidig *et al.*'s (1978) conclusion (built into the model of Heyvaerts *et al.*, 1977) that a combination of newly emerging flux and shear is most favourable for flare occurrence. Also Gaizauskas (1979) observed a shear relaxation combined with, or caused by, a new emerging flux region. This and some other shear relaxations have been seen even in connection with small subflares.

On the other hand, some Crimean maps (e.g. Zvereva and Severny, 1970) show sheared fields at the photospheric level even after flare occurrences. Thus, either the relaxation is much more pronounced at higher altitudes than in the photosphere, or we interpret (at least in some events) altitude-dependent shear erroneously as a time-dependent phenomenon (cf. Section 2.2.1).

A different kind of rearrangement of magnetic field has been reported by Brùzek (1975, 1979) and Neidig (1979), who found patterns of parallel Hα fibrils changed into a pattern of vertical or radial fibrils after a flare or subflare. This observation also may reflect a relaxation process, but the changes generally lag the energy release (Neidig) and are not associated directly with the main part of the flare (Bruzek).

C. Filament activation Flares originate close to the line of zero longitudinal magnetic field, $H_{\parallel}=0$, dividing the magnetic polarities. This zero line is also identical with the position of quiescent dark filaments, which begin to form in active regions during a later phase of their development. The filaments then stay on the disc for many weeks after the sunspots in the active region have disappeared and the surrounding plage has faded. Therefore, it is not surprising that these filaments, when present, are greatly disturbed if a flare occurs in close vicinity to the $H_{\parallel}=0$ line along which they extend.

What may be surprising, however, is the fact that such a filament disturbance starts prior to the flare, preceding it by between a few minutes and one hour. About 45% of major flares are preceded by such filament activations (Smith and Ramsey, 1964; Bruzek, 1969) and flares that occur in old regions without sunspots are quite generally preceded by filament disruptions (Dodson and Hedeman, 1970). Thus, considering that quiescent dark filaments do not exist in relatively young active regions, in which many flares occur, one is inclined to suspect that the magnetic field changes which lead to the preflare activation essentially precede all major flares. In any case, these magnetic field changes obviously precede all two-ribbon flares (cf. Section 2.1.4).

The activation of a filament clearly demonstrates that the stabilizing agent, i.e., the magnetic field in the active region, becomes unstable, and that this destabilization starts long before the onset of the optical flare. However, quite a number of the activated filaments do not disappear completely. Many of them begin to reappear again during the decay phase of the flare, or they resume their former stable shape within several hours (Bruzek, 1969). This shows that the gross configuration of the magnetic field does not undergo any strong permanent change. Rust *et al.* (1975) have demonstrated one event when the preflare destabilization of a filament was most probably caused by a newly emerging flux. This may well be the case in many other events, but hardly in old decaying regions in which extensive spotless two-ribbon flares still can and do occur. There new flux emergences appear to be very rare and the filament must be destabilized by another trigger: either by a disturbance propagating from another active event (Dodson *et al.*, 1972) or by internal instability.

2.3 CHROMOSPHERIC FLARE SPECTRA

The analysis of chromospheric flare spectra has been described in detail in SVE 49–81 REF; here we will only summarize the results obtained.

2.3.1 Characteristic features of the chromospheric flare spectrum

Flare radiation in the optical region and in the UV up to the Lyman limit of hydrogen is emitted predominantly in spectral lines. At the very beginning of a flare, emission in the visible spectral region first appears in the lowest lines of the Balmer series and in the H and K lines of singly ionized calcium, and slight absorption is seen in the D_3 line of the He I. Very soon, broad wings begin to appear in the Balmer lines and, as soon as the wings are formed, higher lines of the Balmer series also become visible (Švestka *et al.*, 1961). As soon as the Balmer series excitation is sufficiently strong for lines higher than H_{10} to be seen in the spectrum, the emission in metal lines also appears, mostly in quite a regular sequence of appearance (Blaha *et al.*, 1962; Banin *et al.*, 1967). Absorption in the D_3 line weakens. The next development of the flare is characterized by increasing intensity in the cores and further extension of the wings of the Balmer lines. This manifests itself in the Hα time-development curves by the steep rise in the flash phase (cf. Figure 2.2) and it is accompanied by increasing intensity of the metal lines. In contrast to the greatly broadened hydrogen lines, however, the metal emission lines remain sharp and narrow. The absorption D_3 line disappears and is replaced by emission in the flash phase of the flare (Waldmeier, 1955; Švestka *et al.*, 1962). As soon as the D_3 emission appears, other He lines, too, start to be visible in the flare spectrum (Švestka *et al.*, 1962). In the IR, however, the λ 10830 Å helium line seems to be in emission from the flare beginning (Rust and Bridges, 1975). In limb flares, and occasionally in the brightest disk flares, the λ 4686 line of He II can also be seen in the flash phase. H15 or H16 are usually the highest Balmer lines clearly visible in spectra of large solar flares on the disc. Above the limb, Balmer lines up to H30 can be observed on rare occasions (Jefferies and Orrall, 1961a). In smaller flares the development stops before it reaches this fully developed phase.

In the UV and EUV range the most prominent chromospheric emission comes from the hydrogen Lyman lines and continuum and from the He II Lyman series. The hydrogen Lα/Hα flux ratio is close to one, with a large scatter of individual values (Zirin, 1978b; Canfield *et al.*, 1980b). The He II Lyman-line at 304 Å is anomalously strong both in the quiet Sun and in flares. Whereas it is formed primarily by collisional excitation in the quiet Sun (Avrett *et al.*, 1976), it appears to be produced largely by recombination in solar flares (Raymond *et al.*, 1979); therefore, the line comes from lower temperature gas in the flare (of the order of 10^4 K) than in the quiet atmosphere ($\sim 10^5$ K).

Observations of flares on the limb (SVE 54/55 REF), as well as profiles of EUV lines (Cheng, 1978), show that non-thermal motions are small in the chromosphere, but increase significantly through the transition layer into the corona.

Continuum flare emission will be discussed in Section 2.3.5.

2.3.2 Physical characteristics of the chromospheric flare

The hydrogen Balmer lines are broadened predominantly by the Stark effect in combination with Doppler broadening (SVE 56–58 REF); in the Lyman series, because of its large optical thickness, the Doppler broadening prevails (De Feiter and Švestka, 1975). From the Stark-broadened higher lines of the Balmer series one can deduce the mean electron density in chromospheric flares observed on the disk (SVE 59–63 REF), which ranges from 4.4×10^{19} m^{-3} (4.4×10^{13} cm^{-3}) in the maximum phase of a large proton flare (De Feiter and Švestka, 1964) to $< 7 \times 10^{18}$ m^{-3} (7×10^{12} cm^{-3}) in a flare knot observed by Stepanyan (1969). The chromospheric electron density variation during

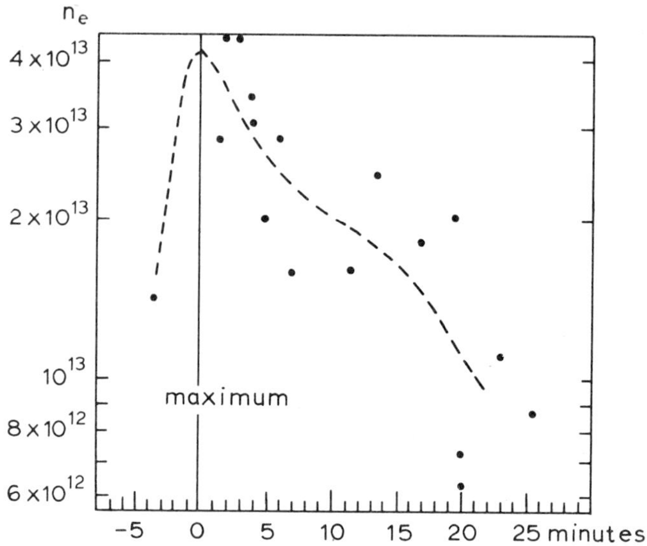

FIGURE 2.4 The chromospheric electron density variation during flare development, as composed from measurements of five flares of importance $\geqslant 2$. (After Fritzová and Švestka, 1967.)

flare development, as composed from measurements of five flares of importance >2n, is shown in Figure 2.4.

Recently, Spicer and Davis (1975) and Bakshi and Kalman (1976) suggested that Stark broadening might be increased (for the same n_e) in plasma which is driven unstable by shock waves or energetic particle beams which generate turbulent fields. This effect might influence density determinations in the flare kernels during the impulsive phase. The highest n_e values, however, have been found in the maximum of Hα flares, which occurs after the impulsive phase, and almost all other n_e values have been determined during the flare decay.

It would certainly be of interest to measure the line profiles in the flare kernels, which has not been done so far (because one does not know where to place the spectrograph slit in advance, and the impulsive phase is so short). W. D. Davis (1977) tried to determine the electric field in flares using the Stark effect exhibited by neutral helium atoms. He believes to have seen electric field strengths as high as 7×10^4 V m^{-1}.

For chromospheric flares seen above the limb other methods of n_e determination have to be used (SVE 63–66 REF). A list of the n_e values thus determined is shown in Figure 2.5. Summarizing, one can say that the determinations in limb flares, though much less accurate than in flares on the disc, clearly show that the electron density decreases with height. The structure of individual flares, however, may differ greatly in higher atmospheric layers. In the spray-type flare of 7 March, 1959, n_e was found to decrease continuously

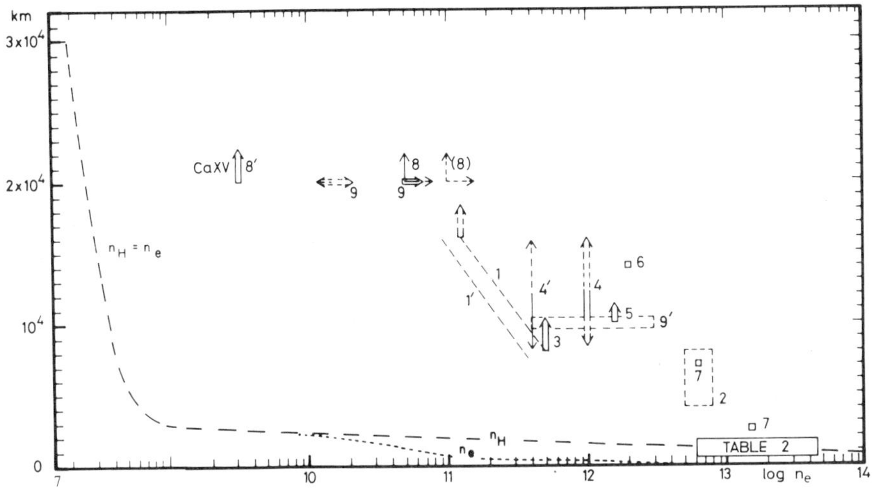

FIGURE 2.5 A summary of the n_e values measured in limb flares. A comparison with the decrease of hydrogen density (n_H) and electron density (n_e) with height in the quiet solar atmosphere shows that the density in flares is generally about two orders of magnitude greater. (After Švestka, 1972, and de Feiter, 1974).

with height, and at 16 000 km it was a factor of 4 lower than at 7700 km (Jefferies and Orrall, 1961a; Fritzová and Švestka, 1967) — number 1 in Figure 2.5. In the loop-type (obviously two-ribbon) flare of 9 June, 1959, however, n_e remained constant over a height difference of 14,000 km (Jefferies and Orrall, 1961b; Fritzová and Švestka, 1967 — number 4).

De Feiter (1974) projected into Figure 2.5 the n_H and n_e dependence on height for the quiet solar atmosphere. The n_e in flares is close to the quiet Sun n_H in the chromosphere, but exceeds by one or two orders of magnitude the $n_H = n_e$ values at greater altitudes. Thus the flare density must be 10–100 times greater than that of the quiet atmosphere, and the ionization should be low in the flaring chromosphere. This contradicts the result reported by Severny (1977) that Salyut 4 observations of Lα and Lβ line indicate 100 times higher ionization in the Lyman-series-producing flare than in the chromosphere. As De Feiter and Švestka (1975) showed, the determination of n_e from the Lyman lines alone is very uncertain.

If n_2 denotes the number of two-quantum hydrogen atoms per m^3 and z the geometrical flare thickness, the most common value of n_2z in large flares is close to 3×10^{19} m^{-2}, this value varying by factors of about 0.5 to 2 in individual flare events. The flare optical thickness is found to be significantly smaller in those parts of flares that penetrate over sunspots (SVE 67–69 REF).

The electron temperature in chromospheric flares rises from about 4900 K near the temperature minimum (the first recognizable excess in T above the preflare conditions, cf. Canfield *et al.*, 1980) through some 2–6 $\times 10^4$ K in the He II flare region (Raymond *et al.*, 1979), before T steeply rises into the transition layer. The metals emit at $T_e < 7000$ K, while from the Balmer series one deduces $T_e < 10{,}000$ K in flares on the disk (SVE 69–71 REF). Above the limb T_e in the Balmer lines and continuum is found between 10^4 and 2×10^4 K.

The He I emission corresponds to $T_e < 23{,}000$ K (Banin, 1965), but T_e determinations for He I regions are uncertain (SVE 70 REF). He II lines in flares appear to be produced largely by recombination at astonishingly low temperatures of 1–2 $\times 10^4$ K (Raymond *et al.*, 1979). However, He II can be emitted at temperatures as high as 3×10^5 K (Zirin, 1964; Shine *et al.*, 1975). The Lyman series emits at 1–3 $\times 10^4$ K, with 2×10^4 K as the most typical T_e value (Machado *et al.*, 1979). For the formation of Lyman continuum, Machado and Noyes (1978) found T_e between 8000 K and 9000 K.

Much better, high-resolution observations of the Lyman lines in flares are needed, in order to be able to deduce the hydrogen density, and the ionization degree, in flares. Presently the degree of ionization in the chromospheric flare appears to be astonishingly low. At the flare of August 7, 1960, e.g., for which $n_e \simeq 3 \times 10^{19}$ m^{-3} was found from the Stark broadening of high Balmer lines, the ionization degree $n_e/(n_e + n_0)$ was 0.59 at $T_e = 10^4$ K, and only 0.016 at $T_e = 7500$ K (SVE 72). This ionization has been computed under the assumption of a detailed balancing between the hydrogen levels 1 and 2; i.e. for $b_1 = b_2$, where b are the deviation factors from the LTE. If $b_1 \neq b_2$, then necessarily $b_1 > b_2$ and hence n_0 increases; therefore, the deduced low ionization degree is an upper limit to the ionization in the chromospheric part of the flare.

From studies of Lyman continuum in flares, Machado and Noyes (1978) find that the b_1 is smaller in flares than in the quiet Sun. The level of Lyman continuum formation is lowered in flares by a factor of 60 to 200 in the mass column density, with the high values probably corresponding to the flare kernels.

2.3.3 Shell or filamentary structure of chromospheric flares

By several different methods one can find and confirm that the geometrical thickness z of a chromospheric flare is astonishingly small (SVE 72–76 REF). The values of z vary from a few hundred to only 15 km; from infrared observations of subflares Hudson (1975) finds $z < 10$ km, and above spots Machado and Seibold (1973) have found z as low as ~ 3 km.

There are two ways these extremely small z values van be interpreted. First, one can assume that this is the actual thickness of a shell-like emitting flare layer in the chromosphere; if the primary source of the flare is in the corona or transition layer, flare energy must be brought downwards into the chromosphere, predominantly by conduction (Švestka, 1973), and in that case it could be deposited in a very thin layer of the dimensions we have deduced from the flare spectra (Shmeleva and Syrovatsky, 1973; Canfield and Athay, 1974). This 'shell model' of the flare has been also advocated by Machado and Rust (1974).

However, there are several objections to this interpretation:

(a) Wherever streams of energetic particles propagate downwards, they dominate the energy input, and in this case the energy deposition is expected to occur over a wide range of altitudes (Brown, 1974). In the case of white-light emission, e.g., significant energy deposit occurs throughout the chromosphere down to its bottom. Therefore, at least in impulsive kernels, the thickness of the chromospheric flare is certainly larger than the small values of z we have obtained. This implies an inhomogeneous structure in all flares that show an impulsive phase.

(b) Even in many flares which do not show white-light emission, the atmosphere is affected down to the temperature minimum region (cf. Canfield *et al.*, 1980a). Thus, at least in some parts of the flare, energy must be deposited all throughout the chromosphere.

(c) Machado and Seibold (1973) have found z above sunspot umbrae equal to or less than ~ 3 km. This certainly cannot be interpreted as the actual thickness of the layer where energy is deposited above sunspots. On the other hand, filamentary structure in the region of flare penetration over sunspot umbrae and the dilution of radiation associated with it can reasonably explain the absurdly small (and obviously fictitious) geometrical thickness.

Thus we come to the other interpretation of the small values of z, namely, an inhomogeneous and filamentary structure for flares. This is the original explanation of the small z values, first proposed by Suemoto and Hiei (1959) and later adopted by many other authors. In reality, we probably meet with a greatly inhomogeneous 'filamentary shell structure' which combines both these pictures together.

2.3.4 The chromospheric line asymmetry

One of the most puzzling problems in chromospheric flare spectra is the asymmetrical shape of flare emission lines. The asymmetry does not change the position of the center of the line, but only makes one of the wings stronger (SVE 76–80 REF). We probably encounter here some sort of internal motion in the flare region.

Very often at the onset of a flare a blue asymmetry is observed (Figure 2.6), i.e., the short-wave wing is more intense (Švestka *et al.*, 1962; Severny, 1968; Švestka, 1968). This blue asymmetry, however, disappears within one or just a few minutes, and subsequently a strong red asymmetry sets in which reaches

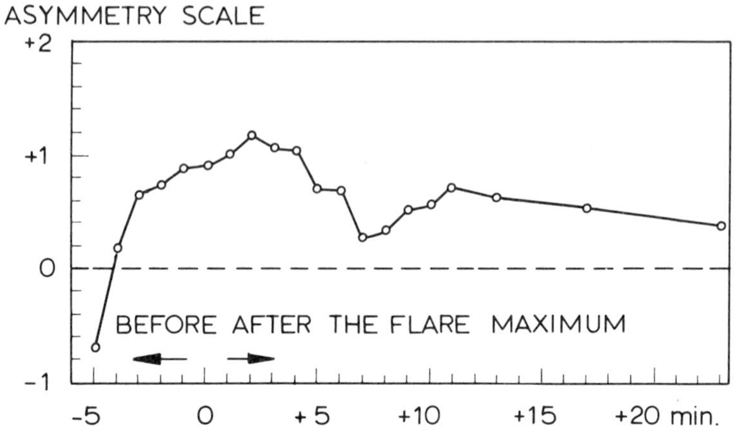

FIGURE 2.6 Mean time dependence of the Hα line asymmetry on the flare development (time of maximum Hα linewidth = 0), constructed from a set of 244 spectra of 92 flares. The asymmetry scale is qualitative: 0 = no, 1 = weak, 2 = moderate, 3 = strong asymmetry. Plus sign means an intensification of the long-wave wing, i.e., a red asymmetry. (After Švestka *et al.*, 1962.)

its maximum approximately at the time of the flare maximum in Hα light and persists for a few minutes after this phase, as one can see in Figures 2.2 and 2.6. Note in these figures that the asymmetry does not show any relation to the impulsive phase (when down-streaming particles might possibly cause it), which occurs prior to the Hα maximum.

Figure 2.2 clearly illustrates that the red asymmetry increases not only at the maximum, but also at all other times of secondary brightenings. Obviously, any newly brightened flare element shows a red asymmetry for a few minutes (Martin, 1979b). In two-ribbon flares, e.g., the asymmetry is seen for a few minutes at newly formed outer edges of the expanding flare ribbons (where, apparently, a new loop was formed in the corona, rooted in the newly brightened, and asymmetrical, region) (Švestka *et al.*, 1979b).

If the asymmetry is explained by motion, then the highest velocity is found for hydrogen (~ 100 km s^{-1}), smaller velocities are deduced from helium and Ca II H and K lines (a few tens of km s^{-1}), whilst the velocity in metals is close to 10 km s^{-1} (Jefferies *et al.*, 1959; Fritzová, 1960). Banin (1965a) interpreted this velocity variation as being due to the difference in height of formation of the different lines in the solar atmosphere. In such a case, the velocity decreases with depth in the flare, which indicates a downward slowing motion in the flare region. On the other hand, the deduced velocities may indicate conservation of momentum. Then, as Hyder (1967) has pointed out, this differentiation might also be due to some kind of a mass effect (e.g. if the radiating atoms are set in motion after collisions with particles in an accelerated stream).

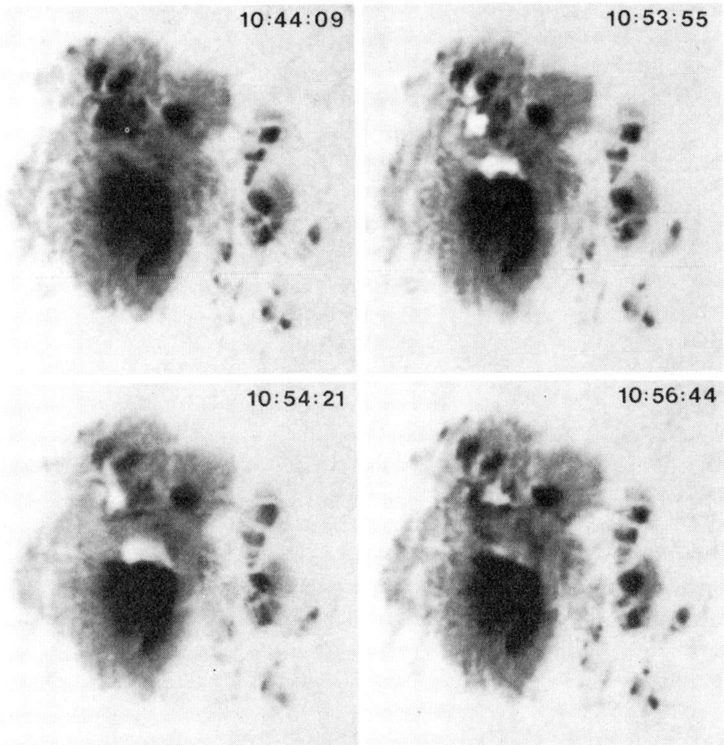

FIGURE 2.7 The white-light flare of 11 July, 1978. (After Dezsö *et al.*, 1980.)

2.3.5 Continuous emission

Continuum flare emission in the visible spectral range is a very rare event (SVE 82–86 REF). At shorter wavelengths, of course, increases in Lyman continua of H and He are common (Linsky *et al.*, 1976; Machado and Noyes, 1978); in some flares above the limb even the Balmer continuum can be observed (SVE 54–55 and 81 REF). A strange phenomenon is the enhancement of EUV continuum between 1400 and 1960 Å (Cheng and Kjeldseth Moe, 1978; Cook and Brueckner, 1979) which indicates a flare-associated temperature increase very deep in the atmosphere, near the temperature minimum. This is in agreement with an analysis of the wings of the Ca II-K line (Machado and Linsky, 1975) and indicates a depression of the temperature minimum region in flares by some two scale-heights below its preflare level (Machado *et al.*, 1978; Canfield *et al.*, 1980a).

In the visible spectral range we very rarely see short-lived continuous emission in major disk flares (SVE 82 REF). An example of such a white-light flare is shown in Figure 2.7 (also see SVE Figure 34). According to McIntosh and Donnelly (1972) one can expect some 5 to 6 white-light flares per year near the maximum of solar activity.

Typically, the white-light emission appears as two, and sometimes more, bright patches that occur very close to (or in) penumbral regions of sunspots (examples: De Mastus and Stover, 1967; Rust and Hegwer, 1975; Figure 2.7). This kind of emission exactly coincides with the impulsive phase of the flare (i.e. with hard X-ray and impulsive EUV and microwave bursts) and one can interpret it as enhancement at the feet of one or more loops which bridge the zero line (Švestka, 1970; McIntosh and Donnelly, 1972; Rust and Bar, 1973; Slonim and Korobova, 1975). The white-light emission patches lie within the bright Hα flare ribbons (De Mastus and Stover, 1967; Švestka, 1970; Slonim and Korobova, 1975; Dezsö, 1980). Whereas according to Rust and Hegwer (1975) the white-light patches on August 7, 1972 did not move with the ribbons, Dezsö *et al.* (1980) detected such a motion on 11 July 1978.

A completely different type of white-light emission has been described by Machado and Rust (1974) and Rust and Hegwer (1975) in the flare of 1972, August 7. After the stationary and short-lived bright white-light knots disappeared, the whole edge of the moving Hα ribbon became visible in white light, moving with a velocity of ~ 40 km s^{-1}, very close to that determined by Krivský *et al.* (1973) for the Hα ribbon (~ 50 km s^{-1}). Occurrence of this type of continuous emission did not coincide with the impulsive phase (which coincided with the occurrence of white-light knots a few minutes earlier). Rust and Hegwer suppose that the continuum wave they observed was not unique; it was detected because a cinematic film was available for the first time.

Still another kind of white-light emission has been reported by Zirin and Tanaka (1973) and Zirin (1979) who observed in four major flares short-lived flashes when looking at the Sun through a broad band centered at 3835 Å. The lifetime of these blue flashes is about 10 sec; they occur in pairs at opposite sites of the $H_{\|}=0$ line and coincide with pulses of impulsive microwave and hard X-ray bursts.

The exact coincidence of the white-light emission with impulsive burst shows clearly that this emission is a product of impinging energetic particles. Presently, however, there is an uncertainty as to which kind of particles is the producer of this emission. Whereas Švestka (1970) suggested >20 MeV protons penetrating into the uppermost photospheric layers, possibly with a contribution of relativistic electrons (Najita and Orrall, 1970), Hudson (1972) suggests non-relativistic electrons depositing their energy at a height between 700 and 1500 km above the base of the chromosphere; Machado (1976) shows that the white-light emission might be due to a huge amount of material formed at $T_e \simeq 2 \times 10^4$ K in the region where the Lyman-α line is formed. As Hudson and Ohki (1975) emphasize, observations in the infrared may contain information about the dimensions of the white-light source and its density, thus deciding between the chromospheric or photospheric origin of this emission.

Orrall and Zirker (1976) suggested that hydrogen Lα profiles from solar flares should show detectable asymmetry due to effects of nonthermal proton beams impinging on the solar chromosphere from above. Canfield and Cook (1978) checked this at the flare of 1973 August 9 when the Lα line was seen during the impulsive phase, and could not detect any statistically significant line asymmetry. They deduced from it, and from the simultaneous microwave burst, that the flux of >20 keV protons injected into the chromosphere was less than 2×10^{-2} that of electrons of the same energy range. However, many free parameters are involved in this analysis, and the spatial resolution in Skylab might not have been high enough to resolve the effect of the proton stream. In any case, at higher energies the proton/electron ratio appears to be quite different: according to Ramaty *et al.* (1980) it is about 10^2 at 10 MeV. Thus, in flares with hard energy spectra, which is a typical characteristic of white-light flares (Najita and Orrall, 1970), the number of protons seems to be sufficient for producing the white-light continuum (Ramaty *et al.*, 1980; Zirin 1979). This interpretation is also supported by the close time coincidence of white-light emission with the γ-ray line enhancement shown in Figure 2.17 in Section 2.9.4 C.

2.4 THERMAL FLARES IN SOFT X-RAYS

As soon as an active region forms on the solar surface, temperature and density in the atmosphere increase, and this increase manifests itself by an enhanced flux in soft X-rays. The enhancements bridge the $H_{\parallel}=0$ line, obviously being composed of loops.

This structure of the X-ray active region undergoes both gradual (Krieger *et al.*, 1972; Parkinson, 1973) and fast brightness variations. The fast variations are often, but not always, associated with chromospheric flares; in the integrated flux of X-rays (most commonly observed, since X-ray imaging of the Sun is rare) these enhancements are called *soft X-ray bursts* (Figure 2.8).

2.4.1 Soft X-ray bursts

Essentially all flares seen in Hα also produce a brightening in soft X-rays (Thomas and Teske, 1971). The most intense soft X-ray enhancements are always flare-associated, but as one proceeds to weaker events, the association becomes worse. Many smaller events are associated with non-flaring disparition brusques (Webb *et al.*, 1976; Howard and Švestka, 1976; Rust and

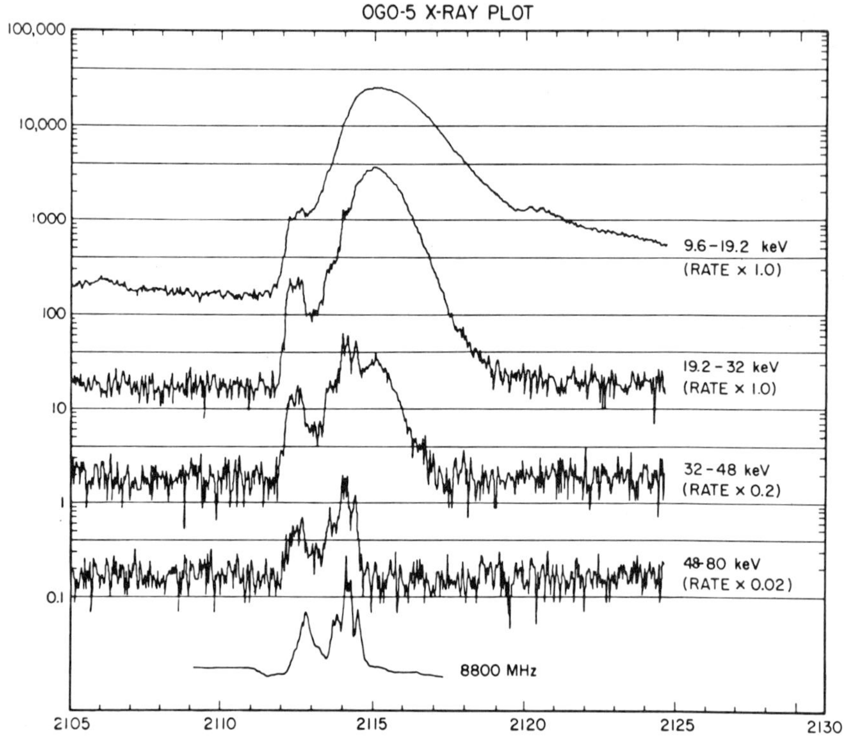

FIGURE 2.8 Records of the flare emission at different X-ray energies and on microwaves (after Zirin *et al.*, 1971). That part of the X-ray emission which shows great similarity to the 8800 MHz record is the hard X-ray burst (from 21:12.0 through 21:14.7 UT). The remaining X-ray emission is the soft X-ray burst; however, most of this emission abote 19.2 keV is due to pile-up of lower-energy quanta (Kane and Hudson, 1970).

Webb, 1977), and X-ray imaging reveals many other sources of X-ray enhancements, which usually are not detected in the integral X-ray flux: flaring bright points (Golub *et al.*, 1974, 1975), brightenings of loops interconnecting active regions (Švestka *et al.*, 1977; Švestka and Howard, 1979), coronal features brightened by travelling disturbances (Rust and Švestka, 1979), and other brightenings of loops for reasons we do not yet understand.

A typical flare-associated soft X-ray burst (Figure 2.8) has a time profile roughly similar to the time development of the Hα intensity in the brightest point of the flare and, similarly, is characterized by a fast rise and much slower decay. Both the Hα and soft X-ray emission start and peak at about the same time (Datlowe *et al.*, 1974); Thomas and Teske (1971) found the soft X-ray burst to start and peak by about 2 minutes earlier, whereas according to Falciani *et al.* (1977) it starts by about 2 minutes later.

By using observations of an X-ray burst in two different wavelength bands, one can compute the temperature T_e and emission measure

$$M = \int n_e^2 \mathrm{d}V$$

in the X-ray source of mean electron density n_e and volume V(SVE 128–134 REF). The peak temperature usually exceeds 1×10^7 K and can reach values up to 4×10^7 K in major flares. The characteristic value of the maximum emission measure is of the order of 10^{55} m^{-3} (10^{49} cm^{-3}) for flares and of 10^{54} m^{-3} for subflares. However, because of obvious inhomogeneities in flares, the deduced T_e and M values vary with the selected wavelength bands, as they refer to different parts of the flare (SVE 132–134 REF).

All the results show that generally there is a fast rise both in temperature and emission measure in the initial phase of the burst. However, whilst T_e peaks sharply in the few minutes after the flare onset and decays approximately exponentially after that, the increase in emission measure, though slowing down at about the time of the T_e peak, continues to increase further and reaches its maximum value much later in the burst development. Usually the emission measure stays fairly constant for ten minutes or more, irrespective of the simultaneous fast decrease in T_e and falls only in the later phase of development of the burst. The time sequence of maxima always keeps to the schedule 'temperature–flux–emission measure' (SVE Figure 50a, b).

The fact that the emission measure still rises while temperature is already decreasing is not easy to interpret. Increasing emission measure implies that either n_e, or V, or both, increase. In many flares (and in compact flares in particular) we do not see in X-ray images any apparent expansion. Therefore, it must be the electron density that is increasing. This can happen either through contraction or through bringing additional plasma into the radiating volume. Contraction would lead to heating, but the contrary is observed. Hence, new matter must be added to the radiating volume. The most realistic assumption seems to be that the flare slowly penetrates downwards (i.e. pushes the transition layer between chromosphere and corona deeper into the chromosphere, as some observations, e.g. by Machado and Noyes (1978), indeed indicate). The lower layers are dense and thus they offer the most efficient way to increase M with only a slight increase of V. Alternatively, or in combination with it, material may flow into the upper parts of a loop from lower layers (see Chapter 5).

2.4.2 The soft X-ray flares

Soft X-ray filtergrams from Skylab have established that the basic structural form of the thermal X-ray plasma in flares is that of closed loops or arches (Vorpahl *et al.*, 1975; Kahler *et al.*, 1975; Moore *et al.*, 1980). The X-ray arches bridge the magnetic neutral line between opposite-polarity regions of the chromospheric flare (Vaiana and Giacconi, 1969; Widing and Cheng, 1974; Pallavicini *et al.*, 1975; Vorpahl, 1976). EUV and XUV spectroheliograms have confirmed that the highest-temperature ($T = 10^7$ K) plasma is situated over the neutral line and between the lower-temperature components of the flare, which are on opposite sides of the neutral line (Neupert *et al.*, 1974; Widing and Cheng, 1974; Cheng and Widing, 1975). Thus, the observations show that the thermal X-ray plasma is contained in the upper portions of closed magnetic field arches rooted in the chromospheric flare.

The X-ray arches in a flare generally are clustered together in the form of a dome or arcade. Particularly during the early stages of a flare, a bright core is often embedded in a larger, more diffuse arch or set of loops (Vaiana and Giacconi, 1969; Vorpahl *et al.*, 1975; Petrasso *et al.*, 1975; Pallavicini *et al.*, 1975). The core connects the brightest parts of the chromospheric flare. In addition, there may be one or more small bright X-ray knots, sometimes located near the ends of the loops (Kahler *et al.*, 1975). Some X-ray knots are similar to Hα kernels observed in some flares, and may be small regions of emerging bipolar magnetic flux (Kahler *et al.*, 1976; Petrasso *et al.*, 1979a). In other cases, however, the knots are not spatially coincident with the Hα kernels and could be the highest parts of the brightened loop structures (Petrasso *et al.*, 1975; Falciani *et al.*, 1977). Most of them are clearly not identical with the sites of the impulsive flare bursts.

The X-ray flare evolves in size, shape and structure by individual loops and knots increasing and decreasing in brightness. Both the core loops and the knots tend to be brightest early in the flare. In general, the lifetime of component structural features and the lifetime of the flare as a whole scale with size: the greater the length-scale the longer the lifetime.

As mentioned in Section 2.1.4, the Skylab observations have revealed that there are two different kinds of flare: a *compact flare*, which is a sudden brightening of a stationary loop, or system of loops, and a *two-ribbon flare* in which new flare loops are sequentially formed. This distinction is based both on an analysis of X-ray flare images on the disk (Moore *et al.*, 1980) and on observations of X-ray flares above the limb (Pallavicini *et al.*, 1977). Whereas most flares and subflares observed on the Sun seem to be of the compact type, essentially all major phenomena are two-ribbon flares.

2.4.2.*A* *Compact* (*simple-loop*) *flares* In compact flares we see brightenings of loops that do not show any apparent expansion, rise, or other kind of motion. In Hα, the brightened footpoints of the flare stay in the same position until they decay.

On the limb, the compact flares have smaller volume ($\sim 10^{20}$–10^{21} m^3) and lower height (mostly $< 10^4$ km) than the two-ribbon flares (Pallavicini *et al.*,

1977). They have faster rise and decay times, shorter duration ($\sim$tens of minutes), and are located often very low at the base of an active region. They do not appear to be associated with filament disruption (which is a characteristic feature of the two-ribbon flares), nor with white-light coronal transients (which are consequences of the filament disruptions).

A detailed study of compact flares has been made by Moore *et al.* (1980), who also summarized results obtained by Kahler *et al.* (1976), Dere *et al.* (1977) Cheng (1978), Underwood *et al.* (1978), and Hiei and Widing (1970). They have found that the thermal X-ray arch in compact flares consists of several loops or a continuum of loops which span a range of cooling times. The longer, outer loops cool more slowly than the shorter, inner loops. Thus, in the decay phase the temperature structure of the flare arch is similar to that in the large two-ribbon flares: cooler on the inside, hotter on the outside. The entire flare arch appears to brighten in place with no detectable increasing separation of the foot points or continued expansion of the envelope of loops. That is, there is no obvious progressive formation of larger external loops in the decay phase as for the large two-ribbon flares.

There is a downflow of cool material in the decay phase of compact flares, similar to that in the two-ribbon flares. In distinct contrast to large two-ribbon flares, however, there is no strong evidence for continued heating in the decay phase of compact flares. Compact flares also do not show enhanced brightness at the top of the loops as the two-ribbon flares do (Pallavicini *et al.*, 1977). They are mostly bright along their entire structure with a tendency to be brighter in the lower sections of loops.

Estimates of density in compact flares generally result in values one order of magnitude higher than in the two-ribbon flares: 10^{17}–10^{18} m^{-3} (10^{11}–10^{12} cm^{-3}) in the corona and transition layer.

2.4.2.*B* *Two-ribbon flares* During the Skylab period several flares clearly belonged to the other category of flaring phenomena, the two-ribbon flares. These were, in particular, the flares on 29 July, 1973 (Moore *et al.*, 1980; Nolte *et al.*, 1979; Petrasso *et al.*, 1979b; Moore and LaBonte, 1979), September 7, 1973 (Withbroe, 1978; Moore, *et al.*, 1979), June 15, 1973 (Pallavicini *et al.*, 1975; Widing and Dere, 1977; J. B. Smith *et al.*, 1977a; Krall *et al.*, 1978) and the limb event on 13/14 August 1973 (Vorpahl *et al.*, 1977; McCombie and Rust, 1979; Cheng, 1980). The results of all the studies of these events can be summarized as follows.

Prior to the flare, a dark Hα filament is seen at the flare location, which is under an arcade of closed magnetic field lines. The magnetic field in the chromosphere and in the filament is strongly sheared across the $H_{\parallel}=0$ line This filament becomes activated; i.e. motions inside its structure are enhanced and parts of it begin to rise; the speed of rise increases, and eventually the filament disrupts into the corona. The time between the onset of activation and the filament disruption can be a few minutes, up to more than an hour. Martin (1979b) could clearly recognize first enhanced motions up to three hours before the filament disruption in some other events.

As soon as the final filament disruption starts, bright flare patches begin to be seen in the chromosphere which merge and subsequently form two bright ribbons running parallel to the $H_{\|}=0$ line (the original site of the filament). These Hα patches are at the footpoints of hot coronal loops visible in soft X-rays.

The initial distance of the Hα ribbons from the $H_{\|}=0$ line can be much less than the height of the filament, thus indicating that the first flare loops were formed or enhanced low in the atmosphere below the original filament site. In other cases the ribbons start to be formed quite far from the $H_{\|}=0$ line so that already the first flare loops obviously extend high into the corona. The first X-ray flare loops and Hα brightenings usually appear in the vicinity of the steepest magnetic field gradient in the flare region.

The newly formed loop system then grows in size by the heating of new plasma on progressively higher loops. The outer, newer loops are always hotter than the inner, older loops. After some time the innermost loops cool so much that they become visible (at $T<20{,}000$ K) in the Hα light. We begin to see the well-known 'loop prominences' associated with major two-ribbon flares (Bruzek, 1969; SVE 233–237 REF., Figures 7.1, 7.2 and 7.3 in this book). The Hα loops are always rooted at the inner edge of the Hα bright ribbons, which shows that the Hα brightening is produced at the footpoints of the newly brightened hot X-ray loops. As new progressively higher loops are formed, their footpoints (i.e. the two bright ribbons) move progressively farther from the $H_{\|}=0$ line (cf. Figures 7.4, 7.6, and 7.7 in this book; SVE 42–48 REF and Figure 19).

In spotless regions this motion is uninhibited. The speed of the loop-rise and ribbon-expansion decreases with time, but it continues for many hours. Contrary to that, in well-developed sunspot groups the expansion of the ribbons stops after the ribbons enter regions of strong longitudinal field, i.e. sunspots (SVE Figure 23). The loops are still growing, but apparently only in the vertical direction, with fixed footpoints at the peripheries of sunspots (SVE Figure 24b).

In the flare of 29 July 1979 the growth of loops was definitely established for more than 10 hours (Figure 7.6), and the total duration of the X-ray enhancement lasted in various events for one to three days! In the flare of 29 July 1973 the altitude of the loops exceeded 100,000 km (Figures 7.5 and 7.6), and on August 14, 1973 loops could be seen up to an altitude of 180,000 km. Since the Hα loops are the cool remnants of hot loops which were previously seen in X-rays at the same altitude, one can estimate the cooling time for each individual loop, or group of loops, from the time difference between the first appearance of the hot and cold loops (Figure 7.6). This cooling time is definitely much shorter than the life-time of the whole loop-system. On July 29, 1973, e.g., the cooling time rose from 10 minutes to a few hours, but it never exceeded 4 hours, whereas the loops were seen growing for at least 10 hours. Therefore, the loop arcade is obviously maintained and grows through the continual formation of new hot loops at its outer boundary.

All these observations are in very good agreement with Kopp and Pneuman's (1976) supposition that the originally closed magnetic field opens as the filament disrupts, and subsequently the field lines reconnect giving rise to the loops we see to be formed at progressively larger altitudes. This Kopp and Pneuman theory is explained more in detail in Chapter 7 of this book.

In most events on Skylab we did not see the earliest phase of the event in X-rays, but on 15 June 1973 the temperature at the top of the early loops was about 2.5×10^7 K. Even hours after the flare onset the temperature at the top of newly formed loops is close to or exceeds 5×10^6 K. Along the loops temperature decreases downwards, and the lower parts of visible X-ray loops are by $1–3 \times 10^6$ K cooler than the top. Density, on the other hand, seems to be constant along the whole loop, possibly with an enhancement in a condensation at the top. In the early phase of loop formation this electron density exceeds 10^{17} m^{-3} (10^{11} cm^{-3}) (cf. Figure 7.17); it is generally of the order of 10^{16} m^{-3} in most of the loops, and slightly less than 10^{16} m^{-3} in very late phase of the loop-system development. Different loops vary in density (or alternatively, the spatial density of elementary thin loops along the line-of-sight is variable) and the brightest Hα patches clearly correspond to the footpoints of loops with the highest density.

In Hα one can directly see the sequential formation of new, progressively higher loops (e.g. Bruzek, 1969). Each individual loop starts as a rapid brightening of a knot above existing loops; the knot grows for several minutes and eventually it flows downward along the magnetic field lines.

The fall of cool material from the corona to the chromosphere is characteristic of loop prominences (Martin, 1979a; Engvold *et al.*, 1979). However, the amount of material observed in the prominence system (10^{12}–10^{13} kg) is too large to be explained by loop build-up from the material in the ambient corona (Kleczek, 1964). Therefore, evaporation of some chromospheric material into the loops after their formation seems to be necessary.

This, of course, is difficult to observe in the X-ray loops; in one case, Albregtsen and Engvold (1979) observed an upward rising disturbance in a cool Hα loop, with a speed of about 200 km s^{-1}.

As we will see in Section 2.8, a two-ribbon flare is the most powerful manifestation of the filament disruption. In many other cases, when a quiescent filament disrupts in very old active region remnants or far from any activity, we do not see any chromospheric flare as the consequence of the disruption (Webb *et al.*, 1976; Howard and Švestka, 1977). Still, the process appears to be the same: an opening, and subsequent reconnection, of the magnetic field lines (Rust and Webb, 1977). Two-ribbon flares, in contrast to the compact flares, are sources of coronal transients, type II and IV radio bursts, and high-energy particles recorded in space. Coronal transients (Munro *et al.*, 1979) and slow-mode waves (Rust and Švestka, 1979) can apparently accompany even some filament disruptions that are not associated with chromospheric flarings.

2.4.3 Flare spectrum in X-ray region

The radiation in the soft X-ray region is composed of continuum and emission in lines of highly ionized ions. The continuous emission consists of two parts: free–free (bremsstrahlung) and free–bound (radiative recombination) transitions. Below about 5×10^6 K also the two-photon decay of the 2s states of hydrogen-like and helium-like ions makes a contribution (Tucker and Koren, 1971), but above 2×10^7 K the bremsstrahlung fully dominates in the formation of the continuum spectrum.

The line emission results from downward radiative transitions in highly ionized, almost completely stripped ions, following the occupation of an excited level either by recombination or by inelastic collisions. Collisional excitation is prevalent, but dielectronic recombination can make a very significant contribution in some spectral lines (SVE 116 REF).

The most prominent lines in the X-ray region are formed in helium-like ions by transitions to the ground $1s^2\ ^1S$ term from $1s2p\ ^1P$ (resonance line), $1s2s\ ^2S$ (forbidden line), and $1s2p\ ^3P$ (intercombination line). These three lines form strong complexes in the X-ray spectrum for Fe XXV, Ca XIX, Ar XVII, S XV, Si XIII, Al XII, Mg XI and O VII, which can be resolved into the three components if the resolving power is good enough. Weak lines of the same type also seem to be present for Mn XXIV, Cr XXIII, Ti XXI, and K XVIII, as well as for Ni XXVII, Na X, and Ne IX, where these lines are blended with other possibly important lines of iron. In addition, for most abundant elements higher lines to the ground term are observed, possibly up to $1s^2$–$1s5p$ for Fe, Si, and Mg.

Next most important lines in the X-ray region are the lines in the Lyman series in hydrogen-like ions. The Ly-α line is clearly observed for Fe XXVI, Ca XX, S XVI, Si XIV, Al XIII, Mg XII (particularly strong), and O VIII, most probably for Na XI and Ne X, where it is blended, and possibly for Ar XVIII. Lyman lines up to Ly-γ are clearly seen for sulphur, silicon and probably magnesium (blended).

Lines of a third type, which are fairly weak, but very important from the theoretical point of view, are produced through innershell transitions in ions with three or more electrons. Apart from these basic lines, which fully dominate the spectrum below 10 Å, many other lines are observed at longer wavelengths. Most originate in various states of ionization of the iron atom.

A list of the lines observed below 22 Å is given in SVE 124–127 REF. A few additional silicon lines were detected by Parkinson *et al.* (1978). Other lists of lines were prepared by Neupert *et al.* (1973), and Bergman *et al.* (1973). Particular attention has been paid (and will be paid at the Solar Maximum Mission satellite) to the line complex of iron (Fe XXV and lower stages of ionization) between 1.86 and 1.90 Å; it is supposed that this line group can be a very sensitive diagnostic tool not only for the thermal plasma, but also for non-thermal streams of particles expected to occur in the flare flash phase. The best measurements of this line complex, with a resolution of 4×10^{-4} Å, have been made by Mandelstam's group in Moscow (Grineva *et al.*, 1973; Mandelstam, 1978, 1979; Korneev *et al.*, 1979).

2.5 IMPULSIVE HARD X-RAY AND MICROWAVE BURSTS

2.5.1 Hard X-ray bursts

De Jager (1965) was the first to suggest that one should distinguish two components in solar X-ray bursts: one thermal, or quasi-thermal, which we have discussed in the preceding chapter, and another one of impulsive, non-thermal character, probably caused by directed streams of accelerated electrons, with non-Maxwellian velocity distribution.

This has been confirmed in later years, as X-ray measurements have been extended to higher energies. However, as Kane (1969) showed and many others confirmed, the non-thermal component is not present in all X-ray bursts; Datlowe *et al.* (1974) have found detectable hard X-ray emission only at 2/3 of soft X-ray bursts with peak fluxes above 10^7 photons $(m^2\ s\ keV)^{-1}$, and when also smaller flares are considered, the impulsive component is clearly visible only in less than 15% of all the flare brightenings (Vorpahl, 1972). Thus the impulsive component either does not develop in many flares or it is so weak (or short-lived) that one cannot recognize it in the record.

Characteristic properties of the hard X-ray bursts have been summarized most recently by Kane *et al.* (1980); also see SVE 143–151 REF and §8.3. A typical example is shown in Figure 2.8 (see also Figures 8.2–8.4).

When the impulsive component is present, it occurs in the initial phase (flash phase) of the flare for tens of seconds to a few minutes, and it extends to much higher energies than the thermal X-ray burst. The thermal burst usually ceases to be discernible above 10 or 20 keV. In contrast to it, the short-lived impulsive burst of non-thermal nature can be seen up to energies of hundreds of keV in strong events.

It is usually difficult to recognize the impulsive component in X-ray records below 10 keV. Nevertheless, Kahler and Kreplin (1971) have shown positively that the impulsive component extends in energy downwards as far as ~ 3 keV. On the other hand, the highest energy ever recorded in a hard X-ray burst was about 3 MeV in the flares of 1972, August 4 and 1978, July 11 (cf. Section 2.9.4.C). However, such a high energy is quite exceptional. In most flares, the burst ceases to be recognizable below 100 keV, and only in major flares does the emission extend over this limit.

The hard X-ray burst always starts either simultaneously with the soft X-ray enhancement (Figure 2.8) or after the soft X-ray onset (see example in SVE, Figure 54a). The time difference between the soft X-ray and hard X-ray onsets may be more than 8 min (Anderson and Mahoney, 1974; Spangler and Shawhan, 1974). Therefore, the impulsive non-thermal burst evidently is not the source of the whole flare phenomenon as many authors have been suggesting. There are also several lines of evidence that the site of the impulsive burst is relatively small, much smaller than the coronal condensation which gives rise to the soft thermal bursts (SVE 144–146 REF; Zirin, 1978; Neidig, 1978).

Detailed analyses of hard X-ray bursts have been performed by Kane and Anderson (1970), Kane (1971, 1973, 1974), Peterson *et al.* (1973), Datlowe *et al.* (1974), Elcan (1978), and Crannell *et al.* (1978). The results, generally referring to small flares or subflares, can be summarized as follows:

1) The hard X-ray burst occurs during the flash phase and peaks usually 0.5 to 3 min before the flare maximum observed in Hα light.

2) The X-ray flux increases and decreases are roughly exponential, with *e*-folding rise times of 2 to 5 s and *e*-folding decay times of 3 to 10 s at 40 keV. The rise time is always shorter or equal to the decay time and quite a number of short simple hard X-ray bursts are symmetric (Crannell *et al.*, 1978). This is not true for the correlating microwave bursts which generally have longer decay time.

3) With higher time resolution, many bursts can be decomposed into several short-lived spikes ('elementary flare bursts' after Van Beek *et al.* (1974)), with the rise and decay times as low as 1 to 2 s (Frost, 1969; Anderson and Mahoney, 1974; Hurley and Duprat, 1977; Crannell *et al.*, 1978). According to De Jager and De Jong (1978), for each one flare the individual elementary bursts have similar durations, varying from 4s to 24s in the 8 flares they studied.

4) The X-ray spectrum which fits the observations at the burst maximum is of the power-law form

$$\frac{dJ(E)}{dE} = CE^{-\gamma} \text{ photons m}^{-2} \text{ s}^{-1} \text{ keV}^{-1}, \tag{79}$$

with $2.5 < \gamma < 5.0$ (median $\gamma \simeq 4.0$) in the range 10 keV $< E < E_0$, and much steeper for $E > E_0 \simeq 60$ to 100 keV. γ may be as high as 8 (Peterson *et al.*, 1973), but almost never below 2.5 (Kane, 1971, 1973), which appears to be an actual limit to the hardness of the spectrum, even in major flares (Hoyng *et al.*, 1976).

5) According to Kane and Anderson (1970) and Kane (1973), the spectrum hardens as the flux increases and softens during the decay time, so that the hardest spectrum is obtained at the time of the peak flux. A similar pattern has been found by Elcan (1978) and Crannell *et al.* (1978); this appears to be a characteristic feature of the impulsive phase, but it is not observed in all bursts (Datlowe *et al.*, 1974; Hoyng *et al.*, 1976).

In more important flares, the impulsive hard X-ray burst is sometimes followed by a second-phase gradual hard X-ray emission (Frost and Dennis, 1971; Hoyng *et al.*, 1976; Hudson, 1978) of much longer duration, but so far observations of this second phase have been rare (see §8.3.2).

Some interesting information has been obtained from flares that produced hard X-ray bursts behind the solar limb. Kane and Donnelly (1971) estimated that the hard X-ray emission in a large burst of 1969, March 30, extended above 25,000 km over the photosphere. Roy and Datlowe (1975) found 25 over-the-limb events with a non-thermal component and concluded from it that at least some of the hard X-ray emission must come from considerable heights (of the order of 10^4 km) in the corona. More recently Hudson (1978) detected hard X-rays from a flare that was at least 25° beyond the limb, which gives a minimum occultation height of 66,000 km. This, however, was obviously the gradual 'second-phase' burst we mentioned above, because it lasted more than 40 minutes.

On the other hand, for most flares behind the limb the hard X-rays are not seen (Roy, 1975), or their data are consistent with most of the hard X-rays being produced at a low altitude (McKenzie, 1975; Kane *et al.*, 1980). Thus, as it seems, both thick-target and thin-target emissions can produce the impulsive burst, possibly with varying importance of one or the other mode in different flares. For example, it has been suggested, as observations indicate, that in two-ribbon flares the acceleration process is accomplished during magnetic reconnection at the top of newly formed loops (Švestka *et al.*, 1979b). Then a (possibly small) fraction of accelerated electrons is trapped in the newly formed loop, producing (very efficiently) hard X-ray emission in the thick-target mode. A huge number of other accelerated electrons escape from the loop trap, and produce (less efficiently) hard X-ray emission in the thin target mode.

There are three observations which indicate, in spite of the occasional observed extension of hard X-ray emission to large altitudes, that a thick-target process should be present (cf. SVE 159–161 REF):

a) As soon as we accept the interpretation that the Hα kernels and the white-light flares are produced by accelerated particles streaming radially from the corona into the chromosphere, the same streams should also produce the hard X-rays (cf. Kane, 1973b).

b) Preliminary results of the hard X-ray imaging aboard the SMM satellite indicate a thick target process (Hoyng *et al.*, *Astrophys. J. Lett.*, in press, and unpublished results).

c) Tindo *et al.* (1972), Tindo and Somov (1978) and Mandelstam (1979) claim that their measurements of polarization of flare X-rays are in agreement with bremsstrahlung produced by directed beams of accelerated electrons streaming in predominantly radial directions.

Nevertheless, this does not seem to be the case in all flares (Tindo *et al.*, 1976). Sometimes either the electron stream flow greatly deviates from the radial direction or the stream does not develop at all.

Brown (1972) and Petrosian (1973) have pointed out that radial streams of electrons should produce a brightening of the hard X-ray sources from solar center to the limb, which has not been confirmed by observations (no limb-brightening in Pizzichini *et al.*, 1974 and Datlowe *et al.*, 1974, 1977; small one in Kovrizhnikh *et al.*, 1976). However, as Tindo and Somov (1978) pointed out, if Compton scattering of the primary flare emission by the photosphere is taken into account, one may obtain no substantial dependence of burst intensity on the heliocentric distance at all, even with a strongly anisotropic electron distribution.

Compton back-scattering on the photosphere could be used to estimate the height of the hard X-ray emission, as suggested by Brown *et al.* (1975), Bai (1978), and Bai and Ramaty (1978), by measuring the position or time shift between the original and reflected emissions. The hard X-ray imaging on the SMM satellite (and later aboard Space Shuttle) can be used for such measurements. It also can definitely decide about the role of the thick-target and thin-target mode in various flares.

An interesting, and so far unique phenomenon, was observed by Hoyng *et al.* (1976) during the long-lasting hard X-ray bursts in the γ-ray flares of 4 and 7 August 1972. Several clearly identified temporal features in the hard X-ray flux occurred progressively later at higher energy channels. According to Bai and Ramaty (1979), the delay time grows slowly to about 5s as energy increases from $\sim$30 through 150 keV, and it jumps to $\sim$15–20s for energies between 150 and 280 keV. Bai and Ramaty suggest that the small delay at lower energies is due to the increase of the electron energy loss time with electron energy, whereas the increased delay above 150 keV manifests second-step acceleration. Thus the second-step acceleration must set in within less than 20 seconds after the first acceleration step, which also appears to be in agreement with observations of γ-rays and white-light emission (cf. Section 2.9.4.C).

2.5.2 Microwave bursts

Impulsive microwave bursts, i.e., fast radio enhancements on frequencies above ~ 1000 MHz, occur simultaneously with hard X-ray bursts and have similar characteristics (see §8.5). There is a very good time correlation between the hard X-ray bursts and impulsive microwave bursts, which are often quite similar, even in the fine details of the time-development curves (Figure 2.8). The correlation is usually best at very high frequencies (Anderson and Winckler, 1962; Parks and Winckler, 1971), and very hard X-ray bursts (above 50 keV) are most similar to radio bursts at frequencies of 1 GHz or more (Arnoldy *et al.*, 1968). According to Crannell *et al.* (1978), in some cases the microwave burst can be slightly delayed behind the X-ray burst. In the vast majority of cases, the impulsive microwave burst is short, with a duration ranging between 1 and 10 min. Very short bursts are usually simple, but the longer ones often show complex fluctuations which are more pronounced at lower frequencies. Quite often, the burst has a long-lived tail, which is called the post-burst increase and is probably of thermal origin. In some 10% of events, however, the impulsive burst itself is longer, with a duration up to one hour (Takakura, 1967), and usually with very complex structures. In such complex and long-lived cases only the onset part of the burst shows a good time coincidence and similarity with the associated hard X-ray burst, whilst the later flux variations have no counterpart in the hard X-ray region. These long-live microwave bursts are mostly associated with prolonged emission at lower frequencies up to the metric range, and form a part of a type IV burst (cf. Section 2.7.1).

The spectrum of the microwave bursts is a broad-band continuum, with maximum intensity generally at about 3 to 10 GHz (Takakura, 1967), the maximum tending to shift to higher frequencies for very strong bursts; in particular, for strong type IV bursts associated with particle emission in space, the maximum is found between 5 to 70 GHz, with the mean value at about 17 GHz (Croom, 1971).

The intensity of a radio burst is commonly given in standard flux units (sfu) of 10^{-22} W m^{-2} Hz^{-1}. Whilst quasi-thermal microwave bursts, associated with soft X-rays (SVE 140–142 REF), usually have their maximum flux below 40 sfu, the maximum flux of impulsive bursts is often much higher and it can exceed 10^4 sfu in outstanding events.

All strong microwave bursts are associated with flares. Bursts of lower intensity (<100 sfu) sometimes appear without reported Hα flares, but almost all can be identified with flare-like Hα brightenings in the chromosphere. The same is true for the hard X-ray bursts. The correlation with reported flares appears to be much better when bursts at millimeter wavelengths are considered, where essentially all bursts are associated with flares (Shimabukuro, 1968; Croom and Powell, 1971). The reverse association is weaker: only about 60% of flares of importance >1 are accompanied by bursts at 10 GHz (Kundu, 1965), and for smaller flares and subflares the correlation percentage is very small. This association, however, is different in different active regions. In some of them, microwave bursts do not occur at all. In others, in particular where a magnetically complex configuration forms, microwave bursts accompany almost all flares, and even many subflares (Švestka and Simon, 1969; Švestka, 1971a; Matsuura and Nave, 1974).

Many authors have tried to determine the center-limb variation of the microwave burst intensity (SVE 174 REF) but the results are fairly confused and contradictory.

The sizes of impulsive microwave bursts are smaller than those of Hα flares and thermal microwave emission. Kundu *et al.* (1974) observed a microwave burst of 3.7 cm, with 40% of the energy being released within a spherical volume of only 2″ diameter. The size increased to $\sim 5''$ in a later phase of the flare. At 11 cm the dimensions were about 5 times larger. Quite recently, by using the VLA antenna at 4.9 GHz, Marsh *et al.* (1979) found a thermal component in small flares of size $\sim 10''$, whereas four weak impulsive bursts had sizes between 2″ and 4″.

Ogawa and Kawabata (1975), using the 35 GHz interferometer at Nagoya, found that the position of the microwave radio source in the flare of August 2, 1972, coincided with an Hα kernel. Kundu *et al.* (1976) found a coincidence of the microwave burst core with a soft X-ray flare kernel.

In several sources Tanaka *et al.* (1967; Tanaka and Enomé, 1970) clearly distinguished a double structure roughly coinciding with sunspots of opposite magnetic polarity. This implies that the microwave burst may be produced inside the feet of a loop-like magnetic structure which stretches between the opposite polarities. This loop may be also the seat of the hard X-ray burst, of the EUV burst and, in particularly strong events, of the white-light emission as well.

Generally, the microwave bursts are partially circularly polarized and the sense of polarization reverses between 2 and 4 GHz. The polarization behavior of the microwave bursts (rather complicated, cf. SVE 176 REF) can be understood if one assumes that all sources have a double structure and the radio emission is produced by gyro-synchrotron radiation.

2.5.3 Quasi-periodic flux variations

Frost (1969) and Parks and Winckler (1971) have detected simultaneous intensity variations in the hard X-ray and microwave flux which appear to have a quasi-periodic structure. In the case studied by Parks and Winckler, five successive maxima were observed in the flux of >20 keV and >50 keV X-rays, with a periodicity of 16 s. In the microwave records these periodic variations were best seen at high frequencies, above 1 GHz. In the case studied by Frost, three successive maxima were observed in the flux of 28 keV X-rays and >2800 MHz radio records, with a period of about 35 s. Periods ranging from 20 to 100 seconds with the majority in the range of 30 to 50 seconds have been found by Lipa (1978).

Janssens *et al.* (1973) observed quasi-periodic pulsations typically with periods of 10 to 20 s in 7 microwave bursts at ~ 3000 MHz. Their observations are similar to those in hard X-rays. Quasi-periodic fluctuations of intensity and polarization have also been reported by Erjushev and Tsvetkov (1973) at 3.15 cm, with periods from 6 to 320 s. Kaufmann and co-workers (Kaufmann, 1976; Kaufmann *et al.*, 1977) have found periodic oscillations at 7 GHz with periods of 4.7 and 17 s, respectively. Quasi-periodic changes at 2.8 GHz with periods between 6 and 600 s were detected by Cliver *et al.* (1976). Similar long periods, 200–900 s, were reported in 1–8 Å X-rays by Jakimiec and Jakimiec (1974).

In fact, however, the small number of pulses (five in a typical event) does not prove any statistically significant 'periodicity'. Anderson and Mahoney (1974) observed two hard X-ray bursts composed of many 'flashes' (cf. SVE Figure 57) and could not find any periodicity in their occurrence. Hoyng *et al.* (1976) could find an indication of periodical variations only in one out of eleven events they studied.

In the flare of August 2, 1972, Zirin and K. Tanaka (1973) observed fast flashes along the zero line in a sunspot penumbra (cf. SVE Figure 35). A comparison of the time-variations of the hard X-ray flux (Van Beek *et al.*, 1974) with Zirin and Tanaka's photographs suggests an association of the flashes in both wavelength regions. Any X-ray peak, when associated with an optical counterpart, had a tendency to be keyed to the brightening of a particular point or pair of points and not to a repeated brightening of the same place (Kane *et al.*, 1980). This suggests that the impulsive phenomenon may consist of an energy release which occurs only once at a given location, but can be repeated at several locations, in a quasi-periodic sequence of hard X-ray or microwave 'elementary bursts'.

Also the spatially-resolved microwave observations made by Enomé and H. Tanaka (1973) and multiple spikes of EUV emission observed by Thomas and Neupert (1975) showed that for several flares the individual components of the microwave or EUV burst occurred at different sites within the active region. However, Karpen *et al.* (1978) suggest that both periodical variations and 'travelling bursts' are possible. They have observed two kinds of multiple impulsive burst events during which the microwave turnover frequency (where the source becomes optically thick) and spectral shape remain the same from peak to peak, and others during which the turnover frequency and spectral shape change significantly. They suppose that in the first case the emission can be characterized by a constant magnetic field and therefore a single source region, in which case the (quasi-periodic) multiplicity may be due to modulation of the emission process; in the second case, the individual spikes appear to originate in regions of different magnetic-field strengths, corresponding to separate source regions that flare sequentially. In that case the quasi-periodic occurrence of the elementary bursts is either a chance coincidence, or it is due to equidistant distribution of the loops that subsequently flare, after being excited by a travelling trigger-disturbance. Vorpahl (1976) and Zirin (1978a) have reported observations of such sequential brightenings in flares.

2.6 ENERGY TRANSFER THROUGH THE TRANSITION LAYER

So far we have discussed the hot flare plasma in the corona, as seen in soft X-rays, and its effects upon the chromosphere — the chromospheric flare. In between is the flaring transition layer, seen in the EUV lines.

As in the X-ray region, the EUV flare emission appears to have two components: one being of apparently non-thermal and impulsive nature (see §8.8), and the other of quasi-thermal and gradual character (SVE 162 REF). The impulsive EUV burst occurs simultaneously at all temperatures from 10^4 to 10^6 K (Emslie and Noyes, 1978) and coincides in time with the hard X-ray and impulsive microwave burst, often even in detailed time structure. It also coincides with white-light flare occurrence in the rare cases when this is observed (McIntosh and Donnelly, 1972). The EUV burst life-time is much shorter than the Hα brightening or the soft X-ray burst, but it is generally longer than the life-time of the accompanying non-thermal X-ray burst.

The impulsive EUV flare radiation rises simultaneously with the net 10 to 1030 Å flash deduced from the SFD data, which thus can be used as a substitute for direct EUV measurements if these are missing (Donnelly *et al.*, 1973). Such data indicate that the relative strength of impulsive EUV bursts, when compared both with flares and hard X-ray bursts, decreases towards the solar limb (Donnelly, 1971; Kane and Donnelly, 1971; Donnelly and Kane, 1978). This suggests that the EUV burst originates in sources embedded deep in a cool absorbing atmosphere.

Under the assumptions that the impulsive EUV bursts are produced by downward streaming electrons in a thick-target model (§§8.9, 8.10), the electron power deposited, as deduced from the contemporary hard X-rays, is far greater than the EUV luminosity (Emslie *et al.*, 1978; Donnelly and Kane, 1978). Thus, either most of the power deposited is conducted away to the optical flare, or only a fraction, less than 10%, of the X-ray emitting electrons is injected downwards.

2.6.1 The EUV spectrum

The EUV emission appears in many spectral lines, ranging from neutral hydrogen and oxygen to highly ionized ions of metals. The amount of energy radiated in the strong EUV resonance lines of O VI and Mg X is typically 10^{-1} to 10^{-3} times the energy radiated in Hα.

According to Hall and Hinteregger (1969; Hall, 1971) and Donnelly *et al.* (1973), the lines formed in the chromosphere and in the transition region zone dominate during the impulsive phase. The coronal emission lines, on the other hand, tend to dominate during the gradual, quasi-thermal phase, which peaks later. Lists of EUV lines in flares can be found in SVE 163 REF (284–1394 Å), Dere (1978) (171–630 Å), and Cohen *et al.* (1979) (1000–1900 Å).

Emission in the EUV spectrum originates within a wide range of temperatures. The lowest temperature is deduced from the Lyman series of hydrogen, where the colour temperature of hydrogen Lyman continuum is only 7500–9000 K (Noyes, 1973; Machado and Noyes, 1978). In the Lyman series of helium, Linsky *et al.* (1976) found a colour temperature of about 24,000 K in the Lyman continuum of a flare. The brightness temperature at the head of the Lyman continuum was about 41,000 K, both these last values being some 10,000 K higher than T_c deduced from the quiet Sun spectrum. Various spectral lines then trace the region from 10^4 through 10^6 temperatures, as one can see, e.g., in Figure 2.9. Whereas the height of the Lyman lines of hydrogen can be estimated to be 3600 ± 2400 km above the base of the chromosphere, the Fe XVI radiation, at $T \simeq 3 \times 10^6$ K, comes from some $15{,}200 \pm 2300$ km altitude (Simon and Noyes, 1972).

Brueckner (1975b) reported non-thermal Doppler broadening of EUV lines in flares with characteristic velocities of ~ 70 km s^{-1}. Non-thermal motions can be observed also before flares, but the broadening rather characterizes active regions prolific in flares than the immediate preflare phase (Brueckner *et al.*, 1975). The existence of turbulent motions in flares, up to speeds of 160 km s^{-1} in Fe XXIII and XXIV lines, but usually in the range between 30 and 80 km s^{-1} at lower temperatures, has been confirmed by Cheng (1977, 1980). According to Doschek and Feldman (1978) turbulent velocities are higher in moving components than in stationary features. In Fe XIX and XXI lines Doschek *et al.* (1975) observed random velocities of 60 to 70 km s^{-1} in the early phase of the 15 June 1973 flare, decreasing to some 30 to 40 km s^{-1} twenty minutes later. In the impulsive phase of the same flare, Doschek *et al.*

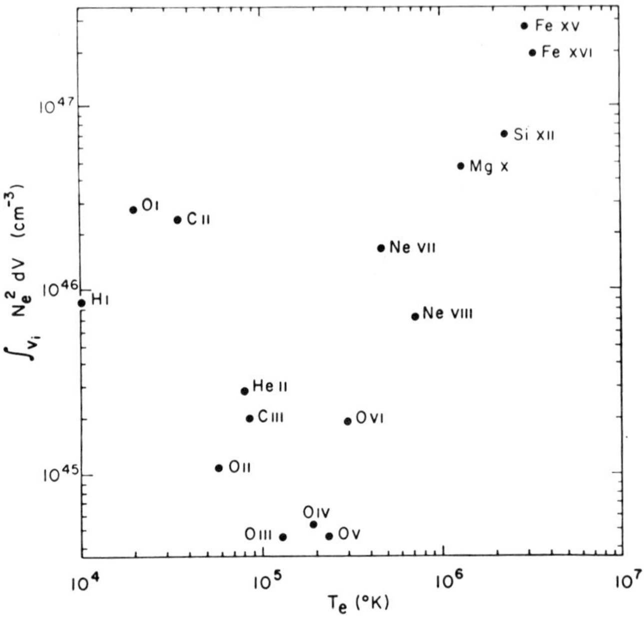

FIGURE 2.9 Peak emission measure (deduced from different EUV lines) versus electron temperature. The minimum at about 2×10^5 K and the increase in emission measure both toward lower and higher *T* values indicate the existence of a narrow transition layer in flares. (After Wood and Noyes, 1972.)

(1977) and Cheng (1977) observed plasma moving towards the observer for about 2 minutes, with velocities as high as 60–80 km s^{-1}. Contrary to that, the transition zone lines in the flare of 5 September 1973 showed a red-shifted component during the initial phase of the flare (Cheng, 1978), with speed of ~ 50 km s^{-1}, and red asymmetry was seen by Cheng (1980) in two other flares. According to Brueckner (1975a, b), asymmetry towards short or long wavelengths appears in the transition layer lines without any particular preference. If so, then this behaviour is quite different from chromospheric lines, in which the asymmetry (except the earliest phase, cf. Figure 2.6) is predominantly towards the red.

2.6.2 The EUV flare morphology and density

Observations of flares in monochromatic images, produced in the various EUV lines, confirm the conclusions which have been drawn from the X-ray images of flares (Widing and Cheng, 1974; Neupert *et al.*, 1974; Cheng and Widing, 1975; Dere *et al.*, 1977; Widing and Dere, 1977; Levine, 1978; Moore *et al.*, 1980). In two-ribbon flares one finds as a rule in the early phase a hot core

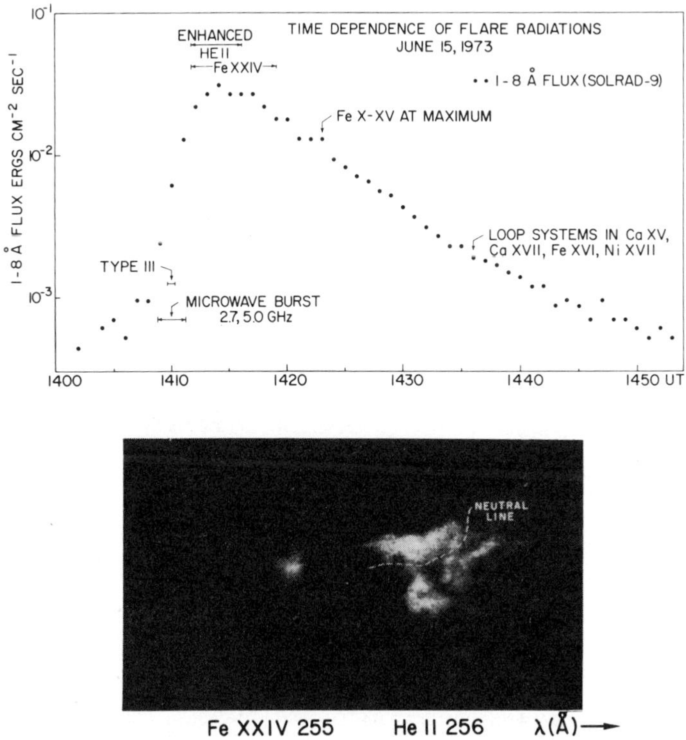

FIGURE 2.10 Upper part shows the soft X-ray emission of the two-ribbon flare of June 15, 1973, and a description of the emissions observed. He II and Fe XXIV are bright at flare maximum, but in different spatial locations, as one can see in the lower part of the figure. This lower part shows monochromatic images of the flare in the Fe XXIV (λ 255 Å) and He II (λ 256 Å) lines, obtained with the NRL slitless objective-grating spectrograph on Skylab. While the hot Fe XXIV emits on the top of the 'flare loop' above the $H_{\|}=0$ line, the 'cool' He II is seen in the loop feet. (After Widing and Cheng, 1974; *Astrophysical Journal*, University of Chicago Press. Copyright the American Astronomical Society.)

(represented, e.g., by the Fe XXIV line at $\sim 1.6 \times 10^7$ K) centered over the $H_{\|}=0$ line, while the cooler material (represented by the He II, Ne VII, and Fe XIV–Fe XVI lines) shows the familiar two-ribbon structure co-spatial with the Hα flare (Figure 2.10). As the flare cools the hot cloud disappears and the region between the ribbons is filled gradually with emissions from ions of lower ionization temperature, exhibiting loop structures.

Great interest has been paid to determinations of densities in the flaring transition region, from the corona (at 10^6–10^7 K) down to the chromosphere (10^4–10^5 K). One can estimate the density from the emission measure if the radiating volume is known, or from the cooling time if the contribution of the conductive cooling is known (SVE 138 REF), but both these methods may involve serious errors because of the insufficient spatial resolution of the observed images. There is only one method which does not depend on the spatial resolution and that is the intensity comparison of properly selected density-sensitive lines in the flare spectrum. Their ratio should yield reliable results on the densities in that part of the flare from which the lines are emitted.

The whole story started when Gabriel and Jordan (1969) pointed to the interesting fact that collisional interchange can take place between the levels ^3S and ^3P in the helium-like ion so that the relative intensities of the intercombination $^3P \rightarrow {}^1S_0$ and forbidden $^3S_1 \rightarrow {}^1S_0$ lines are density-dependent. Unfortunately, the method did not work, mainly because of simultaneous dependence on temperature (SVE 139–140 REF). However, there are other iso-electric sequences, where the resulting n_e values appear to be more reliable.

Munro *et al.* (1971) have shown that the ratio of intensities from certain pairs of lines from ions of the beryllium iso-electronic sequence (e.g., C III, O V) is sensitive to the electron density through the influence of a metastable state. From O V lines, 6 min after a flare maximum, Noyes (1973) thus found $n_e = 5 \times 10^{17}$ m^{-3}, which is 5 times the preflare value, 30 times that in a typical nonflaring active region, and 150 times that in the quiet Sun. Generally, Noyes reports increases by factors of 5 to 10 in the transition zone for small flares, also confirmed by increases in the emission measure. In a flare of Hα importance 1b, Donnelly and Hall (1973) found $n_e > 10^{18}$ m^{-3} (10^{12} cm^{-3}) from the C III lines. By using ions in other iso-electronic sequences, Purcell and Widing (1972) estimated the density from the ratio of the Fe XIV lines at 211 and 219 Å, applying Blaha's (1971) theoretical computations of the dependence of the ratio value on electron density. Four minutes after the maximum of a 2b importance flare they found $n_e = 3 \times 10^{16}$ m^{-3} (3×10^{10} cm^{-3}) in the hot region where Fe XIV was emitted. Similar results were also found from Fe XV lines by Cowan and Widing (1973). As one can see from Figure 2.9, these densities correspond to flare regions with significantly higher temperatures of $\sim 3 \times 10^6$ K, as compared with $\sim 3 \times 10^5$ K in O V and $\sim 8 \times 10^4$ K in C III lines.

In more recent years, many other ions, mainly in the Be and B iso-electronic sequences, have been used for density estimates in flares (cf. Dupree, 1978, for a review). Whereas Vernazza and Mason (1978) give n_e as low as 4.0×10^{15}–1.0×10^{16} m^{-3} from B-like ions (Si X), all the other results have led to significantly higher densities. Cheng (1980), also from B-like ions, gives $n_e = 6 \times 10^{17}$–2×10^{18} m^{-3} near the maximum phase of three flares (Si III and O IV lines). From Ca XVII, Doschek *et al.* (1977) have deduced $n_e = 5 \times 10^{17}$ m^{-3}; from Fe IV (near 10^6 K) Feldman *et al.* (1978) give n_e at least as high as 2×10^{17} m^{-3}; and from O IV (near 10^5 K) Feldman *et al.* (1977) and Feldman and Doschek (1978) find densities from 10^{17} up to $> 10^{19}$ m^{-3}. These highest values are the

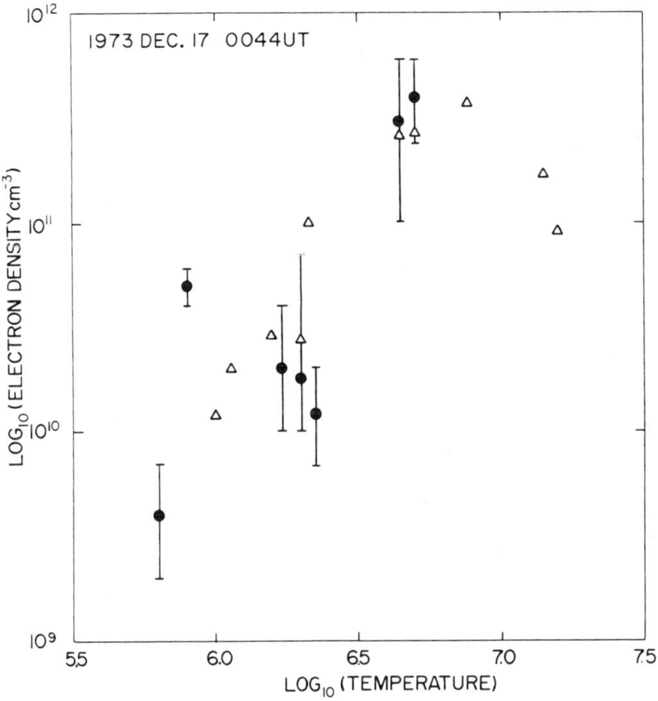

FIGURE 2.11 Electron density as a function of electron temperature for the December 17, 1973, compact flare. A dot, ●, gives n_e from line intensity ratios; a triangle, Δ, gives n_e from the total line intensity and apparent size of the emitting region for that wavelength. (After Dere *et al.*, 1979.)

same as electron densities deduced from the Balmer series (near 10^4 K) in the chromosphere (cf. Section 2.3.2). The enhanced densities are apparently related to very small volumes, 10^{14}–10^{17} m^3.

Dere *et al.* (1979) have studied ions Fe IX through Fe XV, Ni XI through Ni XVII, and members of the Be, B, C, and Ni iso-electric sequences for two compact flares. They have found that the electron density rises from $\sim 10^{16}$ to 5×10^{17} m^{-3} for the temperature rise of 10^6 to 10^7 K. An example of their result is shown in Figure 2.11. Generally, as we said already in Section 2.4.2.A, the density in compact flares is higher than in the large loops of the two-ribbon flares (Moore *et al.*, 1979), but apparently it can decline very fast: Dere and Cook (1979) report a decline in density from 3×10^{18} to 1.3×10^{17} m^{-3} within 3 minutes at $T < 3 \times 10^5$ K; this estimate was made from emission measure-values, and higher-temperature plasma showed different density variations, namely, a rise near 2×10^6 K and a constant value at still higher temperature.

Measurements in the early phase of a major two-ribbon flare are not available; one can see from Figure 7.17, however, that n_e in the onset phase of the 1973 July 29 two-ribbon flare could significantly exceed 10^{17} m^{-3} (10^{11} cm^{-3}) and values near 10^{18} m^{-3} at the very onset of the flare (almost half-an-hour earlier than the first measurement in Figure 7.18) cannot be excluded. Thus the maximum densities in the compact and two-ribbon flares may not differ so much as the present results indicate.

One has to be aware of the fact that some of these density determinations can still be uncertain. For some line-ratios the effects of temperature variations can surpass the effect of density changes; according to Brown and Nakagawa (1978) plasma turbulence can invalidate the density determination; and mass flow through a temperature gradient can change the character of line emission; according to Raymond and Dupree (1978) a downflow of material in the transition region can lead to an underestimate of n_e. Thus, if mass-motions vary with temperature (as some observations indicate), even the n_e variation with T, shown in Figure 2.11, may be misleading.

2.6.3 Energy transfer

It is now generally believed that the site of the primary flare energy release is in the corona (Švestka, 1973; Canfield *et al.*, 1980). Thus the flare phenomena occurring lower in the atmosphere must come about as a result of transfer of energy from the primary energy release site in the corona. Opinions, however, vary greatly on the basical physical processes that carry out this transfer. It may be fast particle energy transport, mass motion, conduction, or radiation.

Many authors have assumed that streams of non-thermal particles accelerated in an active region drive all other flare phenomena and are thus the exclusive mode of energy transport in the flare (SVE 95 REF). This idea, however, contradicts several observed facts.

a) In many flares we do not see any evidence for a non-thermal acceleration process and thus for the presence of suprathermal particles in the flare region.

b) In those flares where particle acceleration is observed (in the form of hard X-ray, EUV, and impulsive microwave bursts) it often occurs several minutes after the onset of the thermal flare (soft X-rays) and even after the first brightening in the Hα light.

c) The energy contained in the particles producing the non-thermal burst is in many flares definitely smaller than the total energy of the flare.

Therefore, as Takakura (1973), Švestka (1973), and Somov and Syrovatsky (1974) have emphasized, the basic flare phenomenon should be a thermal flare; by this we mean that all the kinds of radiation we observe can be interpreted as due to a simple temperature rise and decrease. Only in some flares do we also observe non-thermal effects, and then it is always for a short time of only tens of seconds or a few minutes during the flash phase. There are indications that particle acceleration is accomplished for a longer time than just the duration of the impulsive phase. The polarization in hard X-rays may be enhanced even late in the flare (Tindo *et al.*, 1976; Mandelstam, 1979), and in two-ribbon flares acceleration to low energies can continue for hours (Švestka *et al.*, 1979b). Still, the chromospheric emission must be predominantly due to other effects, mainly heat conduction (see SVE 95–107 for a detailed discussion of this problem) in combination with mass motions (Craig and McClymont, 1976).

A serious complication of this picture has been the discovery of a slight overheating of the temperature-minimum region in flares, detected by Machado and Linsky (1975) from the wings of the Ca II lines. This overheating was later confirmed by Cook and Brueckner (1979) from enhancements of the Si I photoionization continua in the ultraviolet, and by Cook (1979) from intensities of Al I autoionization lines. According to Machado *et al.* (1978) the flare temperature minimum is depressed some two scale-heights below its preflare level, and there is no obvious mechanism which might produce the observed enhancement if the primary flare source is in the corona.

A completely new mode of energy transfer, namely an illumination of the chromospheric levels by the X-ray flare, was suggested by Somov (1975, 1976) and further studied by Hénoux and Nakagawa (1977a, b; 1978). According to Machado (1978) there are emission halos in chromospheric hydrogen images which can be interpreted in this way, in particular for the compact flares. However, wherever heat conduction is appreciable (i.e. at footpoints of loop structures) it dominates over X-ray heating.

Hénoux (1979) has tried to explain even the bright Hα ribbons in two-ribbon flares by X-ray illumination. However, the sharp edges of some of the ribbons (Machado, 1978) and the transient red asymmetry seen at the outer edges of newly formed ribbons (Švestka *et al.*, 1979b) clearly show that the transfer must be due to inflow of material following magnetic field lines.

2.7 METRIC RADIO BURSTS

Following Wild and McCready (1950) and Boischot (1957), radio bursts at lower frequencies are classified as types I through IV. Types II and IV are typically, and type III often, associated with flares.

2.7.1 Type IV bursts

As we mentioned in Section 2.5.2, some impulsive microwave bursts are accompanied by long-lived radio enhancements at lower frequencies. Such an increase in continuum intensity over practically the whole frequency spectrum is called the type IV burst. It consists of a microwave component, IVμ, decametric component, IV dm, and two metric components: moving (IVmA) and stationary (IVmB) type IV bursts (SVE Figure 68, 181–183 REF). In the last few years (after significant contributions to the problems of mass ejecta from the Sun in Culgoora and aboard Skylab) some authors tend to call only the moving metric component as the type IV burst, denoting the stationary component simply as 'continuum'.

The moving type IV component is observed as a rule on frequencies below 200 MHz (SVE 183 REF), and far the most data on it come from the Culgoora radioheliograph on 80 and 160 MHz (Wild *et al.*, 1968; Labrum and Smerd, 1968; Kai, 1970; Sheridan, 1970; Riddle, 1970; Smerd and Dulk, 1971; Dulk and Altschuler, 1971; Wild and Smerd, 1972; McLean, 1974; Kai, 1978; Robinson, 1978; Kai, 1979). Usually it starts a few minutes after the onset of type IVμ burst and the onset delay time increases with decreasing frequency. A typical type IVmA burst lasts several tens of minutes and its source moves outward (not necessarily radially) with a speed between 200 and 1500 km s^{-1}. According to Kai (1979) only less than 15% of these bursts have velocities in excess of 1000 km s^{-1} (almost equal to the fast MHD shock velocity), whereas most bursts have velocities averaging around 400 km s^{-1}. The source ascends to some maximum height in the corona, which usually amounts to 2 or 3, but sometimes more than 5 or even 10, solar radii. Sometimes the source ceases to be visible while still rising; in other cases it stops at some altitude or moves slowly downward with decreasing intensity (SVE 183 REF). The radiation is emitted within a wide cone of angles (unlike type IVmB), and it may be (especially in the late phase of the event) circularly polarized.

The two-dimensional observations with the Culgoora radioheliograph revealed that the term 'moving type IV burst' involves a spectacular variety of configurations that might signify several different physical phenomena. The following three types have been recognized:

a) *Magnetic arch.* An example of this is the event shown in Figure 2.12. Of the 25 moving type IV sources observed at Culgoora before 1973, 6 cases might belong to this type (SVE 183–185).

b) *Advancing front.* This type of burst, first described by Kai (1970), also has an arch shape, but it differs from the 'magnetic arch' in three aspects: it seems to be closely related to a type II burst; it has a large extent perpendicular to the direction of motion; and it is only weakly polarized. One can suppose that we observe in this case an advancing shock front which first produced a type II burst. Most (or all) of the type IVs moving with velocity in excess of 1000 km s^{-1} belong to this type.

c) *Isolated plasmoid.* This term was introduced by Smerd and Dulk

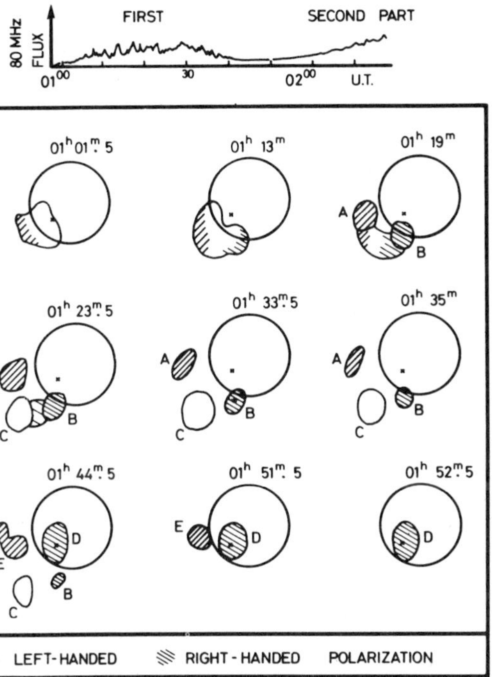

FIGURE 2.12 Moving (A, B, C) and stationary (D) sources of the type IV bursts on 22 November, 1968, as recorded by the 80 MHz radio-heliograph at Culgoora. C was unpolarized and located near the top of the magnetic arch, A and B were circularly polarized in the opposite sense and located in the arch legs. The position of the associated flare is marked with a cross.

Above: record of the integrated 80 MHz flux showing the 'two parts' of the type IV, corresponding to the moving and stationary sources, respectively. (After Wild, 1969.)

(1971) for a uniform radial motion of a 'blob' source to heights of many solar radii, and it also covers many cases when the source is observed to break up into a number of sources moving with similar velocities. About 70% of the moving type IV bursts belong to this category. As the 'blob' moves, the degree of polarization is mostly increasing to very high values of 80 to 90%. This type of moving type IV usually continues to move without any obvious deceleration until it fades.

The moving type IV bursts have been ascribed to gyro-synchrotron emission from energetic electrons trapped in a magnetic cloud ejected from the flare region. However, the sense of circular polarization in the moving bursts is often found contrary to what would be expected from the generally accepted synchrotron hypothesis according to the underlying magnetic field (SVE 186–187; Kai, 1978). Thus either the IVmA bursts are plasma radiation or, which seems to be more likely, the radio-emitting cloud is permeated by a strong magnetic field of opposite polarity to that of the ambient magnetic field (Smerd and Dulk, 1971; Dulk and Altschuler, 1971; Kai, 1978). Since the moving type IV bursts are closely associated with two-ribbon flares and erupted filaments in general (cf. Section 2.4.2.B), one can consider the disrupting filaments as the optical counterparts of the IVmA sources. The filaments follow a greatly sheared field along the $H_{\parallel}=0$ line. The question then arises why and how the polarity of the magnetic field within a moving source is related to the polarity of those magnetic fields that dominate the region of disruption (Kai, 1978).

The stationary type IV burst (IVmB, for example source D in Figure 2.12) may follow type IVmA, but it also may occur without any preceding moving burst. Its duration is usually one or several hours, and if the burst lasts for a long time it often transforms to a 'continuum storm' after one or two hours of its duration. Sometimes this continuum storm may further develop into an ordinary type I noise storm, with a duration of many hours or even days (Pick, 1961; Boischot and Pick, 1962). The emission of the stationary component is directive; due to this, the IVmB component is missing in flares that occur close to the solar limb. Total absence of type IV burst from sources on or behind the limb can be due to the heavy absorption of metric radio waves in the solar corona. The source of the IVmB emission may be either plasma waves or synchrotron radiation (SVE 189 REF) and possibly both these processes may be involved in different phases of the continuum development (Kai, 1975).

At decimeter wavelengths the continuum enhancement is usually characterized by large variability and complexity. At low decimetric and upper metric frequencies series of peculiar and often pulsating fine structures are observed in some type IV bursts (SVE 189 REF, Figure 72). The pulsations, usually with periods between 1 and 3 s, characteristically occur during the late part of a type II burst which started prior to the type IV.

The origin of the pulsations has been attributed by Rosenberg (1970) to synchrotron radiation from electrons trapped in a flux tube undergoing radial oscillations from standing MHD waves. Rosenberg's mechanism was incorporated in a model proposed by McLean *et al.* (1971) to account for the association with the type II bursts. An ascending MHD shock front (which also produces the type II burst) intercepts a discrete flux tube that loops high in the corona. When the shock reaches the top of the loop its front becomes essentially aligned with the flux tube over a length which is large compared to the tube diameter. Thus the tube is excited into radial oscillations as Rosenberg proposed it.

There are some very attractive features of this model (cf. SVE 190–191, Figure 73). However, Pick and Trottet (1978) observed a pulsating structure associated with a moving type IV burst, which disagreed with this interpretation, and Tapping (1978) even reports pulsations in a noise storm that occurred without any flare or type II association.

2.7.2 Type II bursts

A typical type II burst is characterized by an abrupt cut-off in frequency; above the cut-off no radiation is detected, while below the cut-off the radio event starts with a fairly high intensity. It usually has the form of two relatively narrow bands (of several MHz width) with a frequency ratio of approximately 2:1, one being the fundamental band and the other the second harmonic. The starting cut-off frequency of the fundamental is commonly close to or below 100 MHz. The bands drift to lower frequencies at a rate of some 100 kHz up to 1 MHz s^{-1}, and become invisible between 20 and 30 MHz, which corresponds to heights of 2 to 3 solar radii. The total duration of a type II burst is typically between 5 and 10 min (SVE 195 REF).

If one assumes that the emission occurs near the plasma frequency, i.e., near the level where the coronal electron density reduces the refractive index to zero, the frequency drift corresponds to ascending velocities of between 500 and 5000 km s^{-1}, with the most common values between 1000 and 1500 km s^{-1} (Maxwell *et al.*, 1964). This speed is several times greater than both the sonic and the Alfvén velocity in the corona, and therefore the disturbance can be attributed to a shock wave. The very large dimensions of the type II burst source indicate a quasi-spherical shock expanding outward from the flare and exciting plasma oscillations of a given frequency from regions where the shock front crosses the appropriate plasma level (McLean, 1967; Wild, 1968; SVE 201–202 REF).

More than 90% of type II bursts are clearly associated with flares (Swarup *et al.*, 1960; Maxwell and Thompson, 1962). The missing <10% can be ascribed to flares behind the limb so that probably all type II bursts are flare-associated. The type II burst often occurs in association with a type IV burst and both these radio bursts are associated with mass ejections from two-ribbon flares and disrupting filaments (cf. Section 2.4.2.B).

Typically, the type II burst originates before the Hα maximum during the flare flash phase (Krivský, 1976). The distribution of time intervals between the maximum of the impulsive microwave (or hard X-ray) burst and the observed onset of a type II burst on metric waves peaks at 2 minutes (Švestka and Fritzová-Švestková, 1974). This corresponds to a shock wave speed of about 1500 km s^{-1} if we assume that the shock originated at the time of microwave burst maximum. Thus the type II burst appears to originate at the time of the non-thermal acceleration process that gives rise to the hard X-ray and impulsive microwave burst.

As we shall see in Section 2.9.1.A, type II bursts seem to be closely related to the second-step acceleration of pre-accelerated particles in flares.

Some direct and indirect measurements (SVE 195–197 REF) show continuing propagation of the type II-burst-producing-shock through interplanetary space (see also Section 2.8.2).

2.7.3 Type III bursts

The type III burst is characterized by a short duration and very fast frequency drift from high to low frequencies. The bursts are identified over a great range of frequencies: from several hundred MHz, which corresponds to the base of the corona, down to tens of kHz, corresponding to heights of about 200 solar radii, i.e., the distance of the Earth (SVE 208–209 REF; Kurt *et al.*, 1977).

The generally accepted explanation of type III bursts is that proposed long ago by Wild *et al.* (1954). Each individual burst is radiation from an agent that moves outward through the corona and excites plasma oscillations of progressively diminishing frequency as the stream passes through coronal plasma of diminishing electron density. Observations in interplanetary space and near the Earth have confirmed beyond any doubt that the exciting agent is a stream of accelerated electrons as was proposed by Wild *et al.* (1959) and by de Jager (1960, 1962) long ago. Frank and Gurnett (1972) and Lin *et al.* (1973) were the first to succeed in correlating a type III burst, propagating through space, with direct measurement of the onset of a stream of non-relativistic electrons near the Earth. According to Lin *et al.*, the electron spectrum peaked at >100 keV at the type III burst onset and at ~ 15 keV at the burst maximum. From simultaneous observations of hard X-ray and type III bursts, Kane (1972) has estimated that the total number of >20 keV electrons responsible for a group of intense type III bursts is $<10^{34}$.

Type III bursts are the most frequent discrete radio events on the Sun. They occur either singly or in groups, and there are days when more than 50 such bursts or groups of bursts are recorded on the Sun; prolonged storms of weak bursts sometimes occur over periods of the order of one hour (SVE 202–204).

In contradistinction to the bursts of type II and IV, which are essentially always associated with filament disruptions and two-ribbon flares, type III bursts appear to be predominantly active-region-associated rather than flare-associated phenomena. In some active regions the type III bursts occur very often (sometimes only for one or two days), while in others they do not appear at all. If type III bursts appear in an active region, they tend to occur when a flare is formed, but many of them also occur independently of flares. Statistics show that some 70% of type III bursts occur without any event classified as a flare or subflare in the optical region (SVE 205; Kuiper, 1973). Kane (1972) has found that $\sim 32\%$ of the impulsive (flare-associated) hard X-ray bursts are associated with type III bursts; the converse correlation is only $\sim 25\%$ for the most intense type III burst and considerably less for weaker events. However, in the rare cases when such an association occurs, the time correlation between the two emissions is invariably very good (cf. SVE Figure 80).

The associated flare events are usually subflares. In the case that a type III (or mostly a group of type III bursts) is associated with a major radio event, the type III bursts usually occur at the very beginning of the event, prior to type II and type IV bursts.

Recently several authors have tried to detect the optical chromospheric counterpart of type III bursts by studying high-resolution Hα photographs (SVE 206–208 REF). Zirin and Lazareff (1975) claim that many flares (15 out of 21) occurring in front of a rapidly moving spot produced type III bursts, while flares in other locations in the same active region were much less productive in type III activity (3 out of 13). Labonte (1976) associated a set of type III bursts with a homologous set of Hα flares located at the limb-ward edge of a spot group. These observations are somewhat similar to Axisa's (1974) statement that flares located outside the general bipolar pattern of the active region are characterized by a higher type III burst association than those taking place inside of it.

Axisa *et al.* (1973), who have studied 44 groups of type III bursts, claim to find a typical chromospheric activity associated with these radio events, namely, absorbing features extending along the $H_{\|}=0$ line at the border of an active region and moving both downward and upward in the atmosphere. According to these authors most of the type III-associated flares are also related to an absorbing feature of this type. Mercier (1973, 1973) and Kane *et al.* (1974) found some supporting evidence for this association, while Priest and Heyvaerts (1974) even offered a theoretical interpretation for the "neutral line absorbing features associated with type III radio bursts", in terms of newly emerging magnetic flux, penetrating an area of opposite polarities. Nevertheless, we must not forget that we compare here two phenomena that occur very frequently on the Sun so that the probability of a chance coincidence is very high.

Summarizing, our present knowledge indicates that type III bursts might occur mostly in the peripheral parts of active regions, often where a dark filament with distinct motions extends (or extended before) from the active region to the surroundings along the $H_{\|}=0$ line. But we still do not know what makes some active regions extremely prolific in type III bursts, while other regions (looking very similar) do not produce them at all.

Some type III bursts are followed at metric wavelengths by a short-lived (1–3 min) broadband continuum, which is called a type V burst. The III + V bursts appear more likely to be associated with flares (Švestka and Fritzová-Švestková, 1974; Robinson, 1977; Steward, 1978) and with electrons directly recorded in space (Švestka, 1979) than simple type III's. Therefore, a type V burst occurrence is probably the manifestation of a rich stream of electrons, of which a fraction is trapped in the corona and becomes visible either through synchrotron radiation (Wild *et al.*, 1959) or through plasma waves (Weiss and Steward, 1965).

2.8 FLARE EJECTA

Besides the chromospheric (optical) flare proper a number of other phenomena are observed in the optical range in the chromosphere and corona that are obviously related to the flare event:

a) Activation and disruption of filaments (or prominences when seen on the limb) preceding the optical flare, as we discussed in Sections 2.2.4.C and 2.4.2.B.

b) Coronal transients seen in white light in the corona, following the flare.

c) Blast waves (or similar kinds of disturbance) travelling both into the corona as well as along the solar surface, originating close to the onset of the flare.

d) Surges and sprays.

2.8.1 Activation and disruption of filaments

As we mentioned in Sections 2.2.4.C and 2.4.2.B, most two-ribbon flares are preceded by an activation and disruption of a dark filament which extends along the $H_{\parallel}=0$ line in the active region. It represents condensed cool material, with a temperature close to 8000 K (Rust *et al.*, 1980).

As a matter of fact, dark filaments are the most common features seen in Hα light on the solar disk (or as bright prominences when projected above the solar limb). These features are observed inside active regions, in old remnants of active regions, as well as far from any existing centers of activity, up to latitudes pretty close to the solar poles. They always extend along the zero lines of the longitudinal magnetic field, $H_{\parallel}=0$, and this alignment is so perfect that they can be used as a means for mapping the large-scale distribution of magnetic field on the Sun (see, e.g., McIntosh *et al.*, 1976).

Whereas some dark filaments can survive without any obvious alteration for weeks, many others are subject to powerful changes. These can be enhanced internal motions (e.g., Valnícek, 1968; Bruzek, 1969); or damped vertical oscillations, which move the filament image alternately out of and into the passband of a narrow Hα filter (the observers speak then about a "winking" filament; see, e.g. Ramsey and Smith, 1966); or, ultimately, a complete disappearance of the filament which begins to rise with increasing velocity and finally disrupts and vanishes (the so-called "disparition brusque").

The filament disappearance can occur in active regions as well as in quiescent filaments far from any center of activity. The disappeared filament can be (but need not be) restored again after a few hours or days and the same filament (or filaments in very much the same position) can disrupt several times during the period of its existence (see, e.g., Engvold *et al.*, 1976).

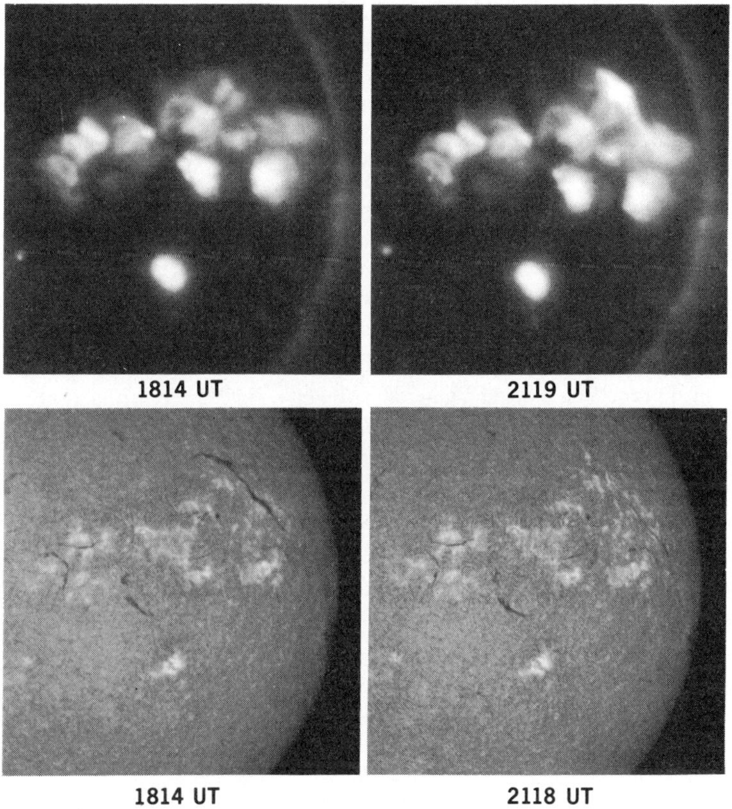

FIGURE 2.13 'Disparition brusque' (DB) of a quiescent filament as seen in the Hα line (below) and in soft X-rays (above). Note that only a very small brightening accompanied the DB in the chromosphere. In the corona, on the other hand, the DB produced a very strong enhancement in brightness; the filament, completely invisible in X-rays prior to its activation, became an outstanding feature when rising into the corona. (After Rust, 1976.)

Most disrupted filaments cause a strong heating in the overlying corona, which can be observed in soft X-rays as a brightening starting at about that time when the filament ceases to be visible in the Hα (Figure 2.13). This coronal enhancement is seen at filament disappearances both inside and outside active regions, corresponding to temperature enhancements to at least 6×10^7 K (Sheeley *et al.*, 1975; Smith *et al.*, 1977, 1978). The response of the chromosphere, however, is not always the same (cf. e.g., Howard and Švestka, 1977). Far from any active center, in weak magnetic fields, the filament just disappears and no other change can be seen in the Hα line in the chromosphere (Figure 2.13). In old remnants of activity the filament disappearance is accompanied by small chromospheric brightenings seen in Hα as scattered dots on both sides of the disappeared filament. Usually, the stronger the field, the more striking is the chromospheric brightening and in active regions it forms two continuous bright Hα ribbons on both sides of the $H_{\parallel}=0$ line along which the disappeared filament extended before (Figures 2.7.4 and 2.7.8). These bright chromospheric enhancements following filament disappearance in active regions are then called *two-ribbon flares*, which we described in detail in Section 2.4.2.B. The chromosphere just glows in these events. Still, in principle the Hα after-effects of the filament disappearance are the same, though much more powerful than in an old region without spots.

An instructive example of a filament activation was described by Schmahl and Hildner (1977; also see Rust *et al.*, 1980). Using movies taken in the green coronal line, Bruzek and DeMastus (1970) have found a case of coronal arch expansion preceding a flare. This case showed a very close similarity to what is seen in Hα quiescent filaments (cf. Section 2.4.2.B). About $1\frac{1}{2}$ h before the flare a bright hill appeared at the base of a system of stable coronal arches, and it expanded with a relatively small velocity of <5 km s^{-1} for more than 1 h. The expansion velocity began to increase some 15 min before the flare onset, at a rate indicative of an explosion. Shortly before the arch disappeared high in the corona (6 min before the flare onset), the expansion velocity increased to about 100 km s^{-1}.

For interpretation of the effects of a filament activation see Section 2.4.2.B.

2.8.2 Coronal transients

The white-light coronagraph on Skylab detected a large number of coronal disturbances, generally called *coronal transients* (MacQueen *et al.*, 1974). Their basic characteristic is that the changes in coronal structure appear fast, on a time-scale of a few tens of minutes. Some of them may reflect just rearrangements of the coronal magnetic field, but at least 77 such events have been clearly identified as mass ejections coming from the lower parts of the solar atmosphere (Munro *et al.*, 1979).

Many coronal transients come from sources behind the solar limb. Moreover, because all the transients were recorded on Skylab only above the altitude of 2 solar radii, some cannot be uniquely associated with events observed in the chromosphere. However, for 25 mass-ejection coronal transients Munro *et al.* believe to be able to make a definite association with chromospheric phenomena, and find that eruptive prominences (i.e. filament disruptions) were definite sources of 19 of them. In another one, the early phase of activity was not seen, but the subsequent appearance of Hα loops clearly indicates that a filament disruption preceded it. Thus, 20 coronal transients (i.e. 80% of all observed events) were clearly associated with the activation and disruption of a filament (in these cases mostly seen as a prominence on the limb).

One has to realize that a disruption of the magnetic field — which is illustratively demonstrated by the rise and disappearance of a filament — can also occur before any visible Hα filament (or prominence) could condense and cool. In such a case we cannot see any filament activation prior to the flare, but still the magnetic configuration and its instability are the same. Thus the remaining 5 events do not necessarily mean that the chromospheric source of the coronal transient differed from the other 20 sources in which the field disrupted. This is particularly true for two other sources with associated type II and type IV radiobursts, which are typically related to two-ribbon flares. Some of the associations, moreover, still might be erroneous. Thus it appears very likely that the source of *all* mass-ejection coronal transients is the process of disruption of a filament.

Out of the 20 events of Munro *et al.* (1979) in which filament disruption occurred without any doubt, only 5 were associated with flares. However, all the other 15 cases were observed on the limb, where a flare easily can be missed if the source is just on, or only slightly behind, the limb. Nevertheless, it is improbable that this would be true for all 15 events, the more so since *six of these filament disruptions occurred outside active regions.* Therefore, Munro *et al.*'s associations indicate not only that the real source of the coronal transients is the process of filament disruption, but also that this source-disruption need not be necessarily followed by a flare.

This is also in agreement with previous, less extensive, studies made by DeMastus *et al.* (1973) and Gosling *et al.* (1974), who already stated that coronal transients are more frequently associated with eruptive prominences than with flares. According to Webb *et al.* (1978) the coronal transients coming from sources outside active regions (i.e., not associated with flares) tend to occur on the borders of evolving coronal holes.

The velocities of mass-ejection coronal transients vary within wide limits (Gosling *et al.*, 1976), from ~ 100 to > 1200 km s^{-1}, with $\bar{v} = 470$ km s^{-1} as an average. Typically, flare-associated events travel faster ($\bar{v} = 775$ km s^{-1}) than events associated solely with eruptive prominences ($\bar{v} = 330$ km s^{-1}) Thus the filament disruptions which are followed by a flare apparently produce faster (i.e., more energetic) coronal transients than the disruptions without flares. The leading edges of the ejecta move outward with constant or increasing speeds.

Metric type II or IV radio bursts are associated only with events moving faster than 400 km s^{-1}. All but two events studied by Gosling *et al.* and moving faster than 500 km s^{-1} produced either a type II or IV radio burst or both. Since a type II burst manifests the presence of a shock, this result means that, as soon as the speed of the mass-ejecta propagation exceeds some 400 or 500 km s^{-1}, the MHD bow wave which moves in front of the coronal mass ejection becomes a shock.

Jackson and Hildner (1978) have reported that transients are surrounded by broad regions of slightly enhanced coronal density that move outward at approximately the same speed as their underlying transients. Since these regions arrive earlier at a given point in the corona than their associated transients, they have been called *forerunners*. Jackson (see Rust *et al.*, 1980) has inferred that a forerunner is in motion prior to the onset of its associated Hα surface activity, but this has yet to be confirmed.

The mass of a coronal transient is usually of the order of 10^{12} kg, up to 2×10^{13} kg (Rust *et al.*, 1980), and the material appears to come from the low corona; the associated eruptive prominence does not supply the majority of the mass in the coronal transient (Schmahl and Hildner, 1977). The energy involved in a coronal transient can be as high as 10^{24} J (10^{31} ergs) (Rust *et al.*, 1979). The most energetic events are those associated with flares and type II and IV radio bursts; however, even those without flares still carry energy of the order of 10^{22} J.

Several authors have tried to establish the mutual relations between coronal transients, metric radio bursts, and ejections of chromospheric material seen in Hα (Riddle *et al.*, 1974; Stewart *et al.*, 1974a, b; Gergeley and Kundu, 1974; Robinson and MacQueen, 1975; Gosling and Dulk, 1975; Poland and Munro, 1976; Kosugi, 1976). The results tend to converge to the following picture.

Ejection of material from the Sun, which manifests itself as a disrupted filament or a spray in Hα light, is seen in the white-light corona as a coronal transient. Around it we may see the less dense forerunner, and in front of it an MHD bow wave, which makes itself only indirectly visible through its effects upon nearby coronal structures. If the speed and energy of the ejection exceeds a certain limit (in such a case the ejection, as a rule, is associated with a flare), the MHD bow wave becomes a shock. We then can see it, far in front of the coronal transient, on metric radiowaves as a type II burst. We also begin to see a moving type IV burst, which coincides with the white-light emission but begins and peaks before the arrival of the most densely populated transient loops or condensations at any given altitude. A stationary type IV continuum source may appear at lower altitudes during the time when the moving type IV begins to decay high in the corona. The ejected chromospheric material, though possibly partially heated, only manifests the associated changes in the magnetic field configuration, but does not contribute in any decisive way to the mass of the coronal transient.

A comparison of kinetic energy density in a coronal transient (Dulk *et al.*, 1976) or an eruptive prominence (Steward *et al.*, 1978) with the magnetic energy density deduced from associated radio bursts has led to the conclusion that magnetic forces provide the principal mechanism for acceleration of the transient material from the Sun.

The reader is referred to Rust *et al.* (1980) for modelling of coronal transients.

2.8.3 Flare waves

It has been known for several decades that some flares initiate various kinds of disturbances in distant regions of the solar disc, after time intervals which correspond to a propagation speed of the disturbing agent of the order of 1000 km s^{-1}. At the present time there are four different kinds of evidence for the existence of these fast-mode flare waves (SVE 225–229 REF).:

a) A fast-moving diffuse front is seen in emission in the Hα-line center and in absorption (or absent) in the Hα wings, propagating away from the flare.

b) Progressive, short-lived brightenings of small points appear in the chromosphere in places passed by the wave.

c) An abruptly initiated activation or oscillation of a filament which stood in the way of the passing wave is occasionally seen, in some cases eventually leading to a "disparition brusque".

d) A type II radio burst demonstrates the existence of a shock wave propagation through the corona (Section 2.7.2).

Many authors also believe in the existence of "sympathetic" flares in distant active regions, when one flare is supposed to cause another. However, statistical analyses (Fritzová, 1959; Smith and Smith, 1963; Fritzová *et al.*, 1976) show clearly that the occurrence of sympathetic flares, if real at all, must be a very rare event, and it apparently belongs to the above-mentioned category (c). Fritzová *et al.* (1976) found a slight excess in the statistical occurrence of small (<20 min) intervals between successive flares for regions closer than $\sim 30°$; however, we meet here more likely with "sympathetic active regions of common subphotospheric origin" than with the individual "sympathetic flares".

A study of the time difference between the flare and the onset of activation of a distant filament yields propagation velocities between 400 and 2200 km s^{-1} (45 cases) with a mean value of 880 km s^{-1}. Measurements of the motion of 15 visible wave-fronts yielded velocities between 440 and 1100 km s^{-1}, with a mean value of 600 km s^{-1} (S. Smith and K. Harvey, 1971). Thus we have supersonic velocities of the order of the Alfvén velocity in the solar corona.

The travelling wave fronts are usually semicircular, extending across a sector of 60 to 120° (S. Smith and Harvey, 1971), and they are sometimes called "Moreton waves". However, this term may also involve sprays projected on the disc. The wave-front seems to be a pure wave effect, which causes a depression and successive relaxation of the fine structure of the chromosphere (Dodson and Hedeman, 1964; Bruzek, 1969). The spicules and fibrils are forced to move downward for about 40 s with a velocity amplitude of 6 to 10 km s^{-1} in the leading wave-front and afterward they return more slowly to their former positions.

The front usually becomes visible only at a distance of more than 100,000 km from the place of its origin. This makes it difficult to establish the exact place and time of its onset, as it is necessary to extrapolate the velocity curve back along the time coordinate. However, even when errors of a few minutes cannot be excluded, this extrapolation clearly indicates that the waves originate very close to the flare onset, during the flash phase (Smith and Harvey, 1971).

Some flare waves can be followed on the disk up to distances exceeding 500,000 km and they propagate with fairly constant speed (Moreton, 1964; Dodson and Hedeman, 1968). Some of them are restricted to a rather narrow cone of propagation (much less than 90°), while others (SVE Figure 86) have a well-developed semi-circular shape.

Some flare waves seem to be well correlated with type II bursts (Lin and Anderson, 1967; Kai, 1970; Harvey *et al.*, 1974; Pintér, 1978). On the other hand, a statistical analysis of this correlation carried out by S. Smith and K. Harvey (1971) for 50 flare waves, leads to the result that only 36% of waves were associated with a type II burst. This percentage is higher after 1963 (>50%), probably due to more complete and more sensitive radio spectrum observations. Even this rather loose correlation, however, indicates a common origin of the visible wave fronts and type II bursts, because visibility conditions of these two phenomena are different (Uchida, 1974). Thus one can expect that sometimes only one or the other phenomenon is observed, even when they are both produced by the same wave effect.

The fact that a travelling flare wave activates distant filaments has been established by many authors (SVE 230 REF). However, this activation often takes only the form of filament oscillations ("winking"), without a definite disruption of the filament. As it seems, the disruption, when due to a travelling disturbance, is caused by another, slow mode, of the wave.

The existence of slow wave disturbances which propagate at about the normal velocity of sound, has been suspected for many years by Bruzek (1952, 1958), Öhman and Öhman, 1953; Öhman *et al.* (1962), and Yajima (1971). From activations of distant filaments a long time after the flare onset, these authors deduced hypothetical velocities of these waves between 60 and 200 km s^{-1}.

A synoptic movie made from soft X-ray pictures taken aboard Skylab made it possible to detect slowly moving disturbances of similar kind, causing sequential brightenings of distant stationary (pre-existing) coronal structures (Rust and Švestka, 1979; Figure 2.14a). The disturbance starts with a speed of $\sim$400 km s^{-1}, but is damped to 50–60 km s^{-1} after one hour, and to $\sim$10 km s^{-1} after 4 or 5 hours of propagation. In extreme events the effects of the disturbance can be followed for more than 18 hours up to a distance in excess of 600,000 km.

JUNE 12-13, 1973

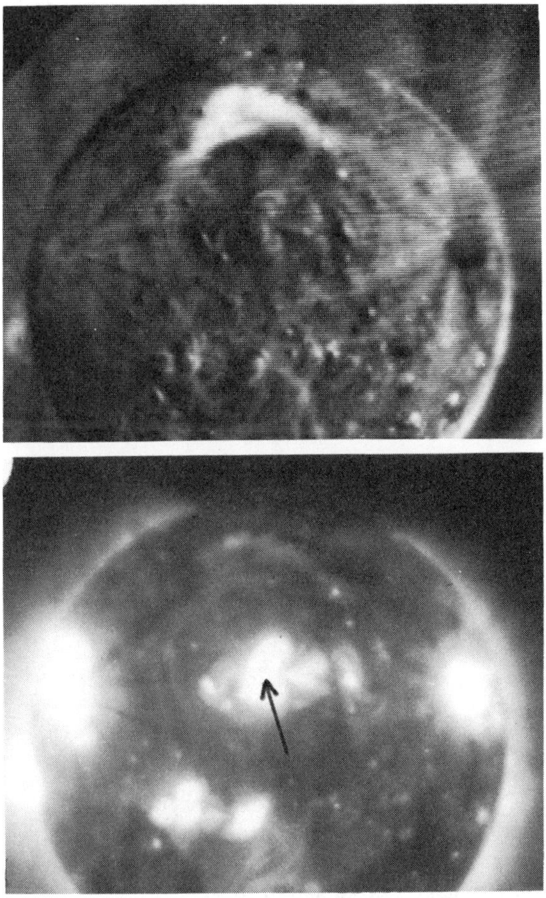

FIGURE 2.14a Example of a coronal brightening in soft X-rays produced by a slowly moving disturbance originating in a disrupted filament. The arrow points to the position of the filament. The upper picture has been obtained by subtracting the X-ray photograph below from a preceding one, on which the brightening was not yet seen. The brightness variation is enhanced in the subtraction. (AS & E photographs.)

This disturbance indeed appears to be identical with the slow-wave mode hypothesized long ago by the Hα observers. If one plots in one graph the mean velocities deduced from the X-ray pictures by Rust and Švestka, and from the Hα picture by Bruzek, Öhman, and Yajima, the fit is extremely good (Figure 2.14b).

However, there is one very significant difference between the Hα and X-ray results: Bruzek, Öhman, and Yajima supposed that this mode of disturbance

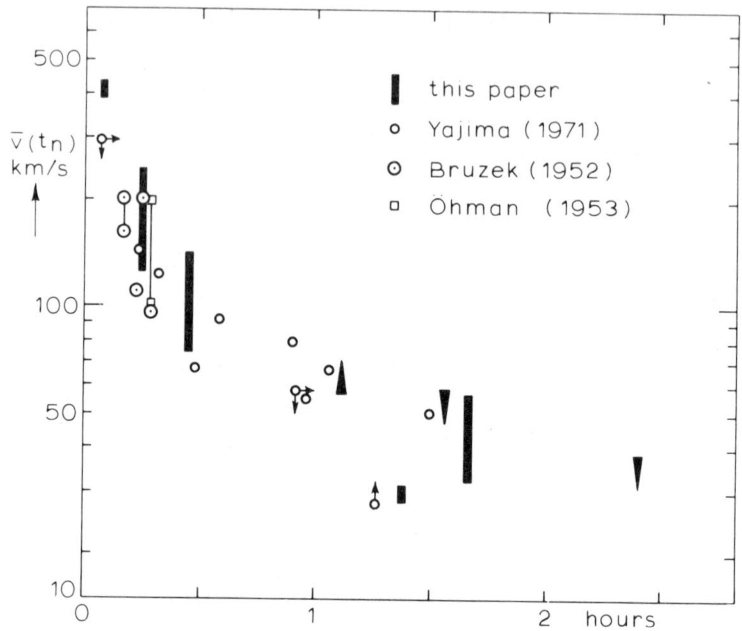

FIGURE 2.14b Mean speeds of slowly moving disturbances, as deduced from the X-ray corona pictures (bars) and from Hα observations (circles) at times Δt after filament disruption or flare onset. (After Rust and Švestka, 1979.)

originated in flares, particularly in major flares; indeed, all sources of the Hα disturbances shown in Figure 2.14b were flares. In X-rays, however, this is not true. In at least three cases the source of a slowly moving X-ray disturbance (i.e. of the slow-mode wave) was an activated filament that was not associated with any reported flare. Two major flares that were sources of this kind of disturbance in X-rays were also preceded by powerful filament activations. Therefore, we must conclude that the source of the slow-wave disturbance is most closely associated with the eruption of a filament, and when the disturbance follows a flare, it is only because a filament activation preceded it.

Thus disk observations of the corona in soft X-rays have confirmed what the limb observations of the white-light corona indicate, namely, that coronal disturbances can originate in disrupted filaments that are not associated with any reported (i.e. big enough) chromospheric flare. The fact that the Hα observers always found a flare to be the source of the slow-wave mode is simply due to an effect of selection. They correlated observed disparitions brusques always with flares and never with other disparitions brusques, for which patrol observations are not available.

2.8.4 Surges and sprays

When a flare starts, it is quite often accompanied by gas ejections which manifest themselves as moving prominences on the limb or Doppler-shifted dark filaments on the disc. We meet here with a great variety of events but generally can divide them into three groups: *eruptive prominences*, about which we talked in Section 2.8.1., *surges*, in which the gas goes up and down again (sometimes along the same magnetic lines of force), and *sprays*, in which most of the ejected gas does not return to the active region of its origin.

The distinction between sprays and eruptive prominences on the limb, and sprays and Moreton waves on the disk, is not easy. Generally sprays are faster than other eruptions on the limb, and more compact (cloud-like) than other Moreton waves on the disk (SVE 223–225 REF). They tend to reach the characteristic "escape" velocities within a minute or two of the start of the phenomenon, whereas eruptive prominences start ascending slowly and accelerate to high velocities only at a late phase (Valnícek, 1964; E. Smith, 1968). However, when many events are studied, there seems to be a continuous transition between these two types of eruption (Tandberg-Hanssen, 1979; Rust *et al.*, 1980). As a matter of fact, all these high-speed phenomena are so complex and variable that it appears almost impossible to classify them properly.

In contradistinction to that, surgest are distinct phenomena which can be observed very often on the Sun. A typical surge has a shape of a straight or slightly curvilinear spike, which grows upward from the chromosphere with velocities typically between 50 and 200 km s^{-1}. After reaching the maximum height (from 20,000 up to 200,000 km) the material is usually seen to descend, as it seems along the original trajectory. Lifetimes of these spikes are mostly between 10 and 30 min. Surges have a strong recurrence tendency and the surge ejections often repeat themselves at intervals of the order of 1 h at apparently exactly the same place (SVE 221 REF).

Surges need not be associated with flares, but essentially all surges show chromospheric brightenings in the initial phase near their footpoints which Roy (1973) identified with moustaches (Ellerman bombs). However, as Bruzek (1972) emphasizes, there are many moustaches which do not give rise to any detectable surge, hence a moustache appearance does not necessarily imply the surge occurrence.

The surge sources are usually situated close to the edges of sunspot penumbrae (Giovanelli and McCabe, 1958), according to Rust (1968) above small, changing, satellite sunspots which represent small "islands" of polarity reversals near a large spot (also see Rust *et al.*, 1980).

Some surges can be seen in white light (Becker, 1959; Roy, 1976) because of scattered photospheric radiation by free electrons. From that, one can deduce electron densities in surges between 10^{17} and 10^{18} m^{-3}. The highest electron density occurs simultaneously with the phase of upward acceleration. In contrast to other prominence eruptions, surges do not heat in any significant way the corona (Rust *et al.*, 1977; Rust *et al.*, 1980).

In spite of the fact that many surges appear without flares, their association with flares is high (SVE 221–222 REF). Apparently, the surge-producing instability is often accomplished in active regions that flare.

2.9 PRODUCTION AND EFFECTS OF HIGH-ENERGY PARTICLES IN FLARES

In many (and possibly all) flares particles are accelerated to suprathermal energies of tens and hundreds of keV. Evidence for this is the occurrence of apparently nonthermal impulsive X-ray and microwave bursts in the flash phase of flares (cf. Sections 2.5, 8.3 and 8.5). In some flares, particle acceleration to much higher energies is also accomplished, to tens and hundreds of MeV, apparently through a second-step acceleration (§§8.21–8.26) of the particles pre-accelerated (§§8.11–8.20) during the impulsive phase. (Observations relevant to particle acceleration are summarized in §§8.1–8.10.)

The accelerated particles are atomic nuclei and electrons, and we get information on them in three different ways:

a) Through their effects in the solar atmosphere: >20 keV electrons produce the hard X-ray and microwave bursts, as well as the type III bursts on metric radio waves; mildly relativistic electrons are the source of type IV bursts; and >30 MeV protons and α-particles produce γ-ray emission on the Sun.

b) Through direct records of the particle "prompt events" in space, seen at various spacecraft shortly after a flare (§8.2).

c) Through effects of the particles in the atmospheric layers of the Earth: >10 MeV protons produce polar-cap absorption (PCA) effects in the polar cap ionosphere (see, e.g., Hultquist, 1969), and >500 MeV protons cause transient increases (GLE events) in the cosmic-ray flux (as recorded by the neutron monitors) at the surface of the Earth. The very rare flares that cause the GLE's are often called "cosmic ray flares", and those that produce the PCA's (or high proton flux on spaçecraft) are sometimes called "proton flares".

2.9.1 Proton events

2.9.1.*A* *The source flares* Ellison *et al.* (1961) were the first to draw attention to the fact that all cosmic-ray flares have the typical two-ribbon shape described in Section 2.4.2.B, with the ribbons embedded within large sunspots. The same is true for almost all proton flares and for the vast majority of other flares that produce enhanced flux of atomic nuclei. For example, in the *Catalog of Solar Particle Events, 1956–1969* (Švestka and Simon, 1976), for fifty proton events one could safely identify the flare source (this has been done by Dodson and Hedeman) and find Hα pictures and other basic characteristics of the flare (this has been done by Martres). Out of the 50 identified sources, 45 appeared to be two-ribbon flares. Thus one can reasonably suppose that whereas both the compact flares and two-ribbon flares are seats of the primary acceleration to ~100 keV, only in two-ribbon flares can particles be accelerated to much higher energies.

In the radio range, a proton flare is characterized by the occurrence of a type II radio burst (Švestka and Fritzová, 1974) and by a strong type IV burst (SVE 241–242 REF). The type II demonstrates the presence of a shock; it is feasible to suppose that this shock is related to a second-step acceleration process which accelerates, in turbulent fields, to much higher energies some of the particles that were pre-accelerated during the impulsive phase. Some accelerated particles are trapped in coronal magnetic fields and when the density of mildly relativistic electrons is large enough electrons become visible through synchrotron radiation in the radio continuum as a type IV burst.

Besides, the moving type IV (cf. Section 2.8.2) indicates the presence of expanding or fast-moving coronal transients above the flare that probably have their origin in the same instability which instigated the acceleration process. According to Kahler *et al.* (1978) all prompt proton events in space are associated with mass-ejection coronal transients.

Proton flares are also accompanied as a rule by intense hard X-ray bursts. Strong events often have multiple maxima and usually a fairly long duration of several minutes. This complexity and long duration of hard X-ray bursts may be considered a characteristic of the most energetic proton flares.

2.9.1.*B* *The source active regions* The active regions in which proton flares occur are characterized by several specific properties:

a) The region is always of a magnetically complex type, i.e., some mixing of magnetic polarities occurs. As Martres (1968) has shown, such active regions occur when a new region forms within the remnants of an old one.

b) Of course, only some of the complex sunspot groups produce proton flares. A necessary condition is a substantial increase of the gradient of the magnetic field in the active region (Gopasyuk *et al.*, 1962; Severny, 1964, 1969b) manifested by a close approach of large spots of opposite magnetic polarity (Avignon *et al.*, 1963). Such a configuration is particularly favorable for the production of energetic particles (Warwick, 1966) if a common penumbra is formed embracing spots of different polarities (the so-called δ-configuration of a sunspot group). According to Sakurai (1972) high gradients are most likely to occur in a sunspot group with rotating motion. Therefore, the sunspot groups which show a reversed polarity distribution appear to be very productive in proton flares (see examples in McIntosh, 1969; Sawyer and Smith, 1970; Zirin and Tanaka, 1973).

c) As a rule, a magnetic field gradient increase in an active region is associated with an intensification and a hardening both of the microwave and X-ray spectrum. In the microwave range the usual maximum close to 10 cm is submerged in the greatly increased flux at shorter wavelengths, so that the microwave spectrum flattens and eventually the maximum shifts toward centimeter wavelengths (Tanaka and Kakinuma, 1964; Tanaka *et al.*, 1969; Tanaka and Enomé, 1975).

For other characteristics of the active regions capable of producing proton flares see SVE 247 REF.

2.9.1.*C* *Time occurrence* Some active regions produce quite a number of proton flares, of which several can be very powerful. An analysis of 27 pairs of successive major proton flares which originated in identical regions shows that the shortest time interval between them was 13 h and the median interval was 55 h. Thus, obviously, some tens of hours (and about two days on the average) are necessary to restore a magnetic situation capable of producing a new major proton flare, and to build up the energy necessary for this process.

Generally, the number of particle events and their importance increase with increasing sunspot number in the solar cycle. During 1 or 2 years near the minimum of solar activity, flare-associated particle events are extremely rare. As the sunspot number rises, particle events occur more frequently, but, somewhat surprisingly, the increase stops before the solar maximum is reached. During solar cycles 18 and 19 the cosmic-ray flares appeared to avoid the cycle maxima completely (Carmichael, 1962; Obayashi, 1964) and the same tendency has been found for strong PCA events (Švestka, 1966; Sakurai, 1967; Hakura, 1974). The highest number of particle events and the strongest of them occurred about 1 year before and 2 or 3 years after the maximum of the sunspot cycle.

In the cycle No. 20 the rule of "maximum-avoidance" appeared to be broken, since three cosmic-ray flares were observed very close to the cycle maximum (Dodson and Hedeman, 1969b; McCracken and Rao, 1970; Švestka, 1970b) and the number of proton flares did not decrease during the maximum of the cycle. Thus, since this maximum was much lower than the two preceding ones, one may conclude that the most important activity phenomena on the Sun do not avoid the phase of the solar maximum, but simply the periods when the sunspot number is extremely high. Even in this rather low cycle a series of very outstanding events occurred again in August 1972, i.e., more than 3 years after the cycle maximum (Section 1.2.1).

A plausible explanation might be that the active region does not find a "quiet" time long enough to build up the conditions necessary for the production of what is actually a big event, if the general activity of the Sun is too high. Due to many sources of disturbance, the flare mechanism is triggered before the conditions for a large event can develop.

2.9.1.*D* *Low-energy proton streams* Van Hollebeke *et al.* (1975) have studied 185 solar particle events recorded on IMPs 4 and 5. They were able to determine the parent flare for 68% of the events and believe that most of the unidentified increases occurred in the invisible hemisphere of the Sun. This explanation is acceptable, since, according to Fritzová and Švestka (1971), the percentage of proton sources behind the solar limb should be ~38% for weak proton events. However, Van Hollebeke *et al.* based their statistics on observations of relativistic electrons (with energies from 0.5 to 1.1 MeV), which obviously show a very high correlation with flares. When the statistic is based on low-energy protons only (also counting the events when relativistic electrons are missing), one finds many cases that do not seem to be flare-associated at all. (Compare the *Catalog of Solar Particle Events*, which comprises more than 700 cases (Švestka and Simon, 1976).)

So far, we do not know where these, apparently non-flare, protons originate. Some, of very low energy, might be accelerated in non-flaring disparition brusques (Švestka *et al.*, 1979b), which occur quite often on the Sun. According to Webb *et al.* (1978) their occurrence is most frequent along borders of growing coronal holes and this might explain why some magnetic sector boundaries carry behind them significant amounts of low-energy particles whereas other sectors are essentially "empty" (Švestka *et al.*, 1976). However, there is strong evidence that a significant part of the low-energy particle streams is not accelerated at the Sun at all, their acceleration being accomplished in interplanetary space. This originates from the McDonald *et al.* (1976) observations that the average density of recurring MeV proton streams at heliocentric distances of >3 AU is an order of magnitude greater than those observed at 1 AU; also, the relative abundances in these co-rotating streams indicate that the source population for them is the solar wind and not flare ejections (McGuire *et al.*, 1978). These particles may be accelerated in shocks travelling or co-rotating through the interplanetary medium (Pesses *et al.*, 1978; Pahner and Gosling, 1978).

2.9.1.*E* *The energy spectrum* The energy spectra of particles observed in space represent the combined effects of acceleration, coronal transport, release into interplanetary medium and interplanetary propagation. As a result of the propagation conditions, the observed spectra may vary considerably from one event to another. Therefore, it is very difficult to deduce from the space data the real spectrum near the acceleration site. Generally, at least at higher energies, the spectrum can be approximated by a differential power law,

$$\frac{dN}{dE} = CE^{-\gamma},$$

or in the form of the integral power law,

$$N(>E) = KE^{-\gamma'},$$

where

$$\gamma' = \gamma - 1.$$

According to the most thorough study carried out by Van Hollebeke *et al.* (1975), for proton energies between 20 and 80 MeV $2.5 < \gamma' < 3.7$ for 90% of all the events in which the protons had direct access to the Earth.

At lower energies, the spectrum becomes flatter and thus the power-law approximation fails to fit the whole spectrum. Of course, partial energy ranges can still be approximated by power laws with different values of the exponent γ, as many authors often do when only a small energy range is observed by them.

Good fit with the spectrum is also found if an exponential law in rigidity, P, is assumed (Freier and Webber, 1963),

$$\frac{dN}{dP} = \frac{dN_0}{dP_0} \exp[-P/P_0(t)].$$

Here P_0 is the characteristic rigidity, variable in time, which characterizes the hardness of the spectrum, similar to γ in the power law. (For energies >30 MeV, $P_0 \approx 200$ MV approximately corresponds to $\gamma \approx 2$.) However, this approximation also no longer fits for proton energies below ~20 MeV.

At higher energies, on the other hand, above 100 MeV, the spectrum appears to be steeper. (For both propagation and storage effects see SVE 256–264 REF.)

2.9.2 Heavier nuclei

Even though observations of nuclei of $Z \geqslant 2$ have been made only in a limited number of events, there seems to be good evidence that these heavier particles are always present in the flux from proton flares (§8.7).

Let us first suppose that atoms of all elements in the flare are completely stripped during acceleration and propagation. In such a case, since $A/Z \simeq 2$ for all nuclei with $A \geqslant 4$, α-particles and all heavier nuclei moving with the same

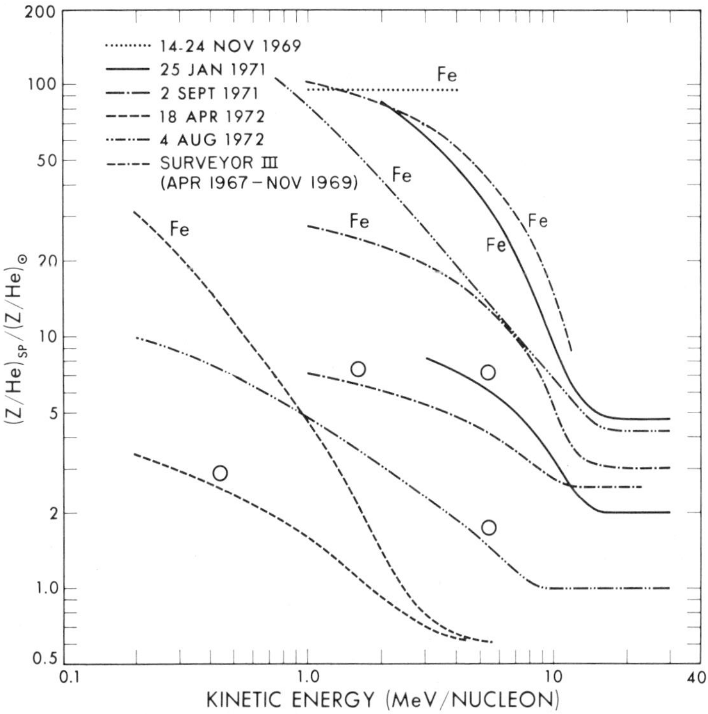

FIGURE 2.15 Energy dependence of the abundance ratios Fe/He and O/He in various flares, normalized to the solar abundance ratios (i.e., $(Z/He)_{\odot} = 1$). Note that the energy where the abundance of Fe and O begin to be enriched varies within wide limits. (After Crawford *et al.*, 1975; *Astrophysical Journal*, University of Chicago Press. Copyright the American Astronomical Society.)

velocity have essentially the same rigidity; thus they also have the same gyroradii and behave identically during propagation. Consequently, the charge spectrum for $Z \geqslant 2$ can be considered as a faithful reproduction of the chemical composition of the particles at the point of acceleration and abundance ratios of different nuclei in the particle flux should be the same in all flares.

This expectation has been confirmed in several flares by the comparison of the helium and medium nuclei spectra. These two spectra actually can be brought to a very good fit by multiplying the flux of the medium nuclei ($6 \leqslant Z \leqslant 9$) by a factor which appears constant ($\simeq 58 \pm 5$) within the limitations of experimental error (Biswas and Fichtel, 1965; Durgaprassad *et al.*, 1968; Bertsch *et al.*, 1969, 1972). However, this is true only for particle energies in excess of ~15 MeV per nucleon (Figure 2.15). At lower energies the particle flux is enriched in heavy elements by an amount that increases with charge number Z, decreases with energy E, and varies from flare to flare (reviews in Price, 1973, and Crawford *et al.*, 1975; SVE 264–265 REF). Quite exceptionally, the enhancement of solar heavy nuclei can extend even far above 25 MeV per nucleon (Bertsch *et al.*, 1974; Bertsch and Reames, 1977; Dietrich and Simpson, 1978).

Under the assumption that the energy spectra are the same for all nuclei above ~15 MeV, so that they become identical when mulitplied by constant factors, these factors give us the relative abundances of the different species. Abundances thus obtained are very similar to those in the solar photosphere and corona. This indicates that large deviations of the high-energy particle population from average solar abundances do not occur. On the other hand, as Figure 2.15 shows, at 1 MeV nucleon^{-1} the O/He abundance ratio can exceed that above 15 MeV by a factor up to 10, and the Fe/He abundance ratio by a factor up to 25.

Scholer *et al.* (1978) have found in two events that the Fe/O ratio at the same energy per nucleon (~1 MeV) decreased during the events by about one order of magnitude, whereas the C/O ratio was time-independent. This indicates that while carbon and oxygen are essentially fully stripped, iron nuclei are only partially ionized. Also all the interpretations of the abundance enhancement of heavy nuclei at low energies demand that the ions do not enter the acceleration process completely stripped, as we assumed at the beginning of this section. On the other hand, according to Mogro-Campero and Simpson (1972a, b) even iron nuclei are fully stripped when recorded near the Earth on OGO 5 in the energy range from 25 to 61 MeV nucleon^{-1}.

The situation is different when we compare α-particles (or other heavier nuclei) with protons. Since the rigidities of protons and α-particles of the same velocity differ by a factor of two, the protons propagate differently. This is actually observed, and the helium/proton abundance ratio α/P varies greatly from flare to flare as well as during the development of a single event. It generally decreases with increasing energy per nucleon, and this decrease has as its consequence that the helium spectrum is always softer than the spectrum of protons (Lanzerotti, 1973).

2.9.3 Electrons

2.9.3.*A* *Relativistic electrons* Relativistic electrons, first detected by Cline and McDonald (1968), generally accompany prompt proton events (SVE 279–280 REF). However, the flux of relativistic electrons is always much lower than that of protons of the same energy, probably because of rigidity-dependent injection into the acceleration process and due to the heavy synchrotron losses that the electrons suffer before leaving the Sun. This is probably also the reason why the number of flare-associated relativistic-electron events is smaller than that of proton events. According to Datlowe (1971) the ratio of electrons to protons in the 12 to 45 MeV range varies between 10^{-2} to 5×10^{-6}, with a median value $\approx 3 \times 10^{-4}$. However, Lanzerotti and Maclennan (1971) have found the flux of >1.1 MeV electrons still comparable to proton flux at the same energy, hence a decrease in electron number should occur only at higher energies. Simnett (1974a, b) compared the 0.3 to 0.9 MeV electron flux with >11 MeV proton flux in more than 30 events. The ratio of these two fluxes varied within a very wide range from ~ 0.2 to >10 (see also §8.2.2).

There are also relativistic electron events in which no protons are recorded at all. It seems that in such cases any flare-association is indeed missing. Many of them are observed during quiet time periods (McDonald, 1970; McDonald *et al.*, 1972; L'Heureux *et al.*, 1972) and can be explained as being of galactic origin.

The energy spectra of prompt relativistic electron events below 10 MeV fit well to the power law, with $\gamma = 3.0 \pm 0.4$ (Simnett, 1972a, b); beyond 10 MeV the spectrum seems to be somewhat steeper (Datlowe, 1971; Simnett, 1972a, b).

A serious problem is raised when time-delays of relativistic electrons are considered (Simnett, 1971, 1974a). A 660 keV electron moves with velocity 0.9c; thus if the length of a typical spiral field-line is assumed to be ~ 1.3 AU, electrons of this energy should arrive at the Earth within 12 min. After making allowance for the 8 min which light needs for propagation to the Earth, the theoretical delay should be 4 min. But the delay times observed are tens of minutes. This indicates some delay in the release of the relativistic electrons from the Sun (unless the late onset is due to the very weak flux of these kinds of particles).

2.9.3.*B* *Non-relativistic electrons* Flares that produce non-relativistic electrons are very common phenomena on the Sun. During six and one-half years, 1964 to mid-1970, 230 electron events have been recorded in the vicinity of the Earth, with a threshold of 10^5 electrons m^{-2} s^{-1} sr^{-1} above 40 keV. Many more events have been recorded above 20 keV, but there the survey is not complete, since long-lived streams of solar particles obscure the weak events.

The occurrence of the electron events, however, is very unevenly distributed in time. This is due to the fact that some particular active regions tend to generate many flares that accelerate electrons (up to 15 or 20), whilst other regions do not produce any electron events at all. This behavior is similar to that of the production of the radio type III bursts (cf. Section 2.7.3) and indeed, non-relativistic electron events in space are identical with strong type III radio events on the Sun (SVE 270–273 REF). These type III bursts which produce electrons recorded in space are essentially always associated with flares (mostly subflares) and with microwave bursts that occur at the same time as the type III burst.

Similar to proton flares, the flares producing observable electron events are concentrated in the western hemisphere of the Sun. The concentration of the electron events around the footpoint of the Sun–Earth magnetic field-line is very strong; according to Lin (1974) the cones of emission are $<60°$ wide.

The intensity of the electron flux at the Earth ranges from the threshold limit up to $\sim 10^8$ electrons m^{-2} s^{-1} sr^{-1}. A typical moderate electron event produces a few times 10^6 electrons m^{-2} s^{-1} sr^{-1}. Most rise-times are in the range of 5 to 50 min, whilst the decrease takes many hours. The total number of electrons above 40 keV can be estimated as 10^{32} to 10^{33} (Lin and Hudson, 1971; Alvarez *et al.*, 1972), which does not exceed the number needed for a type III burst. Holt and Ramaty (1969), Lin and Hudson (1971) and Kane and Lin (1972) estimate that less than 1% of the electrons accelerated in the flare gain access to interplanetary space.

Essentially all proton events also produce non-relativistic electrons. On the other hand, electron events without detectable protons are quite common and are often called "pure electron events". It is possible that these events represent a phenomenon essentially different from proton flares. For example, the acceleration may be rigidity-dependent in pure electron (PE) events, but energy-dependent when protons are also observed (PR events).

In both types of electron event the power law fits the spectrum down to below ~ 5 keV, and sometimes even to ~ 2 keV. Whereas, however, the number of >20 keV electrons is estimated at $\sim 10^{33}$ in PE events, it can be as large as $\sim 10^{36}$ in large proton events. Sarris and Shawhan (1973) find that in the microwave radio range the PE events are associated mostly with peak fluxes $<5 \times 10^{-5}$ W m^{-2} and a short decay-time of the radio burst, whereas the PR's are preceded mostly by peak fluxes $>5 \times 10^{-5}$ W m^{-2} and the microwave bursts have considerably longer decay-times.

However, all these differences between PE and PR events can also be explained as being simply due to a difference in intensity of the acceleration process. In that case protons should always be present in the flux, but we simply do not record them for a very simple reason. While 40 keV electrons travel to the Earth in ~ 20 min, 40 keV protons need about 14 h to cover the same distance. Therefore, the time-association of the two species at very low energies is very uncertain.

The fact that the power law fits the electron spectrum down to below 5 keV without any obvious flattening can be used to estimate the altitude at which the electrons are accelerated. For the event of 27 April, 1972, Lin (1973) has estimated that the number of ambient ions the electrons met along their path to the Earth must have been $<3.5 \times 10^{23}$ m^{-2}. For hydrogen this is <600 kg m^{-2} of material traversed. This leads to an estimate that the acceleration could not take place below the coronal layer where $n_i \simeq 2 \times 10^{15}$ m^{-3} (2×10^9 cm^{-3}). This is the density in the low corona and Lin places it in an active region about 30,000 km above the photosphere. Since no bend has been detected so far in the electron spectra at low energies, this is in fact the minimum altitude of the electron acceleration in the flare region.

More recently, Lin *et al.* (1978) have succeeded in determining the electron spectrum for several events from the 10 keV range to about 10 MeV (Figure 2.16). A break at ~ 100 keV is clearly evidenced in these spectra. Above this energy, an unbroken power law can be seen up to the highest observed energies.

2.9.4 Nuclear reactions in flares

So far, we have discussed explicitly only the effects caused by electrons in the solar atmosphere, through bremsstrahlung, gyro-synchrotron radiation, or plasma waves. Perhaps jointly with protons, the electrons also produce the white-light flare emission. But there is only one effect in solar flares that is due solely to atomic nuclei, and that is the production of nuclear reactions (see §8.6). The most common nuclear reactions one can expect (Lingenfelter and Ramaty, 1976; Cheng, 1972; de Feiter, 1974; Ramaty *et al.*, 1974) can be summarized as follows:

1) The production of neutrons in reactions initiated either by high energy protons or α-particles. Neutrons can be produced through interaction of these nuclei with the most abundant elements, but far the most important appear to be the reactions between the hydrogen and helium nuclei themselves. Nuclei must be accelerated to more than 30 MeV to accomplish these reactions. A great part of the neutrons produced in these reactions escape from the Sun, and they should be observable near the Earth shortly after the acceleration process. The remaining neutrons are captured by ambient ^{1}H and ^{3}He nuclei in the photosphere. While captures on ^{3}He do not lead to photon emission, captures on protons produce deuterons and the 2.23 MeV gamma-ray line.

2) The production of positrons emitted from radioactive nuclei. These positron emitters originate again through interactions of accelerated protons and α-particles with the ambient medium. The thresholds for these reactions range from ~ 3 MeV nucleon^{-1} for producing ^{11}C through protons on ^{14}N, or ^{15}O through α-particles on ^{12}C, to 41.4 MeV for producing ^{10}C through protons on ^{16}O. The emitted positrons partially escape from the source region, while those remaining on the Sun slow down through collisions and finally combine with an electron to form a pair of γ-ray photons of 0.511 MeV photon energy.

3) The production of nuclei in excited states; these are isotopes which either are abundant in the solar atmosphere or are easily produced in nuclear reactions of energetic protons and α-particles with the abundant constituents.

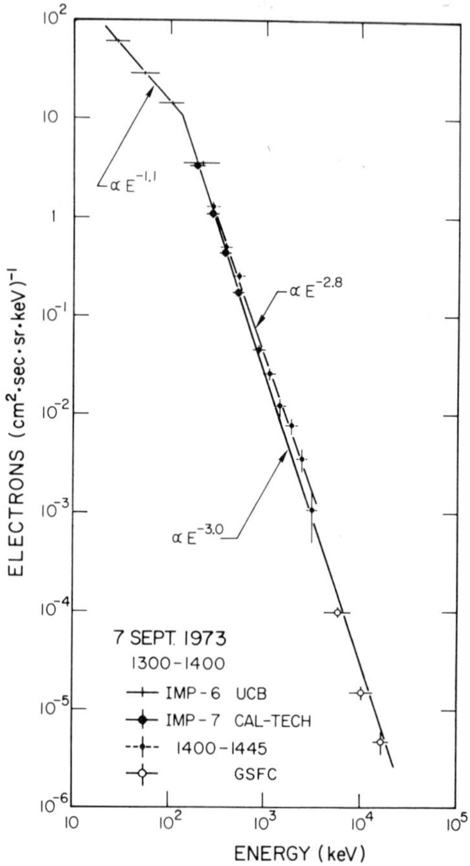

FIGURE 2.16 Differential electron energy spectrum observed in the interplanetary medium from the 7 September, 1973, flare. The spectrum is constructed from the flux maximum in each energy channel and compiles from 3 different detector systems as shown by the different symbols. (After Ramaty *et al.*, 1979.)

De-excitation of these nuclei to the ground state takes place by the emission of a γ-ray photon of energy corresponding to the energy of the excited level. Unlike 1) and 2) above, this emission is prompt, whereas in both 1) and 2) the emission is delayed due to relatively long half-life times of the emitters.

4) Production of ^{2}H (deuterium), ^{3}H (tritium) and ^{3}He as secondary products of interactions between accelerated protons and α-particles (threshold $\sim$25–30 MeV per nucleon) and of some other nuclear reactions.

Thus, summarizing, production of deuterium, tritium, helium-3, neutrons, positrons and γ-rays is to be expected on the Sun when protons and α-particles are accelerated in flares to energies exceeding $\sim$30 MeV per nucleon.

2.9.4.*A* *Accelerated isotopes of hydrogen and helium* The abundance ratio $\Gamma(^2H/^1H)$ in proton flares has been found to be four times higher than the upper limit found by Epstein and Taylor (1970) in the solar wind, 16 times higher than the upper limit found by Grevesse (1969) in quiet regions of the Sun, and it is even above the current estimate for the 'cosmic' abundance of ^2H (Reeves *et al.*, 1973). Anglin *et al.* (1973a, b) conclude that these increased values are consistent only with their production in nuclear interactions associated with flares, even when the yield certainly varies greatly from flare to flare, being very sensitive to the energy spectra of the bombarding nuclei.

While the increase in ^2H and ^3H abundances becomes statistically significant only as an average in a great set of flare data, the high yield of ^3He can be shown even for individual flares. However, the flares with a high yield of ^3He are rare (Dietrich, 1973; Anglin *et al.*, 1973a; Garrard *et al.*, 1973; Serlemitsos and Balasubrahmanyan, 1975). Ramaty *et al.* (1980) list less than 40 events with $\Gamma(^3He/^4He) > 0.1$ between 1968 and 1976.

One very peculiar feature of the ^3He-rich events is their lack of correlation with ^2H and ^3H enhancements (Ramaty and Kozlovsky, 1974; Anglin, 1975; Serlemitsos and Balasubrahmanyan, 1975). This is strange, because the hydrogen isotopes are thought to be the products of ^4He fragmentation: $H + {}^4He \rightarrow {}^3He + {}^2H$. On the other hand, the ^3He-rich events are always associated with enhancements of Fe nuclei, whereas the opposite is not true (Anglin *et al.*, 1977; Zwickl *et al.*, 1977).

2.9.4.*B* *Neutrons and the 2.23 MeV line* Biermann *et al.* (1951) were the first to point out that protons accelerated on the Sun could interact with the solar atmosphere and produce nuclear reactions with the emission of neutrons. Detailed calculations of this process (SVE 286–288 REF) indicate that after flares with very hard spectra ($\gamma \gtrsim 200$ in the power-law in energy of $P_0 \lesssim 2$ MV in an exponential rigidity spectrum) observable flux of neutrons should be detected near the Earth (exceeding the threshold sensitivity of the neutron detectors, $\sim 10^2$ neutrons m^{-2} s^{-1}). The neutrons should be looked for as soon as the flare brightens, since they are produced during the impulsive phase and move fast without any magnetic trapping through the corona and interplanetary space.

Attempts to detect neutrons from the Sun have been made by many authors in the past decade (Chupp, 1971; SVE 288 REF), all with a negative result. However, as a matter of fact, no measurements have been carried out on balloons or spacecraft after a flare from which an observable neutron flux should be expected. Therefore, Kirsch (1973) has tried to detect solar neutron emission by analyzing neutron monitor data of several mountain stations on the surface of the Earth. He finds 4×10^2 neutrons m^{-2} s^{-1} as the highest upper limit to the neutron flux at the Earth even in the case of the flare of 1967,

TABLE 2.III.
Observations of the 2.23 MeV line in flares.

Date	Source of data	Peak flux (photons m^{-2} s^{-1})	Reference
1972 Aug. 4	OSO-7	2200 ± 200	Chupp *et al.* (1973)
1972 Aug. 7	OSO-7	460 ± 13	Chupp *et al.* (1973)
1977 Nov. 22	Prognoz	$1500 \pm ?$	Chambov *et al.* (1978)
1978 July 11	HEAO-1	8300 ± 240	Hudson *et al.* (1979)

May 23, for which the expected flux (Švestka, 1971b) was two orders of magnitude higher.

According to Wang and Ramaty (1975) > 10 MeV positrons should also be detected in interplanetary space following major flares.

Direct observation of neutrons in space is one possibility for detecting their production in solar flares. Another is observation of the γ-line at 2.23 MeV, produced when neutrons are captured by protons in the solar atmosphere. This line was actually observed in three or four flares as Table 2.III (Hudson, 1979) demonstrates.

In the two well-observed events, on August 4, 1972, and July 11, 1978, the γ-ray flux at 2.23 MeV was only slightly delayed behind the hard X-rays and microwaves (cf. Figure 2.17 and SVE Figure 118). The delay was 94 ± 30 s in the July flare and < 180 s in the August flare. Some delay in the production of the 2.23 MeV γ-line is to be expected due to the finite neutron capture time in the photosphere. According to Wang and Ramaty (1974) this delay is between 30 s and 3 min, depending on the neutron energy, the density which the neutrons traverse, and the ^{3}He population, since ^{3}He "kills" the neutrons without emitting any photon. Thus a comparison of the γ-rays and hard X-rays in Figure 2.17 gives convincing proof that < 1 MeV electrons and > 30 MeV protons are accelerated either by the same (one-step) acceleration process, or that the second-step acceleration (giving rise to > 30 MeV protons) immediately follows the process of pre-acceleration.

2.9.4.*C* *Other γ-ray lines and continuum* Other γ-ray lines so far observed in the flare spectra are the 0.511 MeV line produced by positrons and the 4.43 MeV and 6.14 MeV lines due to ^{12}C and ^{16}O nuclei de-excitation. While the positron line is also delayed (Wang and Ramaty, 1975), the two other lines are prompt. The γ-ray flux at 1 AU (photons m^{-2} s^{-1}) was given by Chupp *et al.* (1973), for the August flare, as $(7 \pm 1.5) \times 10^2$ in the 0.511 MeV line, and $(3 \pm 1) \times 10^2$ in either of the two de-excitation lines, and by Hudson *et al.* (1979), for the July flare, as 1500 ± 600 in the 4.43 MeV line.

The observed flux ratio 4.43 MeV/2.23 MeV (0.14 ± 0.06 in the August and 0.18 ± 0.09 in the July flare) can be used to deduce the spectrum of the

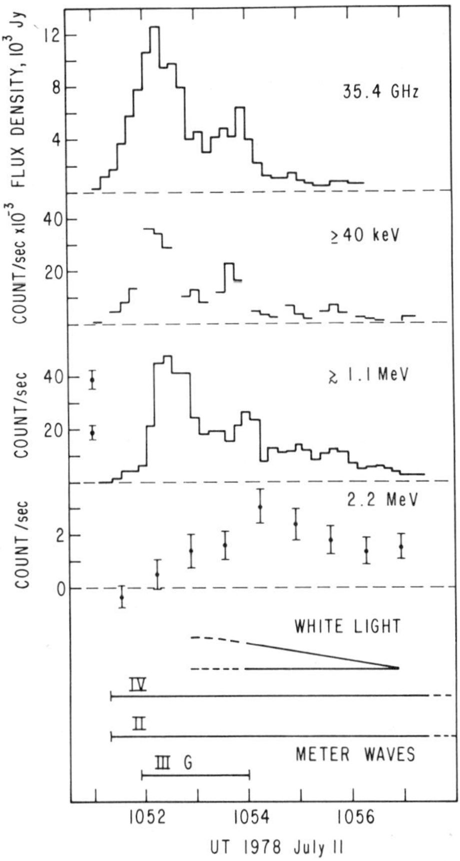

FIGURE 2.17 A comparison of time profiles of the microwave impulsive burst (35.4 GHz), hard X-rays (> 40 keV), γ-ray continuum (> 1.1 MeV), the neutron γ-ray line at 2.2 MeV, and the white-light emission (Debrecen) in the γ-ray flare of July 11, 1978. (After Hudson *et al.*, 1979.)

accelerated protons (Ramaty *et al.*, 1975). Unfortunately, the fact that many neutrons can decay on ^{3}He in the photosphere without emitting a photon makes these computations rather uncertain.

There are many other lines resulting from de-excitations of nuclear levels that might be seen in flares (cf. Ramaty *et al.*, 1980). An important line is at 15.11 MeV from ^{12}C de-excitation; the 15.11 MeV/4.44 MeV flux ratio could be another sensitive diagnostic of accelerated particle spectra (Crannell *et al.*, 1979).

During both the γ-ray flares of August 4, 1972, and July 11, 1978, an enhanced γ-ray continuum was also recorded extending well above 1 MeV. The continuum spectrum could be described as a power law, $dJ/dE \simeq 10^4$

$E^{-1.7}$ photons m^{-2} s^{-1} MeV^{-1} in the August flare (Suri *al.*, 1975) and $dJ/dE \simeq 8.4 \times 10^4\ E^{-3}$ photons m^{-2} s^{-1} MeV^{-1} in the July flare (Hudson *et al.*, 1979) in the 1 to 5 MeV range. This continuum might be due to a merging of unresolved gamma-ray lines (Ramaty *et al.*, 1977). However, the steep spectrum in the July event indicated that an appreciable bremsstrahlung contribution exists at energies at least up to 3.1 MeV (Hudson *et al.*, 1979).

2.9.5 The proton/electron ratio

Because of the fact that protons produce very few observable effects in the solar atmosphere, the proton/electron ratio is still poorly known. At high energies (>10 MeV) flare-associated protons in interplanetary space greatly exceed the flux of electrons in the same energy range, as we mentioned in Section 2.9.3.A. The γ-rays indicate that even in the solar atmosphere the flux of 10 MeV protons is about 100 times stronger than that of 10 MeV electrons (Ramaty *et al.*, 1980). Analysis of the observations indicates that this ratio increases towards higher energies, but it decreases significantly as we go down to and below 1 MeV. According to Canfield and Cook (1978) the flux of protons injected into the chromosphere at energies below ~ 100 keV is less than 0.02 that of electrons in the same energy range.

This latter estimate, deduced from the lack of asymmetry in the Lyman-α line as observed on Skylab on 9 August, 1973, is uncertain by at least one order of magnitude. Nevertheless, a significant decrease in the proton/electron ratio with decreasing energy seems to be well indicated.

This variation of proton/electron ratio should not be due to propagation effects, because the processes of propagation and storage are not significantly dependent on particle rigidity or energy (Ma Sung *et al.*, 1975). It may be caused by differences in energy loss, or it may be the result of two different stages of acceleration. According to Bai and Ramaty (1976, 1979) the γ-ray and hard X-ray spectra and time-shifts on August 4, 1972, indicate indeed that relativistic electrons and relativistic nuclei are accelerated by a mechanism which is different from the mechanism which accelerates ~ 100 keV electrons in flares.

On the other hand, the small time-shift of the γ-ray emission on 11 July, 1978, with respect to hard X-rays and white-light emission (Figure 2.17) shows that, if such two different mechanisms do exist, the two acceleration processes must be almost simultaneous. On 11 July, the γ-ray integral flux lagged behind the impulsive phase hard X-rays (>40 keV) by only ~ 10 seconds (Hudson *et al.*, 1979).

This problem is also related to the origin of white-light flares. If the white-light emission is caused by protons, as Švestka (1970a) and Najita and Orrall (1970) suggested, both the 2.23 MeV line and the white light should be excited by nuclei of about the same energy (>30 MeV/nucleon). Because of the expected delay of the 2.23 MeV line by >30 seconds (Wang and Ramaty, 1974), Figure 2.17 does not exclude this possibility. However, the white-light emission is obviously produced at footpoints of closed flare loops (cf. Figure 2.7); hence, particles of >30 MeV energies must be captured in a high enough number inside the loops. A model accelerating the pre-accelerated particles in a shock high in the atmosphere does not fit this picture; whereas a model accelerating the particles in the reconnection process at the top of the loops (Švestka *et al.*, 1979) can hardly accelerate particles to such very high energies.

References

Albregsen, F. and Engvold, O., *IAU Colloquium* No. 44, p. 246.(1979).
Alvarez, H., Haddock, F. T., and Lin, R. P., *Solar Phys.* **26**, 468 (1972).
Anderson, K. A. and Mahoney, W. A., *Solar Phys.* **35**, 419 (1974).
Anderson, K. A. and Winckler, J. R., *J. Geophys. Res.* **67**, 4103 (1962).
Anglin, J. D., *Astrophys. J.* **198**, 733 (1975).
Anglin, J. D., Dietrich, W. F. and Simpson, J. A., in *Symp. on High Energy Phenomena on the Sun*, NASA SP-342, p. 315 (1973a).
Anglin, J. D., Dietrich, W. F. and Simpson, J. A., *Astrophys. J. Letters* **186**, L41 (1973b).
Anglin, J. D., Dietrich, W. F. and Simpson, J. A., 15*th International Conf. Cosmic Rays* **5**, 43 (1977).
Arnoldy, R. L., Kane, S. R. and Winckler, J. R. *IAU Symp.* **35**, 490 (1968).
Avignon, Y., Martres, M. J. and Pick, M., *Compt. Rend. Acad. Sci. Paris* **256**, 2112 (1963).
Avrett, E. H., Vernazza, J. E. and Linsky, J. L. *Astrophys. J.* **207**, L199 (1976).
Axisa, F., *Solar Phys.* **35**, 207 (1974).
Axisa, F., Martres, M. J., Pick, M., and Soru-Escaut, I., *Solar Phys.* **29**, 163 (1973).
Bai, T., *Solar Phys.* **59**, 141 (1978).
Bai, T. and Ramaty, R., *Solar Phys.* **49**, 343 (1976).
Bai, T. and Ramaty, R., *Astrophys. J.* **219**, 705 (1978).
Bai, T. and Ramaty, R., *Astrophys. J.* **227**, 1072 (1979).
Baker, D., *Flare Classification Based upon X-ray Intensity*, AIAA Paper 70-1370, Huntsville, Ala. (1970).
Bakshi, P. and Kalman, G., *Solar Phys.* **47**, 307 (1976).
Banin, V. G., *Izv. Krymsk. Astrofiz. Obs.* **33**, 118 (1965a).
Banin, V. G., *Izv. Krymsk. Astrofiz. Obs.* **34**, 252 (1965b).
Becker, U., *Z. Astrophys.* **48**, 189 (1959).
Bell, B. and Glazer, H., *Smithson. Contrib. Astrophys.* **3**, 4 (1959).
Bertsch, D. L. and Reames, D. V., *Solar Phys.* **55**, 491 (1977).
Bertsch, D. L., Fichtel, C. E. and Reames, D. V., *Astrophys. Letters* **157**, L53 (1969).
Bertsch, D. L., Fichtel, C. E. and Reames, D. V., *Astrophys. J.* **171**, 169 (1972).
Bertsch, D. L., Biswas, S. and Reames, D. V., *Solar Phys.* **39**, 479 (1974).
Bierman, L., Haxel, O. and Schulter, A., *Z. Naturforsch.* **6a**, 47 (1951).
Biswas, S. and Fichtel, C. E., *Space Sci. Rev.* **4**, 705 (1965).
Blaha, M., *Solar Phys.* **17**, 99 (1971).
Boischot, A., *Compt. Rend. Acad. Sci. Paris* **244**, 1326 (1957).
Boischot, A. and Clavelier, B., *Astrophys. Letters* **1**, 7 (1976a).
Boischot, A. and Clavelier, B., *Ann. Astrophys.* **31**, 445 (1967b).
Boischot, A. and Pick, M., *J. Phys. Soc. Japan* **17**, A II, 203 (1962).
Brown, J. C., *Solar Phys.* **26**, 441 (1972).
Brown, J. C., *Solar Phys.* **36**, 371 (1974).
Brown, J. C. and Nakagawa, Y., *Astrophys. J. Letters*, in press (1978).
Brown, J. C., van Beek, H. F. and McClymont, A. N., *Astron. Astrophys.* **41**, 395 (1975).

Brueckner, G. E., *IAU Symp.* **68**, 105 (1975a)
Brueckner, G. E., *IAU Symp.* **68**, 135 (1975b).
Brueckner, G. E., Patterson, N. P. and Scherrer, V. E., *Solar Phys.* **47**, 127 (1976).
Bruzek, A., *Z. Astrophys.* **31**, 111 (1952).
Bruzek, A., *Z. Astrophys.* **44**, 183 (1958).
Bruzek, A., *Solar Phys.* **2**, 451 (1976).
Bruzek, A., in C. de Jager and Z. Švestka (eds.), *Solar Flares and Space Research*, p. 61 (1969).
Bruzek, A., *Solar Phys.* **42**, 215 (1975).
Bruzek, A., *Solar Phys.* **61**, 35 (1979).
Canfield, R. C. and Athay, R. G., *Solar Phys.* **34**, 193 (1974).
Canfield, R. C. and Cook, J. W., *Astrophys. J.* **225**, 650 (1978).
Canfield, R. C. and Fisher, R. R., *Astrophys. J.* **210**, L149 (1976).
Canfield, R. C., Brown, J. C., Brueckner, G. E., Cook, J. W., Craig, I. J. G., Doschek, G. A., Emslie, A. G., Hénoux, J. C., Lites, B. W., Machado, M. E., and Underwood, J. H., in P. A. Sturrock (ed), *Proc. of Skylab Workshop on Solar Flares*, Ch. 6, in press (1979a).
Canfield, R. C., Cheng, C. C., Dere, K. P., Dulk, G. A., McLean, D. J., Robinson, R. D., Schmahl, E. J. and Schoolman, S. A., in P. A. Sturrock (ed), *Proc. of Skylab Workshop on Solar Flares*, Appendix A, in press (1979b).
Carlqvist, P., *Solar Phys.* **7**, 377 (1969).
Catalano, C. P. and Van Allen, J. A., *Astrophys. J.* **185**, 335 (1973).
Chambou, G., Hurley, K., Niel, M., Talon, R., Vedrenne, G., Likine, O. B., Kouznetsov, A. V. and Estouline, I. V., in T. L. Cline and R. Ramaty (eds): *Gamma Ray Spectroscopy in Astrophysics*, NASA-TM-79619 (1978).
Cheng, C. C., *Space Sci. Rev.* **13**, 3 (1972b).
Cheng, C. C., *Solar Phys.* **55**, 413 (1977).
Cheng, C. C., *Solar Phys.* **56**, 205 (1978).
Cheng, C. C., *Solar Phys.*, in press (1980).
Cheng, C. C. and Kjeldseth Moe, O., *Solar Phys.* **59**, 361 (1978).
Cheng, C. C. and Widing, K. G., *Astrophys. J.* **201**, 735 (1975).
Chupp, E. L., *Space Sci. Rev.* **12**, 486 (1971).
Chupp, E. L., Forrest, D. J., Higbie, P. R., Suri, A. N., Tsai, C. and Dunphy, P. P., *Nature* **241**, 5388 (1973).
Cline, T. L. and McDonald, F. G., *Solar Phys.* **5**, 507 (1968).
Cliver, E. W., Hurst, M. D., Wefer, F. L. and Bleiweiss, M. P., *Solar Phys.* **48**, 307 (1976).
Coffey, H. E. (ed.), WDC-A Report UAG–28, NOAA, Boulder (1973).
Cohen, L., Feldman, V. and Doschek, G. A., *Astrophys. J. Supplements*, in press (1979).
Cook, J. W., *Astrophys. J.*, in press (1979).
Cowan, R. D. and Widing, K. G., *Astrophys. J.* **180**, 285 (1973).
Cook, J. W. and Brueckner, G. E., *Astrophys. J.* **227**, 645 (1979).
Craig, I. J. D. and McClymont, A. N., *Solar Phys.* **50**, 133 (1976).
Crannell, C. J., Frost, K. J., Mätzler, C. H., Ohki, K. and Saba, J. L., *Astrophys. J.* **223**, 620 (1978).
Crannell, C. J., Crannell, H. and Ramaty, R. *Astrophys. J.* **229**, 762 (1979).
Crawford, H. J., Price, P. B., Cartwright, B. G. and Sullivan, J. D., *Astrophys. J.* **195**, 213 (1975).
Croom, D. L., *Solar Phys.* **19**, 152 (1971).
Croom, D. L. and Powell, R. J., *Solar Phys.* **20**, 136 (1971).
Datlowe, D., *Solar Phys.* **17**, 436 (1971).
Datlowe, D. W., Hudson, H. S. and Peterson, L. E., *Solar Phys.* **35**, 193 (1974).
Datlowe, D. W., O'Dell, S. L., Peterson, L. E. and Elcan, M. J., *Astrophys. J.* **212**, 561 (1977).
Davis, W. D., *Solar Phys.* **54**, 139 (1977).
De Feiter, L. D., *Space Sci. Rev.* **16**, 3 (1974).
De Feiter, L. D. and Švestka, Z., *Bull. Astron. Inst. Czech.* **15**, 117 (1964).
De Feiter, L. D. and Švestka, Z., *Solar Phys.* **41**, 415 (1975).
De Jager, C., *Space Res.* **1**, 628 (1960).
De Jager, C., *Space Sci. Rev.* **1**, 487 (1962).
De Jager, C., *Ann. Astrophys.* **28**, 283 (1965).
De Jager, C., *Solar Phys.* **2**, 327 (1967).
De Jager, C., *Solar Phys.* **50**, 133 (1975).

De Jager, C. and De Jonge, G., *Solar Phys.* **58**, 127 (1978).
DeMastus, H. L. and Stover, R. R., *Publ. Astron. Soc. Pacific* **79**, 615 (1976).
DeMastus, H. L., Wagner, W. L. and Robinson, R. D., *Solar Phys.* **31**, 449 (1973).
Dere, K. P., *Astrophys. J.* **221**, 1062 (1978).
Dere, K. P. and Cook, J. W., *Astrophys. J.* **229**, 772 (1979).
Dere, K. P., Horan, D. M. and Kreplin, R. W., *Astrophys. J.* **217**, 976 (1977).
Dere, K. P., Mason, H. E., Widing, K. G. and Bhatia, A. K., Skylab/ATM Preprint, E. O. Hulburt Center for Space Research (1979).
Dezsö, L., Gezstelyi, L., Kondás, L., Kovács, A. and Rostás, S., *Solar Phys.*, in press (1980).
Dietrich, W. F., *Astrophys. J.* **180**, 955 (1973).
Dietrich, W. F. and Simpson, J. A., *Astrophys. J.* **225**, L41 (1978).
Dizer, M., *Solar Phys.* **10**, 416 (1969).
Dodson, H. W. and Hedeman, E. R., *Astrophys. J.* **110**, 242 (1949).
Dodson, H. W. and Hedeman, E. R., *Astron. J.* **65**, 51 (1960).
Dodson, H. W. and Hedeman, E. R., in *AAS-NASA Symp. on Physics of Solar Flares*, p. 15 (1964).
Dodson, H. W. and Hedeman, E. R., *Solar Phys.* **4**, 229 (1968).
Dodson, H. W. and Hedeman, E. R., *Solar Phys.* **9**, 278 (1969b).
Dodson, H. W. and Hedeman, E. R., *Solar Phys.* **13**, 401 (1970).
Dodson, H. W. and Hedeman, E. R., WDC-A Report UAG–14 (1971).
Dodson, H. W., Hedeman, E. R. and Miceli, M. R., *Solar Phys.* **23**, 360 (1972).
Donnelly, R. F., *Solar Phys.* **20**, 188 (1971).
Donnelly, R. F. and Hall, L. A., *Solar Phys.* **31**, 411 (1973).
Donnelly, R. F. and Kane, S. R., *Astrophys. J.* **222**, 1043 (1978).
Donnelly, R. F., Wood, A. T. and Noyes, R. W., *Solar Phys.* **29**, 107 (1973).
Doschek, G. A. and Feldman, U., *Astron. Astrophys.* **69**, 11 (1978).
Doschek, G. A., Feldman, U., Dere, K. P., Sandlin, G. D., Van Hooiser, M. E., Brueckner, G. E., Purcell, J. D. and Tousey, R., *Astrophys. J.* **196**, L83 (1975).
Doschek, G. A., Feldman, U. and Rosenberg, F. D., **Astrophys. J. 215**, 329 (1977a).
Doschek, G. A., Feldman, U. and Dere, K. P. *Astron. Astrophys.* **60**, L11 (1977b).
Dulk, G. A. and Altschuler, M. D., *Solar Phys.* **20**, 438 (1971).
Dulk, G. A., Smerd, S. F., MacQueen, R. M., Gosling, J. T., Magun, A., Stewart, R. T., Sheridan, K. V., Robinson, R. D. and Jacques, S., *Solar Phys.* **49**, 369 (1976).
Dupree, A. K., in *Advances in Atomic and Molecular Physics*, Vol. 14 (1978).
Durgaprassad, N., Fichtel, C. E., Guss, D. E. and Reames, D. B., *Astrophys. J.* **154**, 307 (1968).
Elcan, M. J., *Astrophys. J.* **226**, L99 (1978).
Ellison, M. A., McKenna, S. M. P. and Reid, J. H., *Dunsink Obs. Publ.* **1**, 1 (1960).
Ellison, M. A., McKenna, S. M. P. and Reid, J. H., *Dunsink Obs. Publ.* **1**, 53 (1961).
Emslie, A. G. and Noyes, R. W., *Solar Phys.* **57**, 373 (1978).
Emslie, A. G., Brown, J. C. and Donnelly, R. F., *Solar Phys.* **57**, 175 (1978).
Engvold, O., McKim Malville, J. and Rustad, B. M. *Solar Phys.* **48**, 137 (1976).
Engvold, O., Jensen, E. and Andersen, B. N., *Solar Phys.* **62**, 331 (1979).
Enome, S. and Tanaka, H., in R. Ramaty and R. G. Stone (eds), *High Energy Phenomena on the Sun*, NASA SP-342, p. 78 (1973).
Epstein, S. and Taylor, H. P., *Geochim. Cosmochim. Acta* **2**, Suppl. 1, 1085 (1970).
Erjushev, N. N. and Tsvetkov, L. I., *Izv. Krymsk. Astrofiz. Obs.* **48**, 85 (1973).
Falciani, R., Giordano, M., Rigutti, M. and Roberti, G., *Solar Phys.* **54**, 169 (1977).
Feldman, U. and Doschek, G. A., *Astron. Astrophys.* **65**, 215 (1978).
Feldman, U., Doschek, G. A. and Widing, K. G. *Astrophys. J.* **219**, 304 (1978).
Fokker, A. D., *Solar Phys.* **2**, 315 (1967).
Frank, L. A. and Gurnett, D. A., *Solar Phys.* **27**, 446 (1972).
Freier, P. S. and Webber, W. R., *J. Geophys. Res.* **68**, 1605 (1963).
Fritzová, L., *Bull. Astron. Inst. Czech.* **10**, 145 (1959).
Fritzová, L., *Bull. Astron. Inst. Czech.* **11**, 177 (1960).
Fritzová, L. and Švestka, Z., *Solar Phys.* **2**, 87 (1976).
Fritzová, L. and Švestka, Z., *Solar Phys.* **17**, 212 (1971).
Fritzová-Švestková, L., Chase, R. C. and Švestka, Z., *Solar Phys.* **48**, 275 (1976).

Frost, K. J., *Astrophys. J. Letters* **158**, L159 (1969).
Frost, K. J. and Dennis, B. R., *Astrophys. J.* **165**, 655 (1971).
Gabriel, A. H. and Jordan, C., *Monthly Notices Roy. Astron. Soc.* **145**, 241 (1969).
Gaizauskas, V., *IAU Colloquium*, **44**, 272 (1979).
Garrard, T. L., Stone, E. C. and Vogt, R. E., in R. Ramaty and R. G. Stone (eds), *High Energy Phenomena on the Sun*, NASA SP-342, p. 341 (1973).
Gergely, T. E. and Kundu, M. R., *Solar Phys.* **34**, 433 (1974).
Giovanelli, R. G., *Astrophys. J.* **89**, 555 (1939).
Golub, L., Krieger, A. S., Silk, J. K., Timothy, A. F. and Vaiana, G. S., *Astrophys. J. Letters* **189**, L89, L93 (1974).
Golub, L., Krieger, A., Silk, J. K., Timothy, A. F. and Vaiana, G., *IAU Symp.* **68**, 23 (1975).
Gopasyuk, S. I., Ogir, M. B., Severny, A. B. and Shaposhnikova, E. F., *Izv. Krymsk. Astrofiz. Obs.* **29**, 15 (1962).
Gosling, J. T. and Dulk, G. A., Paper presented at the AAS Meeting, Boulder, Colorado (1975).
Gosling, J. T., Hildner, E., MacQueen, R. M., Munro, R. H., Poland, A. I. and Ross, C. L., *J. Geophys. Res.* **79**, 4581 (1974).
Gosling, J. T., Hildner, E., MacQueen, R. M., Munro, R. H., Poland, A. I. and Ross, C. L., *Solar Phys.* **48**, 389 (1976).
Grevesse, N., *Mem. Soc. Roy. Sci. Liege* **19**, 249 (1969).
Grineva, Y. I., Karev, V. I., Korneev, V. V., Krutov, V. V. Mandelstam, S. L., Vainstein, L. A., Vasilyev, B. N. and Zhitnik, I. A., *Solar Phys.* **29**, 441 (1973).
Hagen, J. P. and Neidig, D. F., *Solar Phys.* **18**, 305 (1971).
Hakura, Y., *Solar Phys.* **39**, 493 (1974).
Hall, L. A., *Solar Phys.* **21**, 167 (1971).
Hall, L. A. and Hinteregger, H. E., in C. de Jager and Z. Švestka (eds), *Solar Flares and Space Research* (1969).
Harvey, K. L. and Harvey, J. W., *Solar Phys.* **47**, 233 (1976).
Harvey, K. L. and Martin, S. F., *Solar Phys.* **32**, 389 (1973).
Harvey, K. L., Martin, S. F. and Riddle, A. C. *Solar Phys.* **36**, 151 (1974).
Harvey, K. L., Harvey, J. W. and Martin, S. F., *Solar Phys.* **40**, 87 (1975).
Hénoux, J. C., Paper presented at the SMY Meeting of IAU Comm. 10 in Montreal (1979).
Hénoux, J. C. and Nakagawa, Y., *Astron. Astrophys.* **57**, 105 (1977a).
Hénoux, J. C. and Nakagawa, Y., *Solar Phys.* **53**, 279 (1977b).
Hénoux, J. C. and Nakagawa, Y., *Astron. Astrophys.* **66**, 385 (1978).
Heyvaerts, J., Priest, E. R. and Rust, D. M., *Astrophys. J.* **216**, 123 (1977).
Hiei, E. and Widing, K., *Solar Phys.* **61**, 407 (1979).
Holt, S. S. and Ramaty, R., *Solar Phys.* **8**, 119 (1969).
Howard, R. and Švestka, Z., *Solar Phys.* **54**, 65 (1977).
Hoyng, P., Brown, J. C. and Van Beek, H. F., *Solar Phys.* **48**, 197 (1976).
Hudson, H. S., *Solar Phys.* **24**, 414 (1972).
Hudson, H. S., *Solar Phys.* **45**, 69 (1975).
Hudson, H. S., *Astrophys. M.* **224**, 235 (1978).
Hudson, H. S., Contribution to Workshop on Particle Acceleration, La Jolla, January 1979 (1979).
Hudson, H. S., Bai, T., Gruber, D. E., Matteson, J. L., Nolan, P. L. and Petterson, L. E., *Astrophys. J.*, in press (1979).
Hultqvist, B., in C. de Jager and Z. Švestka (eds), *Solar Flares and Space Research*, p. 215 (1969).
Hurley, K. and Duprat, G., *Solar Phys.* **52**, 107 (1977).
Hyder, C. L., *Solar Phys.* **2**, 267 (1967).
Jackson, B. V. and Hildner, E., *Solar Phys.* **60**, 155 (1978).
Jakimiec, J. and Jakimiec, M., *Astron. Astrophys.* **34**, 415 (1974).
Janssens, T. J., White, K. P., III and Broussard, R. M., *Solar Phys.* **31**, 207 (1973).
Jefferies, J. T., Smith, E. V. P. and Smith, H. J., *Astrophys. J.* **129**, 146 (1959).
Kahler, S. W. and Kreplin, R. W., *Astrophys. J.* **168**, 531 (1971).
Kahler, S. W., Krieger, A. S. and Vaiana, G. S., *Astrophys. J. Letters* **199**, L57 (1975).
Kahler, S. W., Petrasso, R. D. and Kane, S. R., *Solar Phys.* **50**, 179 (1976).
Kahler, S. W., Hildner, E. and Van Hollebeke, M. A. I., *Solar Phys.* **57**, 429 (1978).

Kai, K., *Solar Phys.* **11**, 310 (1970).
Kai, K., *Solar Phys.* **45**, 217 (1975).
Kai, K., *Solar Phys.* **56**, 417 (1978).
Kai, K., *Solar Phys.* **61**, 187 (1979).
Kane, S. R., *Astrophys. J. Letters* **157**, L139 (1969).
Kane, S. R., *Astrophys. J.* **170**, 587 (1971).
Kane, S. R., *Solar Phys.* **27**, 174 (1972).
Kane, S. R., *in Symp. on High Energy Phenomena on the Sun*, NASA-SP 342, p. 55 (1973).
Kane, S. R., *Space Sci. Lab. Series* **14**, Issue 76 (1974).
Kane, S. R. and Anderson, K. A., *Astrophys. J.* **162**, 1003 (1970).
Kane, S. R. and Donnelly, R. F., *Astrophys. J.* **164**, 151 (1971).
Kane, S. R. and Hudson, H. S., *Solar Phys.* **14**, 414 (1970).
Kane, S. R. and Lin, R. P., *Solar Phys.* **23**, 457 (1972).
Kane, S. R., Kreplin, R. W., Martres, M. J., Pick, M. and Soru-Escaut, I., *Solar Phys.* **38**, 483 (1974).
Kane, S. R., Crannell, C. J., Datlowe, D., Feldman, U., Gabriel, A., Hudson, H. S., Kundu, M. R., Mätzler, C., Neidig, D., Petrosian, V. and Sheeley, N. R., in P. A. Sturrock (ed), *Proc. of Skylab Workshop on Solar Flares*, Chapter 5, in press (1979a).
Kane, S. R., Anderson, K. A., Evans, W. D., Klebesadel, R. W. and Laros, J., Preprint, Space Sci. Lab., UC Berkely (1979b).
Karpen, J. T., Crannell, C. J. and Frost, K. J., NASA TM-79663 (1978).
Kaufmann, P., *Solar Phys.* **50**, 197 (1976).
Kaufmann, P., Piazza, L. R. and Raffaelli, J. C., *Solar Phys.* **54**, 179 (1977).
Kirsch, E., *Solar Phys.* **28**, 233 (1973).
Kleczek, J., *Bull. Astron. Inst. Czech.* **4**, 9 (1953).
Kleczek, J. in *AAS-NASA Symp. on Physics of Solar Flares*, p. 77 (1964).
Korneev, V. V., Krutov, V. V., Mandelstam, S. L., Urnov, A. M., Zhitnik, I. A., Kononov, A. Y., Sylwester, B. and Sylwester, J., *Solar Phys.* **63**, 319 (1979).
Kosugi, T., *Solar Phys.* **48**, 339 (1976).
Kovrizhnikh, O. M., Savenko, I. A. and Chupova, L. M., *Solar Phys.* **50**, 447 (1976).
Krall, K. R., Reichmann, E. J., Wilson, R. M., Henze, W. and Smith, J. B., *Solar Phys.* **56**, 383 (1978).
Krieger, A. S., Paolini, F., Vaiana, G. S. and Webb, D., *Solar Phys.* **22**, 150 (1972).
Krivský, L., *Bull. Astron. Inst. Czech.* **27**, 374 (1976).
Krivský, L., Valnícek, B., Böhme, A., Fürstenberg, F., and Krüger, A., Paper presented at the *7th Regional Consultation on Solar Physics*, Starý Smokovec, Czechoslovakia (1973).
Krüger, A., *Geodät, Veröff. II* **10**, 41 (1969).
Kuiper, T. B., *Solar Phys.* **33**, 461 (1973).
Kundu, M. R., *Solar Radio Astronomy*, Interscience Publishers, New York (1965).
Kundu, M. R., Velusamy, T. and Becker, R. H., *Solar Phys.* **34**, 217 (1974).
Kundu, M. R., Alissandrakis, C. E. and Kahler, S. W., *Solar Phys.* **50**, 429 (1976).
Künzel, H., *Astron. Nachr.* **285**, 271 (1960).
Kurt, V. G., Logachev, Y. I. and Pissarenko, N. F., *Solar Phys.* **53**, 157 (1977).
Labrum, N. R. and Smerd, S. F., *Proc. Astron. Soc. Australia* **1**, 140 (1968).
Lanzerotti, L. J., in *Symp. on High Energy Phenomena on the Sun*, NASA-SP 342, p. 427 (1973).
Lanzerotti, L. J. and Maclennan, C. G., in Proceedings of *COSPAR Symp. on November* 1969 *Solar Particle Event*, p. 85 (1971).
Levine, R. H., *Solar Phys.* **56**, 185 (1978).
L'Heureux, J., Fan, G. Y. and Meyer, P., *Astrophys. J.* **171**, 363 (1972).
Lin, R. P., in *Symp. on High Energy Phenomena on the Sun*, NASA-SP 342, p. 439 (1973).
Lin, R. P., *Space Sci. Rev.* **16**, 189 (1974).
Lin, R. P. and Anderson, K. A., *Solar Phys.* **1**, 446 (1976).
Lin, R. P. and Hudson, H. S., *Solar Phys.* **17**, 412 (1971).
Lin, R. P., Evans, L. G. and Fainberg, J., *Space Sci. Lab. Series* **14**, Issue 26 (1973).
Lin, R. P., Mewaldt, R. A., Stone, E. C., Vogt, R. E., Paizis, C. and Van Gollebeke, M. A. I., *Bull. Amer. Phys. Soc.* **23**, 4-AJ7 (1978).
Lincoln, J. V. and Leighton, H. I., WDC-A Report UAG-21, NOAA, Boulder (1972).
Lingenfelter, R. E. and Ramaty, R., *High Energy Nuclear Reactions in Astrophysics*, W. A. Benjamin, New York, p. 99 (1967).

Linsky, J. L., Glackin, D. L., Chapman, R. D., Neupert, W. M. and Thomas, R. J., *Astrophys. J.* **203**, 509 (1976).
Lipa, B., *Solar Phys.* **57**, 191 (1978).
MacCombie, W. J. and Rust, D. M., *Solar Phys.* **61**, 69 (1979).
Machado, M. E., *Solar Phys.* **49**, 91 (1976).
Machado, M. E., *Solar Phys.* **60**, 341 (1978).
Machado, M. E. and Linsky, J. L., *Solar Phys.* **42**, 395 (1975).
Machado, M. E. and Noyes, R. W., *Solar Phys.* **59**, 129 (1978).
Machado, M. E. and Rust, D. M., *Solar Phys.* **38**, 499 (1974).
Machado, M. E. and Seibold, J. R., *Solar Phys.* **29**, 75 (1973).
Machado, M. E., Emslie, A. G. and Brown, J. C., *Solar Phys.* **58**, 363 (1978).
Machado, M. E., Avrett, E. H., Vernazza, J. E. and Noyes, R. W., *Astrophys. J.*, in press (1979).
MacQueen, R. M., Eddy, J. A., Gosling, J. T., Hildner, E., Munro, R. H., Newkirk, G. A., Poland, A. I. and Ross, C. L., *Astrophys. J.* **187**, L85 (1974).
Malville, J. M. and Smith, S. F., *J. Geophys. Res.* **68**, 3181 (1963).
Mandelstam, S. I., in *Uspechi Sovetskogo Sojuza v Issledovanii Kosmitsheskogo Prostranstva*, 1967–1977, Moscow, p. 209 (1978).
Mandelstam, S. L., Paper presented at COSPAR Meeting in Bangalore, India (1979).
Marsh, K. A., *Solar Phys.* **59**, 105 (1978).
Marsh, K. A., Zirin, H. and Hurford, G. J., *Astrophys. J.* **228**, 610 (1979).
Martin, S. F., *Solar Phys.*, in press (1979a).
Martin, S. F., unpublished (1979b).
Martres, M. J., *IAU Symp.* **35**, 25 (1968).
Martres, M. J. and Pick, M., *Ann. Astrophys.* **25**, 4 (1962).
Martres, M. J. and Soru-Escaut, I., *Solar Phys.* **53**, 225 (1977).
Martres, M. J., Michard, R. and Soru-Iscovici, *Ann. Astrophys.* **29**, 249 (1966).
Martres, M. J., Michard, R., Soru-Iscovici, I. and Tsap, T., *IAU Symp.* **35**, 318 (1968a).
Martres, M. J., Michard, R., Soru-Iscovici, I. and Tsap, T., *Solar Phys.* **5**, 187 (1968b).
Martres, M. J., Soru-Escaut, I. and Rayrole, J., *IAU Symp.* **43**, 435 (1971).
Ma Sung, L. S., Van Hollebeke, M. A. I. and McDonald, F. B., *Proc. Int. Cosmic Ray Conf.* **14**, 5, 1767 (1975).
Matsuura, O. T. and Nave, M. F. F., *Solar Phys.* **16**, 417 (1971).
Maxwell, A. and Thompson, A. R., *Astrophys. J.* **135**, 138 (1962).
Maxwell, A., Defouw, R. J. and Cummings, P., *Planetary Space Sci.* **12**, 435 (1964).
McCracken, K. G. and Rao, U. R., *Space Sci. Rev.* **11**, 155 (1970).
McDonald, F. B., in V. Manno and D. E. Page (eds), *Intercorrelated Satellite Observations Related to Solar Events*, p. 34 (1970).
McDonald, F. B., Cline, T. L. and Simnett, G. M., *J. Geophys. Res.* **77**, 2213 (1972).
McDonald, F. B., Teegarden, B. J., Trainor, J. H., Von Rosenvinge, T. T. and Webber, W. R., *Astrophys. J.* **203**, L149.
McGuire, R. E., Von Rosenvinge, T. T. and McDonald, F. B., *Astrophys. J.* **224**, L87 (1978).
McIntosh, P. S., WDC-A Report UAG-5, p. 14 (1969).
McIntosh, P. S. and Donnelly, R. F., *Solar Phys.* **23**, 444 (1972).
McIntosh, P. S., Krieger, A. S., Nolte, J. T. and Vaiana, G., *Solar Phys.* **49**, 57 (1976).
McKenzie, D. L., *Solar Phys.* **40**, 183 (1975).
McLean, D. J., *Proc. Astron. Soc. Australia* **1**, 47 (1967).
McLean, D. J., in G. Newkirk (ed), "Coronal Disturbances", *IAU Symp.* **57**, 301 (1974).
McLean, D. J., Sheridan, K. V., Stewart, R. T. and Wild, J. P., *Nature* **234**, 140 (1971).
Mercier, C., *Solar Phys.* **33**, 177 (1973).
Mercier, C., *Solar Phys.* **39**, 193 (1974).
Michard, R., *IAU Symp.* **43**, 359 (1971).
Mogro-Campero, A. and Simpson, J. A., *Astrophys. J. Letters* **171**, L5 (1972a).
Mogro-Campero, A. and Simpson, J. A., *Astrophys. J. Letters* **177**, L37 (1972b).
Moore, R. L. and Datlowe, D. W., *Solar Phys.* **43**, 189 (1975).
Moore, R. L. and LaBonte, B. J., *IAU Symp.* **91**, in press (1979).
Moore, R. L., Tang, F., Bohlin, J. D. and Golub, L., *Astrophys. J.* **218**, 286 (1977).

Moore, R. L., McKenzie, D. L., Švestka, Z., Widing, K. G., Antiochos, S. K., Dere, K. P., Dodson-Prince, H. W., Hiei, E., Krall, K. R., Krieger, A. S., Mason, H. E., Petrasso, R. D., Pneuman, G. W., Silk, J. K., Vorphal, J. A. and Withbroe, G. L., in P. L. Sturrock (ed), *Proceedings of Skylab Workshop on Solar Flares*, Chapter 8, in press (1979).
Moreton, G. E., *Astrophys. J.* **69**, 145 (1964).
Moreton, G. E. and Severny, A. B., *Solar Phys.* **3**, 282 (1968).
Munro, R. H., Dupree, A. K. and Withbroe, G. L., *Solar Phys.* **19**, 347 (1971).
Munro, R. H., Gosling, J. T., Hildner, E., MacQueen, R. M., Poland, A. I. and Ross, C. L., *Solar Phys.* **61**, 201 (1979).
Najita, K. and Orrall, F. Q., *Solar Phys.* **15**, 176 (1970).
Neidig, D. F., *Solar Phys.* **54**, 165 (1977).
Neidig, D. F., *Solar Phys.* **57**, 385 (1978).
Neidig, D. F., *Solar Phys.* **61**, 121 (1979).
Neidig, D. F., Demastus, H. L. and Wiborg, P. H., AFGL-TR-78-0194, *Environmental Research Papers*, No. 637 (1978).
Neupert, W. M., Thomas, R. J. and Chapman, R. D., *Solar Phys.* **34**, 349 (1974).
Nolte, J. T., Gerassimenko, M., Krieger, A. S., Petrasso, R. D. and Švestka, Z., *Solar Phys.* **62**, 123 (1979).
Noyes, R. W., in *Symp. on High Energy Phenomena on the Sun*, NASA-SP 342, p. 231 (1973).
Obayashi, T., *Space Sci. Rev.* **3**, 79 (1964).
Ogawa, H. and Kawabata, K. A., *Solar Phys.* **40**, 159 (1975).
Ohki, K. and Hudson, H. S., *Solar Phys.* **43**, 405 (1975).
Öhman, Y. and Öhman, N., *Observatory* **73**, 203 (1953).
Öhman, Y., Lindgren, A. and Lindgren, U., *Arkiv. Astron.* **3**, 121 (1962).
Orrall, F. Q. and Zirker, J. B., *Astrophys. J.* **208**, 618 (1976).
Pallavicini, R., Vaiana, G. S., Kahler, S. W. and Krieger, A. S., *Solar Phys.* **45**, 411 (1975).
Pallavicini, R., Serio, S. and Vaiana, G. S., *Astrophys. J.* **216**, 108 (1977).
Palmer, I. D. and Gosling, J. T., *J. Geophys. Res.* **83**, 2037 (1978).
Parker, E. N., *Astrophys. J.* **201**, 494 (1975).
Parkinson, J. H., *Solar Phys.* **28**, 137 (1973).
Parkinson, J. H., Wolff, R. S., Kestenbaum, H. L., Ku, W. H. M., Lemon, J. R., Long, K. S., Novick, R., Suozzo, R. J. and Weisskopf, M. C., *Solar Phys.* **60**, 123 (1978).
Parks, G. K. and Winckler, J. R., *Solar Phys.* **16**, 186 (1971).
Pesses, M. E., Van Allen, J. A. and Goertz, C. K., *J. Geophys. Res.* **83**, 553 (1978).
Peterson, L. E., Datlowe, D. W. and McKenzie, D. L., in *Symp. on High Energy Phenomena on the Sun*, NASA-SP 342, p. 132 (1973).
Petrasso, R. D., Kahler, S. W., Krieger, A. S., Silk, J. K. and Vaiana, G. S., *Astrophys. J.* **199**, L127 (1975).
Petrasso, R., Gerassimenko, M. and Nolte, J., *Astrophys. J.* **227**, 299 (1979a).
Petrasso, R. D., Nolte, J. T., Gerassimenko, M., Krieger, A. S., Krogstad, R., Seguin, F. H. and Švestka, Z., *Solar Phys.* **62**, 133 (1979b).
Petrosian, V., *Astrophys. J.* **186**, 291 (1973).
Pick, M., *Ann. Astrophys.* **24**, 183 (1961).
Pick, M. and Trottet, G., *Solar Phys.* **60**, 353 (1978).
Pinter, S., in *Proceedings COSPAR Symposium B, Tel Aviv*, p. 35 (1978).
Pizzichini, G., Spizzichino, A. and Vespignani, G. R., *Solar Phys.* **35**, 431 (1974).
Poland, A. I. and Munro, R. H., *Astrophys. J.* **209**, 927 (1976).
Price, P. B., in *Symp. on High Energy Phenomena on the Sun*, NASA-SP 342, p. 377 (1973).
Priest, E. R. and Heyvaerts, J., *Solar Phys.* **36**, 433 (1974).
Purcel, J. D. and Widing, K. G., *Astrophys. J.* **176**, 239 (1972).
Ramaty, R. and Kozlovsky, B., *Astrophys. J.* **193**, 729 (1974).
Ramaty, R., Kozlovsky, B. and Lingenfelter, R. E., *Space Sci. Rev.* **18**, 341 (1975).
Ramaty, R., Kozlovsky, B. and Suri, A. N., *Astrophys. J.* **214**, 617 (1977).
Ramaty, R., Colgate, S. A., Dulk, G. A., Hoyng, P., Knight, S. W., Lin, R. P., Melrose, D. B., Paizis, C., Orrall, F., Shapiro, P. R., Smith, D. F., and Van Hollebeke, M., in P. A. Sturrock (ed), *Proceedings of the Skylab Workshop on Solar Flares*, Chapter 4 (1980).
Ramsey, H. E. and Smith, S. F., *Astron. J.* **71**, 197 (1966).

Raymond, J. C. and Dupree, A. K., *Astrophys. J.* **222**, 379 (1978).
Raymond, J. C., Noyes, R. W. and Stopa, M. P., *Solar Phys.* in press (1979).
Reeves, H., Andouze, J., Fowler, W. A. and Schramm, D. N., *Astrophys. J.* **179**, 909 (1973).
Ribes, E., *Astron. Astrophys.* **2**, 316 (1969).
Riddle, A. C., *Solar Phys.* **13**, 448 (1970).
Riddle, A. C. and Sheridan, K. V., *Proc. Astron. Soc. Australia* **2**, 62 (1971).
Riddle, A. C., Tandberg-Hanssen, E. and Hansen, R. T., *Solar Phys.* **35**, 171 (1974).
Robinson, R. D., *Solar Phys.* **55**, 459 (1977).
Robinson, R. D., *Solar Phys.* **60**, 383 (1978).
Robinson, R. and MacQueen, R. M., Paper presented at the AAS Meeting, Boulder, Colorado (1975).
Rosenberg, J., *Astron. Astrophys.* **9**, 159 (1970).
Roy, J. R., *Solar Phys.* **26**, 418 (1972).
Roy, J. R., *Solar Phys.* **28**, 95 (1973).
Roy, J. R., *Search for Solar Hard X-ray Bursts from Flares behind the Limb* from the Utrecht TD-IA Spectrometer Data, unpublished (1975).
Roy, J. R., *Solar Phys.* **48**, 149 (1976).
Roy, J. R. and Datlowe, D. W., *Solar Phys.* **40**, 165 (1975).
Rust, D. M., *Solar Phys.* **33**, 205 (1973a).
Rust, D. M., Sacramento Peak Obs. Contr. No. 209 (1973b).
Rust, D. M., *Solar Phys.* **47**, 21 (1976).
Rust, D. M. and Bar, V., *Solar Phys.* **33**, 445 (1973).
Rust, D. M. and Bridges, C. A., *Solar Phys.* **43**, 129 (1975).
Rust, D. M. and Hegwer, F., *Solar Phys.* **40**, 141 (1975).
Rust, D. M. and Roy, J. R., Sacramento Peak Obs. Contr. No. 221 (1974).
Rust, D. M. and Švestka, Z., *Solar Phys.* **63**, 279 (1979).
Rust, D. M. and Webb, D. F., *Solar Phys.* **54**, 403 (1977).
Rust, D. M., Nakagawa, Y. and Neupert, W. M., *Solar Phys.* **41**, 397 (1975).
Rust, D. M., Webb, D. F. and MacCombie, W., *Solar Phys.* **54**, 53 (1977).
Rust, D. M., Hildner, E., Dryer, M., Hansen, R. T., McClymont, A. N., McKenna-Lawlor, S. M. P., McLean, D. J., Schmahl, E. J., Steinolfson, R. S., Tandberg-Hanssen, E., Tousey, R., Webb, D. F. and Wu, S. T., in P. A. Sturrock (ed), *Proc. of Skylab Workshop on Solar Flares*, Chapter 7, in press (1979).
Sakurai, K., *Rep. Iono. Space Res. Japan* **21**, 29 (1967).
Sakurai, K., *Solar Phys.* **23**, 142 (1972).
Sarris, E. T. and Shawhan, S. D., *Solar Phys.* **28**, 519 (1973).
Sawyer, C. and Smith, S. F., WDC-A Report UAG-9, p. 9 (1970).
Schmahl, E. and Hildner, E., *Solar Phys.* **55**, 473 (1977).
Scholer, M., Hovestadt, D., Klecker, B., Gloeckler, G. and Fan, C. Y., *J. Geophys. Res.* **83**, 3349 (1978).
Serlemitsos, A. T. and Balasubrahmanyan, V. K., *Astrophys. J.* **198**, 195 (1975).
Severny, A. B., *Izv. Krymsk. Astrofiz. Obs.* **22**, 12 (1960).
Severny, A. B., in *AAS-NASA Symp. on Physics of Solar Flares*, p. 95 (1964).
Severny, A. B., *Nobel Symp.* **9**, 71 (1968).
Severny, A. B., *Annals IQSY* **3**, 11 (1969a).
Severny, A. B., in C. de Jager and Z. Švestka (eds). *Solar Flares and Space Research*, p. 38 (1969b).
Severny, A. B., *Solar Phys.* **53**, 285 (1977).
Sheeley, N. R., Bohlin, J. D., Brueckner, G. E., Purcell, J. D., Scherrer, V. E., Tousey, R., Smith, J. B., Speich, D. M., Tandberg-Hanssen, E., Wilson, R. M., Deloach, A. C., Hoover, R. B. and McGuire, J. P., *Solar Phys.* **45**, 377 (1975).
Sheridan, K. V., *Proc. Astron. Soc. Australia* **1**, 376 (1970).
Shimabukuro, F. I., *Solar Phys.* **5**, 498 (1968).
Shine, R., Gerola, W. and Linsky, J. L., *Astrophys. J.* **202**, L101 (1975).
Shmeleva, O. P. and Syrovatsky, S. I., *Solar Phys.* **33**, 341.
Simnett, G. M., *Solar Phys.* **20**, 448 (1971).
Simnett, G. M., *Solar Phys.* **22**, 189 (1972a).

Simnett, G. M., *Space Res.* **13**, 745 (1972b).
Simnett, G. M., *Space Sci. Rev.* **16**, 257 (1974a).
Simnett, G. M., *Solar Phys.* **34**, 377 (1974b).
Simon, G. W. and Noyes, R. W., *Solar Phys.* **22**, 450.
Simon, P. and McIntosh, P. S., in P. S. McIntosh and M. Dryer (eds), *Solar Activity Observations and Predictions*, p. 343 (1972).
Slonim, Y. M. and Korobova, Z. B., *Solar Phys.* **40**, 397 (1975).
Smerd, S. F. and Dulk, G. A., *IAU Symp.* **43**, 616 (1972).
Smith, E. v. P., *Nobel Symp.* **9**, 137 (1968).
Smith, H. J. and Smith, E. v. P., *Solar Flares*, The Macmillan Co., New York (1963).
Smith, J. B., Wilson, R. M. and Henze, W., *Astrophys. J.* **216**, L79 (1977a).
Smith, J. B., Speich, D. M., Wilson, R. M., Tandberg-Hanssen, E. and Wu, S. T., *Solar Phys.* **52**, 379 (1977b).
Smith, J. B., Speich, D. M., Wilson, R. M. and Reichmann, E. J., in *Proceedings COSPAR Symposium B*, Tel Aviv, p. 3 (1978).
Smith, S. F. and Harvey, K. L., in C. J. Macris (ed.), *Physics of the Solar Corona*, p. 156 (1971).
Smith, S. F. and Ramsey, H. E., *Z. Astrophys.* **60**, 1 (1964).
Somov, B. V., *Solar Phys.* **42**, 235 (1975).
Somov, B. V., *Trudy Fiz. Inst. P.N. Lebedeva* **88**, 127 (1976).
Somov, B. V. and Syrovatsky, S. I., *Solar Phys.* **39**, 415 (1974).
Spangler, Z. R. and Shawhan, S. D., *Solar Phys.* **37**, 189 (1974).
Spicer, D. S., *Solar Phys.* **53**, 305 (1977).
Spicer, D. S. and Davis, J., *Solar Phys.* **43**, 107 (1975).
Stepanyan, N. N., *Astron. Zh.* **46**, 580 (1969).
Stewart, R. T., McCabe, M. K., Koomen, M. J., Hanssen, J. T. and Dulk, G. A., *Solar Phys.* **36**, 203 (1974a).
Stewart, R. T., Howard, R. A., Hansen, F., Gergely, T. and Kundu, M. *Solar Phys.* **36**, 219 (1974b).
Stewart, R. T., Hansen, R. T. and Sheridan, K. V., *IAU Colloquium* **44**, 315 (1978).
Sturrock, P. A. (ed.), *Proceedings of the Skylab Workshop on Solar Flares*, in press (1979).
Suemoto, Z. and Hiei, E., *Publ. Astron. Soc. Japan* **11**, 185 (1959).
Şuri, A. N., Chupp, E. L., Forrest, D. J. and Reppin, C., *Solar Phys.* **43**, 415 (1975).
Švestka, Z., *Bull. Astron. Inst. Czech.* **17**, 262 (1966).
Švestka, Z., *Nobel Symp.* **9**, 17 (1968).
Švestka, Z., *Solar Phys.* **8**, 400 (1969).
Švestka, Z., *Solar Phys.* **13**, 471 (1970a).
Švestka, Z., *Space Res.* **10**, 797 (1970b).
Švestka, Z., *Report of the Meeting of CESRA*, **15**, (1971a).
Švestka, Z., *Solar Phys.* **19**, 202 (1971b).
Švestka, Z., *Ann. Rev. Astron. Astrophys.* **10**, 1 (1972).
Švestka, Z., *Solar Phys.* **31**, 389 (1973).
Švestka, Z. and Simon, P. (eds), *Catalog of Solar Particle Events, 1956–1969.* D. Reidel Publ. 149 = Švestka (1976), p. 149; SVE 130 REF = Švestka (1976), p. 130 and references therein; SVE Figure 10 = Figure 10 in Švestka (1976).
Švestka, Z. and Fritzová-Švestková, L., *Solar Phys.* **36**, 417 (1974).
Švestka, Z. and Howard, R., *Solar Phys.* **63**, 297 (1979).
Švestka, Z. and Simon, P., *Solar Phys.* **10**, 3 (1969).
Švestka, Z. and Simon, P. (eds), *Catalog of Solar Particle Events*, 1956–1969. D. Reidel Publ. Company, Dordrecht, Holland (1976).
Švestka, Z., Kopecký, M. and Blaha, M., *Bull. Astron. Inst. Czech.* **12**, 229 (1961).
Švestka, Z., Kopecký, M. and Blaha, M., *Bull. Astron. Inst. Czech.* **13**, 37 (1962).
Švestka, Z., Castelli, J. P., Dizer, M., Dodson, H. W., McIntosh, P. S. and Urbarz, H., Paper presented at the CINOF Meeting in Sao Paulo, Brazil (1974).
Švestka, Z., Fritzová-Švestková, L., Nolte, J. T., Dodson-Prince, H. W. and Hedeman, E. R., *Solar Phys.* **50**, 491 (1976).
Švestka, Z., Krieger, A. S., Chase, R. C. and Howard, R., *Solar Phys.* **52**, 69 (1977).
Švestka, Z., Dodson-Prince, H. W., Martin, S. F., Mohler, O. C., Moore, R. L., Nolte, J. T. and Petrasso, R. D., *Solar Phys.*, in press (1979a).

Švestka, Z., Martin, S. F. and Kopp, R. A., *Proc. IAU Symp.* **91**, in press (1979b).
Swarup, G., Stone, P. H. and Maxwell, A., *Astrophys. J.* **131**, 725 (1960).
Takakura, T., *Solar Phys.* **1**, 304 (1967).
Takakura, T., in *Symp. on High Energy Phenomena on the Sun*, NASA GSFC Preprint, p. 179 (1973).
Tanaka, H. and Enomé, S., *Nature* **225**, 435 (1970).
Tanaka, H. and Enomé, S., *Solar Phys.* **40**, 123 (1975).
Tanaka, H. and Kakinuma, T., *Rep. Iono. Space Res. Japan* **18**, 132 (1964).
Tanaka, H., Kakinuma, T. and Enomé, S., *Proc. Res. Inst. Atmospherics* **14**, 23 (1967).
Tanaka, H., Kakinuma, T. and Enomé, S., *Annals IQSY* **3**, 63 (1969).
Tanaka, K., *Solar Phys.* **47**, 247 (1976).
Tanaka, K., *Solar Phys.* **58**, 149 (1978).
Tanaka, K. and Nakagawa, Y., *Solar Phys.* **33**, 187 (1973).
Tandberg-Hanssen, E., in Proceedings US–Indian Workshop on Solar–Terrestrial Physics, Udaipur, in press (1979).
Tang, F., *Solar Phys.* **60**, 119 (1978).
Tapping, K. F., *Solar Phys.* **59**, 145 (1978).
Thomas, R. J. and Teske, R. G., *Solar Phys.* **16**, 431 (1971).
Tindo, I. P., Ivanov, V. D., Valnicek, B. and Livshits, M. A., *Solar Phys.* **27**, 26 (1972).
Tindo, I. P., Shuryghin, A. I. and Steffen, W., *Solar Phys.* **46**, 219 (1976).
Tindo, I. P. and Somov, B. V., in K. van der Hucht and G. S. Vaiana (eds), *New Instrumentation for Space Astronomy*, Pergamon Press, Oxford (1978).
Uchida, Y., *Solar Phys.* **39**, 431 (1974).
Underwood, J. H., Antiochos, S. K., Feldman, U. and Dere, K. P., *Astrophys. J.* **224**, 1917 (1978).
Vaiana, G. S. and Giacconi, R., in D. G. Wentzel and D. A. Tidman (eds), *Plasma Instabilities in Astrophysics*, p. 91 (1969).
Valnícek, B., *Bull. Astron. Inst. Czech.* **15**, 207 (1964).
Valnícek, B., *IAU Symp.* **35**, 282 (1968).
Van Beek, H. F., de Feiter, L. D. and De Jager, C., *Space Res.* **14**, 447 (1974).
Van Hollebeke, M. A. I., Ma Sung, L. S. and McDonald, F. B., *Solar Phys.* **41**, 189 (1975).
Van Tend, W., Thesis, University of Utrecht, The Netherlands (1979).
Vernazza, J. E. and Mason, H. E., *Astrophys. J.* **226**, 720 (1978).
Vorpahl, J. A., *Solar Phys.* **26**, 397 (1972).
Vorphal, J. A., *Solar Phys.* **28**, 115 (1973).
Vorpahl, J. A., *Astrophys. J.* **205**, 868 (1976).
Vorpahl, J. A. and Zirin, H., *Solar Phys.* **11**, 285 (1970).
Vorpahl, J. A., Gibson, E. G., Landecker, P. B., McKenzie, D. L. and Underwood, J. H., *Solar Phys.* **45**, 199 (1975).
Vorpahl, J. A., Tandberg-Hanssen, E. and Smith, J. B., *Astrophys. J.* **212**, 550 (1977).
Wang, H. T. and Ramaty, R., *Solar Phys.* **36**, 129 (1974).
Wang, H. T. and Ramaty, R., *Astrophys. J.* **202**, 532 (1975).
Warwick, C. S., *Astrophys. J.* **145**, 215 (1966).
Webb, D. F., Krieger, A. S. and Rust, D. M., *Solar Phys.* **48**, 159 (1976).
Webb, D. F., McIntosh, P. S., Nolte, J. T. and Solodyna, C. V., *Solar Phys.* **58**, 389 (1978).
Webb, D. F., Cheng, C. C., Dulk, G. A., Edberg, S. J., Martin, S. F., McKenna-Lawlor, S. and McLean, D. J., in P. A. Sturrock (ed.), *Proc. of Skylab Workshop on Solar Flares*, Appendix B, in press (1979).
Weiss, A. A. and Stewart, R. T., *Australian J. Phys.* **18**, 143 (1965).
Widing, K. G. and Cheng, C. C., *Astrophys. J.* **194**, L111 (1974).
Widing, K. G. and Dere, K. P., *Solar Phys.* **55**, 431 (1977).
Wild, J. P., in D. G. Wentzel and D. A. Tidman (eds.), *Plasma Instabilities in Astrophysics*, p. 119 (1968).
Wild, J. P., *Solar Phys.* **9**, 260 (1969).
Wild, J. P. and McCready, L. L., *Australian J. Sci. Res.* **A3**, 387 (1950).
Wild, J. P. and Smerd, S. F., *Ann. Rev. Astron. Astrophys.* **10**, 159 (1972).
Wild, J. P., Roberts, J. A. and Murray, J. D., *Nature* **173**, 532 (1954).

Wild, J. P., Sheridan, K. V. and Noylan, A. A., *Australian, J. Phys.* **12**, 369 (1959).
Wild, J. P., Sheridan, K. V. and Kai, K., *Nature* **218**, 536 (1968).
Withbroe, G. L., *Astrophys. J.* **225**, 641 (1978).
Wood, A. T. and Noyes, R. W., *Solar Phys.* **24**, 180 (1972).
Yajima, S., *Tokyo Astr. Bull.* No. 207 (1971).
Yoshimura, H., Tanaka, K., Shimizu, M. and Hiei, E., *Publ. Astron. Soc. Japan* **23**, 443 (1971).
Zirin, H., *Solar Phys.* **26**, 393 (1972).
Zirin, H., *Astrophys. J.* **140**, 1216 (1964).
Zirin, H., *Solar Phys.* **58**, 95 (1978a).
Zirin, H., *Astrophys. J.* **222**, L105 (1978b).
Zirin, H., Big Bear Solar Observatory Preprint No. 0181 (1979).
Zirin, H. and Lazareff, B., *Solar Phys.* **41**, 425 (1975).
Zirin, H. and Tanaka, K., *Solar Phys.* **32**, 173 (1973).
Zirin, H. and Werner, S., *Solar Phys.* **1**, 66 (1967).
Zirin, H., Pruss, G. and Vorpahl, J., *Solar Phys.* **19**, 463 (1971).
Zvereva, A. M. and Severny, A. B., *Izv. Krymsk. Astrofiz. Obs.* **41**, 97 (1970).
Zwickl, R. D., Roelof, E. C., Gold, R. E., Krimigis, S. M. and Armstrong, T. P., Preprint JHU/APL 77-12.

CHAPTER 3

Current Sheets

E. R. PRIEST

Applied Mathematics Department, The University, St. Andrews, Scotland

3.1 INTRODUCTION

Throughout most of the solar atmosphere, the length-scale L over which the magnetic field B varies is believed to be relatively large, say typically 1000–10,000 km. The corresponding current density

$$j \approx \frac{B}{\mu L} \tag{3.1}$$

is so small that the magnetic field is "frozen to the plasma" (§1.4.1). However, current sheets have widths (l) much smaller than L and possess much larger current densities so that the magnetic field may slip quickly through the plasma. The energy that is released appears partly as plasma kinetic energy due to acceleration by magnetic forces, partly as heat by ohmic dissipation and partly as the energy of fast particles accelerated by the sheet's electric field (§§8.11–8.20).

Most current sheets are likely to be rather transient in nature. Small ones are continually formed throughout the upper atmosphere in response to photospheric motions, and the energy that is liberated in them may well heat the corona. Larger current sheets inside active regions are probably the sites of solar flares.

Suppose the faces of a current sheet have an area of $\mathscr{L}^2$ m^2 and suppose magnetic flux of strength B is carried across the faces into the sheet, where it is converted continuously into heat and plasma kinetic energy. If the field originally occupies two cubes of side $\mathscr{L}$ either side of the sheet, it contains magnetic energy of amount

$$W = \mathscr{L}^3 B^2/\mu.$$

For example, if we take $\mathscr{L} = 20{,}000$ km, and $B = 5 \times 10^{-2}$ T (500 Gauss), the *energy release* W is 2×10^{25} J (2×10^{32} erg), which is typical of a large flare.

Furthermore, suppose the speed V with which magnetic flux is carried into the sheet is 20 km s^{-1} (about one five-hundredth of the Alfvén speed for a number density of 10^{16} m^{-3}). Then the resulting *time for energy release* is

$$\tau = \frac{\mathscr{L}}{V},$$

which has a value of 10^3 sec, characteristic of the flare duration. For a steady-state conversion process, the *width l of the sheet* must be

$$l = \frac{\eta}{V},$$

so that V equals the diffusion speed η/l (§3.3.1); with $T = 10^4$ K, for example, and $V \doteq 20$ km s^{-1}, we find $l = 0.04$ m, which implies a current density $B/(\mu l)$ of 10^6 A m^{-2}.

The above order of magnitude estimates indicate why a current sheet has been a popular candidate for the energy source of a large flare. If one takes instead values of 2,000 km for $\mathscr{L}$ and 2 km s^{-1} for V, W becomes 2×10^{22} J (2×10^{29} erg), typical of a small flare; the time-scale remains as 10^3 sec, while the sheet width becomes 0.4 m and the current density 10^5 A m^{-2}.

The object of the present chapter is to go beyond these order of magnitude considerations in several ways: to demonstrate how current sheets may be formed, both in X-type configurations and closed loops or arcades (§3.2); to describe the behaviour of current sheets in detail (§3.3); and to show the role of current sheets in modern models for solar flares (§3.4, §3.5).

A current sheet may be defined as a non-propagating boundary between two plasmas, with the dominant magnetic field component tangential to the boundary. In other words, it may be regarded loosely as a *tangential discontinuity* without any flow. The tangential field components are arbitrary in magnitude and direction, subject only to the condition that the total pressure be continuous (for transverse equilibrium). In other words,

$$p_2 + \frac{B_2^2}{2\mu} = p_1 + \frac{B_1^2}{2\mu}, \tag{3.2}$$

where subscripts 1 and 2 denote conditions on the two sides of the current sheet. (If the sheet is curved significantly, additional magnetic tension terms need to be included.)

Inside active regions the magnetic field is so strong that for many purposes the plasma pressures p_1 and p_2 outside the current sheet may be neglected. Then (3.2) implies that the magnetic field strength must be the same on both

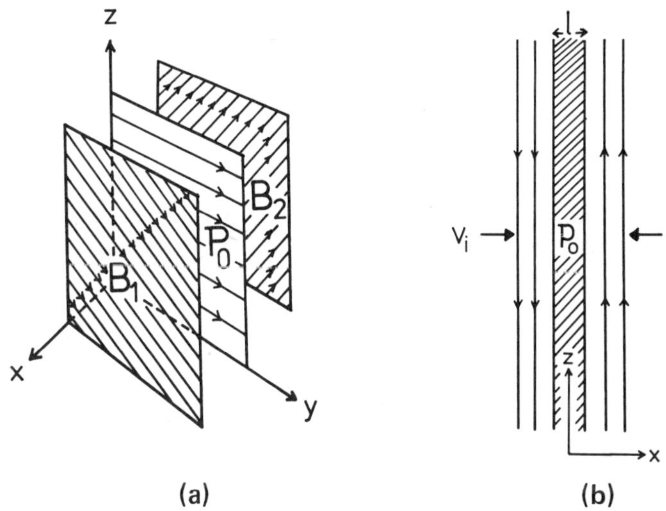

FIGURE 3.1 (a) A current sheet across which the magnetic field rotates from $\mathbf{B}_1$ to $\mathbf{B}_2$. (b) A neutral current sheet, in whose centre the magnetic field vanishes and the plasma pressure is p_0.

sides of the sheet, while the field directions may differ (Figure 3.1a). Suppose the x-axis is taken normal to the plane of the sheet and the y-axis bisects the angle between the two field directions. Then, as one passes through the current sheet, B_y may remain constant, while B_z reverses its direction; at the centre of the sheet, the plasma pressure is enhanced by an amount

$$p_0 = \frac{B_{z1}^2}{2\mu} = \frac{B_{z2}^2}{2\mu}.$$

In the particular case when B_y vanishes we have a *neutral* current sheet (or *neutral sheet*); the magnetic field vanishes completely in the centre of such a sheet and the field reverses its direction from one side of the sheet to the other (Figure 3.1b). It is the *neutral* current sheet that is, for simplicity, most often analysed.

A current sheet is rather like a *shock wave*, in the sense that it may be regarded as a discontinuity separating two regions where the equations of ideal magnetohydrodynamics hold; also, its width and the details of its interior are determined by diffusive processes. However, the similarity ends there. Unlike shocks, current sheets do not propagate; they tend to diffuse away in time and jets of plasma are squirted from their ends at Alfvénic speeds.

3.1.1 Formation of current sheets (summary)

There are four ways in which a current sheet may be formed.

1) The region near an X-type neutral point can collapse and form into a current sheet (§3.2.1).

2) If topologically-separate parts of a magnetic configuration are pushed together, a current sheet can appear at the boundary between them (§3.2.3). This may give rise to a small (simple-loop) flare when either a complex active region field is evolving or new flux is emerging (§3.4.1, §3.5.1).

3) Current sheets may develop when a magnetohydrostatic equilibrium ceases to exist (i.e. non-equilibrium). In particular, as the photospheric footpoints of a complex coronal force-free field move, the coronal field cannot always adjust to a new force-free equilibrium; instead a current sheet forms and magnetic energy is rapidly released, a process called *topological dissipation* by Parker (§3.2.2).

4) When a magnetic configuration is evolving slowly through a series of equilibria, it may reach a point where the equilibrium becomes unstable; during the instability, high magnetic gradients may develop, so that current sheets (or sheaths or filaments) form. The instability may be thermal, ideal or resistive in nature (§3.2.4).

3.1.2 Properties of current sheets (summary)

1) In the absence of flow, a current sheet will *diffuse away* at a speed η/l, where η is the magnetic diffusivity. The magnetic field is *annihilated* and magnetic energy converted into heat by ohmic dissipation (§3.3.1).

2) The region outside a current sheet is effectively "frozen" to the plasma. Plasma and magnetic flux may be brought towards the sheet from the sides at speed v_i, say. If v_i is less than η/l the sheet expands, whereas if v_i exceeds η/l the sheet will become thinner. When $v_i=\eta/l$ a steady state is maintained (Figure 3.1b).

3) The enhanced plasma pressure in the centre of the sheet expels material from the ends of the sheet at the Alfvén speed v_A. Magnetic flux is ejected with the material and so one effect of the sheet is to *reconnect the field lines* (Figure 3.2). The centre of the sheet is an X-type neutral point. In *steady flow*, the rate at which magnetic flux is transported remains constant; in other words, the rate v_iB_i at which flux enters the sheet equals the rate v_AB_0 at which it leaves, where subscripts i and o denote input and output values, respectively. Thus, for subAlfvénic inflow ($v_i<v_A$), the outflow field-strength,

$$B_o=\frac{v_i}{v_A}B_i,$$

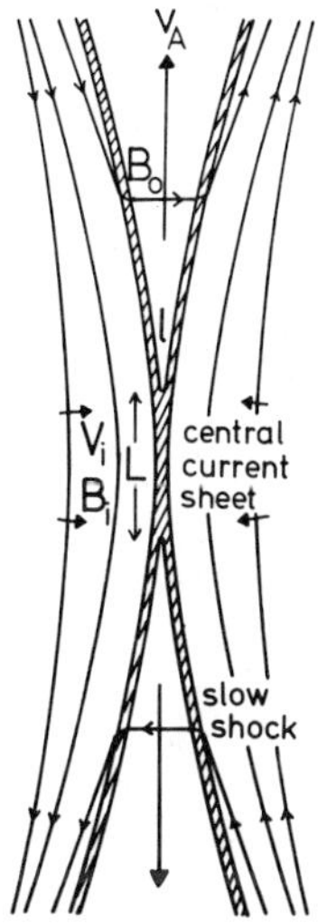

FIGURE 3.2 The reconnection of magnetic field lines by their passage through a current sheet.

is smaller than the inflow field-strength B_i. An important effect of a current sheet is therefore to *convert magnetic energy* into *heat and flow energy.*

4) Pairs of *slow-mode shock waves* propagate from the ends of the current sheet and in a steady flow they remain as standing waves. The dimensions of the central current sheet are given by

$$l=\frac{\eta}{v_i},$$

$$L=\frac{\rho_o v_o}{\rho_i v_i}l,$$

the second equation being an expression of mass conservation. *Magnetic reconnection may occur for a wide range of inflow speeds* v_i; as v_i is varied, so the dimensions of the sheet respond. There is a *maximum inflow speed*, which lies typically between 0.1 v_A and 0.01 v_A and is weakly-dependent on the magnetic Reynolds number (§3.3.4).

5) Consider an equilibrium current sheet with a unidirectional field $B(x)\hat{\mathbf{y}}$ reversing at $x=0$ and a uniform total pressure

$$p(x)+\frac{B(x)^2}{2\mu}=\text{constant}.$$

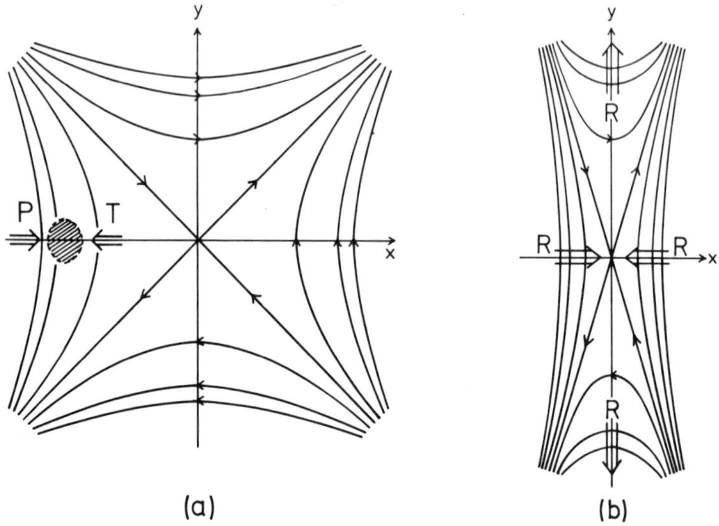

FIGURE 3.3 The magnetic field lines near an X-type neutral point (a) in equilibrium with no current and (b) away from equilibrium with a uniform current.

Such a sheet is subject to the *tearing-mode instability* on a time-scale of typically the geometric mean $(\tau_d \tau_A)^{1/2}$ of the diffusion time τ_d and the Alfvén travel-time τ_A (§3.3.3). In the nonlinear development, one is likely to find (depending on the boundary conditions) either a state of quasi-steady magnetic reconnection or the creation of stationary loops which slowly diffuse away.

3.2 MEANS OF FORMATION

3.2.1 X-type collapse

An X-type neutral point in a magnetic configuration tends to be locally unstable, provided the sources of the field are free to move (Dungey, 1953). This may be demonstrated by considering the equilibrium, current-free field

$$B_x = y, \qquad B_y = x, \tag{3.3}$$

near an X-type neutral point (0, 0). The field lines are the hyperbolae

$$y^2 - x^2 = \text{constant},$$

as shown in Figure 3.3a; in sketching these field lines, we have taken account of the fact that the field strength increases with distance from the origin, so that the hyperbolae are situated successively closer to one another. Any element of plasma, such as the one shown near the x-axis, experiences a resultant magnetic tension T due to the outwardly-curving field lines. It acts outwards from the origin and is exactly balanced by a magnetic pressure force P, which acts inwards because the magnetic field strength weakens as one approaches the origin.

Now suppose the field (3.3) is distorted to the form

$$B_x = y, \qquad B_y = \alpha^2 x,$$

where $\alpha^2 > 1$. The corresponding field lines are given by

$$y^2 - \alpha^2 x^2 = \text{constant},$$

and are sketched in Figure 3.3b. The limiting field lines ($y = \pm \alpha x$) through the origin are no longer inclined at $\frac{1}{2}\pi$, but have closed up a little like a pair of scissors. On the x-axis, the field lines are more closely spaced than those in Figure 3.3a, so that the magnetic pressure has increased; but they also have smaller curvature than before, so that the magnetic tension has increased less than the pressure. The dominance of the pressure therefore produces a resultant force R acting inward. On the y-axis, the field lines have the same spacing as in Figure 3.3a, but they are more sharply curved, so that the pressure force remains the same while the tension force increases; the resultant force R therefore acts outward as shown. These comments are borne out by evaluating the current density

$$j_z = \frac{\alpha^2 - 1}{\mu}$$

and the resulting Lorentz force

$$\mathbf{j} \times \mathbf{B} = -\frac{(\alpha^2 - 1)\alpha^2 x}{\mu}\hat{\mathbf{x}} + \frac{(\alpha^2 - 1)y}{\mu}\hat{\mathbf{y}}.$$

The magnetic force is therefore such as to increase the perturbation by closing up the scissors even more, which means that the equilibrium (3.3) is unstable (provided that conditions at distant boundaries allow such a displacement). As the instability proceeds, α increases and the limiting field lines through the origin close up, so that the current density and ohmic heating j^2/σ also increase.

The instability may be demonstrated formally by showing that the linearized form of the time-dependent equations

$$\frac{\partial \mathbf{B}}{\partial t} = \text{curl}\,(\mathbf{v} \times \mathbf{B}),$$

$$\rho \frac{\mathrm{D}\mathbf{v}}{\mathrm{D}t} = \text{curl}\,\mathbf{B} \times \frac{\mathbf{B}}{\mu},$$

$$\frac{\mathrm{D}\rho}{\mathrm{D}t} = -\rho \boldsymbol{\nabla}.\mathbf{v},$$

possess solutions of the form

$$B_x = B_o(1 + \varepsilon \mathrm{e}^{\omega t})y/l,$$

$$B_y = B_o(1 - \varepsilon \mathrm{e}^{\omega t})x/l,$$

$$v_x = \varepsilon u_o \mathrm{e}^{\omega t} x/l,$$

$$v_y = -\varepsilon u_o \mathrm{e}^{\omega t} y/l,$$

$$\rho = \rho_o,$$

where B_o, l, ρ_o, $u_o(=B_o/(\mu\rho_o)^{1/2})$, $\varepsilon(\ll 1)$ are constants and the growth-rate is

$$\omega = \frac{2u_o}{l}.$$

The two-dimensional collapse of the X-type configuration towards a sheet was later investigated more fully by others, who included both the effect of pressure gradients and the nonlinear time-development. Chapman and Kendall (1963) found an exact incompressible solution, whereas Imshennik and Syrovatsky (1967) discovered a self-similar, compressible solution, which indicates that the collapse takes some finite multiple of the time (l/u_o) for an Alfvén wave to cross the system. Syrovatsky (1966) has included the idea in a flare model, as outlined below (see also §1.4.3).

Limitations of the above local analysis arise from the effect of line-tying at the sources, so Syrovatsky (1966, 1969) attempted an ingenious, analytic, non-local treatment. He starts with the potential field $\mathbf{B}_0 = \text{curl}\ \mathbf{A}_0$, where

$$\mathbf{A}_0 = \tfrac{1}{2}h_0(x_0^2 - y_0^2)\hat{z}, \tag{3.4}$$

and zero subscripts refer to the initial state. He studies the magnetic field evolution that results from an approach of the two sources by a distance δ at a speed v_0 such that

$$v_S \ll v_0 \ll v_A, \tag{3.5}$$

where v_S is the sound speed. The object is to obtain values for the current density j and number density n near the neutral point. The distance

$$r_s = \frac{(2\mu n_0 kT)^{1/2}}{h}$$

at which the gas pressure ($n_0 kT$) and magnetic pressure are equal is assumed to be much smaller than the initial separation (unity) of the sources. Also, the plasma is supposed frozen to the field lines, so (3.5) implies that, at distances larger than r_s, motions along field lines at a speed v_S can be neglected, while Alfvén waves propagate rapidly across the system and set up a series of equilibria. The equilibria are characterised by a vanishing current density and develop by means of displacements perpendicular to the field lines. Thus for $1 \gg r \gg r_s$ the new field is the same as the initial one but with the field lines displaced, as indicated by the dashed arrows in Figure 1.7. The new magnetic vector potential is

$$\mathbf{A} = \tfrac{1}{2} h_0 (x^2 - y^2 + 2\delta)\hat{\mathbf{z}}, \tag{3.6}$$

in terms of the new coordinates (x, y) of a particle originally at (x_0, y_0). Equating (3.4) and (3.6) gives

$$x_0^2 - y_0^2 = x^2 - y^2 + 2\delta, \tag{3.7}$$

while the fact that the displacements are along rectangular hyperbolae normal to the field lines implies

$$x_0 y_0 = xy. \tag{3.8}$$

Thus, for example, the field line which initially crosses the x-axis at $x_0 = (2\delta)^{1/2}$ moves over to the position occupied initially by the separatrix through the origin. Then the equation of mass continuity implies that

$$\frac{n(x, y)}{n_0} = \frac{\partial(x_0, y_0)}{\partial(x, y)},$$

where (x, y) is given in terms of (x_0, y_0) by (3.6) and (3.8). In the limit $r_s^2 \ll r^2 \ll \delta$, it yields

$$\frac{n}{n_0} = \frac{r^2}{2\delta} \ll 1. \tag{3.9}$$

Also, one finds $n/n_0 \gg 1$ near $y = \pm(2\delta)^{1/2}$, giving a strong compression, as indicated by the shading in the figure.

Within a distance r_s of the origin, the above analysis fails because currents are generated and there are significant motions along magnetic field lines. But, if the field lines remain frozen to the plasma, there must be no reconnection and so they become distorted as shown in Figure 1.7. The current density near the neutral point is estimated by supposing that all the magnetic flux originally crossing the x-axis between 0 and $(2\delta)^{1/2}$ becomes compressed into the region $0 \leqslant x \leqslant r_s$ to a strength B_{comp}, say. In other words,

$$[h_0(2\delta)^{1/2}](2\delta)^{1/2} = B_{\text{comp}} r_s,$$

which gives the resulting current density as

$$j \equiv \frac{B_{\text{comp}}}{\mu r_s} = \frac{h_0 2\delta}{\mu r_s^2}. \tag{3.10}$$

Thus, if the original displacement of the sources is so large that $2\delta \gg r_s^2$, we have a large current density by comparison with the typical value h_0/μ and also, by (3.9), a very strong rarefaction at $r = r_s$. The resulting large value for j/n is the basis for Syrovatsky's 'dynamic dissipation' theory for flares, as proposed in (1966) and developed in later papers (Syrovatsky, 1969, 1970, 1972). He explains three stages of the flare process as follows.

Initial or soft phase (characterized by the appearance of local Hα brightenings, filament activation and a rise in soft X-ray emission). This phase is the result of a sufficiently rapid ($v_0 \gg v_s$) movement by a large enough distance δ (see (3.11)) of the sources of a magnetic field which was previously current-free and contained a neutral line N. (Equivalently, the dipole moment of one of the sources may increase in value.) An Alfvén wave is generated by the motion of the sources and, after typically a few minutes, it converges on N. When it reaches the neutral line, a current sheet of length $2\sqrt{\delta}$ develops, as described above (Figure 1.7). Three (shaded) regions of compressed plasma are formed – one just beyond each end of the sheet (due to the squeezing of plasma away from N) and one at N. If δ is large enough, the electron conduction speed $u \equiv j/(ne)$ at a distance $r = r_s$ exceeds the electron thermal speed $(kT_e/m_e)^{1/2}$.

Having substituted for j and n from (3.10) and (3.9), respectively, this condition may be written

$$\delta^2 > (n_0 r_s^4 \mu)/(4h_0)(kT_e/m_e)^{1/2}. \quad (3.11)$$

Ion-sound turbulence develops inside the current sheet. The enhanced electrical resistivity leads to a heating of the plasma, which produces the Hα brightenings and soft X-ray emission. The critical value for u, namely $(kT_e/m_e)^{1/2}$, holds when the electron and ion temperatures are the same ($T_e \approx T_i$); we have used it rather than Syrovatsky's choice of the ion-sound speed $(kT_e/m_i)^{1/2}$, which is relevant only when $T_e > T_i$. If a simple bipolar region surrounded by a uniform horizontal field increases its dipole moment, the resulting current sheet will be horizontal. Syrovatsky identifies the dense regions at the ends of the sheet with the 'two ribbons' of a flare. But this seems unlikely because the dense regions are ejected too rapidly after the initial formation process, namely at the Alfvén speed; also, two lines of sunspots would need to increase their moments simultaneously for the picture to hold in three dimensions.

Explosive, flash, Y-, hard or impulsive phase (describing a simultaneous rapid increase in Hα area and intensity, a movement apart of the two ribbons and a hard X-ray burst). According to Syrovatsky, this is produced by 'dynamic dissipation', which is the direct transfer of energy from the magnetic field to high energy particles. It is said to occur when the current sheet is ruptured, due to either a tearing-mode instability or the attainment by the electron conduction velocity of a relativistic value. A strong impulsive electric field at the site of rupture accelerates particles along the neutral line by the runaway effect.

Particle acceleration is considered to be the primary mechanism. The production of X-ray, optical and radio emission is a secondary effect which is due to the accelerated particles. The Hα emission results from a flow of heat during the soft phase and of highly accelerated particles during the hard phase.

With an initial source separation of 100,000 km and typical values $n_0 = 10^{17}$ m^{-3}, $T = 10^4$ K, $v_s = 10^4$ ms^{-1}, $h_0 = 100$ G m^{-1}, he finds $r_s = 20$ km and an onset for dynamic dissipation if $\delta > 100$ km. The amount of magnetic energy supplied to fast particles is $h_0^2 \delta^2/\mu = 10^{24}$ J; it is contained in the field $B_{comp} = h_0 \delta / r_s = 500$ G occupying an area r_s^2. He claims that 10^{34} electrons can be accelerated to 100 MeV. Ranges for these and other parameters are given by Syrovatsky (1970).

Hot, decay or main phase (during the greater part of which both the flare area and intensity decrease). The hot region, which formed in the previous phase, evolves and cools. Turbulent heating in the current sheet may continue.

The appeal of Syrovatsky's mechanism is the directness of the conversion of magnetic energy to particle energy. Also, when it first appeared, it seemed the only way of releasing energy fast enough, since a quasi-steady current sheet seemed too slow (Parker, 1963), although that is no longer believed to be the case (see §3.3.4). But there have been a number of serious criticisms.

Rather than being stored before the flare, the magnetic energy is assumed by Syrovatsky to be propagated up towards the neutral line from below during the initial phase. But the observations do not indicate large enough changes in the photospheric magnetic field. Also the ion-acoustic turbulence, which occurs in the initial stage when u exceeds the electron thermal speed, would lead to a diffusion of the plasma across the magnetic field efficiently enough to prevent u from becoming relativistic (Smith and Priest, 1972).

But a still more fundamental criticism was given by Anzer (1973), who claims that j/n does not become large enough for dynamic dissipation, since there is neither an extreme density rarefaction nor a strong magnetic field compression. First of all, he points out that the flow field used to calculate n assumes complete reconnection and so is valid only far from the neutral line ($r > (2\delta)^{1/2}$). It is irrelevant near the current sheet. He therefore constructs the density more carefully and concludes that there is either an increase in n or a small decrease, but certainly not a large decrease. Secondly, according to (3.10), the magnetic field strength in the compressed region ($0 < x < r_s$) is

$$B_{\text{comp}} = \frac{2h_0\delta}{r_s},$$

whereas that outside the compressed region ($r_s < x < 1$) is

$$B_{\text{ext}} = h_0(2\delta)^{1/2}.$$

Thus, when (as assumed by Syrovatsky) $2\delta \gg r_s^2$, the ratio $B_{\text{comp}}/B_{\text{ext}} = (2\delta)^{1/2}/r_s \gg 1$ and the resulting restoring force severely limits the magnetic compression. Anzer argues on largely physical grounds, but a more recent numerical simulation by Syrovatsky (1978a) has shown that a quasi-steady current sheet (§3.3.4) does develop and that n can decrease and lead to values of u large enough to trigger turbulence.

3.2.2 Non-equilibrium

Current sheets can be formed when a magnetostatic equilibrium ceases to exist at all, a situation known as *non-equilibrium*. If the photospheric footpoints of a *simple bipolar magnetic field* move slowly, the low-β corona responds by

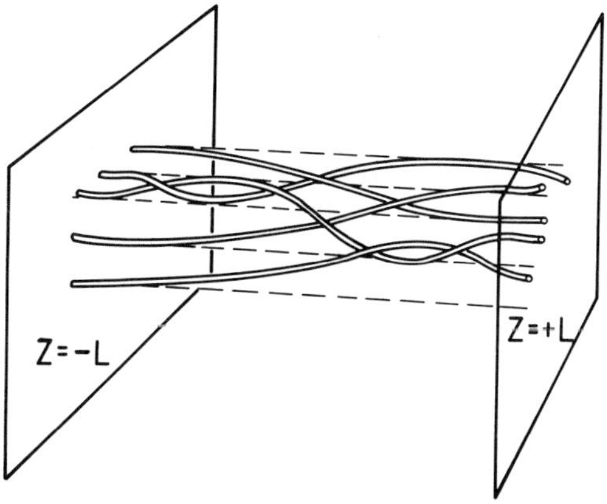

FIGURE 3.4 Schematic drawing of the topology of magnetic tubes of force following a displacement of the ends of the tubes where they intersect $z=+L$. (From Parker, 1972.)

establishing a series of force-free configurations. In general, however, the coronal magnetic field is much more complex than this and contains topologically distinct flux systems. Parker (1972) and Syrovatsky (1978b) have demonstrated that, as the footpoints of such a magnetic field move, the *corona cannot adjust to a new force-free equilibrium* and current sheets are formed instead. These current sheets are themselves not in equilibrium, since they allow a rapid reconnection at some fraction of the Alfvén speed (§3.2.3) and the magnetic configuration reduces to the state of lowest potential energy. Parker referred to such a process as *topological dissipation.* Continual footpoint motion means that the coronal field is all-the-time responding by reconnecting and converting magnetic energy into heat in the process. This may be the means whereby the corona is heated, especially inside active regions, but it may also produce subflares.

Parker (1972) establishes that, *if the pattern of small-scale variations is not uniform along a large-scale field, then the field cannot be in magnetostatic equilibrium.* In other words, equilibrium exists only if the field variations consist of a simple twist extending from one footpoint to another. More complex topologies, such as braided flux tubes with several field lines wrapped around each other, are not in equilibrium. To obtain his result, Parker considers a uniform configuration, with plasma pressure p_0, and magnetic field, $B_0\hat{\mathbf{z}}$, and with the footpoints anchored at the planes $z=\pm L$(Figure 3.4). Suppose that a displacement of the footpoints by at most $\lambda\,(\ll L)$ leads to small deviations $\mathbf{B}_1$ and p_1 from the uniform field and pressure, respectively, so that

$$\mathbf{B}=B_0\hat{\mathbf{z}}+\mathbf{B}_1(x,\ y,\ z),$$

$$p=p_0+p_1(x,\ y,\ z),$$

where $B_1/B_0 \sim p_1/p_0 \ll 1$. Then the equation

$$-\boldsymbol{\nabla} p + (\boldsymbol{\nabla} \times \mathbf{B}) \times \mathbf{B}/\mu = \mathbf{0}$$

for magnetostatic equilibrium becomes, to first order,

$$-\boldsymbol{\nabla}\left(p_1 + \frac{B_0 B_{1z}}{\mu}\right) + \frac{B_0}{\mu}\frac{\partial \mathbf{B}_1}{\partial z} = \mathbf{0}. \tag{3.12}$$

But the basic equation $\boldsymbol{\nabla} . \mathbf{B} = 0$ gives

$$\boldsymbol{\nabla} . \mathbf{B}_1 = 0,$$

and so the divergence of (3.12) implies

$$\nabla^2\left(p_1 + \frac{B_0 B_{1z}}{\mu}\right) = 0. \tag{3.13}$$

Now, variations of $(p_1 + B_0 B_{1z}/\mu)$ extend at most a distance of order λ ($\ll L$) into the region $-L < z < L$ from the end-planes $z = \pm L$, so the solution to Laplace's equation (3.13) is

$$p_1 + B_0 B_{1z}/\mu = \text{constant}, \tag{3.14}$$

save in boundary layers of width λ near $z = \pm L$. However, the z-component of (3.12) is

$$\frac{\partial p_1}{\partial z} = 0,$$

and hence the z-derivative of (3.14) implies

$$\frac{\partial B_{1z}}{\partial z} = 0.$$

Thus, if a field is in magnetostatic equilibrium, its pattern does not vary along the general direction of the field, which establishes Parker's result.

Green (1965) and Syrovatsky (1971) demonstrated that the *slow continuous deformation of a two-dimensional, potential magnetic field containing neutral points leads to the production of neutral current sheets in the perfectly conducting limit*. Consider the initial magnetic field

$$B_x = y, \qquad B_y = x,$$

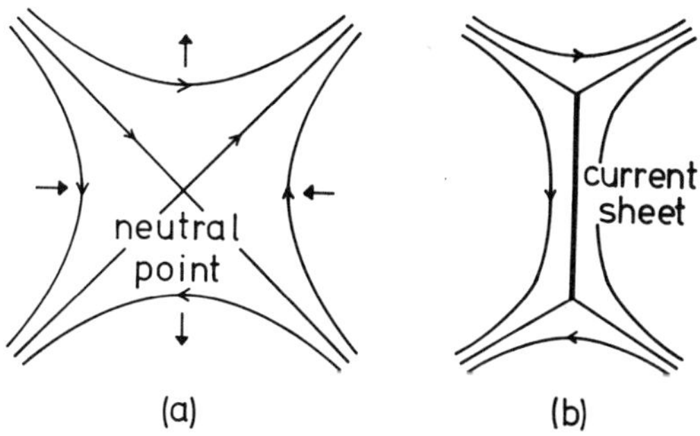

FIGURE 3.5 (a) A potential magnetic field near an X-type neutral point. (b) The field produced by the slow motion indicated in (a) by solid-headed arrows. The plasma is assumed perfectly conducting.

which may be written in terms of the complex variable

$$z = x + iy$$

as

$$B_y + iB_x = z \tag{3.15}$$

and which contains a neutral point at the origin (Figure 3.5). Next, suppose that this configuration is deformed by the imposition of a uniform electric field perpendicular to the z-plane; this drives a motion normal to the magnetic field. If the plasma is regarded as perfectly conducting, the field lines are carried with the motion and Ohm's Law may be written

$$E + v\,B = 0. \tag{3.16}$$

It is clear that the instantaneous plasma speed is determined by (3.16) everywhere except at the neutral point where B vanishes. Furthermore, we suppose that the flow speed and plasma pressure are so small that the configuration passes through a series of equilibria with vanishing Lorentz force, so that, in this two-dimensional situation,

$$0 = j\,B. \tag{3.17}$$

The peculiar role of the neutral point is evident in both (3.16) and (3.17), which imply that a continuous deformation of the original field (3.15) through a series of potential configurations with

$$j = 0$$

is possible everywhere save in the vicinity of the neutral point. At the neutral point itself, (3.17) allows non-zero currents. Indeed, the solution to the problem is that there develops a current sheet, as represented by a cut from $z=-iL$ to $z=+iL$, say, in the complex plane. The resulting magnetic field components are given by

$$B_y+iB_x=(z^2+L^2)^{1/2},$$

and the field is sketched in Figure 3.5. This complex-variable technique (for obtaining the position of the current sheet that forms as oppositely directed fields approach one another) has been extended to the case of approaching bipolar fields that are equal or unequal (§3.2.3). In the latter case, the sheet is curved. All these calculations use the assumption that the plasma remains frozen to the field everywhere, but in practice this approximation fails inside the current sheet. During the approach of the two flux systems, the current sheet bifurcates into two slow magnetoacoustic shocks and reconnection occurs in the manner described in §3.3.4. When the footpoint motion ceases, the magnetic configuration rapidly reduces (over an Alfvén travel-time) to its lowest energy state, namely a potential field.

Syrovatsky (1978b) generalised his previous result for potential fields to show that a *continuous deformation of a force-free field in general leads to the production of current sheets.* It is only for rather simple fields and simple footpoint motions that no current sheets are generated. Suppose a magnetic configuration evolves through a series of force-free states such that

$$(\nabla\times\mathbf{B})\times\mathbf{B}=\mathbf{0}, \tag{3.18}$$

with

$$\nabla\cdot\mathbf{B}=0.$$

Then, in general, this evolution cannot maintain the condition of frozen-in flux, since, given a time-sequence of magnetic fields satisfying (3.18), the solution of the induction equation

$$\frac{\partial\mathbf{B}}{\partial t}=\nabla\times(\mathbf{v}\times\mathbf{B}) \tag{3.19}$$

for $\mathbf{v}\times\mathbf{B}$ requires it to have (in general) a component parallel to $\mathbf{B}$. This is unacceptable since $\mathbf{v}\times\mathbf{B}$ must by definition be normal to $\mathbf{B}$. The difficulty in solving (3.19) for a given magnetic field shows up particularly clearly when a closed magnetic field line C exists. Simply integrate over a surface S bounded by such a C to give

$$\frac{\partial}{\partial t}\iint_S \mathbf{B}\,.\,\mathbf{dS} = \int_C (\mathbf{v}\times\mathbf{B}).\mathbf{ds}. \tag{3.20}$$

Then, for a given sequence of solutions to (3.18), there is in general no reason to suppose that the flux through S should remain constant. This implies that the left-hand side of (3.20) is non-zero, whereas the right-hand side must vanish (since $\mathbf{v}\times\mathbf{B}$ is normal to the field-line curve C).

3.2.3 Forced magnetic evolution

The collapse of an X-type neutral point to form a current sheet has not yet been studied completely. It involves the time-development of a two-dimensional solution of the magnetohydrodynamic equations. For the above analytic solutions (§3.2.1) one needs to impose rather severe simplifying assumptions. Furthermore, during the later stages, when the current density becomes so high that the local magnetic Reynolds number is no longer large, one needs to include resistive effects.

A complementary approach to the formation of current sheets has been taken by Priest and Raadu (1975). They suppose that the magnetic evolution is driven by the boundary conditions, rather than by the local instability near the neutral point. The boundary conditions are assumed to vary so slowly that the configuration passes through a series of equilibria. If the initial equilibrium contains an X-type neutral point, a current sheet forms and slowly grows in length; the surrounding field is assumed to be two-dimensional and potential. The magnetic field sources are two line dipoles and the analysis is a natural extension of the previous work by Green (1965) for two line currents (§3.2.2).

The dipoles have moment $(2\pi D/\mu)$ and are situated initially at $(a_0, 0)$ and $(-a_0, 0)$, as shown in Figure 3.6a. The components (B_{0x}, B_{0y}) of the resulting potential field may be written

$$B_{0y}+iB_{0x}=iD[(z+a_0)^{-2}+(z-a_0)^{-2}]. \tag{3.21}$$

Suppose the sources approach one another at a speed which is so fast that significant reconnection does not take place and that the field passes through a series of equilibrium states, with a current sheet stretching from $P(0, p)$ to $Q(0, q)$ when the dipole separation is $2a$, as shown in Figure 3.6b. Outside the current sheet, the magnetic field is current-free and so potential; its components in the combination (B_y+iB_x) are therefore an analytic function of z, namely

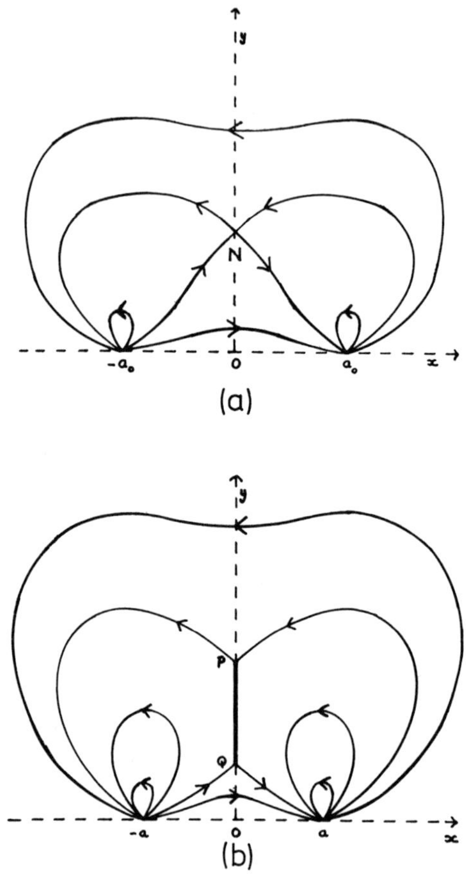

FIGURE 3.6 (a) The current-free (potential) magnetic configuration due to two line dipoles of equal moment, which are situated at $x = \pm a_0$ on the x-axis. An X-type neutral point, where the magnetic field strength vanishes, is located at N. (b) When the dipoles approach one another to $x = \pm a (a < a_0)$ and no reconnection is allowed, the resulting equilibrium field contains a current sheet PQ.

$$B_y + iB_x = \frac{E(z^2 + p^2)^{1/2}(z^2 + q^2)^{1/2}}{(z^2 - a^2)^2}, \tag{3.22}$$

where $E = 4iDa^2(a^2 + p^2)^{-1/2}(a^2 + q^2)^{-1/2}$ is chosen to give the correct behaviour near the sources. The values of p and q are found as functions of a by using the conditions of frozen-in flux, which imply that the fluxes crossing the positive y-axis above and below N in Figure 3.6a must equal those crossing above P and below Q, respectively, in Figure 3.6b. The fluxes can be found by performing the relevant integrals of (3.21) and (3.22) with respect to y at $x = 0$.

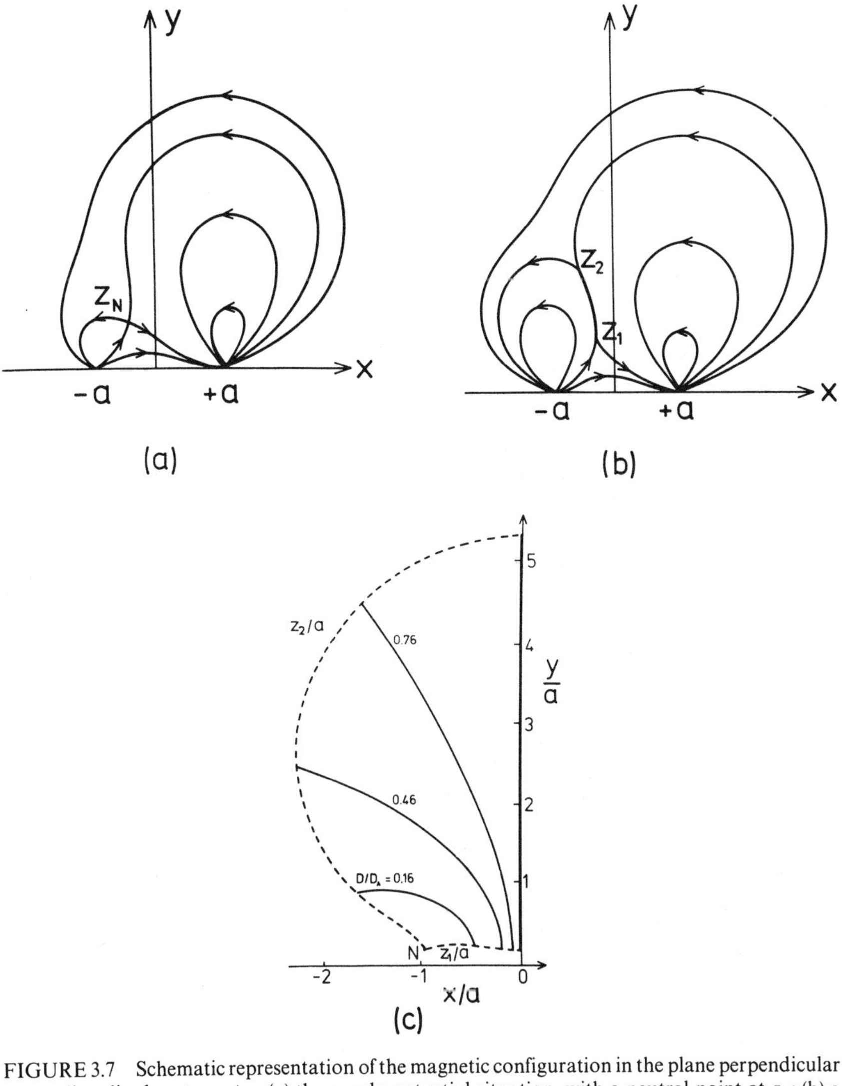

FIGURE 3.7 Schematic representation of the magnetic configuration in the plane perpendicular to two line-dipoles at $x = \pm a$: (a) the purely potential situation, with a neutral point at z_N; (b) a current sheet stretching between z_1 and z_2 after an increase in the moment of the smaller dipole; (c) the locus of the endpoints and a few examples of the shape of the sheet as the moment of the dipole at $x = -a$ increases from 0.01 D_A to D_A.

Tur and Priest (1976) have extended the analysis to include curved current sheets in more realistic configurations. For example, they modelled the current sheet which forms as a small bipolar flux system emerges into and presses up against a much larger bipolar field (Figure 3.7). The initial potential field has components given by

$$B_{0y}+iB_{0x}=\frac{iD_0}{(z+a)^2}+\frac{iD_A}{(z-a)^2},$$

where $2\pi D_A/\mu$ and $2\pi D_0/\mu$ are the moments for the larger ambient dipole and the smaller evolving dipole. The emergence of additional flux is simulated by increasing the strength of the smaller dipole while keeping its position fixed. The resulting magnetic field takes the form

$$B_y+iB_x=\frac{iE[(z-z_1)(z-z_2)(z-z_1^*)(z-z_2^*)]^{1/2}}{(z^2-a^2)^2},$$

where there is a branch cut from z_1 to z_2 and the stars denote complex conjugates. The values of E, z_1, z_2 are determined from the behaviour near $z=\pm a$ and flux-conservation conditions; an example of the resulting current sheet locations is shown in Figure 3.7c. It should be noted that a current sheet also forms if the new flux *evolves* (rather than emerges), so that the distance between the sources decreases while the moments remain constant. The energy is stored in the magnetic field surrounding the current sheet, rather than in the sheet itself; it increases monotonically as the dipole moment increases or the source separation decreases and can easily reach the values appropriate for a flare.

Another model example considered by Tur and Priest is the emergence of new flux (represented by a dipole) into a uniform horizontal field. The initial potential field is

$$B_{0x}-iB_{0y}=b+D_0/z^2.$$

When the dipole moment has increased from $2\pi D_0/\mu$ to $2\pi D/\mu$ the field becomes

$$B_x-iB_y=E[(z^2-z_1^2)(z^2-z_1{}^{*2})]^{1/2}/z^2,$$

where the end-points of the current sheet are z_1 and $z_2=-z_1^*$, as shown in Figure 3.8a. E and z_1 are determined by boundary conditions and flux conservation conditions. The resulting sheet is again curved, as indicated in Figure 3.8b.

One limitation of the above models incorporating complex variable techniques is that they are purely two-dimensional. This has been overcome by Tur (1977) following a suggestion by Raadu (private correspondence) for the case of the approach of symmetric dipoles (Figure 3.6), when in three-

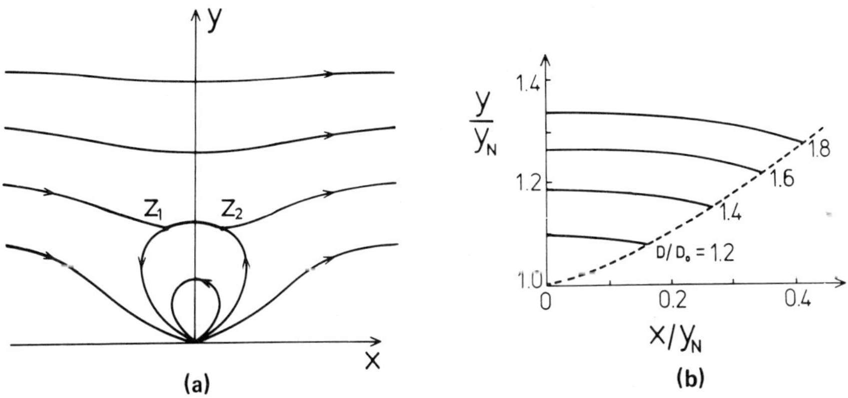

FIGURE 3.8 (a) Schematic representation of the magnetic field due to a dipole emerging into a uniform horizontal magnetic field, with a current sheet joining z_1 and z_2. (b) The position of one endpoint (dashed) and a few locations of the right-hand half of the sheet as the dipole moment increases.

dimensions the current sheet forms an annulus. In spherical polar coordinates (r, θ, φ) the dipoles (of moment $2\pi D/\mu$) lie at a distance α from the origin in the directions $\theta=0$ and $\theta=\pi$ (Figure 3.9a). The initial potential field $\mathbf{B}_0$ is written in terms of a potential Φ_0 as

$$\mathbf{B}_0 = -\nabla\Phi_0,$$

where

$$\Phi_0 = \frac{D\cos\theta_1}{r_1} + \frac{D\cos\theta_2}{r_2} \tag{3.23}$$

and the values r_1, r_2, θ_1, θ_2 (which may be written in terms of r and θ) are indicated in Figure 3.9a. When the two dipoles approach one another and the current sheet forms, the magnetic field is

$$\mathbf{B} = -\nabla\Phi,$$

where

$$\Phi = \Phi_s + \Phi_D,$$

with Φ_s arising from the current sheet and Φ_D from the dipole sources. The solution for Φ_s may be built up as an integral sum of the potentials arising from the individual current loops that may be considered to make up the current

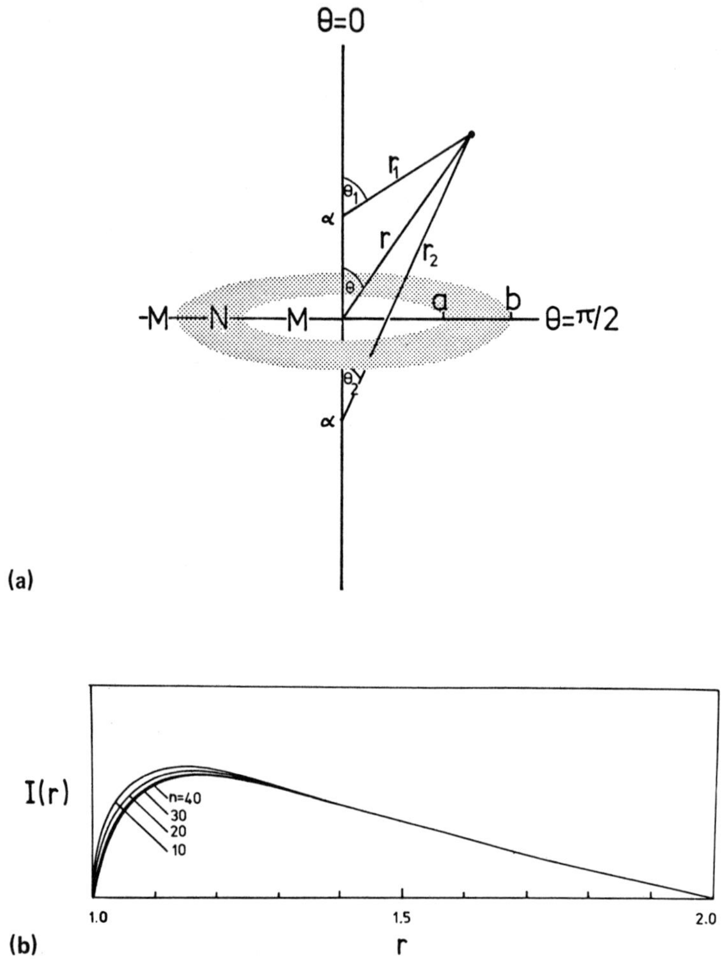

FIGURE 3.9 (a) An annular current sheet of radii a and b situated in the plane $\theta=\frac{1}{2}\pi$, with dipole sources at distances α above and below the plane. (b) The current-density function in an annular current sheet of radii $a=1$ and $b=2$ for $D=1$ and several values of n.

sheet. Thus, the potential at a point (r, θ) due to a circular current loop of radius ρ in the plane $\theta=\frac{1}{2}\pi$ may be written

$$\varphi(\rho)=\begin{cases}\varphi^{(-)}(\rho) & \text{for } r<\rho,\\ \varphi^{(+)}(\rho) & \text{for } r>\rho,\end{cases}$$

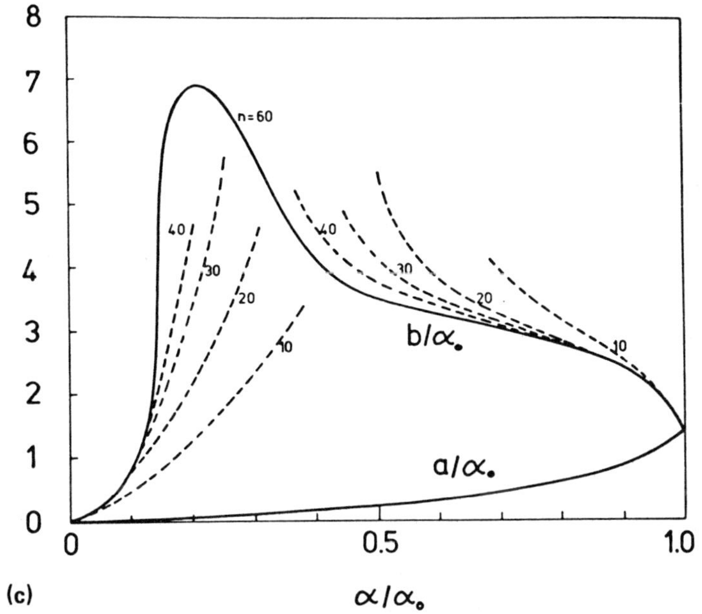

FIGURE 3.9 (c) The loci of the inner and outer radii $r=a$ and $r=b$ as a function of the separation 2α of the dipole sources.

where

$$\varphi^{(-)}(\rho)=2\pi I(\rho)\left[1-\sum_{n=0}^{\infty}(-1)^n\frac{(2n)!}{4^n(n!)^2}\left(\frac{r}{\rho}\right)^{2n+1}P_{2n+1}(\cos\theta)\right],$$

$$\varphi^{(+)}(\rho)=2\pi I(\rho)\sum_{n=1}^{\infty}(-1)^{n+1}\frac{(2n)!}{4^n(n!)^2}\left(\frac{\rho}{r}\right)^{2n}P_{2n-1}(\cos\theta).$$

Then the potential Φ_s due to the current annulus is formed by integrating over ρ from a to b, being careful to split the range of integration at $\rho=r$:

$$\Phi_s=\begin{cases}\displaystyle\int_a^b\varphi^{(-)}(\rho)d\rho, & r<a,\\ \displaystyle\int_a^r\varphi^{(+)}(\rho)d\rho+\int_r^b\varphi^{(-)}(\rho)d\rho, & a<r<b,\\ \displaystyle\int_a^b\varphi^{(+)}(\rho)d\rho, & r>b.\end{cases}\tag{3.24}$$

The unknown function $I(\rho)$ is determined by the boundary conditions on Φ_s at $\theta=\frac{1}{2}\pi$, which may be found by subtracting the boundary conditions on Φ and Φ_D as follows. In the annular region $a\leqslant r\leqslant b$, the net field is tangential, so that

$$\frac{1}{r}\frac{\partial\Phi}{\partial\theta}=0;$$

and from (3.23) the dipolar field has

$$\frac{1}{r}\frac{\partial\Phi_D}{\partial\theta}=\frac{2D(r^2-2\alpha^2)}{(r^2+\alpha^2)^{5/2}}.$$

Hence, subtraction gives

$$\left[\frac{1}{r}\frac{\partial\Phi_s}{\partial\theta}\right]_{\theta=\pi/2}=\frac{-2D(r^2-2\alpha^2)}{(r^2+\alpha^2)^{5/2}}.$$

Application of this condition to Φ_s in (3.24) yields the following integral equation for $I(\rho)$:

$$2\pi\int_a^r\sum_{n=1}^{\infty}(-1)^{n+1}\frac{(2n)!(2n+1)P_{2n+2}(0)}{4^n(n!)^2r}\left(\frac{\rho}{r}\right)^{2n}I(\rho)\mathrm{d}\rho-2\pi\int_r^b\sum_{n=0}^{\infty}(-1)^n$$

$$\frac{(2n)!(2n+1)P_{2n+2}(0)}{4^n(n!)^2r}\left(\frac{r}{\rho}\right)^{2n+1}I(\rho)\mathrm{d}\rho$$

$$=-\frac{2D(r^2-2\alpha^2)}{(r^2+\alpha^2)^{5/2}},\qquad(a\leqslant r\leqslant b).$$

This has been solved numerically (Figure 3.9b); one finds a maximum in the current density function close to the inner boundary $r=a$, because the field is stronger there.

Just as for the two-dimensional problem, flux conservation may be used to determine the inner and outer radii $a(\alpha)$ and $b(\alpha)$ as the dipoles approach one another with a separation 2α. A numerical solution is necessary, with a value of $n=60$ required for convergence (Figure 3.9c); the resulting current sheet is somewhat larger than in the two-dimensional case.

Sakurai and Uchida (1977) have calculated the potential fields from a set of dipoles. Distinct flux systems are separated by a surface known as a separatrix and so, if the polarity of one flux system is reversed, the field lines may occupy the same positions as before, while the separatrix becomes a current sheet separating oppositely-directed fields. This technique for constructing curved current sheets is used for several examples; it applies when there is no reconnection so that the sheets form closed surfaces enclosing each flux system.

3.2.4 Instability

Current sheets can develop during the course of an instability. One such instability is the collapse of an X-type neutral point, as described in §3.2.1, but there are several other relevant instabilities, such as those outlined below. It will be noticed that it is not necessary to have an X-type neutral point in the original equilibrium; current sheets can equally well be produced in magnetic flux loops and arcades.

3.2.4.*A* *Ideal instability* Ideal magnetohydrodynamic instabilities normally grow very fast, on the Alfvén time-scale l/v_A, where l is a typical dimension and v_A is the Alfvén speed. In some cases the current density in one location may grow indefinitely and produce a current singularity or current sheet, within which the ideal equations no longer hold. As an example, the kink instability of a twisted flux tube (§4.4) can, in its nonlinear development (Sakurai, 1976), produce current sheets, as the flux loop twists round on itself.

Furthermore, the instability of a magnetic flux loop (§4.4.8) or coronal arcade (§6.5.3) that is line-tied at its footpoints in the dense photosphere may lead to an eruption of the coronal part of the configuration. This produces an open field structure, which takes the form of a large-scale current sheet.

3.2.4.*B* *Thermal instability* A magnetic configuration that is non-potential, such as a sheared force-free structure, may become unstable in several ways, with the electric current concentrating into filaments. One mechanism is the tearing-mode instability (§3.2.4.C), and another is thermal instability (§8.19.2). Heyvaerts (1974) has described two such types of instability, namely the *Joule mode* and the *antidiffusion mode*. They both cause a uniform electric current to concentrate into many small *current threads* (or filaments) parallel to the magnetic field. The main conditions for their occurrence are that the temperature be about 10^5 K, so that the radiative loss function $Q(T)$ is approximately constant (§1.5), and that the perturbation wavenumber k be both so small that thermal conduction is swamped by Joule heating and also so large that a local stability analysis is valid.

For a disturbance propagating at an angle θ to the ambient magnetic field with $\gamma = 5/3$ and an electrical conductivity σ proportional to $T^{3/2}$, the *Joule mode* has a growth-rate

$$\omega = (\sin^2\theta - \cos^2\theta)\frac{j_0^2}{\sigma p_0}, \tag{3.25}$$

where j_0 is the equilibrium current density and p_0 the equilibrium pressure. It occurs most effectively for $\theta = \frac{1}{2}\pi$ and so forms fine structures aligned along the magnetic field and consisting of hotter current concentrations separated by cooler regions. The mode is, however, restricted to wavenumbers

$$k \gg (\omega\mu\sigma)^{1/2},$$

such that the plasma is not frozen to the field, and another limitation on applicability is that the current density needs to be rather large for the growth-rate to be reasonably high. For example, suppose we adopt $T_0 = 10^5$ K, $N_0 = 10^{15}$ m^{-3}, together with a turbulent conductivity that is a factor of 100 smaller than the Coulomb value and a current density $j_0 = 1$ A m^{-2} (corresponding to a magnetic field change of 100 Gauss (10^{-2} T) over 1,000 km). Then (3.25) gives a growth-rate of only 4×10^{-4} sec^{-1}.

The *antidiffusion mode* causes the magnetic flux to concentrate rather than diffuse and it occurs only for θ close to $\frac{1}{2}\pi$. Its growth-rate is

$$\omega = \frac{k^2}{\mu\sigma(\theta - \frac{1}{2}\pi)^2},$$

so that $\omega \gg k^2/(\mu\sigma)$ and the plasma is almost frozen to the field. A local increase in temperature enhances the electrical conductivity and current density, which in turn produces more heating and drives the instability.

3.2.4.C *Resistive instability* The magnetic field in a current sheet of width l diffuses through the plasma on time-scale

$$\tau_d = \frac{l^2}{\eta},$$

where $\eta = (\mu\sigma)^{-1}$ is the magnetic diffusivity (§3.3.1). During the process of magnetic diffusion, magnetic energy is converted ohmically into heat at the same rate. However, solar values of τ_d are enormous unless l is tiny (§1.4.1). Furth, Killeen and Rosenbluth (1963) showed how the diffusion can drive three distinct instabilities at a much faster rate. These instabilities occur provided the sheet is wide enough that

$$\tau_d \gg \tau_A,$$

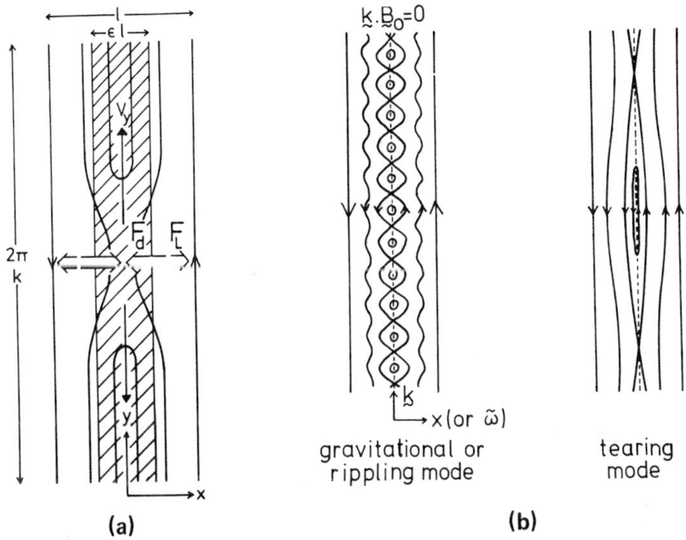

FIGURE 3.10 (a) Resistivity instability in a current sheet for which the driving force F_d exceeds the restoring force F_L. Significant diffusion takes place over a fraction ε of the width of the sheet. Here just one wavelength $2\pi/k$ is shown, but in practice there may be many such features end to end. (b) Small- and long-wavelength resistive instabilities in a current sheet or a sheared magnetic field.

where

$$\tau_A = \frac{l}{v_A}$$

is the time it takes to traverse the sheet at the Alfvén speed $v_A = B_0/(\mu\rho_0)^{1/2}$. They occur on time-scales $\tau_d(\tau_A/\tau_d)^{\delta}$, where $0<\delta<1$ (see below) and have the effect of creating in the sheet many small scale magnetic loops. In other words, resistive instabilities produce current filaments in current sheets (or in any sheared structure); subsequently, the filaments and associated magnetic loops diffuse away, releasing magnetic energy in the process.

During the instability, the magnetic field slips through the plasma in a region of width εl about the centre of the sheet, where the magnetic field vanishes. At the edge of this diffusion region the magnetic field is εB_0, a fraction ε of the value B_0 at the edge of the sheet (Figure 3.10a). If the velocity with which plasma enters the diffusion region is $-v_x\hat{\mathbf{x}}$, the resulting electric current is

$$\mathbf{j} \approx \sigma(\mathbf{v}\times\mathbf{B}) = \sigma v_x(\varepsilon B_0)\hat{\mathbf{z}};$$

and the Lorentz force $\mathbf{F}_L$ which opposes the flow is roughly

$$\mathbf{F}_L \equiv \mathbf{j} \times \mathbf{B} \approx -\sigma v_x (\varepsilon B_0)^2 \hat{\mathbf{x}}, \tag{3.26}$$

(where $v_x < 0$ for $x > 0$). Now, if an instability is to take place, this restoring force must be exceeded by some driving force $\mathbf{F}_d$, say, which has roughly the same magnitude as $\mathbf{F}_L$. The rate of working of $\mathbf{F}_d$ during the instability is therefore

$$-\mathbf{v} \cdot \mathbf{F}_L \approx \sigma v_x^2 (\varepsilon B_0)^2. \tag{3.27}$$

The effect of the driving force is to accelerate plasma (to a speed v_y) a distance $2\pi/k$ along the sheet, given that the instability has a wavenumber $k(\gg l^{-1})$ in this direction. For an incompressible plasma, the equation

$$\nabla \cdot \mathbf{v} = 0$$

implies that v_y is given in terms of the slow inflow speed $-v_x$ by

$$k v_y + (\varepsilon l)^{-1} v_x \approx 0,$$

or

$$v_y \approx -\frac{v_x}{k \varepsilon l}.$$

The rate of increase in kinetic energy over a time $(\omega)^{-1}$ is therefore

$$\omega \rho_0 v_y^2 \approx \frac{\omega \rho_0 v_x^2}{(k \varepsilon l)^2},$$

which, when equated to the work done driving the instability (Equation (3.27)), gives

$$(\varepsilon l)^4 = \frac{\omega \rho_0 l^2}{\sigma k^2 B_0^2},$$

or, in terms of the time-scales τ_d and τ_A,

$$\varepsilon^4 = \frac{\omega}{(kl)^2} \frac{\tau_A^2}{\tau_d}. \tag{3.28}$$

For two of the three resistive instabilities, one can equate (in order of magnitude) the particular driving force $\mathbf{F}_d$ to $-\mathbf{F}_L$ given by (3.26) to produce a relationship between ε, ω and k, which, together with (3.28), determines the dispersion relation for ω as a function of k. (A more rigorous analysis produces the same result.)

Gravitational mode

The *gravitational mode* exists when a gravitational (or equivalent) force $\rho g\hat{\mathbf{x}}$ transverse to the current sheet acts in the presence of a density stratification $\rho_0(x)$. Plasma is transported without compression, so that density changes satisfy

$$\frac{\partial \rho_1}{\partial t}+v_x\frac{\mathrm{d}\rho_0}{\mathrm{d}x}=0$$

for mass continuity; the driving force $\rho_1 g\hat{\mathbf{x}}$ then becomes

$$\mathbf{F}_d=-\frac{v_x}{\omega}\frac{\mathrm{d}\rho_0}{\mathrm{d}x}g\hat{\mathbf{x}}, \tag{3.29}$$

and drives plasma into the sheet if $\mathrm{d}\rho_0/\mathrm{d}x<0$. Equating (3.29) to (3.26) and using (3.28) for ε gives

$$\omega=\left(\frac{(kl)^2\tau_A^2}{\tau_d\tau_G^4}\right)^{1/3},$$

where

$$\tau_G=\left(-\frac{g}{\rho_0}\frac{\mathrm{d}\rho_0}{\mathrm{d}x}\right)^{-1/2}$$

is a gravitational time-scale.

Rippling mode

The *rippling-mode* instability occurs when there is a spatial variation across the sheet in magnetic diffusivity $\eta_0(x)$. This may arise, for instance, from a temperature structure in the basic state. If the diffusivity at a point varies due to advection with the plasma, we have

$$\frac{\partial \eta}{\partial t}+\mathbf{v}\,.\,\boldsymbol{\nabla}\eta=0,$$

and so linear departures η_1 on a time-scale ω^{-1} from the original diffusivity η_0 are given by

$$\eta_1=-\frac{v_x}{\omega}\frac{\mathrm{d}\eta_0}{\mathrm{d}x}. \tag{3.30}$$

This change in diffusivity produces an extra current

$$\mathbf{j}_1=-\frac{\eta_1}{\eta_0}\mathbf{j}_0 \tag{3.31}$$

in Ohm's Law; associated with it is an extra Lorentz force $\mathbf{j}_1\times\varepsilon B_0\hat{\mathbf{y}}$ at the ($x>0$) edge of the diffusion region, where the magnetic field is $\varepsilon B_0\hat{\mathbf{y}}$. Since the original current is, in order of magnitude, $\mathbf{j}_0=B_0/(\mu l)\hat{\mathbf{z}}$, this force may be written (using (3.30) and (3.31)) as

$$\mathbf{F}_d=-\frac{v_x}{\omega\eta_0}\frac{\mathrm{d}\eta_0}{\mathrm{d}x}\frac{\varepsilon B_0^2}{\mu l}\hat{\mathbf{x}}. \tag{3.32}$$

It represents a driving force with the sign shown in Figure 3.10a for $x>0$ provided $\mathrm{d}\eta_0/\mathrm{d}x<0$. Similarly, on the left-hand side of the current sheet ($x<0$), where the magnetic field is $-\varepsilon B_0\hat{\mathbf{y}}$ and $v_x>0$, a driving force of the necessary direction is present provided $\mathrm{d}\eta_0/\mathrm{d}x>0$ there. Thus, if the diffusivity η_0 takes a maximum value in the centre of the sheet, the whole of the sheet is subject to the rippling mode. For a classical diffusivity, η_0 varies as $T_0^{-3/2}$ with temperature T_0, so this would necessitate a temperature minimum at $x=0$. If, on the other hand, the diffusivity increases (or decreases) monotonically through the sheet, the instability takes place only in the left- (right-)hand side of the sheet. Equating (3.32) to (3.26) and using (3.28) for ε determines the growth-rate as

$$\omega=\left[\left(\frac{\mathrm{d}\eta_0}{\mathrm{d}x}\frac{l}{\eta_0}\right)^4\frac{(kl)^2}{\tau_d^3\tau_A^2}\right]^{1/5}$$

It should be noted that, if conduction is efficient enough, diffusivity changes are not advected so that (3.30) fails and the rippling mode is stabilized.

Tearing mode

The third type of resistive instability is the *tearing mode*, which, in contrast to the other two, occurs with a wavelength greater than the width of the sheet ($kl < 1$). Its growth-rate is given by

$$\omega = [\tau_d^3 \tau_A^2 (kl)^2]^{-1/5},$$

for wavenumbers in the approximate range

$$\left(\frac{\tau_A}{\tau_d}\right)^{1/4} < kl < 1.$$

The smallest allowable wavelengths (l) grow in a time $\tau_d^{+3/5} \tau_A^{+2/5}$, whereas the longest wavelength, namely $(\tau_d/\tau_A)^{1/4} l$, is much larger than the width (l) of the sheet and has the fastest growth-rate, namely

$$\omega = (\tau_d \tau_A)^{-1/2}.$$

Finite electrical conductivity effects are important only in a narrow layer with a width $(kl)^{-3/5}(\tau_A/\tau_d)^{2/5} l$. Cross and Van Hoven (1971) give a more accurate expression for the maximum growth-rate, namely

$$\omega = 3.4 \tau_d^{-0.43} \tau_A^{-0.57}.$$

Most tearing-mode calculations assume η is constant, whereas for a Coulomb plasma it decreases with the temperature like $T^{-3/2}$: Cross and Van Hoven (1971, 1973) have taken account of such variations resulting from Ohmic heating and find that the increase in η quenches the tearing mode. (However, the resulting resistivity gradients can destabilise the rippling mode.) The tearing-mode is possibly the most important of the three resistive instabilities, since it has a long wavelength and requires neither a gravitational force nor a resistivity gradient to be excited. It may well be the cause of coronal heating and is possibly relevant to the solar flare (§3.4); it is described in detail in §4.5.

The tearing mode occurs in *any* sheared field not just a neutral sheet, since the analysis is largely unaffected by the addition of a constant magnetic field component normal to the plane of Figure 3.10. Analytic examples of such fields are

$$B_{0y}(x)\hat{\mathbf{y}} + B_{0z}(x)\hat{\mathbf{z}}$$

in plane geometry, and

$$B_{0\phi}(\tilde{\omega})\boldsymbol{\varphi} + B_{0z}(\tilde{\omega})\hat{\mathbf{z}}$$

in cylindrical geometry. Sheared fields are in general resistively unstable at *many* thin sheaths throughout the structure. At *any* particular location specified by the value of x (or $\tilde{\omega}$), the instability has a wavenumber in a direction normal to the field,

$$\mathbf{k} \cdot \mathbf{B}_0 = 0,$$

and in Figure 3.10a the x-y axes refer to the x-k (or $\tilde{\omega}$-k) plane, as indicated in Figure 3.10b. The credit for stressing the importance of tearing modes in general solar fields such as magnetic loops belongs to Spicer.

The presence of a large enough transverse field component *in the plane* of Figure 3.10 is likely to stabilise the plasma to infinitesimal perturbations; a finite perturbation which cancels out B_x locally would probably then be needed to destabilise the configuration. Thus configurations which include a transverse component larger than $R_m^{-1/2} B_0$ (§3.5.1) are likely to be stable to the tearing mode. Another point is that, if the sheet width is less than a collision mean-free path, the magnetohydrodynamic analysis becomes invalid. But, in a collisionless plasma, there is also a type of tearing-mode instability due to the effects of electron inertia in Ohm's Law (see e.g. Furth, 1962; Coppi *et al.*, 1966; Schindler and Soop, 1968). Drake and Lee (1977) distinguish three regimes for the tearing of a current sheet, depending on whether the wavelength (λ) of the perturbation is larger or smaller than some multiple of the electron mean-free-path (λ_e). If

$$\lambda \ll \lambda_e (\lambda_0 / l)^2,$$

one is in the *kinetic* regime, where

$$\lambda_0 = \frac{c}{\omega_p}$$

is the skin depth (typically 5 cm for coronal conditions). Here the growth-rate is

$$\omega = (\lambda_0 / l)^2 \tau_e,$$

where

$$\tau_e = \frac{l}{v_{Te}}$$

is the transit time at the electron thermal speed v_{Te}. The *semi-collisional* regime occurs when

$$\lambda_e \frac{\lambda_0^2}{l^2} \ll \lambda \ll \lambda_e \frac{l}{\lambda_0}\left(\frac{\rho_i}{l}\right)^{3/2},$$

where ρ_i is the ion gyroradius. In this case the growth-rate is

$$\omega = (\lambda_0/a)^{2/3} \tau_e^{-2/3} \tau_d^{-1/3}.$$

Finally, the *collisional* regime applies when

$$\lambda \gg \lambda_e \frac{l}{\lambda_0}\left(\frac{\rho_i}{l}\right)^{3/2}$$

and gives

$$\omega = (kl)^{-2/5} \tau_A^{-2/5} \tau_d^{-3/5};$$

except for extremely thin sheets this is the relevant regime in practice.

As well as depending on the collisional or collision-free character of the current sheet, the tearing growth-rate varies markedly with geometry. In a cylindrical flux tube, Coppi *et al.* (1976) find

$$\omega_{\text{cyl}} \approx \tau_A^{-2/3} \tau_d^{-1/3},$$

which is considerably faster than the maximum plane growth-rate

$$\omega_{\text{plane}} \approx \tau_A^{-1/2} \tau_d^{-1/2}.$$

If there are two surfaces where **k** . **B** vanishes for the same k, the double-tearing mode (Schnack and Killeen, 1979) grows even faster, at a rate

$$\omega_{\text{double}} \approx \tau_A^{-3/4} \tau_d^{-1/4}.$$

For example, suppose the magnetic field strength is 0.01 T (100 G), the temperature is 10^6 K and the density is 10^{16} m^{-3} (10^{10} cm^{-3}); then $\omega_{\text{plane}^{-1}}$, $\omega_{\text{cyl}^{-1}}$ and $\omega_{\text{double}^{-1}}$ are 2×10^4 s, 300 s and 30 s, respectively, for a sheet width (l) of 100 km.

3.3 BEHAVIOUR OF CURRENT SHEETS

3.3.1 Magnetic annihilation in one-dimensional sheets

The behaviour of the magnetic induction **B** is governed by the induction equation

$$\frac{\partial \mathbf{B}}{\partial t} = \text{curl}\,(\mathbf{v} \times \mathbf{B}) + \eta \nabla^2 \mathbf{B} \tag{3.33}$$

once the plasma velocity **v** is known. Physically, it implies that changes in the magnetic field are brought about by either the convection of magnetic field lines frozen into the plasma (if the first term on the right-hand side dominates) or by the diffusion of field lines through the plasma (if the second is dominant). The ratio of the order of magnitude of the two terms on the right gives the magnetic Reynolds number, which may be of order unity for a steady current sheet when the two effects balance one another.

For one-dimensional motion $v(x, t)\,\hat{\mathbf{x}}$ in a one-dimensional current sheet with magnetic field $B(x, t)\hat{\mathbf{y}}$, (3.33) reduces to

$$\frac{\partial B}{\partial t} = -\frac{\partial}{\partial x}(vB) + \eta \frac{\partial^2 B}{\partial x^2}. \tag{3.34}$$

If the plasma velocity is identically zero, this becomes a pure diffusion equation, whose solution may be written in general as

$$B(x, t) = \int G(x - x', t) B(x', 0) \mathrm{d}x',$$

in terms of some initial magnetic profile $B(x', 0)$ and the Green's function

$$G(x - x', t) = \frac{1}{(4\pi\eta t)^{3/2}} \exp\left[-\frac{(x - x')^2}{4\eta t} \right].$$

In particular, when the initial sheet is infinitesimally thin with a piecewise constant magnetic field ($B = +B_0$ for $x > 0$, $B = -B_0$ for $x < 0$), we find

$$B(x, t) = \frac{2B_0}{\pi^{\frac{1}{2}}} \operatorname{erf}\left(\frac{x}{(4\eta t)^{1/2}} \right), \tag{3.35}$$

where the error function is defined by

$$\operatorname{erf}(\xi) = \int_0^{\xi} \mathrm{e}^{-u^2} \mathrm{d}u.$$

The solution (3.35) shows that the magnetic field diffuses away in time at a speed η/l, where the width (l) of the sheet is of order $(\eta t)^{1/2}$ and so increases in time. The resulting magnetic field strength at a fixed position decreases with time and so the field is said, in a rather dramatic fashion, to be "annihilated". During the process, magnetic energy is quite simply converted into heat by ohmic dissipation.

When flows are present, (3.34) is coupled to an equation of continuity

$$\frac{\partial \rho}{\partial t}+\frac{\partial}{\partial x}(\rho v)=0 \tag{3.36}$$

and an equation of motion, which reduces for highly subsonic and subAlfvénic flow speeds to

$$p+\frac{B^2}{2\mu}=\frac{B_e(t)^2}{2\mu}; \tag{3.37}$$

here the edge of the neutral sheet ($x=\pm l(t)$, say) is assumed to be at zero plasma pressure and at a magnetic pressure of $B_e(t)^2/(2\mu)$. In addition, there is an energy equation, but for simplicity we adopt the polytropic form

$$p=K\rho^s. \tag{3.38}$$

The solution to (3.34), (3.36)–(3.38) has not yet been well-studied*, but compressive similarity solutions have been discovered by Priest and Raadu (1975); they take the form

$$B=B_e(t)\sin\frac{\pi x}{2l(t)},$$

$$p=p_0(t)\cos^2\frac{\pi x}{2l(t)},$$

$$v=\frac{x}{2t},$$

$$\rho^s=\frac{p}{K},$$

* See Cheng (1979) *J. Geophys. Res.* **84**, 2129 for a recent numerical treatment.

for $t<0$, where

$$l(t)=\pi\left(\frac{-\eta t}{2-s}\right)^{1/2},$$

$$B_e(t)=\text{const}\times l^{-\gamma/2},$$

$$p_0(t)=\frac{B_e(t)^2}{2\mu}.$$

As t increases from $-\infty$ to 0, so l decreases from ∞ to 0, whereas both B_e, p_0 and $|v|$ increase from 0. As the width of the sheet decreases, so the magnetic field is compressed and at the same time diffuses into the neutral line. The magnetic flux in the sheet decreases like $l^{1-s/2}$, whereas the magnetic energy and ohmic dissipation increase like l^{1-s} and l^{-1-s}, respectively. There is no flow of plasma or electromagnetic energy into the sheet across its moving boundary, but work is done by the external magnetic pressure. Similarity solutions also exist with the width of the neutral sheet increasing in time.

Suppose next that new magnetic flux is carried by a plasma motion towards the plane $x=0$ from both sides at a constant speed v_i. Then a steady-state current sheet of width $l=\eta/v_i$ can exist, in which the outwards diffusion of the magnetic field is balanced by the inwards convection of new flux. In order to prevent a pile-up of matter, the plasma that approaches the null-plane has to turn away and flow out along the sheet, as shown in Figure 3.11. For a steady-state process, it is more convenient to replace (3.33) by Ohm's Law

$$\mathbf{E}+\mathbf{v}\times\mathbf{B}=\eta \text{ curl } \mathbf{B}, \tag{3.39}$$

where curl $\mathbf{E}=\mathbf{0}$. It is supplemented by the equation of motion

$$\rho(\mathbf{v}\,.\,\nabla)\mathbf{v}=-\nabla p+(\text{curl }\mathbf{B})\times\mathbf{B}/\mu, \tag{3.40}$$

where div $\mathbf{B}=0$ and, for incompressible flow, div $\mathbf{v}=0$. A simple exact solution of (3.39) and (3.40) that describes steady magnetic field annihilation in the x-y plane has been found by Parker (1973a) and Sonnerup and Priest (1975) (see also Priest and Sonnerup, 1975). They consider the stagnation-point flow

$$v_x=-kx, \qquad v_y=ky,$$

and the unidirectional magnetic field

$$B_x=0, \qquad B_y=B_y(x),$$

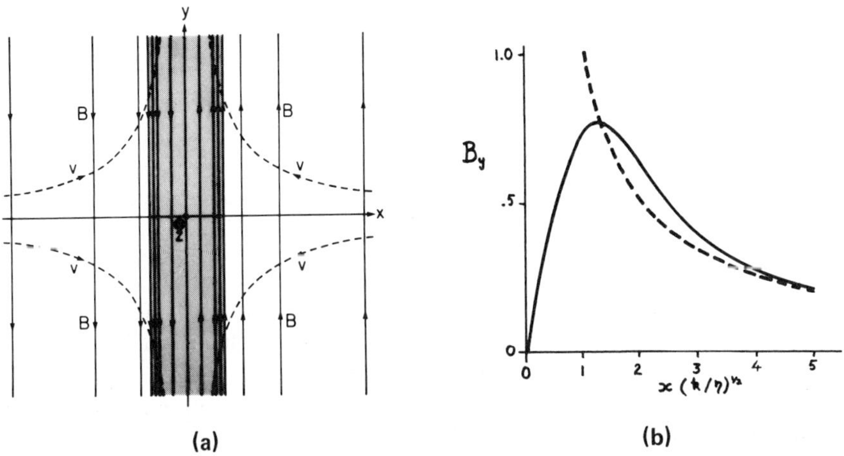

FIGURE 3.11 (a) Oppositely directed magnetic field lines (————) are carried in from the two sides by a stagnation point flow with streamlines shown (- - -). When the field lines enter the current sheet or diffusion region (shaded) they are no longer frozen to the plasma. In the sheet they are annihilated, while the magnetic energy is converted to heat by Joule dissipation. The configuration is a steady one provided the sheet width is such that outwards diffusion of field lines is balanced by their inwards convection. (b) The resulting magnetic field strength as a function of x. It is expressed in units such that the gradient dB_y/dx is unity at the origin. The dashed line represents the solution, $B_y=E(kx)^{-1}$, to Ohm's Law when diffusion is neglected.

where k, E and ρ are constants. Having substituted for these forms, (3.40) simply determines the gas pressure as

$$p=\text{const}-\tfrac{1}{2}\rho v^2-B^2/(2\mu), \tag{3.41}$$

whereas (3.39) reduces to

$$E-kxB_y=\eta\frac{\mathrm{d}B_y}{\mathrm{d}x},$$

which has the solution given in Figure 3.11b.

The above solution has been generalised to include the z-components of both the magnetic field and plasma velocity, by taking

$$\begin{aligned} v_x&=-k_{1x}, & v_y&=k_2y, & v_z&=(k_1-k_2)z,\\ B_x&=0, & B_y&=B_y(x), & B_z&=B_z(x). \end{aligned} \tag{3.42}$$

In this case, the equation of motion again gives (3.41), but the y- and z-components of (3.33) reduce to

$$\eta \frac{d^2 B_y}{dx^2} + k_1 x \frac{dB_y}{dx} + k_2 B_y = 0$$

$$\eta \frac{d^2 B_z}{dx^2} + k_1 x \frac{dB_z}{dx} + (k_1 - k_2) B_z = 0.$$

Their solutions have been presented by Sonnerup and Priest (1975), who demonstrate the existence of solutions which join an arbitrary magnetic field vector at large positive values of x to another arbitrary one at large negative values of x. It is thus possible to bring together fields which are not just antiparallel but may be inclined at any angle. As the fields approach, their directions rotate and their magnitudes change, giving (in general) a magnetic field of non-zero strength at the centre of the current sheet and annihilating part of the magnetic flux in the process.

The above model, like much work on current sheets, deals only with the incompressible situation. The assumption of incompressibility is, of course, made for simplicity, in the hope that the essence of the annihilation process does not depend on compressibility effects. But it would be valuable to extend many current sheet models to include such effects. The other limitation of the exact solution (3.42) is that it is relevant only when the flow speed into the current sheet is much less than the Alfvén speed, so that the length L of the sheet far exceeds its width l. Furthermore, the solution fails near the ends of the sheet, where there are significant transverse components of magnetic field (in the x-direction).

3.3.2 Interchange instability

The possible importance of the interchange instability in current sheets was pointed out qualitatively by Parker (1973b), who suggested that it may enhance the reconnection rate. A plane current sheet, with equal magnetic field magnitudes either side of it, is neutrally stable to interchange modes. But Uchida and Sakurai (1977) analysed curved sheets and showed that they can be unstable. They considered a configuration with circular field lines and a sheet at the radius $r = R$ such that

$$B_{0\phi} = \begin{cases} B_i R/r & \text{for } r < R, \\ -B_e R/r & \text{for } r > R, \end{cases}$$

$$p_0 = \begin{cases} (B_e^2 - B_i^2)/(2\mu) & \text{for } r < R, \\ 0 & \text{for } r > R, \end{cases}$$

The change in potential energy due to a linear perturbation $\boldsymbol{\xi}$ is

$$\delta W = \tfrac{1}{2}\int [\{\gamma p_0(\nabla . \boldsymbol{\xi})^2 + (\boldsymbol{\xi} . \nabla p_0)(\nabla . \boldsymbol{\xi})\} + \{B_1^2 - \boldsymbol{\xi} \times (\nabla \times \mathbf{B}_0) . \mathbf{B}_1\}(2\mu)^{-1}]\, dV,$$

where

$$\mathbf{B}_1 = \nabla \times (\boldsymbol{\xi} \times \mathbf{B}_0).$$

Uchida and Sakurai consider a perturbation

$$\boldsymbol{\xi} = \hat{\boldsymbol{\xi}} \cos\left(\frac{\pi\varphi}{2\cos^{-1}(d/r)}\right) e^{i(kz + \mu\varphi)},$$

which vanishes at the photosphere ($r\cos\varphi = d$) and creates a ripple of wavelength $2\pi/k$ along the (horizontal) z-axis. The expression for δW is minimised with respect to the three components $\hat{\xi}_\phi$, $\hat{\xi}_z$ and $\hat{\xi}_r$ in turn and stability is deduced from the sign of δW. Increasing k or μ (and hence producing ripples with shorter wavelengths) is found to increase stability (due to the effect of the magnetic tension), and so is the addition of an axial field component B_z. However, instability is found to occur for small enough values of d/R (so that line-tying is ineffective) and for large enough values of B_e/B_i (so that the driving force dominates). The growth-rate increases with B_e/B_i and also with kR, so that the smallest wavelengths grow fastest, but there is ultimately a limit at sizes of about 10^2–10^3 m due to viscosity.

The effect of the interchange mode is to create very fine scales in the current sheet and so enhance the natural rate of diffusion; the annihilation and reconnection processes discussed in §3.3.1 and §3.3.3 still apply on larger scales, but with an enhanced eddy diffusivity based on the fine scales.

3.3.3 The growth of reconnection

A simple one-dimensional current sheet, such as discussed in §3.3.1, tends to be subject to instabilities that create two-dimensional structure and so produce a reconnection of the field lines. Two such instabilities that we have already mentioned are particularly important.

a) *The tearing mode* (§3.2.4.C and §4.5). This long wavelength mode leads to the formation of pairs of X- and O-type neutral points and creates long, thin loops on a time-scale $(\tau_d \tau_A)^{1/2}$. It occurs for the following reason. Suppose an equilibrium sheet (Figure 3.12a) is perturbed by squeezing it at some location

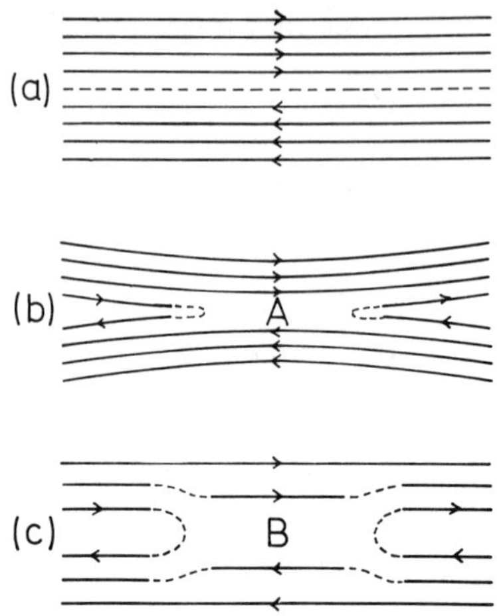

FIGURE 3.12 (a) A neutral sheet in equilibrium. (b) A local squeezing of the sheet at A leads to reconnection. (c) An enhanced diffusion near B leads to reconnection.

A (Figure 3.12b). Then, if there were no diffusion, the field strength would increase locally and the enhanced magnetic pressure would tend to restore the equilibrium. However, the presence of diffusion means that the field near A diffuses more rapidly in response to the squeeze. Hence there are fewer field lines in the sheet at A than at neighbouring locations; this in turn means that some field lines cannot continue to pass between A and so must join with neighbouring ones, as indicated by the dashed lines. The resulting tension forces in the newly-reconnected field lines are such as to continue the perturbation and, provided the wavelength of the original squeezing is long enough, this will dominate the tendency of field lines above and below to straighten out. It is also important to remember that in some cases the tearing mode may not take place; for instance, if fixed boundary conditions are maintained either side of the sheet and the sheet is in a state of minimum energy, then it will remain in stable equilibrium (Birn and Schindler, §6.5).

b) *The rippling mode* (§3.2.4.C). We know that a current sheet of width l tends to diffuse away at a speed η/l (§3.3.1), with the magnetic field strength decreasing in time and the field lines disappearing locally. Suppose then that the magnetic diffusivity near B (Figure 3.12c) is larger than normal, so that the field diffuses more rapidly there. Diagrammatically, this can be represented by having the original eight field lines (Figure 3.12a) decrease to six in the most of the sheet and to only four near B. This implies that only four of the six field lines can continue to be joined to field lines near B and the remainder must link with themselves, as indicated by the dashed lines.

3.3.3.*A* *The causes and results of growth* Reconnection can develop in several distinct ways:

i) it may be generated spontaneously by an instability, such as the tearing mode (Figure 3.12b);

ii) it may be driven from outside if there is a *finite* disturbance (initiated from outside the sheet) of the type shown in Figure 3.12b;

iii) it may be driven locally by a sudden enhanced resistivity at some part of the sheet (Figure 3.12c).

The *nonlinear* results of a growth of reconnection depend crucially on the assumed boundary conditions at large distances, as follows:

i) If the boundary is regarded as rigid so that the flow velocity is held equal to zero, then no new flux can enter the region and (provided the magnetic boundary conditions permit it) the magnetic field ultimately decays away to zero and the magnetic energy is converted into heat. The effect of the reconnection is to create small-scale fields, which can subsequently diffuse away more rapidly than before. In some cases, the one-dimensional analysis of §3.3.1 is relevant here, with the ordinary diffusivity replaced by an eddy value characteristic of the small-scale structure. Furthermore, during the relaxation of the system, a quasi-steady state of reconnection can sometimes ensue (§3.3.4). The nonlinear development of the tearing mode in a fixed region is described in §4.5.3.B. Rutherford (1973) has shown analytically how the third-order Lorentz forces oppose the vortex flows; the exponential growth is replaced by a slower one (proportional to t^2). Also, Dickman *et al.* (1969) and Biskamp and Schindler (1971) have suggested that in a few linear *e*-folding times, all the current filaments coalesce into one. Cross and Van Hoven (1976), on the other hand, have investigated the nonlinear behaviour numerically for a constant η. They find that the instability is saturated after about five linear *e*-folding times. Plasma speeds of up to 15% of the Alfvén speed are produced and the temperature goes up by many orders of magnitude. 92% of the magnetic energy in the sheet is converted to heat, the remainder being split equally between plasma kinetic energy and reconnected magnetic field energy. It would be interesting to make their present calculation more applicable to the solar atmosphere, by adopting different boundary conditions and a more realistic energy equation.

ii) If the flow velocity (and field strength) on the inflowing boundary are held constant, then the configuration evolves to a state of *steady reconnection* (§3.3.4) *based on that inflow speed*. This is relevant when the reconnection is being forced by magnetic evolution (such as emerging flux) far from the reconnection site.

iii) If the boundary conditions are free, in the sense that the inflow speed may adopt any value it likes, then the nonlinear state is likely to approach one of *steady reconnection* (§3.3.4) *at the maximum allowable possible rate.*

3.3.4 Nonlinear phase of steady reconnection

A state of steady reconnection can develop in several ways (§3.3.3) and the main realisation of recent years is that, if oppositely directed fields are being forced together at a prescribed rate, the *plasma will just respond by allowing reconnection at that speed*; this applies up to a certain *maximum rate*, which is typically 0.01–0.1 v_A and depends weakly on the magnetic Reynolds number. If, on the other hand, the system is free to reconnect at any speed, it will tend to choose the maximum rate.

Consider the magnetic configuration around a finite, two-dimensional current sheet or 'diffusion region', where the electric current density is so large that the right-hand side of (3.39) is not small compared with the left side and the field lines can slip through the plasma (Figure 3.13). Outside the diffusion region, the right-hand side of (3.39) is assumed to vanish, so that the magnetic field is effectively "frozen" to the plasma. Magnetic field lines are carried into the diffusion region from the sides (with speed v_i) and out through the top and bottom (with speed v_o). In the process, the field lines are said to have been "reconnected" at the neutral point N, where the magnetic field strength in the plane of the figure vanishes. The outflow magnetic field strength B_o is less than the inflow strength B_i. Thus, for a steady-state, two-dimensional situation, in which the electric field E is constant, the inflow of magnetic energy EB_i, exceeds the outflow EB_o and some of the magnetic field is annihilated. Part of the magnetic energy is converted into heat and part into plasma kinetic energy, since v_o exceeds v_i in value.

For steady, incompressible flow, Sweet (1958) and Parker (1963) derived the following order of magnitude relationships between the input and output parameters:

$$v_o = v_{Ai} \equiv \frac{B_i}{(\mu\rho)^{1/2}}, \tag{3.43}$$

$$v_i = \frac{\eta}{l}, \tag{3.44}$$

$$v_i L = v_o l. \tag{3.45}$$

The first arises because plasma is accelerated away from N by an excess gas pressure of amount B_i^2/μ. The second expresses the balance between inwards convection and outwards diffusion and the last equation is a consequence of

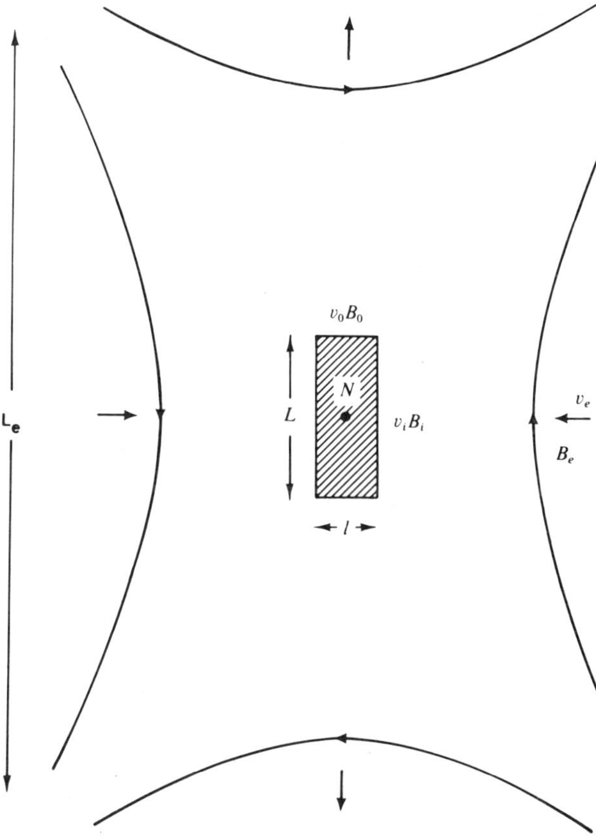

FIGURE 3.13 The magnetic configuration for the reconnection process. Oppositely directed magnetic field lines of strength B_e, frozen to the plasma, are carried towards one another at a speed v_e by a converging flow. They enter a diffusion region (shaded) with dimensions l and L, are reconnected at the neutral point N and are finally ejected from the ends.

mass conservation. For given input values v_i and B_i, (3.43)–(3.45) determine v_o and the dimensions of the diffusion region. In particular, L is much less than the overall dimension L_e for magnetic field variations if

$$v_i \gg (v_{Ai}\eta/L_e)^{1/2}.$$

Such a situation may be referred to as "fast" reconnection and is the one on which most attention has been concentrated. The object has been to determine the details of the flow in the "external region" which surrounds the diffusion region and to find the maximum allowable value for the speed of approach v_e at a distance L_e in terms of the Alfvén speed $v_{Ae} \equiv B_e/(\mu\rho)^{1/2}$ there.

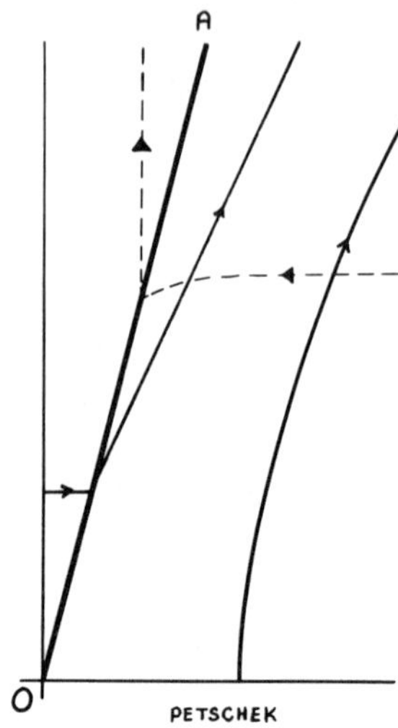

FIGURE 3.14 Magnetic field lines (————) and stream lines (– – –) for the first quadrant of Petschek's mechanism. A finite Alfvén wave is situated at OA. It is generated at the diffusion region, which is here regarded as a point at O, and serves to turn the flow and magnetic field through large angles. In the compressible case OA becomes a slow magnetohydrodynamic shock which may split into two (Petschek and Thorne, 1967).

Petschek (1964) was the first to propose a fast reconnection model, one quadrant of which is shown in Figure 3.14. A slow magnetohydrodynamic shock wave (or finite Alfvén wave in the incompressible limit) remains stationary at the position OA, while plasma carries magnetic flux through it. The direction of the magnetic field vector rotates towards the normal and the strength of the field decreases in the process. When the inflow speed is much less than the Alfvén speed, the angle between OA and the vertical axis becomes very small. Pairs of shock waves are then so close together that they may be regarded loosely as current sheets extending from the central current sheet. Petschek gives a qualitative argument for expecting v_e to possess a maximum value and for the value to depend weakly on the magnetic Reynolds number $R_{me} \equiv v_{Ae} L_e / \eta$.

But Petschek's mechanism was not generally accepted, partly because of the semi-quantitative nature of the analysis, and also because some calculations of Green and Sweet (1966) imply that the field lines to the left of OA are bowed away from O, rather than being straight, which presents a problem for matching to the diffusion region. Even more doubt on the validity of Petschek's analysis was cast when alternative solutions for the external region were put forward by Sonnerup (1970) and Yeh and Axford (1970).

By comparison with Figure 3.14, Sonnerup's model possesses an extra discontinuity OB to the right of OA. The two discontinuities separate the quadrant into three regions, in each of which the magnetic field and plasma velocity are uniform. This enables one to relate the values of variables in each region analytically. However, the discontinuity OB needs to be generated at a corner B in the flow and so the model is not applicable in detail to a current sheet in the solar atmosphere. However, this model does have the great virtue of simplicity and the discontinuity OB may be regarded as a mathematical lumping together of expansion effects in the inflow region. The elegance of the model has meant that it has been possible to generalise it in several ways: Cowley (1974a) has included magnetic field and velocity components normal to the plane of the figure; Yang and Sonnerup (1976) have included compressibility effects, for which OB spreads into a slow expansion fan centred at the external corner in the flow and causing a complicated interaction region ahead of the diffusion region; asymmetry in the opposing inflow regions has been treated by Cowley (1974b) and Mitchell and Kan (1978).

Yeh and Axford's model may be regarded as a generalisation of Sonnerup's. Like Sonnerup's model, it contains the extra discontinuity OB, which is unacceptable physically. But unfortunately in their case OB is not acceptable mathematically either, since it proves to be impossible to join the flow on both sides (Vasyliunas, 1975). The *approach* of Yeh and Axford is, however, a valuable one. They argue that the magnetohydrodynamic variables should be of *self-similar* form on some scale intermediate between the size of the diffusion region and the distance between the magnetic field sources. Having expressed $\mathbf{v}$ and $\mathbf{B}$ in terms of a stream function ψ and a vector potential A, defined by

$$(v_r,\ v_\theta)=\left(\frac{1}{r}\frac{\partial\psi}{\partial\theta},\quad -\frac{\partial\psi}{\partial r}\right),$$

$$(B_r,\ B_\theta)=\left(\frac{1}{r}\frac{\partial A}{\partial\theta},\quad -\frac{\partial A}{\partial r}\right),$$

they look for solutions of (3.39) and (3.40) (with $\eta=0$) in the form

$$\psi=rg(\theta),\qquad A=rf(\theta).$$

They then solve the resulting ordinary differential equations for $g(\theta)$ and $f(\theta)$ subject to the boundary conditions

$$g=\frac{\mathrm{d}f}{\mathrm{d}\theta}=0\qquad\text{on } \theta=0 \text{ and } \tfrac{1}{2}\pi,$$

which imply that the x- and y-axes form streamlines on which $B_r=0$.

Soward and Priest (1977) have tackled the fast reconnection problem in the same spirit, by seeking two-dimensional solutions of (3.39) and (3.40) which are asymptotic in form and so valid at large distances from the diffusion region. But they stipulate that there should be no second discontinuity like OB. The analysis is rather complex and has been summarised by Priest and Soward (1976). They define

$$R = \log_e (r/L) + R_0,$$

where R_0 is a constant, which is chosen for convenience as $\pi/(8M_i) - \log_e M_i$, and $M_i \equiv v_i(\mu\rho)^{1/2}/B_i$. Then they seek series solutions in powers of R for which the Lorentz force dominates the inertial term in (3.40). For technical reasons, they write

$$\begin{aligned} \psi &= rg(R, \theta) \equiv r[R^{-1/2} g_0(\theta) + R^{-3/2}(g_{11}(\theta)\log_e R + g_1(\theta)) + \dots], \\ A &= rf(R, \theta) \equiv r[R^{1/2} f_0(\theta) + R^{-1/2}(f_{11}(\theta)\log_e R + f_1(\theta)) + \dots], \end{aligned} \tag{3.46}$$

where θ is measured from the x-axis, shown vertical in Figure 3.15a. These expansions are substituted into (3.39) and (3.40) and the coefficients of powers of R are equated to zero. The resulting ordinary differential equations for the unknown functions $g_0, f_0, g_{11}, f_{11}, g_1, f_1, \dots$ are solved for $0 < \theta < \frac{1}{2}\pi$, subject to the boundary conditions

$$g = \frac{\partial f}{\partial \theta} = 0 \qquad \text{at } \theta = \tfrac{1}{2}\pi,$$

to all orders. The expansion (3.46) represents only an "outer" solution, since it fails to satisfy the required boundary condition at $\theta = 0$. The "inner" solution, which is valid near $\theta = 0$, takes the form

$$\begin{aligned} \psi &= rG(R, \xi) \equiv r[R^{-1/2} G_0(\xi) + R^{-3/2}(G_{11}(\xi)\log_e R + G_1(\xi)) + \dots], \\ A &= rF(R, \xi) \equiv r[R^{-1/2} F_0(\xi) + R^{-3/2}(F_{11}(\xi)\log_e R + F_1(\xi)) + \dots], \end{aligned} \tag{3.47}$$

where

$$\xi = \frac{\theta}{\Theta(R)},$$

and

$$\theta = \Theta(R) \equiv \Theta_0 R^{-1} + (\Theta_{11} \log_e R + \Theta_1) R^{-2} + \dots$$

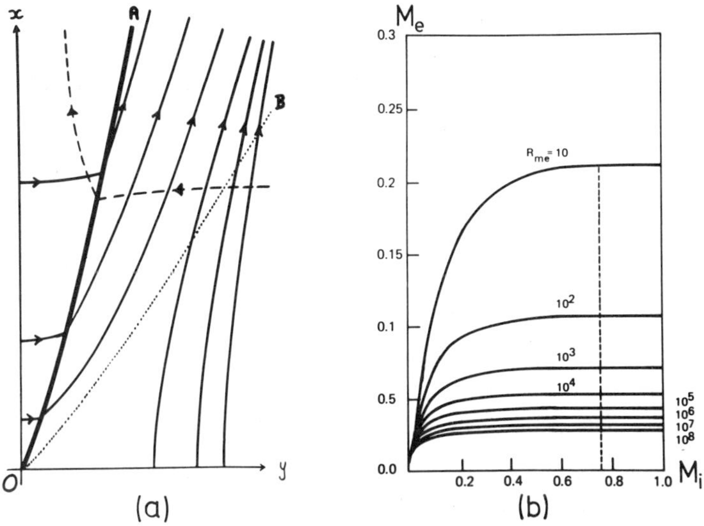

FIGURE 3.15 (a) The modification of Petschek's configuration by Soward and Priest (1977). OA now curves slightly towards the x-axis. OB is a second 'Alfvén line', at which the normal components of the plasma velocity and magnetic field are equal, but the plasma flows smoothly through it without jumps in the tangential components. Any discontinuity at OB would need to be generated externally and so is most unlikely in the solar atmosphere. By comparison with Figure 3.14, extra field lines have been sketched to indicate that, as one approaches O, the field strength decreases along Oy and increases along Ox. (b) The Alfvén Mach number $M_e \equiv v_e/v_{Ae}$, as given by Soward and Priest (1977), is plotted as a function of $M_i \equiv v_i/v_{Ai}$ for a series of values of $R_{me} \equiv v_{Ae}L_e/\eta$. v_A is the Alfvén speed and subscripts e and i refer to values at large distances from the diffusion region and close to it, respectively, as indicated in Figure 3.13. Regardless of the value of R_{me}, M_e has a maximum at $M_i = \pi/4$, as shown by the dashed line.

is the position of the discontinuity OA (Figure 3.15a). Again, they derive equations for the unknown functions in (3.47) and solve them for $0 \leqslant \xi < \infty$, subject to the boundary conditions

$$G = \frac{\partial F}{\partial \xi} = 0 \qquad \text{at } \xi = 0 \text{ (i.e. } \theta = 0).$$

Finally, the inner and outer expansions (3.46) and (3.47) are matched by means of the conditions

$$\lim_{\theta \to 0} g = \lim_{\xi \to \infty} G \quad \text{and} \quad \lim_{\theta \to 0} f = \lim_{\xi \to \infty} F,$$

to all orders. The inner solution exhibits a discontinuity in $\mathbf{v}$ and $\mathbf{B}$ at $\xi = 1$ (i.e. at OA), but it may be chosen to be continuous throughout the remainder of the domain. It transpires that, to the left of OA, the magnetic field behaves like

$R^{-1/2}$, while to the right it behaves like $R^{1/2}$. Thus, as one approaches the diffusion region along the line $\theta = \pi/2$ or moves away along $\theta = 0$, the magnetic field decreases in strength while the plasma speeds up. Furthermore, the position of OA ($\theta = \Theta(R)$) is found to be to lowest order

$$\theta = \frac{\pi}{8R},$$

so that the discontinuity curves slightly, as indicated diagrammatically in Figure 3.15a. The streamlines and magnetic field lines are found to have curvature in the senses shown. In particular, the field lines to the left of OA are convex towards O, which allays the previous fears of Green and Sweet (1966). The figure is exaggerated in the sense that, in practice, OA and OB lie very close to the x-axis.

The significance of the above analysis is that *it puts Petschek's mechanism on a sound mathematical basis as the only workable model for fast reconnection* in the solar atmosphere. Furthermore, when the solution is linked to the diffusion region by means of (3.43)–(3.45), it determines a relation between the Alfvén Mach numbers $M_e \equiv v_e(\mu\rho)^{1/2}/B_e \equiv v_e/v_{Ae}$ and $M_i \equiv v_i(\mu\rho)^{1/2}/B_i$. (They are characteristic of the flow at large distances and at the diffusion region inflow, respectively, as indicated in Figure 3.13.) Figure 3.15b shows the resulting graphs of M_e as a function of M_i, for various values of the magnetic Reynolds number R_{me}. It can be seen that M_e does indeed possess a maximum value, as suggested by Petschek, which is typically a half of what Petschek gave. For instance, the maximum inflow speed v_e (or, loosely, "reconnection rate") varies from 0.2 v_{Ae} when $R_{me} = 10$ to 0.02 v_{Ae} when $R_{me} = 10^6$. The value of the reconnection rate required for the main phase of a large flare is likely to be below this maximum.

A limitation of present reconnection models is that the diffusion region is treated only in an order of magnitude way, with its dimensions given by (3.43)–(3.45) as

$$l = \frac{\eta}{v_i},$$

$$L = \frac{v_{Ai} l}{v_i}.$$

Milne and Priest (1980) have therefore solved the flow equations within the diffusion region, allowing for compressibility and an energy balance. The solutions depend on two parameters,

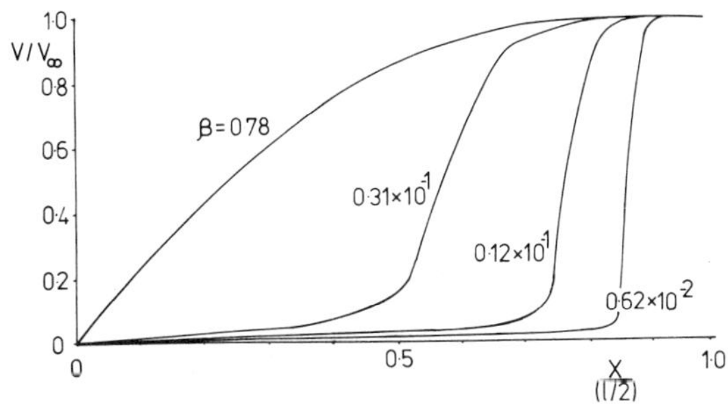

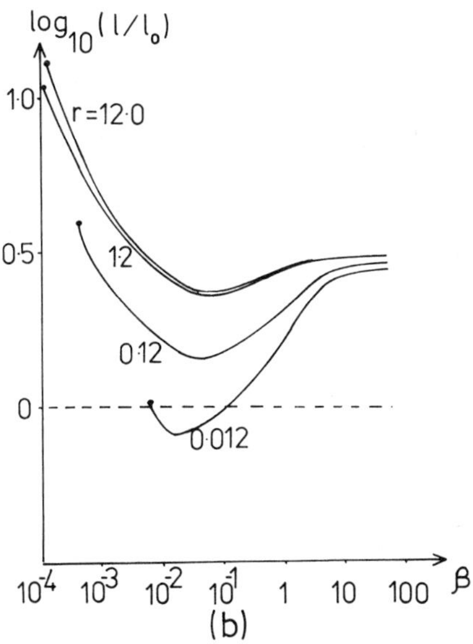

FIGURE 3.16 The model of Milne and Priest (1980) for the central diffusion region. (a) The inflow velocity as a function of distance x from the centre of the sheet for various values of β and $r = 0.012$. (β is the ratio of plasma to magnetic pressure and r is the ratio of radiation to convection.) (b) The width l of the central diffusion region in terms of its order of magnitude value $l_0 = \eta/v_i$.

$$\beta = \frac{2\mu p}{B^2}$$

and

$$r = \frac{\text{radiation}}{\text{convection}}.$$

In Figure 3.16 is plotted the inflow velocity profiles and the sheet width l in terms of the order of magnitude value

$$l_0 = \frac{\eta}{v_i}.$$

It can be seen that the sheet may be a factor of ten thicker than previously predicted. Also, previously any solution was allowed up to a maximum value of v_i, but now the curves are terminated at dots so that there is a β-limitation, with no solution if β is too small. Another feature is the presence at low-β values of steep gradients within the sheet.

3.3.5 Laboratory experiments

There have now been several laboratory experiments on magnetic reconnection and these have been reviewed by Van Hoven (1976), Baum and Bratenahl (1977) and others. The main difficulty for comparison with theory is that the magnetic Reynolds number in the experiments is not very large; also, most of the evolution is strongly time-dependent and it is difficult to resolve any shocks.

The most detailed experiment is that of Bratenahl, Baum and their co-workers (Bratenahl and Yeates, 1970; Baum and Bratenahl, 1974a, b; Baum *et al.*, 1973a, b). It consists of a double-inverse-pinch, whose vacuum field is shown in Figure 3.17a; the dark figure-of-eight is the separatrix which separates cells 1, 2 and 3. The magnetic field is produced by currents in two metal rods at the centres of 1 and 2 and a return current along the outer dark boundary. As the rod current increases, the limiting X-type field lines close up somewhat and reconnection occurs, with flux being transferred from cells 1 and 2 into 3. The plasma is first pre-ionised and then large currents are suddenly passed through the rods; they drive two snowplow fronts (containing antiparallel magnetic fields) towards the neutral line. The resulting current-density contours and field lines are shown in Figure 3.17b. As the magnetic field rises, there is a period of quasi-steady reconnection similar to Petschek's picture, except that the magnetic Reynolds number is not much greater than unity. At about 7 μsec, a sudden increase in resistivity occurs (probably due to turbulence produce by the ion acoustic instability) and the reconnection rate rises; flux crosses much more rapidly into cell 3 and the field relaxes towards its potential form with the X-type field lines ultimately intersecting at $\frac{1}{2}\pi$. In an attempt to understand the influence of boundary conditions, Baum and Bratenahl (1976) have replaced the two rods by flat plates; the result was again a simple reconnection at a single neutral point. They are also attempting to simulate the interaction between two bipolar spot groups.

Dailey *et al.* (1972) have investigated reconnection in an annular current

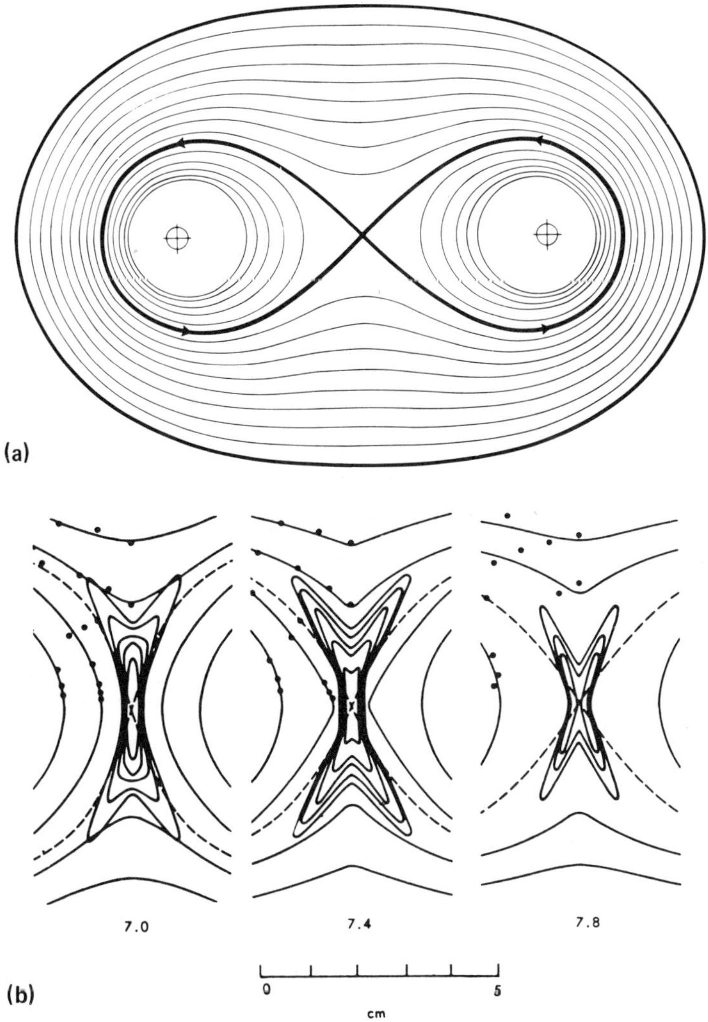

FIGURE 3.17 The double-inverse-pinch experiment of Bratenahl and Baum. (a) The potential field due to two rods. (b) Measured current density contours (solid closed curves) and magnetic field lines at times of 7.0, 7.4 and 7.8 μsec. Slow shocks similar to Petschek's appear to form along the current-density maxima.

sheet. However, again the magnetic Reynolds number is of order unity and so they are probably witnessing ordinary resistive diffusion: there is no evidence for slow Petschek shocks. Anderson and Kunkel (1969) describe an annular Z-pinch, in which a cylindrical current sheet breaks up into a series of filaments. Also, in the hard core theta-pinch of Alidières *et al.* (1968) a cylindrical sheet (with axial field lines) breaks in half.

Ohyabu *et al.* (1972, 1974) study a quadrupole pinch, in which the current sheet breaks first into two and then into three filaments. It may represent a demonstration of the tearing-mode instability and, for some parameters, it exhibits a transition to turbulent conditions.

Quadrupolar fields have also been constructed by Syrovatsky *et al.* (1973), Frank (1976), Bogdanov *et al.* (1975) and Overskei and Politzer (1976). The magnetic Reynolds number is of order unity and a transition to turbulent conditions is demonstrated. An X-type neutral point collapses into a neutral current sheet, which subsequently splits into two parts.

3.3.6 Numerical experiments

There have now been several remarkable numerical experiments, which confirm some aspects of the analytic treatments (§3.3.1–§3.3.3) and in many ways complement the analytic approach. It is difficult to compare with the theory, since often a steady state is not reached and the behaviour is highly nonlinear. Also, one must note carefully what boundary conditions are being assumed since they can affect the evolution markedly. Furthermore, there are numerical problems of covering the diffusion region and shocks adequately and of running the program long enough.

A most complete series of time-dependent, two-dimensional numerical simulations has been carried out by Tsuda, Ugai and Fukao. First of all, Fukao and Tsuda (1973) solved the time-dependent equations of motion and induction

$$\rho \frac{\mathrm{D}\mathbf{v}}{\mathrm{D}t} = \mathbf{j} \times \mathbf{B} - \nabla p + \nu \nabla^2 \mathbf{v},$$

$$\frac{\partial \mathbf{B}}{\partial t} = \nabla \times (\mathbf{v} \times \mathbf{B}) + \eta \nabla^2 \mathbf{B},$$

for an incompressible fluid. The initial flow is of stagnation-point type and the initial magnetic configuration is a one-dimensional current sheet with $\mathbf{B} = \mathbf{B}_y(x)\hat{\mathbf{y}}$. The magnetic Reynolds number R_m is 5,000, although the numerical viscosity in their Friedrichs–Lax scheme reduces the effective value to 50–500. The importance of the experiment is in giving clear evidence of reconnection, since a non-zero B_x is produced. The time-development is consistent with the formation of slow shocks, although the details are obscured by numerical diffusion; furthermore, the mesh size is too large to cover the diffusion region adequately and the experiment is not run long enough to allow a steady state to develop.

Ugai and Tsuda (1977) consider the corresponding compressible problem and assume an energy equation with ohmic heating alone. The equations are solved in conservative form by the two-step Lax–Wendroff scheme. Initially the plasma is at rest in a one-dimensional current sheet with magnetic field

$$\mathbf{B}_x(\mathbf{y}) = \begin{cases} B_0 y/L & \text{for} \quad |y| < L, \\ B_0 & \text{for} \quad y > L, \\ -B_0 & \text{for} \quad y < -L, \end{cases}$$

and the magnetic diffusivity η is assumed to be enhanced by a factor 100 within a circle of radius L about the origin. The ambient magnetic Reynolds number is taken as 10^3. The *resistivity enhancement is found to generate locally a reconnection* of the field lines (Figure 3.18a) and the resulting imbalance of magnetic forces causes an outflow of plasma and magnetic flux along the current sheet, together with an inflow in the y-direction (Figures 3.18b, c). Only the first quadrant is taken as the domain of integration, so that symmetry conditions are imposed on the x- and y-axes. The other boundaries, $y = 2L$ and $x = 2L$, are assumed to be free, so that the plasma can flow freely across them. A region of enhanced current density develops which may be evidence of a slow wave broadened out by numerical diffusion. The current density at the neutral point first decreases because of the diffusion and then increases as the plasma motion builds up; the pressure there decreases because of the expansion. The experiment was run for 7.2 Alfvén travel-times (14.4 L/v_{Ao}), but a steady state was not reached. During this period the magnetic energy in the region decreased in spite of a net inflow of magnetic energy; the thermal energy decreased, mainly due to an outflow of thermal energy, while the kinetic energy increased despite a net kinetic energy outflow.

Three more papers (Tsuda and Ugai, 1977; Ugai and Tsuda, 1979a, b) describe later runs of the same numerical experiment. The flow was found to evolve to a quasi-steady state similar to the Petschek picture after about ten Alfvén travel-times ($20L/v_{Ao}$). A slow wave is present but it is thickened by numerical diffusion to a width of about 0.25 L. The effect of holding the electric field vB constant on the boundary is again to give evolution towards a steady state, which depends on the value of the constant. The steady state is found to represent *reconnection at the maximum rate* calculated by Soward and Priest (1977). The effect of increasing the value of the resistivity is simply to make the diffusion region smaller and change the reconnection rate slightly, in agreement with the analytic results.

Hayashi and Sato (1978) have solved essentially the same set of equations as Ugai and Tsuda with a similar numerical scheme (see also Sato and Hayashi, 1979; Sato, 1979). The magnetic diffusivity is assumed to be non-zero only where the current density exceeds a critical value, in an attempt to model an

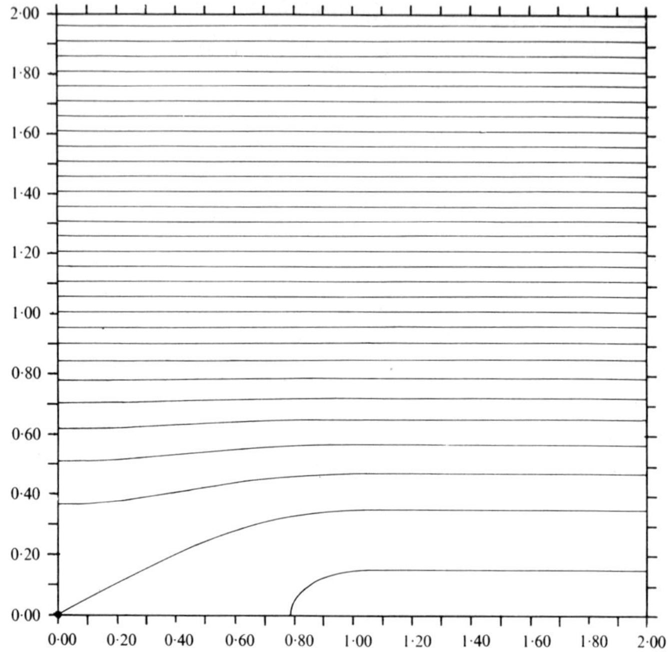
2·00
1·80
1·60
1·40
1·20
1·00
0·80
0·60
0·40
0·20
0·00
0·00
0·20
0·40
0·60
0·80
1·00
1·20
1·40
1·60
1·80
2·00

(a)

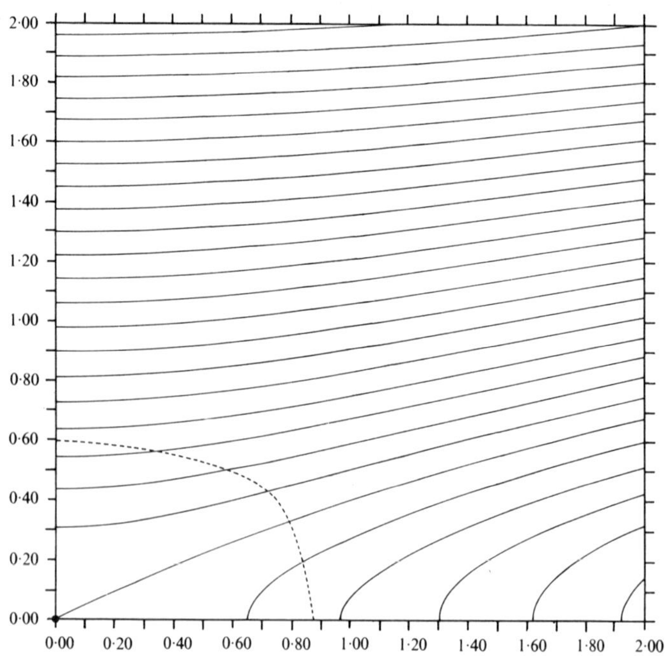
2·00
1·80
1·60
1·40
1·20
1·00
0·80
0·60
0·40
0·20
0·00
0·00
0·20
0·40
0·60
0·80
1·00
1·20
1·40
1·60
1·80
2·00

(b)

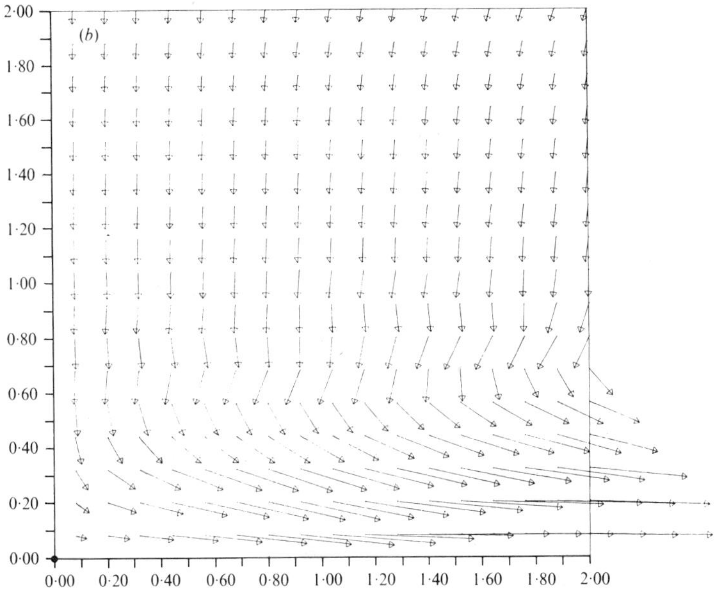

(c)

FIGURE 3.18 The numerical experiment of Ugai and Tsuda (1977). (a) Magnetic field lines in the first quadrant after a time 0.9 L/v_{Ao}, when the hydrodynamic effects are still negligible. (b) Magnetic field lines after a time 14.4 L/v_{Ao}, with a dashed curve indicating the boundary of the diffusion region given by $\sigma|\mathbf{v} \times \mathbf{B}|/j = 10$. (c) The flow pattern after a time 14.4 L/v_{Ao}. An arrow of length 0.07 corresponds to a speed of 0.1 v_{Ao}.

anomalous diffusivity. Symmetry conditions are assumed on the axes of symmetry and free boundary conditions on the outflow boundary, just like Ugai and Tsuda, but on the inflow boundary a fixed mass flux is imposed. They are therefore studying the *response to an imposed inflow* and find that a quasi-steady reconnection is developed after (15–20)L/v_{Ao}; it is of Petschek type, with current layers downstream of the separatrix corresponding to slow shocks (Figure 3.19).

Birn (1980) has performed a numerical experiment on a current sheet with a transverse field, so that the field lines form loops (Figure 3.20a). The aim was to model the geomagnetic tail. R_m is taken as 500 and free boundary conditions are imposed. First of all, the sheet thins (Figure 3.20b) and a pair of X- and O-type neutral points are formed, the latter moving tailwards. Then reconnection develops and approaches a steady state containing two current layers which are interpreted as slow shocks (Figures 3.20c, d).

3.4 ROLE OF CURRENT SHEETS IN FLARE MODELS (SUMMARY)

Most of these flare models are described elsewhere in the book, so here we

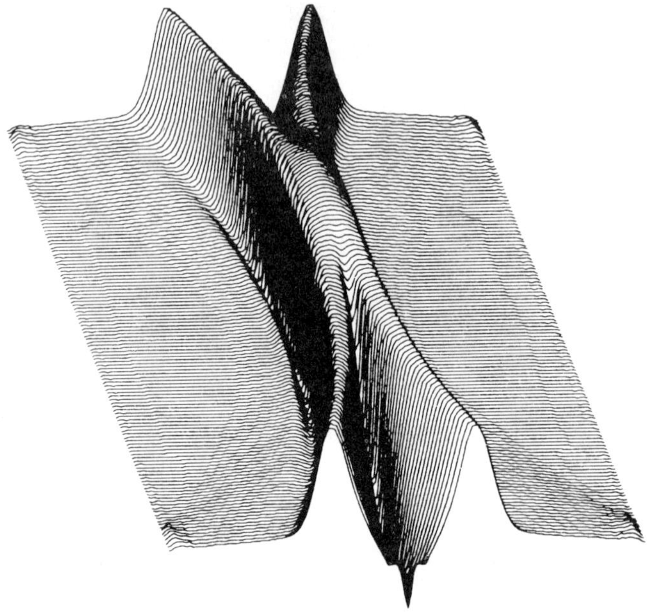

FIGURE 3.19 A three-dimensional display of the current density as a function of position, showing the diffusion region and the four slow shock waves (from Sato and Hayashi, 1979).

simply summarize them; the exception is the emerging flux model, which is discussed in the next section.

3.4.1 Simple-loop flares

There are at present several possible explanations for simple-loop (or compact) flares (§1.6.1):

i) *Emerging flux.* A current sheet is formed at the boundary of newly-emerging flux and the ambient field. As more flux emerges, so it eats its way into the overlying field and the sheet moves higher in the atmosphere. At a critical height the sheet becomes turbulent and the flare is triggered (§3.5). Magnetic energy release occurs in the sheet and in the shock waves which extend from it.

ii) *Thermal non-equilibrium.* The cool core of an active-region loop may lose thermal equilibrium and heat up; current sheets play no part in this theory (§5.4.6).

iii) *(Resistive) kink instability.* When a loop is twisted by more than a critical amount it becomes kink or resistive kink unstable (§4.4, §4.5). If the

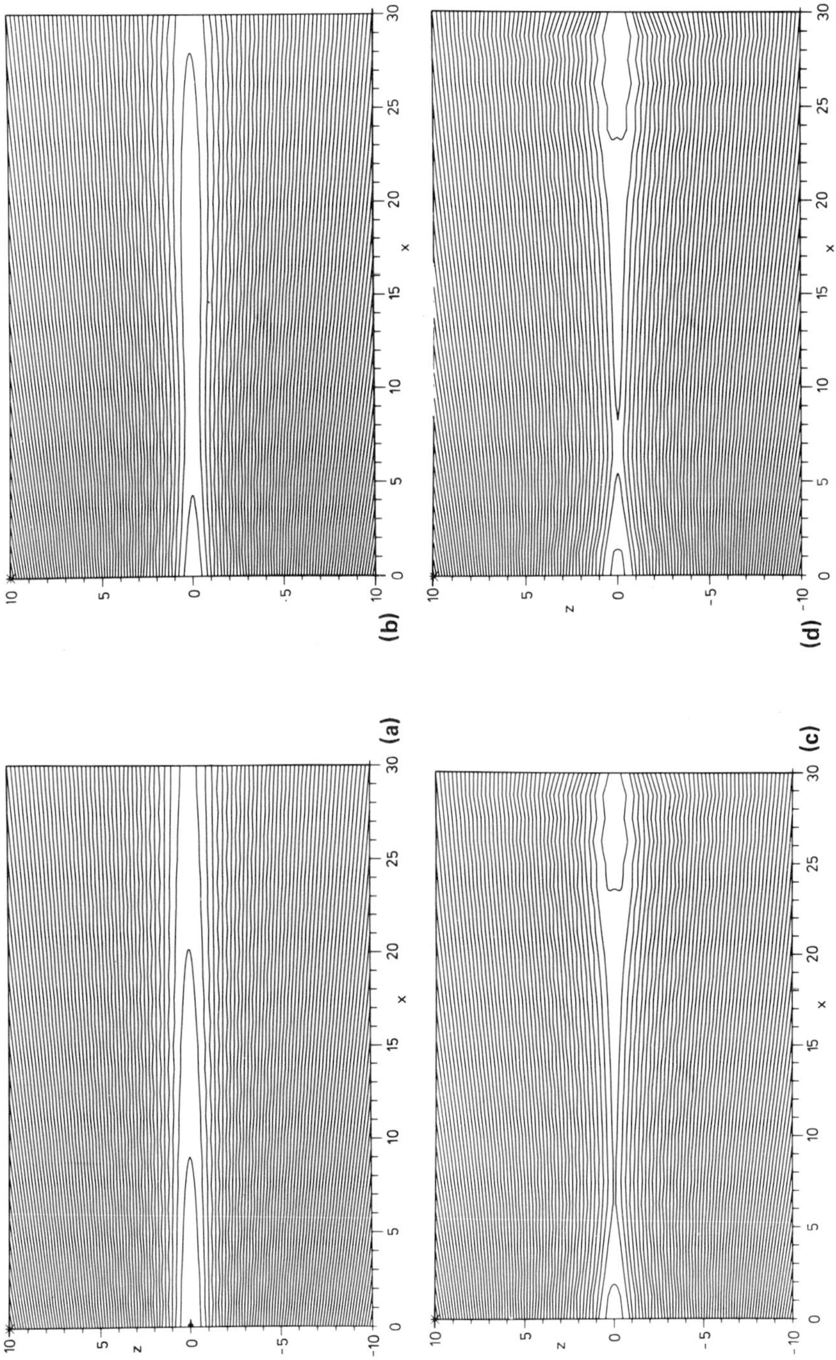

FIGURE 3.20 The numerical experiment of Birn (1980). Magnetic field lines after (a) $t=0$; (b) $t=100\ t_A$; (c) $t=165\ t_A$; and (d) $t=170\ t_A$, where t_A is the Alfvén travel-time.

ideal kink occurs, the loop may become contorted and develop current sheets in the nonlinear development. If the resistive kink takes place, one or several current sheets form at which the magnetic energy is dissipated. This is just a tearing mode in cylindrical geometry, but at present no way of storing the energy and triggering the instability has been demonstrated; one possible solution is the effect of line-tying, which needs to be included in a resistive kink treatment.

3.4.2 Two-ribbon flares

There are essentially two stages to a two-ribbon flare:

i) *The initial filament eruption.* This in turn consists of two phases, namely a *slow* filament rise (which may be due to emerging flux or thermal non-equilibrium (§5.4.6)) and a *fast* rise when the filament ascends much more rapidly. The fast rise may be due to several causes:

a) Emerging flux. The emerging flux creates a small region of enhanced resistivity (§3.5), which allows rapid reconnection to take place in the overlying field. In this case, the emerging flux and its current sheet trigger instability in a much larger configuration.

b) Magnetic instability. When the filament reaches a critical height or the field shear reaches a critical value, the magnetic arcade may become magnetically unstable (§6.5), but current sheets play no part.

c) Tearing-mode instability. As the filament rises, so the underlying field is stretched out into the form of a current sheet, which eventually becomes unstable to the tearing mode when the transverse field component is small enough; the field reconnects and so allows the filament to ascend further.

ii) *Post-flare loops.* When the filament has erupted and dragged the magnetic field lines with it, it leaves behind an open current-sheet configuration. This is unstable to the tearing mode and so reconnection quickly sets in, possibly early in the flare as the filament is rising. The field lines then close back down and create "post"-flare loops in the process (§7.3).

3.5 THE EMERGING (OR EVOLVING) FLUX MODEL

High-resolution ground-based observations have shown that flares tend to occur after the emergence of new flux from below the photosphere (§1.2.1, §2.2.1). Heyvaerts *et al.* (1977) have outlined a flare model based on this supposition. They suggest that the type of flare produced depends on the magnetic environment into which the new flux emerges. Usually, just an X-ray bright point is produced. But emergence near a unipolar sunspot or into a unipolar area near the edge of an active region may give rise to a simple-loop flare. Also, if the new flux appears near the sheared field around an active-region filament, then a two-ribbon flare may result, with the emerging flux triggering the release of stored energy in the much larger overlying field.

It should be noted that the model also applies to a situation of *evolving flux* rather than emerging flux. When distinct flux systems evolve, current sheets form at the interfaces between the systems; thus, horizontal motions of photospheric flux will produce current sheets in just the same way as vertical motions.

It should also be stressed that not all emerging flux can produce a flare. Simple-loop flares are triggered only if sufficient flux emerges that the current sheet occupying the interface between new and old flux reaches a critical height. This height depends on emergence speed and field strength. Two-ribbon flares also need the large-scale field to be highly sheared and so contain the necessary stored energy.

3.5.1 Simple-loop flare

Heyvaerts *et al.* (1977) suggest that *simple-loop flares* occur in three stages as the emerging magnetic flux interacts with the overlying field (Figure 3.21). During the *preflare phase*, steady reconnection occurs in the current sheet formed between the new and old flux. Then, at some critical height, the current sheet finds itself in an unstable state and heats up, seeking a new equilibrium. During this process, the current density may exceed the threshold value for the onset of plasma microinstabilities. This leads to the *impulsive phase*, in which particles are accelerated to high velocities by the resulting large electric fields. The attainment of the critical current density in the sheet is regarded as the trigger for the flare. Finally, in the *main phase*, a new state of steady magnetic field reconnection is reached, the current sheet being a marginally turbulent one.

During the preflare heating phase, the flux, with a magnetic field strength B_∞, emerges at a speed v_∞ from below the photosphere. It presses up against the overlying field and a current sheet forms at the boundary between new and old flux. If v_∞ is larger than a maximum value v_{max}, there is essentially no reconnection. The current sheet grows in length and gains in altitude, as described by Tur and Priest (1976). If v_∞ is smaller than v_{max}, there is continuous reconnection by the Petschek mechanism, but again the current sheet changes its dimensions and moves higher up in the atmosphere. During this process heat is released, the amount depending on v_∞ and B_∞.

The emerging flux continues to reconnect on the basis of a Coulomb resistivity and to rise into regions of increasing temperature and decreasing density until a critical level is reached. At this level the critical current j_{turb} for

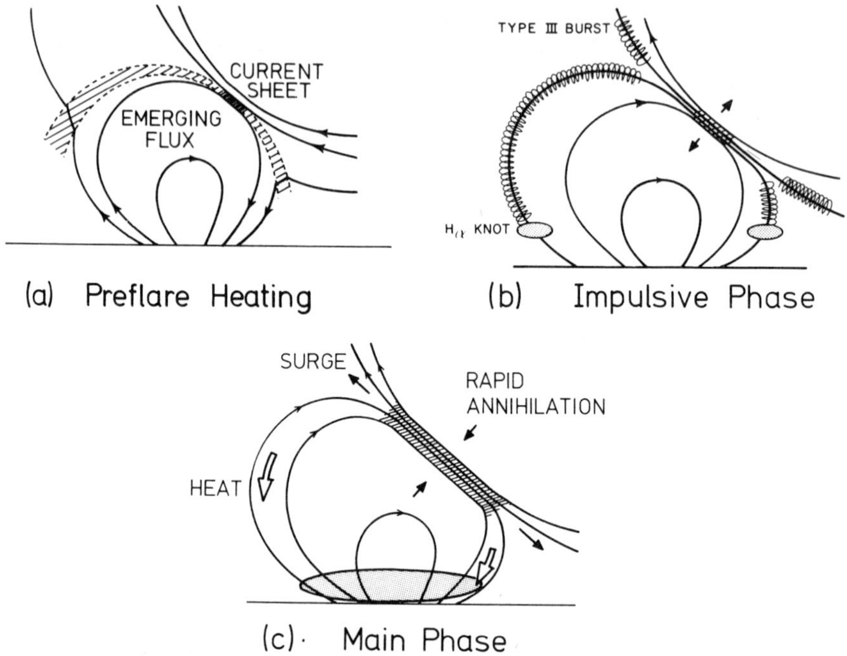

FIGURE 3.21 The emerging flux mechanism for a *small flare* (a simple-loop flare). (a) *Preflare phase.* The emerging flux reconnects with the overlying field. Shock waves (dashed) radiate from a small current sheet and heat the plasma as it passes through them into the shaded region. (b) *Impulsive phase.* The onset of turbulence in the current sheet (when it has reached a critical height) causes a rapid expansion. The resulting electric field accelerates particles, which then escape along field lines and produce an impulsive microwave burst as they spiral. Those that move downwards give rise to hard X-rays by collisional excitation (and possibly Hα knots), while those that escape upwards onto open field lines produce type III radio bursts. (c) *Flash and main phases.* The current sheet reaches a new steady state, with reconnection based on a turbulent resistivity. It is much bigger than before and both heat and particles are conducted down to the lower chromosphere where they produce the Hα flare.

the onset of microinstabilities is reached (so that the plasma becomes turbulent). The condition for the current density to exceed j_{turb} may be written

$$T_c^2 > T_{\text{turb}}^2 \equiv 1.8 \times 10^{16} \; B_\infty / v_\infty,$$

where T_c is the temperature of the plasma inside the current sheet, B_∞ is in tesla and v_∞ is in ms^{-1}. If, for example, we take the typical values $B_\infty = 10^{-2}$ T (100 G) and $v_\infty = 10^3$ ms^{-1}, we find $T_{\text{turb}} = 4.2 \times 10^5$ K. In order to determine the temperature T_c, however, it is necessary to investigate the energy balance inside the sheet, including the effects of mechanical heating, Joule heating, radiative loss and thermal conduction (Heyvaerts and Priest, 1976). The result is that, low down in the atmosphere, there exists a unique solution to the energy

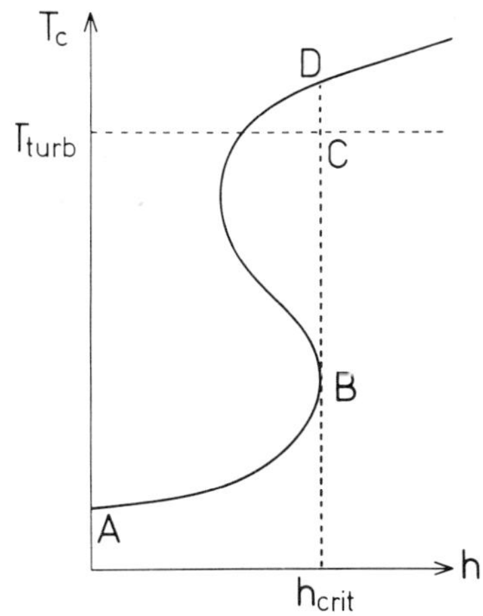

FIGURE 3.22 The thermal equilibrium temperature T_c in the current sheet is shown schematically as a function of height h in the solar atmosphere. As the sheet gains in height, the equilibrium solution moves along AB. When the critical height h_{crit} is attained, there are no neighbouring equilibria and the sheet heats up dynamically along the path BD. But, as the temperature T_{turb} is exceeded, the critical current density for turbulence onset is surpassed.

equation. The main balance is between radiation and Joule heating and T_c is lower than T_{turb} (Figure 3.22). Eventually, a critical height is reached beyond which there is no thermal equilibrium with such a balance! The sheet is therefore in a metastable state! It heats up rapidly, T_{turb} is exceeded and the flare is triggered. The current sheet broadens very rapidly due to the onset of a turbulent resistivity, so that strong electric fields are generated which accelerate particles to high energies and produce the impulsive phase.

The main phase is identified with the enhanced release of energy in a new steady-state reconnection process based on the turbulent resistivity. The amount of energy W released depends on v_∞ and B_∞. If $B_\infty = 10^{-2}$ T and the length of the turbulent sheet is 10,000 km, then $W \approx 10^{22}$ J $(=10^{29}$ erg), characteristic of a small flare. Plasma which is ejected from the top end of the current sheet will be collimated along the ambient field lines and may show up as a tiny surge.

There are four ways in which the flare may be triggered when the sheet reaches a critical height:

i) the critical current density j_{turb} for turbulence onset may be exceeded;

ii) the sheet may lose thermal equilibrium and so heat up and pass the condition for turbulence onset;

iii) as the sheet lengthens, so the transverse magnetic field may decrease until the tearing mode can take place; however, it is unclear whether this will really trigger the flare or just produce a weakly turbulent sheet;

iv) the interchange instability may occur due to the curvature of the sheet (§3.3.2).

The condition for iii) may be estimated as follows. Consider a current sheet of width l and containing a magnetic field $B_y(x)$ which varies between $\pm B_0$. The sheet is subject to the tearing mode on a time-scale

$$\tau_{tmi} = \left(\frac{l(\mu\rho)^{1/2}}{B_0}\frac{l^2}{\eta}\right)^{1/2}.$$

If now a transverse field component $B_t\hat{\mathbf{x}}$ is also present, energy will be transported across the sheet by Alfvén waves on a time-scale

$$\tau = \frac{l(\mu\rho)^{1/2}}{B_t}.$$

Thus the tearing mode will take place as normal when the transverse field is so small that

$$\tau_{tmi} < \tau,$$

i.e.

$$\frac{B_t}{B_0} < \frac{1}{R_m^{1/2}},$$

where $R_m = B_0(\mu\rho)^{-1/2} l/\eta$ is the magnetic Reynolds number.

In studying the thermal evolution of a current sheet rising in the solar atmosphere, Heyvaerts and Priest (1976) considered the thermal equilibrium inside the diffusion region determined by the energy balance

$$E = J + H + K - R, \tag{3.48}$$

where E represents the convective transfer of heat through the sheet; J, H and K represent the Joule heating, mechanical heating and conduction into the sheet, respectively, while R denotes the radiative loss. Each term is approximated and the equation solved for the internal temperature (T_c) of the sheet as a function of the height (h) above the photosphere. A similar analysis by Syrovatsky (1976) predicts qualitatively similar results, although he includes only Joule heating and radiative losses in his model.

Tur and Priest (1978) investigate the current sheet energy balance in more detail as follows. The equations are solved for a wide range of ambient magnetic field strengths (B_∞) and velocities (v_∞) in order to determine their effect on the critical height for the current sheet and to find the temperature at which the sheet becomes turbulent; also, one can distinguish the different ways in which the critical conditions may be reached.

For convenience they divide the solar atmosphere into three regions as follows:

1) Very low down, where the velocity of emerging flux (v_∞) is greater than or roughly equal to the local Alfvén speed (v_A).

2) Intermediate heights, typically the low chromosphere, where $v_{\max} < v_\infty < v_A$, $v_{\max}$ being the maximum allowed reconnection speed (taken as $v_A/40$ approximately). Here a current sheet may be expected to be formed without significant reconnection.

3) Higher up, $v_\infty < v_{\max}$ and the emerging flux reconnects easily with the overlying field.

The order of magnitude approximations of Tur and Priest to the various terms in (3.48) are as follows. The convective term is

$$E = \frac{\gamma}{\gamma - 1}\frac{p}{T}(\mathbf{v}.\boldsymbol{\nabla})T - (\mathbf{v}.\boldsymbol{\nabla})p$$

$$\approx \frac{\gamma}{\gamma-1}\frac{p_c}{T_c}\frac{v_\infty}{l}(T_c - T_\infty) - \frac{v_\infty}{l}(p_c - p_\infty),$$

where subscripts c and ∞ denote conditions inside and outside the sheet, respectively, v_∞ is the ambient inflow speed, l is the current-sheet width, and total pressure balance between inside and outside gives

$$p_c = p_\infty + \frac{B_\infty^2}{2\mu}.$$

The Joule heating term is

$$\mathrm{J} = \frac{j^2}{\sigma},$$

where

$$j \approx \frac{B_\infty}{\mu l}.$$

The radiative cooling and heating terms are taken as

$$R = n_c^2 Q(T_c),$$

$$H = n_c n_\infty Q(T_\infty).$$

Finally, the conductive term is approximated by

$$K = \nabla . (\kappa \nabla T)$$

$$\approx \kappa_{\|} \frac{T_\infty - T_c}{L^2} + \kappa_\perp \frac{T_\infty - T_c}{l^2},$$

where L is the sheet length.

In the non-reconnecting regime, the sheet width and length are approximated by

$$l = \frac{\eta}{v_{\max}},$$

and, from the results of Tur and Priest (1976),

$$L = 2.38(h - h_1).$$

For the reconnecting regime, one puts

$$l = \frac{\eta}{v_\infty},$$

and, from mass continuity,

$$L = l \frac{v_{\text{out}}}{v_\infty} \frac{\rho_c}{\rho_\infty},$$

where the outflow speed is

$$v_{\text{out}} = \frac{B_\infty}{(\mu \rho_c)^{1/2}}.$$

Equation (3.48) may therefore be solved for the sheet temperature T_c in terms of T_∞, n_∞, B_∞ and v_∞. A model atmosphere relating n_∞ and T_∞ to height h is

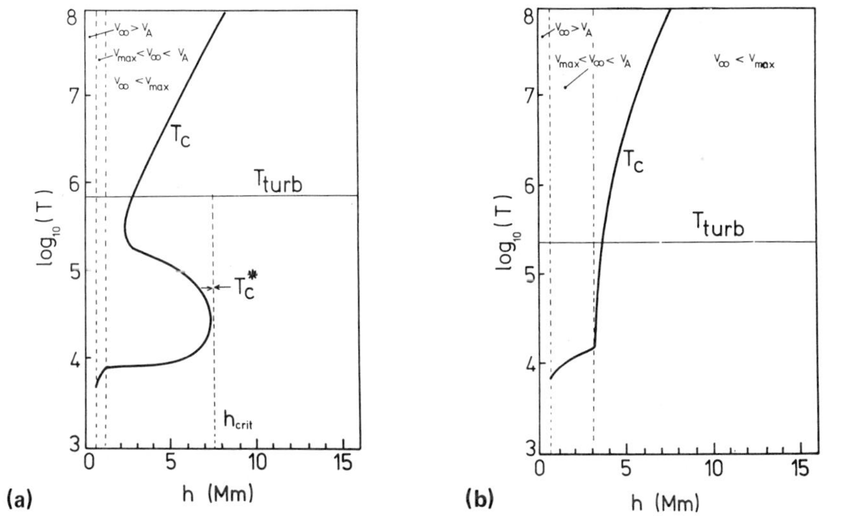

FIGURE 3.23 The variation of the internal current-sheet temperature T_c with height h for (a) $B_\infty = 100$ Gauss, $v_\infty = 700\ \mathrm{ms}^{-1}$ and (b) $B_\infty = 100$ Gauss, $v_\infty = 7\ \mathrm{kms}^{-1}$.

chosen and then T_∞ is plotted as a function of h for various values of v_∞ and B_∞ (Figure 3.23).

The internal temperature T_c may exceed the critical value (T_{turb}) for the onset of turbulence in two rather different ways. Firstly, as the current sheet gains in height, T_c may increase monotonically in the neighbourhood of T_{turb}, and the critical value will be slowly surpassed (Figure 3.23b), the sheet passing from a classically conducting to an unstable state on a time-scale determined by the speed at which it rises. On the other hand, if the graph of T_c becomes triple-valued (Figure 3.23a), the threshold for instability may be reached on a much shorter time-scale as the sheet undergoes a dynamic transition from one stable part of the curve to another. In this latter case, the dynamics of the increase in T_c between points B and C (Figure 3.22) presents a more complex problem than the simple energetics discussed in the preceding sections. Clearly, the sheet will never attain the new equilibrium position D on the upper branch of the curve, since it becomes turbulent before reaching this point.

As the temperature T_{turb} is exceeded near C and the plasma becomes turbulent, there are several effects. First of all, the magnetic diffusivity increases to a turbulent value, which makes the width of the sheet increase and the current density decrease. This in turn alters the energy balance. The convective flux E increases while the Joule heating J decreases. Also, the thermal conduction coefficient, $\kappa_\parallel$ and $\kappa_\perp$ change to turbulent values; the coefficient parallel to the magnetic field decreases, while the coefficient across the field increases. The net imbalance in the energy equation (3.48) means that the temperature T_c must rise, both when the equilibrium $T_c(h)$ curve is monotonic and when it is triple-valued. One result of this temperature increase is that the radiation falls and another is that the critical current density j_{turb} falls in value, which means that, in spite of the decrease in current density j as the sheet broadens, the critical conditions $j = j_{\text{turb}}$ for the onset of microinstabilities can still hold. In fact, a fully-turbulent value for η means in practice that a steady sheet would be too wide for the condition $j \geqslant j_{\text{turb}}$ to hold, so the resulting steadily reconnecting sheet is marked by a *marginally* turbulent diffusivity η^*, whose value, somewhere between the classical and fully-turbulent cases, is determined by the condition $j = j_{\text{turb}}$.

When the current sheet is at height h_{crit} and the point C is reached in Figure 3.22, the plasma starts becoming turbulent, but it is not in equilibrium and so continues heating until a new marginally turbulent equilibrium is reached, indicated in Figure 3.23a by the small arrows that locate the temperature T^*. There is even the possibility of oscillations about the marginally turbulent equilibrium. This new equilibrium is determined by Equation (3.48), with E, J, H, K, R, n_c, l, Lgiven as in the pre-turbulent situation, but now η is replaced everywhere by the unknown marginally turbulent diffusivity η^* and the thermal conduction coefficients $\kappa_{\|}$, $\kappa_{\perp}$ are replaced also by marginally turbulent values $\kappa_{\|}^*$, $\kappa_{\perp}^*$. The introduction of a new variable η^* demands the addition of an extra equation, namely

$$j = j_{\text{turb}},$$

which reduces to

$$\frac{1}{\mu\eta^*} = 1.8 \times 10^{13} \frac{B_\infty}{v_\infty T_c^{1/2}}.$$

The new thermal conduction coefficients are given by

$$\kappa_{\perp}^* = \kappa_{\|} \frac{\eta}{\eta^*},$$

since the product $\kappa_{\|}\eta$ is independent of collision frequency, and

$$\kappa_{\perp}^* = \kappa_{\perp} \frac{\eta^*}{\eta},$$

since both $\kappa_{\perp}$ and η are in practice proportional to collision frequency.

The existence of two distinct types of graph for $T_c(h)$ indicates that there are two different ways in which the solar flare may be triggered. Flux emerging with a high velocity or into a relatively low ambient field will have a single-valued temperature function, or at least one which crosses T_{turb} on a stable branch. Alternatively, if the emerging flux finds itself in a region of high magnetic field strength, the current sheet so formed will become unstable in its transition from the cool branch of the $T_c(h)$ curve to the upper hotter one. This time the trigger may be expected to be more violent, perhaps being more likely to produce an impulsive rather than a thermal flare. Observations by Rust and Roy (1974) tend to support these ideas. In their paper they report on their intensive study of two particular flares during the "Campaign for Integrated Observations of Solar Flares" (CINOF). The X-ray burst associated with the first event was thermal in nature, and displayed a slow rise to maximum Hα flux. This flare was observed in a region of relatively weak field strength. In contrast, the second flare exhibited an impulsive phase, characterised by a large hard X-ray burst and occurred in a region of high magnetic field strength in a compact sunspot group. Despite the very different nature of the two flares, both were reported to have been triggered by the emergence of new magnetic flux. Vorpahl (1973) too has noted the tendency for slowly-rising flares to be thermal and fast-rising ones to produce non-thermal particles. Many flares may be categorised in this way, and Zirin (1974) notes the distinction between the two types in his classification of their physical properties.

Some general conclusions may be drawn from the results of Tur and Priest (1978) for a wide range of values of v_∞ and B_∞.

i) If B_∞ and v_∞ increase in the same proportions, the graphs of $T_c(h)$ do not vary to any great extent.

ii) The critical height at which the sheet becomes unstable varies considerably. As either v_∞ decreases or B_∞ increases, h_{crit} increases in value, often giving a site for the flare onset well into the corona. For B_∞ and v_∞ close to the dividing line between single- and triple-valued solutions, h_{crit} is typically 7.5×10^6 m above the photosphere. The variation in h_{crit} is shown in Figure 3.24i as a function of v_∞ for several values of B_∞.

iii) The internal temperature of the current sheet at the critical point of those curves that are triple-valued does not vary significantly with either v_∞ or B_∞. In most cases it lies between 15,000 and 20,000 K.

iv) For large field strengths and low velocities, the sheet length L exceeds the upper realistic value of 10^7 to 10^8 m, and the present theory may become invalid.

v) In the analysis, it was assumed that the Buneman microinstability was responsible for the onset of turbulence in the sheet. However, one must point out that other microinstabilities exist with an even lower temperature threshold; this has the effect of changing the quantitative details but not the qualitative features of the model.

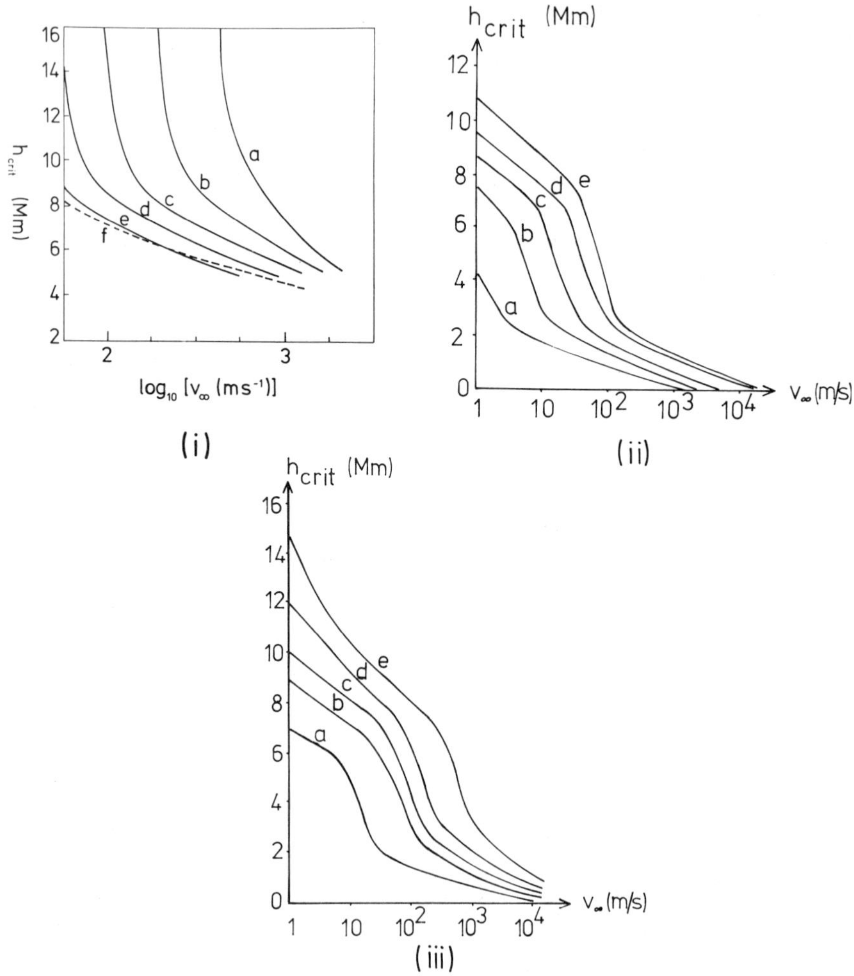

FIGURE 3.24 The variation of the critical height h_{crit} for simple-loop-flare onset with the emergence speed v_∞ for several values of the ambient magnetic field B_∞ (in tesla). (i) Order of magnitude solutions of Tur and Priest for: (a) $B_\infty = 10^{-1.5}$; (b) $B_\infty = 10^{-1.75}$; (c) $B_\infty = 10^{-2}$; (d) $B_\infty = 10^{-2.25}$; (e) $B_\infty = 10^{-2.5}$; (f) B decreasing with height. (ii) Solutions of Milne and Priest for a classical diffusivity and (a) $B_\infty = 10^{-1}$; (b) $B_\infty = 10^{-1.5}$; (c) $B_\infty = 10^{-2}$; (d) $B_\infty = 10^{-2.5}$; (e) $B_\infty = 10^{-3}$. (iii) As for (ii) but with a turbulent diffusivity that is a 100 times larger than the classical value.

The above analysis has treated the diffusion region in an order of magnitude way, but Milne and Priest (1980) have repeated the analysis with the improved model for a diffusion region described in §3.3.4. The resulting values of h_{crit} for a classical diffusivity η are shown in Figure 3.24ii, from which it can be seen that the variation with B_∞ has the opposite trend to Figure 3.24i so that h_{crit}

decreases in value as B_∞ increases. This is because of the low-β limitation (§3.3.4), which makes equilibrium solutions impossible when β is too low, as indicated by the dots in Figure 3.16b. If the possibility of a low-β limitation had not been included, the B_∞-variation would have been similar to Figure 3.24i. The predicted height for the onset of flare activity is rather low in this model (typically upper chromospheric heights for $v_\infty \geqslant 10^2$ ms^{-1}). However, there is evidence that the solar atmosphere is in a weakly turbulent state with the diffusivity enhanced by a factor of at least 100. (Heavy ions appear in the solar wind in nearly coronal abundances, which seems unlikely unless the Coulomb collision-rate is enhanced by turbulence (Benz, private communication); for example, Benz and Gold (1976) suggest the presence of Whistler turbulence.) The effect of this is to increase the length-scales l_0 and L_0 by a factor 100 over their previous values and to increase the critical height in the emerging flux model. Figure 3.24iii shows that equilibrium may then persist into the middle of the transition region for significantly higher values of the inflow velocity than in the non-turbulent model.

Further work, both theoretical and observational, is needed to set the emerging flux model on a firmer footing. Theoretically, the dynamics of the current sheet need to be followed during its search for a new equilibrium state. Observationally, more measurements of emerging-flux magnetic and velocity fields prior to flare are urgently required.

3.5.2 Two-ribbon flare

For a large two-ribbon flare, the production of a small turbulent region by the above process stimulates the release of stored magnetic energy from the much larger overlying force-free field. Observations of two-ribbon flares indicate the following stages, as shown diagrammatically in Figure 1.3.

i) New magnetic flux emerges near or directly beneath an active-region filament (Rust and Roy, 1974; Zirin, 1974). The photospheric field strength may be as much as 1,500 G and the rate of emergence of each polarity is about 10^{19} Mx h^{-1} (Martin and Harvey, 1973). The speed of emergence of the top of the loops may well be the same as that for an emerging arch-filament system, namely 1–10 km s^{-1} (Bruzek, 1969). The filament often becomes disturbed with motions away from the emergence site (Rust *et al.*, 1975).

ii) Hα knots of emission appear near the two ends of the filament and the emerging field region.

iii) Hα brightening occurs next in the form of two ribbons, which move apart, one being on each side of the filament. Sometimes the filament remains, though somewhat disturbed, but usually it rises and disappears, perhaps untwisting in the process. Work by Vorpahl (1973) on a particular active sunspot group showed that 15 out of 16 flares occurred near and were preceded by the emergence of new magnetic flux. The flares were larger and more frequent when there was emergence close to a sunspot, when filament reorientation was occurring and when the emerging flux was most vigorous.

In a complex active region, the magnetic field appears to depart greatly from a potential configuration. Evidence for this is the fact that filaments and fibrils in filament channels run largely parallel to the boundary between positive and negative fields on the solar surface. It is the energy stored in nonpotential fields that is probably the source of energy for a two-ribbon flare. The free energy contained in a force-free field, for example, is certainly adequate to account for the 10^{25} J (10^{32} erg) released in a large flare (Tanaka and Nakagawa, 1973). Furthermore, after the flare is over and the field has relaxed, photospheric motions of the magnetic footpoints can again build up force-free fields and store more energy, ready to be released as an homologous flare. For instance, a shear flow parallel to the axis of the cylinder with speeds of 100 ms^{-1} would move a frozen-in footpoint a distance 10^7 m in a day and so build up appreciable force-free currents.

An idealisation of the typical overall magnetic configuration around the filament is a cylindrical dome, as shown in Figure 3.25 and discussed in §6.4. The field lines make a smaller angle in crossing the polarity inversion line than do those of the corresponding potential field, and the amount of free energy (i.e. energy in excess of that stored in the potential field) increases with the degree of shear. The central portion of the structure, where the filament is located, may even possess twisted field lines (Figure 6.13a).

If the plasma is frozen to the field lines, the force-free configuration is in some cases in stable equilibrium, with the topological connection of each field line prescribed. However, the plasma does possess a finite, albeit large, electrical conductivity, which permits the tearing mode instability (§3.2.4.C, §3.3.3) to take place on a time-scale of order

$$\tau_{tmi} = \tau_d^{1/2} \tau_A^{1/2},$$

where $\tau_A = d/v_A$ is the Alfvén travel-time and τ_d the diffusion time. With $B_\infty = 0.01$ tesla (100 G), $n_\infty = 10^{15}$ m^{-3}, $T = 10^6$ K, $d = 10^7$ m, we find $\tau_A = 1$ s, so

$$\tau_{tmi} = 3 \times 10^6 \text{ sec.}$$

Under normal quiet conditions, with a classical electrical conductivity, the field thus relaxes back to equilibrium over the course of a few weeks, which is much longer than the day that it takes photospheric motions to build up the force-free field.

For a two-ribbon flare the fast release of the excess energy in the force-free

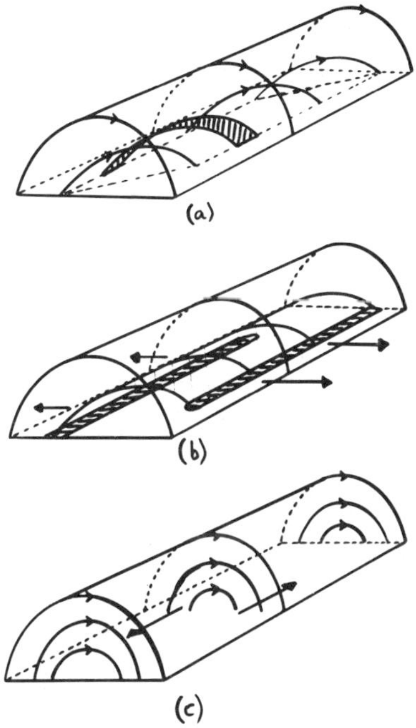

FIGURE 3.25 The cylindrical dome-like magnetic configuration for a two-ribbon flare. (a) New flux is able to emerge most easily near the centre of the dome. During the impulsive phase of Figure 3.21, particles escape from the current sheet along field lines and produce Hα knots where the lines meet the chromosphere. The overlying sheath of field lines, which links the current sheet, becomes filled with turbulence (shown shaded). It may well be sheared much more than indicated. (b) *Either* the enhanced reconnection causes an eruption and the two ribbons are formed as the field closes back down *or* occasionally there may be no eruption; in that case successive overlying cylindrical sheets become subject to a tearing-mode instability, based on a turbulent conductivity. Heat travels down along a particular cylindrical sheet and gives rise to the Hα ribbons, which move apart as the overlying structure becomes infected. (c) After the flare is over, a current-free configuration remains and shearing photospheric motions can proceed to store more excess magnetic energy in new force-free fields.

field may be made possible by both the spread of turbulence (§8.11) into the overall structure and also the presence of the large (say 10^7 m long) turbulent current sheet, which provides a local increase in the magnetic diffusivity by a factor of 10^4 or 10^5, say. Tsuda and Ugai (1977) have examined the effect of such a local diffusivity increase on a two-dimensional field and find that rapid reconnection occurs after some multiple of the Alfvén travel time-scale (§3.3.6). We expect a similar effect for our force-free field.

For the two-ribbon flare model, we envisage the following sequence of events (Figure 3.25).

i) *Preflare Heating.* The new flux emerges and continuously reconnects with part of the overlying prominence field. Heat is generated by the shock waves which radiate from a small, classically resistive current sheet (§3.3.3) and produces the slowly increasing thermal X-ray intensity that often precedes the flash phase. The reconnected field lines take on the relevant potential configuration, while the overlying force-free structure continuously adjusts to the presence of the emerging flux on the Alfvén time-scale. Due to untwisting of the outer layers of the prominence field, some heating may occur along the filament, and the prominence may start to rise slowly (Figure 1.3a). Material may move along the filament away from the site of the emerging flux.

ii) *Impulsive Phase*. When the current sheet reaches such a height that a neighbouring equilibrium no longer exists (point B in Figure 3.22), it begins to heat up dynamically (moving along BC). During this dynamic change, the critical current density for turbulence onset is surpassed (point C) and a regime of weak ion-sound turbulence is produced. The resulting anomalous resistivity causes the current sheet to expand rapidly. An induced electric field is therefore generated, which forces the electrons in the current sheet to behave in a runaway manner and/or to generate high-frequency turbulence. Fast particles are thus accelerated either directly or stochastically. They escape along field lines that link the sheet and, where these field lines intersect the low chromosphere, intense knots of Hα emission are produced, as indicated in Figure 1.3b. Typically, they first appear near the ends of the filament and near the points of emerging flux. The initial acceleration takes a fraction of a second; if the conditions for fast-particle production are not attained, then the impulsive phase is absent, but the presence of the turbulent current sheet still enables the flash phase to occur. (Note that impulsive particle acceleration may possibly continue in a much larger volume while the overlying force-free structure destabilises; this deserves more study, but may explain the much larger number of fast particles and the longer-lasting impulsive phase that are observed in some great events.)

The question of particle acceleration is most important and is discussed at some length in Chapter 8. We note here that in an emerging flux situation there are several possibilities for particle acceleration. A certain number is accelerated in the neutral sheet separating new and old flux, both in the impulsive expansion of the sheet at the onset of turbulence and in the reconnection process. But this is nowhere sufficient to explain the enormous number accelerated in large two-ribbon flares. In that case there may be acceleration in the shocks that propagate from the flash point and in the untwisting field around the rapidly erupting filament; also, the particles that propagate from the flash point (the current sheet) may generate turbulence and trigger the tearing mode at many locations throughout the large-scale structure (§8.11). Finally, there will be some acceleration during the large-scale reconnection that takes place in the main phase as the open field closes back down. Not all the fast particles need to originate in the corona, because some chromospheric particles may first be evaporated into the corona and then accelerated to high energies.

iii) *Flash (or Explosive) Phase.* Turbulence is generated not only in the current sheet but also along flux tubes into which fast particles are injected from the sheet. The resulting local increase by several orders of magnitude in the magnetic diffusivity destabilises the overlying force-free field, which may contain much more excess magnetic energy than the emerging flux. The instability is evidenced by a dramatic increase in filament activation. The filament suddenly erupts much more quickly than before, sometimes untwisting as it does so. The excess magnetic energy is released as mass motion and also as heat and particle energy, which are ducted down to the chromosphere to produce two ribbons of Hα emission (Figure 1.3c). The erupting filament and its surrounding magnetic arcade show up as a mass ejection of cold and hot material, preceeded by a bow shock, in the manner outlined by Sturrock (1980).

iv) *Main Phase.* The eruption of the filament and its largely frozen-in magnetic field leaves behind either an open magnetic structure or a closed one. If an open field is produced, we follow Kopp and Pneuman (1976) in suggesting that, during the main phase, it slowly closes back down toward a potential structure, as shown schematically in Figure 7.13. The point P is a Y-point (or cusp) of the field and slowly rises as successive field lines close down. From P there trail two slow magnetohydrodynamic shock waves, whose effect is to bring to rest plasma which is travelling outward along the newly closed flux tube. Behind the shocks, there is a loop-shaped region, which contains hot stationary material and is inclined slightly to the field lines. Inside that region lies another loop-shaped region, which Kopp and Pneuman identify with a post-flare loop-prominence system; it contains plasma which has had time to cool radiatively and so to start falling along field lines under gravity. Heat is conducted from the shocked plasma down to the chromosphere, where it produces two ribbons of Hα emission, which move outward as P rises. The leading edge of a ribbon is also heated by fast particles that are accelerated either in the shocks or during the reconnection at P. The condensing plasma falls to the trailing edge of a ribbon. On the disk, one observes loops of material linking the ribbons, which would be described as post-flare loops if observed on the limb. (An alternative explanation [Priest, 1976] of the main phase requires that the flash [or explosive] phase leaves a closed cylindrical dome-like structure [with successively higher field lines sheared to one another], rather than a basically open structure. The erupting filament would then be regarded as originally sitting on top of the closed structure rather than being embedded in it. During the main phase, successively higher cylindrical sheets would be destabilised due to the presence of the localised region of enhanced resistivity. The feet of such a sheet would give the two outward-moving Hα

ribbons, and the observed rotation of the post-flare loops would correspond to the shear in the structure. More observational information on the detailed field structure around the filament and above the post-flare loops is needed before one can decide whether the loops result from the closing down of an open configuration or the spreading of a disturbance through a closed one.)

v) *Recovery Phase.* After the flare is finished, the magnetic configuration that remains is largely (if not completely) a potential one, since most of the energy associated with the currents has been released. Shear motions parallel to the filament channel can then build up new force-free fields in the structure, in preparation for the emergence of more new flux to trigger another flare.

It will be noticed that the model has similarities with that of Sturrock (1974), who suggests that, during the preflare phase, a force-free field with excessive shear erupts into an open configuration like that of Figure 7.13. Particles are then accelerated at the impulsive phase by a tearing-mode instability in the open sheet, and the flash and main phases occur as the point P rises. By comparison, in our model it is the emergence of new flux which initiates both the filament eruption and the impulsive phase, and the subsequent closing down of field lines is a purely main-phase phenomenon.

3.6 CONCLUSION

Current sheets are of central importance for solar flare models since they represent regions of greatly enhanced magnetic dissipation and are a possible site for particle acceleration, *either* by direct electric fields *or* by time-varying fields as current sheets explode *or* in regions between the shocks that emanate from current sheets. Magnetic reconnection has several effects. It allows for a change of magnetic topology and prevents the build-up of excessive quantities of magnetic energy by allowing a magnetic configuration to reduce to its lowest-energy state. Magnetic energy is converted into heat and kinetic energy, most of this taking place in slow shock waves when the process is quasi-steady.

The theory for the behaviour of current sheets has progressed greatly during the past ten years. Although often developed for a strict neutral line, it holds more generally for the limiting line which separates two flux systems. It is now possible to describe analytically the annihilation of fields which are not simply antiparallel, but may be inclined at any angle. Moreover, the only acceptable model for fast reconnection is Petschek's, which is now on a firm mathematical basis and gives a maximum reconnection rate of between a tenth and a hundredth of the Alfvén speed. Also, the nonlinear development of the tearing-mode instability is being understood.

Solar flare models are now becoming more sophisticated in an attempt to explain more of the observations. This means, for example, that the configuration studied by Sweet (with a current sheet between equal dipoles) is no longer thought to be completely relevant. Furthermore, the flare is not thought to be triggered initially in the open current sheet that resembles the geomagnetic tail. Indeed, it is clear that the geomagnetic substorm is not identical with the solar flare: there may be similarities in the overall theme, such as the presence of an instability leading to reconnection and the importance of both macroscopic and microscopic processes; however, the physical configurations and parameter ranges are not the same and the basic instabilities are now believed to be different.

There are many aspects of current-sheet behaviour which remain to be understood, such as the two-fluid and non-fluid aspects which are summarised by Sonnerup (1979) and Schindler and Birn (1978), especially in relation to the geomagnetic tail. We have begun to include the effects of compressibility, but, for instance, a compressible version of the Soward and Priest (1977) calculation has not yet been completed. There is a need for more numerical simulations, with configurations and an energy balance that are more appropriate to the solar atmosphere. Also, one would like to analyse the effect of line-tying on the resistive kink instability; it is probably similar to the effect on the ideal kink mode (e.g. Hood and Priest, 1979 — see §4.4.8).

References

Alidières, M., Aymar, R., Jourdan, P., Koechlin, F. and Samain, A., *Plasma Phys.* **10**, 841 (1968).
Anderson, O. A. and Kunkel, W. B., *Phys. Fluids* **12**, 2099 (1969).
Anzer, U., *Solar Phys.* **30**, 459 (1973).
Baum, P. J. and Bratenahl, A., *J. Plasma Phys.* **11**, 93 (1974a).
Baum, P. J. and Bratenahl, A., *Phys. Fluids* **17**, 1232 (1974b).
Baum, P. J. and Bratenahl, A., *Solar Phys.* **47**, 331 (1976).
Baum, P. J. and Bratenahl, A., *J. Plasma Phys.* **18**, 257 (1977).
Baum, P. J., Bratenahl, A. and White, R. S., *Phys. Fluids* **16**, 226 (1973a).
Baum, P. J. and Pollack, J., *J. Appl. Phys.* **44**, 163 (1973b).
Benz, A. and Gold, T., *Astron. and Astrophys.* **55**, 229 (1976).
Birn, J., *J. Geophys. Res.* (1980).
Biskamp, D. and Schindler, K., *Plasma Phys.* **13**, 1013 (1971).
Bogdanov, S., Tokarevskaya, N. P., Frank, A. G. and Khodzhaev, A. Z., *Sov. J. Plasma Phys.* **1**, 71 (1975).
Bratenahl, A. and Yeates, C. M., *Phys. Fluids* **13**, 2696 (1970).
Bruzek, A., *Solar Phys.* **8**, 29 (1969).
Chapman, S. and Kendall, P. C., *Proc. Roy. Soc.* **A271**, 435 (1963).
Coppi, B., Laval, G. and Pellat, R., *Phys. Rev. Letters* **16**, 1207 (1966).
Coppi, B., Galvao, R., Rosenbluth, M. N., Rutherford, P. H., *Fiz. Plasmy* **2**, 961 (1976).
Cowley, S. W. H., *J. Plasma Phys.* **12**, 319 (1974a).
Cowley, S. W. H., *J. Plasma Phys.* **12**, 341 (1974b).
Cross, M. A. and Van Hoven, G., *Phys. Rev.* **A4**, 2347 (1971).
Cross, M. A. and Van Hoven, G., *Bull. Am. Phys. Soc.* **18**, 1328 (1973).
Cross, M. A. and Van Hoven, G., *Phys. Fluids* **19**, 1591 (1976).
Dailey, C. L., Davis, H. A. and Lovberg, R. H., *Bull. Am. Phys. Soc.* **17**, 1011 (1972).

Dickman, D. O., Morse, R. L. and Neilson, C. W., *Phys. Fluids* **12**, 1708 (1969).
Drake, J. F. and Lee, Y. C., *Phys. of Fluids* **20**, 1341 (1977).
Dungey, J. W., *Phil. Mag.* **44**, 725 (1953).
Frank, A. G., in *Neutral Current Sheets in Plasmas* N. W. Basov (ed) Consultants Bureau, New York (1976).
Fukao, S. and Tsuda, T., *Planet. and Space Sci.* **21**, 1151 (1973).
Furth, H. P., *Nuclear Fusion Suppl.* **1**, 169 (1962).
Furth, H. P., Killeen, J. and Rosenbluth, M. N., *Phys. Fluids* **6**, 459 (1963).
Green, R. M., *IAU Symp.* **22**, 398 (1965).
Green, R. M. and Sweet, P. A., *Astrophys. J.* **147**, 1153 (1966).
Hayashi, T. and Sato, T., *J. Geophys. Res.* **83**, 217 (1978).
Heyvaerts, J., *Astron. and Astrophys.* **37**, 65 (1974).
Heyvaerts, J. and Priest, E. R., *Solar Phys.* **47**, 223 (1976).
Heyvaerts, J., Priest, E. R. and Rust, D. M., *Astrophys. J.* **216**, 123 (1977).
Hood, A. W. and Priest, E. R., *Solar Phys.* **64**, 303 (1979).
Imshennik, V. S. and Syrovatsky, S. I., *Soviet Phys. JETP* **25**, 656 (1967).
Kopp, R. A. and Pneuman, G. W., *Solar Phys.* **50**, 85 (1976).
Martin, S. and Harvey, K., *Solar Phys.* **32**, 398 (1973).
Milne, A. and Priest, E. R., *Solar Phys.*, submitted (1980).
Mitchell, H. G. and Kan, J. R., *J. Plasma Phys.* **20**, 31 (1978).
Ohyabu, N. and Kawashima, N., *J. Phys. Soc. Japan* **33**, 496 (1972).
Ohyabu, N., Okamura, S. and Kawashima, N., *Phys. Fluids* **17**, 2009 (1974).
Overskei, D. O. and Politzer, P. A., *Phys. Fluids* **19**, 603 (1976).
Parker, E. N., *Astrophys. J. Suppl. Ser.* **8**, 177 (1963).
Parker, E. N., *Astrophys. J.* **174**, 499 (1972).
Parker, E. N., *J. Plasma Phys.* **9**, 49 (1973a).
Parker, E. N., *Astrophys. J.* **180**, 247 (1973b).
Petschek, H. E., in *AAS-NASA Symp. on Phys. of Solar Flares*, p. 425 (1964).
Petschek, H. E. and Thorne, R. M., *Astrophys. J.* **147**, 1157 (1967).
Priest, E. R., *Solar Phys.* **47**, 41 (1976).
Priest, E. R. and Raadu, M. A., *Solar Phys.* **43**, 177 (1975).
Priest, E. R. and Sonnerup, B. U. O., *Geophys. J. Roy. Astron. Soc.* **41**, 405 (1975).
Priest, E. R. and Soward, A. M., *IAU Symp.* **71**, 353 (1976).
Rust, D. M. and Roy, J. R., AFCRL Environmental Research Papers No. 474 (1974).
Rust, D. M., Nakagawa, Y. and Neupert, W., *Solar Phys.* **41**, 397 (1975).
Rutherford, P. H., *Phys. Fluids* **16**, 1903 (1973).
Sakurai, T., *Publ. Astron. Soc. Japan* **28**, 177 (1976).
Sakurai, T. and Uchida, Y., *Solar Phys.* **52**, 397 (1977).
Sato, T., *J. Geophys. Res.* **84**, 7177 (1979).
Sato, T. and Hayashi, T., *Phys. Fluids* **22**, 1189 (1979).
Schnack, D. D. and Killeen, J. *Nuclear Fusion* **19**, 877 (1979).
Schindler, K. and Birn, J., *Phys. Rep.* **47**, 109 (1978).
Schindler, K. and Soop, M., *Phys. Fluids* **11**, 1192 (1968).
Smith, D. F. and Priest, E. R., *Astrophys. J.* **176**, 487 (1972).
Sonnerup, B. U. O., *J. Plasma Phys.* **4**, 161 (1970).
Sonnerup, B. U. O., *Solar System Plasma Phys.* C. F. Kennel, L. J. Lanzerotti and E. N. Parker (eds.) **3**, 45 (1979).
Sonnerup, B. U. O. and Priest, E. R., *J. Plasma Phys.* **14**, 283 (1975).
Soward, A. M. and Priest, E. R., *Phil. Trans. Roy Soc. Lond.* **A284**, 369 (1977).
Sturrock, P. A., in *Flare-related magnetic field dynamics* Y. Nakagawa and D. M. Rust (eds.), NCAR, Boulder p. 187 (1974)
Sturrock, P. A., *Solar Flares*, Colo. Ass. Univ. Press, Chapter 9 (1980).
Sweet, P. A., *Nuovo Cimento Suppl.* **8**, Ser X, 188 (1958a).
Sweet, P. A., *IAU Symp.* **6**, 123 (1958b).
Syrovatsky, S. I., *Soviet Astron.* **10**, 270 (1966).
Syrovatsky, S. I., in *Solar Flares and Space Res.* C. de Jager and Z. Švestka (eds.), p. 346 (1969).

Syrovatsky. S. I., in *Solar Terrestrial Phys.* E. R. Dyer (ed.), p. 119 (1970).
Syrovatsky, S. I., *Soviet Phys. JETP* **33**, 933 (1971).
Syrovatsky, S. I., *Comments Astrophys. and Space Sci.* **4**, 65 (1972).
Syrovatsky, S. I., *Soviet Astron. Letters* **2**, 35 (1976).
Syrovatsky, S. I., private communication (1978a).
Syrovatsky, S. I., *Solar Phys.* **58**, 89 (1978b).
Syrovatsky, S. I., Frank, A. G. and Khodshaev, A. Z., *Sov. Phys. Tech. Phys.* **18**, 580 (1973).
Tanaka, K. and Nakagawa, Y., *Solar Phys.* **33**, 187 (1973).
Tsuda, T. and Ugai, M., *J. Plasma Phys.* **18**, 451 (1977).
Tur, T. J., Ph.D. Thesis, St. Andrews University (1977).
Tur, T. J. and Priest, E. R., *Solar Phys.* **48**, 89 (1976).
Tur, T. J. and Priest, E. R., *Solar Phys.* **58**, 181 (1978).
Uchida, Y. and Sakurai, T., *Solar Phys.* **51**, 413 (1977).
Ugai, M. and Tsuda, T., *J. Plasma Phys.* **17**, 337 (1977).
Ugai, M. and Tsuda, T., *J. Plasma Phys.* **21**, 459 (1979a).
Ugai, M. and Tsuda, T., *J. Plasma Phys.* **22**, 1 (1979b).
Van Hoven, G., *Solar Phys.* **49**, 95 (1976).
Vasyliunas, V. M., *Rev. Geophys. Space Phys.* **13**, 303 (1975).
Vorpahl, J. A., *Solar Phys.* **28**, 115 (1973).
Yang, C. K. and Sonnerup, B. U. O., *Astrophys. J.* **206**, 570 (1976).
Yeh, T. and Axford, W. I., *J. Plasma Phys.* **4**, 207 (1970).
Zirin, H., *Vistas in Astron.* **16**, 1, (1974).

CHAPTER 4

Simple-loop flares: magnetic instabilities

GERARD VAN HOVEN

Department of Physics, University of California, Irvine, California 92717

The era of rocket and spacecraft observations of the Sun has provided an entirely new view of the structure and evolution of the solar atmosphere. It is now clear, particularly since the extended series of Skylab flights, that the lower corona is quite non-uniform. In active regions, the strong ambient magnetic fields collimate and confine the emissive plasma into a myriad of loops and arcades. These features are observed to evolve slowly, with a time-scale much longer than the relevant hydromagnetic or Alfvén period and, therefore, must be considered generally to be stable. A conspicuous exception is the sporadic flare activity of these loops, which is believed to be due to localized departures from infinite-conductivity behavior.

4.1 INTRODUCTION

In this chapter, we will describe the current state of knowledge about the structure and stability of coronal loops. We will begin, in Section 4.2, with an exposition of the observations and data analyses which describe the two main physical aspects of atmospheric features, the plasma and the magnetic field. The high temperature of the solar corona dictates that the plasma be observed mainly above the terrestrial atmosphere in the extreme ultraviolet (EUV) or X-ray spectral range. The magnetic field cannot, at present, be observed in coronal features, and must be theoretically extrapolated from solar surface measurements.

An important conclusion of the EUV (Foukal, 1976) and X-ray (Krieger, 1977) observations of coronal loops is that they appear to be stable, with some examples maintaining their identities for periods longer than a day. One of the major aims of this chapter is to describe the steps necessary, and the progress made, in the theoretical explanation of this empirical fact.

It is indeed difficult, from a plasma-physics viewpoint, to understand the behavior of these relatively long-lived, large-scale loops. This is because of the known tendency for cylindrical and toroidal laboratory-plasma structures to be unstable on the very short magnetohydrodynamic (MHD) time-scale

(Newcomb, 1960; Goedbloed, 1971b). This time, 1–10 sec, is a factor $S^{-1/2}$ smaller than the resistive flaring time and is *even smaller* than the observed quiescent period between flares. (Here S is the magnetic Reynolds number, which is typically $>10^9$ for large-scale coronal structures — see §4.5.)

The challenge, then, is to find a model for the plasma/field structure of such a loop which can explain its manifest stability. The ambient physical conditions of high plasma temperature and large spatial scale indicate that one can generally use an ideal, infinite-conductivity ($S \gg 1$), MHD description. Such an MHD model has several aspects, which may be treated seriatim.

First, one must devise an equilibrium configuration of plasma and magnetic field, as described in Section 4.3.

Second, one must investigate the infinite-conductivity stability (Bernstein *et al.*, 1958) of this configuration. That is, are small perturbations stable (or oscillating), or do they grow in time? This aspect, which is determined by a linearization of the MHD equations in terms of a displacement from the equilibrium state, restricts the usable range of parameters found in step one. Examples of such calculations, which provide stability during the period when quiescent loops are observed, are given in Section 4.4.

The *third* step, if one finds ideal MHD stability, is to consider slower instabilities due to the local effects of finite resistivity. The primary process of this type is resistive magnetic tearing (Furth *et al.*, 1963; Van Hoven, 1979), which decouples the plasma from the field in certain singular layers, and allows the release of magnetic energy which is observed as a solar flare. Such processes are treated in Section 4.5. The final section (4.6) contains a brief assessment of the present state of coronal loop flare models.

4.2 PHYSICAL CONDITIONS

Although coronal condensations have been observed in high-temperature optical-line radiations for more than twenty years, quantitative measurements of their plasma properties must generally be made in the EUV or in X-rays, which do not penetrate the terrestrial atmosphere. The first hard-radiation images of coronal loops were obtained by a rocketborne X-ray filter telescope (Vaiana *et al.*, 1968), and the subsequent Skylab flights of 1973–74 confirmed that such loops were *the* characteristic atmospheric feature of the flare-producing active regions of the Sun (Kahler *et al.*, 1975; Vorpahl *et al.*, 1975).

4.2.1 Plasma parameters

A full determination of the physical parameters describing the loop plasma requires spatial and spectroscopic images in lines which cover the temperature

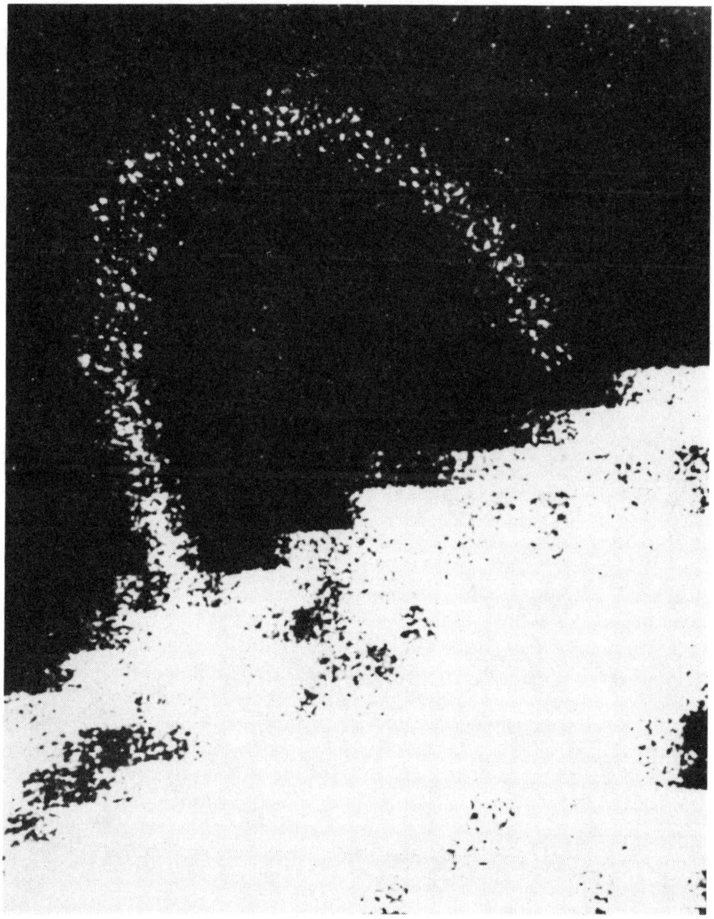

FIGURE 4.1 A cool coronal loop, observed in a rastered display of the emission from an EUV line of OIV (Foukal, 1976).

span from the photosphere to the flaring corona. At the present time, the lower end of this range (*v.* Figure 4.1) is covered by EUV lines, with good temperature discrimination but poor density sensitivity (Foukal, 1975). The higher end is covered by X-ray filtergrams (*v.* Figure 4.2), with poor temperature differentiation, but the ability to diagnose plasma pressure in some cases (Krieger, 1977). The resulting picture of a typical coronal structure is thus a composite one.

In this chapter, we will concentrate our attention on *preflare, active-region loops*. For this subclass, the electron number density n_e can range from $10^{14.7}$ to $10^{16.6}$ m^{-3} ($10^{8.7}$ to $10^{10.6}$ cm^{-3}), with the span 10^{15} to 10^{16} m^{-3} (10^9 to

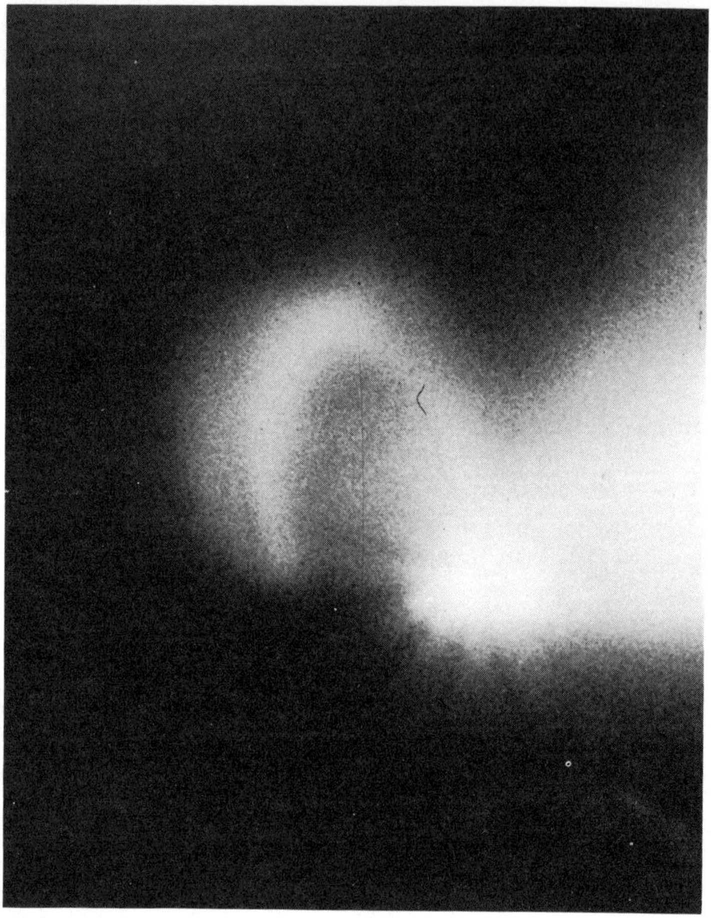

FIGURE 4.2 A hot coronal loop, as seen in broadband soft X-rays (courtesy A. S. Krieger).

10^{10} cm^{-3}) more commonly quoted. Electron temperatures T_e between 10^4 and $10^{6.5}$ K have been stated in the literature, with the range 10^5 to $10^{6.3}$ K well documented.

The spatial distribution of these quantities is less clearly determined. By virtue of its designation as a loop, such a structure is longer than it is wide. The *ratio* R/a of the global radius of curvature R to the transverse radius a is larger than 10 usually. Loop lengths (πR) lie between 10^4 and 10^5 km, with the shorter loops exhibiting higher densities and pressures. In spite of the fact that they extend beyond a gravitational scale-height in some cases, the loops exhibit a remarkable axial uniformity, both in width and brightness (Foukal, 1976; Krieger, 1977).

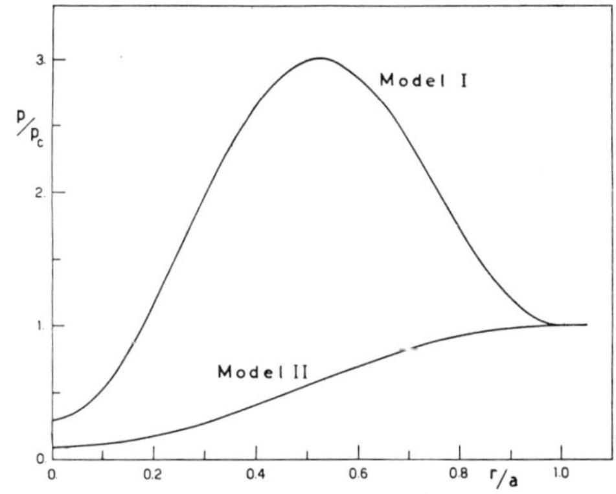

FIGURE 4.3 Models for transverse pressure profiles which qualitatively correspond with loop observations (Chiuderi *et al.*, 1977).

4.2.2 Transverse plasma profiles

The *transverse* variation of n_e and T_e is not well established, partly because of problems of spatial resolution, and partly because the range of temperatures encountered requires the coordinated analysis of data from different instruments. Gabriel and Jordan (1975) and Foukal (1975) have analyzed the observed structure of large, relatively cool loops and have shown them to have *relative* temperature minima $T_e < 10^6$ K on their axes; Foukal finds an annular shell structure with a monotonic variation and $T_e < 10^5$ K as a central value. There is less agreement about the variation of number density, Gabriel and Jordan (1975) showing a drop toward the axis and Foukal (1975) showing near constancy or a slight rise.

As we will see in Section 4.4, an important determining influence for hydromagnetic stability of the plasma itself is the *pressure profile*. In the cool portions of the loops this variation is found from the product of temperature and density. For the large loops analyzed in the EUV, as shown in Figure 4.1, the overall profile appears to exhibit a rise from the external ambient coronal value of $p_c \approx 10^{-1.5}$ N m^{-2} to an off-axis pressure maximum of 5–15 p_c, as shown schematically in Model I of Figure 4.3, then a drop to an axial value either greater or less than the coronal pressure, depending upon the presently undetermined central temperature. Foukal's (1975, 1978) diagnosis is also consistent with an alternative monotonic profile, as shown in Model II of the figure.

When coronal loops are observed in X-rays, as exemplified by Figure 4.2, none of the cool plasma described above is visible. What can be observed through some of the available broadband filters is a response which is directly proportional to $p(r)$, at an ideal spatial resolution approaching 1500 km. In the 2–3 × 10^6 K range to which the filters respond, the loop profiles appear to exhibit an internal pressure maximum (Krieger, 1977) of less than a factor of ten larger than the external coronal value. However, the "point spread function" of X-ray telescopes tends to widen the image of intense sources, which would obscure the existence of a narrow cool core in such a loop. Krieger (1977) states that the existence of a high-temperature core can be specifically excluded and that the few available spectra indicate a mixture of hot and cool material. Although the X-ray data cannot verify the regular coaxial transverse variation seen in the EUV, it also does not contradict such an empirical model.

4.2.3 Magnetic fields

The greatest deficiency in our observational knowledge of the structure of coronal loops is the local, small-scale, form of the confining magnetic field. The primary reason is the absence of reliable direct measurements of the field in the upper solar atmosphere. A secondary fact in this regard is that a number of pieces of evidence indicate that magnetic forces dominate pressure (and gravitational) forces in active-region loops. In such a physical situation, the strength and detailed spatial variation of the magnetic field are the nearly exclusive determinants of loop stability. We will expand on these points in what follows.

Our present information on coronal magnetic fields is the result of a four-step process. *First*, measurements are made of the two opposite, circularly-polarized, Zeeman components of spectral lines originating from the cooler surface layers of the Sun. Subsequent subtraction of such images quantifies the strength of the line-of-sight field (averaged within the limitations of spatial resolution) in the photospheric or chromospheric stratum where the lines originate. *Second*, finite regions are selected for further analysis, usually (for "closed" lower coronal features) with vanishing net line-of-sight flux. These areas must not be too far from disk center, so that they will not suffer from perspective errors. The surface values are then expanded in a two-dimensional Fourier series in this region. *Third*, based on the Fourier amplitudes as a boundary-condition description, the atmospheric magnetic field is computed (Nakagawa, 1974; Levine, 1975). The algorithm commonly used assumes a force-free form for the field (*v*. Section 4.3), which satisfies the Helmholtz equation $(\nabla^2+\alpha^2)\mathbf{B}=0$. The parameter α/μ measures the proportionality between the current density and the field (including the current-free limit when $\alpha=0$) and, for feasible calculations, is taken to be a constant. *Fourth*, and finally, spatial plots of the field lines are compared with images of coronal plasma structures to choose, *a posteriori*, the constant value of α providing the best spatial correspondence. An example of such a comparison has been shown by Levine (1976), in which he makes the point that different parts of a complex region require different values of α for a reasonable match of fields with emissive features.

A crucial point to be made about such estimates of the coronal magnetic field is that, with present capabilities, they merely indicate the average global form of the field. This is reflected in values of $\alpha < 10^4\ \mathrm{km}^{-1}$, which can only give the coarse orientation of the field lines. However, the fact that coronal loops appear along a selected few of these flux tubes indicates some local differences in force balance (to support the observed pressure gradients) and in current density (to heat and confine the emissive plasma). This evidence means that the local field structure must have a smaller scale and increased shear as compared to that indicated by the global value of α. Methods to measure such local field orientations and strengths are not currently available.

4.2.4 Relative influence of plasma and magnetic field

One important plasma-physics parameter is $\beta \equiv 2\mu\, p/B^2$, the ratio of plasma to magnetic pressure. If $\beta \ll 1$ then the ambient plasma does not have a strong influence on the structure of coronal features, which is determined mainly by the magnetic field (*v.* Section 4.3).

It is generally believed that the $\beta \ll 1$ condition holds for the global, overall corona. However, due to the difficulties in determining the coronal field in specific, small-scale structures (as described above), this condition is not well established for coronal loops.

Since flaring coronal loops maintain their form through more than ten-fold increases in both density and temperature, pressure-balance arguments indicate that the preflare loop must have had a β value of at most 10^{-2}–10^{-1}. However, such simple considerations also provide very large estimates of the coronal magnetic field at extended heights in those cases when the actual plasma pressures (up to $n = 10^{18}\ \mathrm{m}^{-3}$ and $T = 10^7$ K (Krieger, 1978)) can be determined. Values of 0.02 T (200 G) at 50,000 km altitude are difficult to understand when compared with the usual global-field calculations. These considerations indicate that one should keep an open mind about the relative effect of plasma-pressure gradients in small-scale structures.

4.3 EQUILIBRIUM MAGNETIC FIELDS

The first step in a magnetohydrodynamic model of a coronal loop is the formulation of a stationary structure for the plasma and magnetic field. These physical quantities are related by the equilibrium MHD equations, which ignore flows and time variations.

The momentum-transport equation (1.2) then assumes the form

$$0 = -\nabla p + \mathbf{j} \times \mathbf{B}, \tag{4.1}$$

which states a balance between the pressure-gradient and magnetic forces. (Gravitational effects have been ignored by empirical constraint, as described in Section 4.2.1.) In cases where the pressure-gradient term is negligible, the magnetic configuration is called *force-free* and has **j** parallel to **B**. When (4.1) is solved for $\mathbf{j}_\perp$ and the result is inserted in Ampere's Law (1.5), one obtains the *equilibrium equation*

$$\begin{aligned}\nabla \times \mathbf{B} &= \mu(\mathbf{j}_\parallel + \mathbf{j}_\perp) \\ &= \alpha(\mathbf{r})\mathbf{B} + \mathbf{B} \times \nabla p / B^2,\end{aligned} \tag{4.2}$$

where the parallel and perpendicular subscripts refer to the direction of the magnetic field, and $\alpha(\mathbf{r})$ is the usual notation for the constant of proportionality of the aligned component. We see that perpendicular currents result only from significant pressure *gradients* across the field.

In the remainder of this section, we will solve Equation (4.2) in several different limits, beginning with the simplest case of the *constant-α, force-free field*, in which perpendicular currents can be ignored. This is the assumption used in the calculation of the global coronal field (as described in Section 4.2.3), although we will apply it to a smaller-scale variation. Such local current enhancements give rise to the confining fields which make coronal loops visible in the surrounding active-region atmosphere. We will also restrict our attention to cylindrical symmetry. That this approximation is allowable for the treatment of narrow coronal loops will be justified in Section 4.4.5.

4.3.1 Constant-α (linear), force-free fields

When α is uniform in space, the curl of Equation (4.2) gives the Helmholtz equation, as written in Section 4.2.3, and the equilibrium field has a periodic character. The axially-symmetric solution in cylindrical geometry, with only an r-variation, is given by (Lundquist, 1951)

$$B_z(r) = B_0 J_0(\alpha r), \qquad B_\theta(r) = B_0 J_1(\alpha r), \tag{4.3}$$

where B_0 is a common amplitude and the J_i are the cylindrical Bessel functions. Such fields steadily rotate with increasing r, as will be shown by example in Section 4.3.3.

The constant-α model has added importance as the minimum-energy state under certain constraints (Woltjer, 1958). This fact has been suggested by Taylor (1975) as the reason for the spontaneous generation of reversed fields in the toroidal z-pinch (Christiansen and Roberts, 1978). Finally, when resistive behavior is included, the preservation of the force-free condition with time requires the constancy of α (Jette and Sreenivasan, 1969).

4.3.2 Variable-α (nonlinear), force-free fields

A disadvantage of the Lundquist field (4.3) for coronal-loop applications is that the axial field continues to alternate its direction with increasing radius, a feature which is not observed in line-of-sight magnetograms. To eliminate this mathematical artifact, one must use force-free equilibria in which $\alpha(r)$ decreases with radius.

A commonly used model for this class of fields is that proposed by Gold and Hoyle (1960), which has constant *pitch* $[2\pi r B_z(r)/B_\theta(r)]$ and thus, by definition, no magnetic *shear*. (This circumstance has certain undesirable side effects when stability is considered, as will be described in Section 4.4.6.) The form of this field is

$$B_z = B_0(1+\tilde{\mu}^2 r^2)^{-1}, \qquad B_\theta = B_0 \tilde{\mu} r(1+\tilde{\mu}^2 r^2)^{-1}, \tag{4.4}$$

where $2\pi\tilde{\mu}^{-1}$ is the field-line pitch distance. Equation (4.4) indicates the value

$$\alpha(r) = 2\tilde{\mu}(1+\tilde{\mu}^2 r^2)^{-1},$$

so that the current density goes to zero as the radius increases. A clear treatment of this and several other constant-parameter fields is given in Goedbloed (1971a), and other analytical models are described in Ferraro and Plumpton (1966), Alfvén and Fälthammar (1963), Anzer (1968), Chiuderi *et al.* (1980), and Hood and Priest (1979).

However, as we will see in Section 4.4.6, stability calculations (with the exception of the constant-α case) must necessarily be performed numerically. Thus, it is convenient to have a simple and flexible form which can describe a range of physical situations. One that has been used by Chiuderi *et al.* (1980) has

$$\alpha(r) = \begin{cases} \alpha_0 & 0 \leqslant r \leqslant r_0, \\ \frac{1}{2}\alpha_0\left[1+\cos\frac{\pi}{\delta}(r-r_0)\right] & r_0 < r < r_0+\delta, \\ 0 & r_0+\delta \leqslant r < \infty, \end{cases} \tag{4.5}$$

which has the radial profile shown in Figure 4.4, along with the field

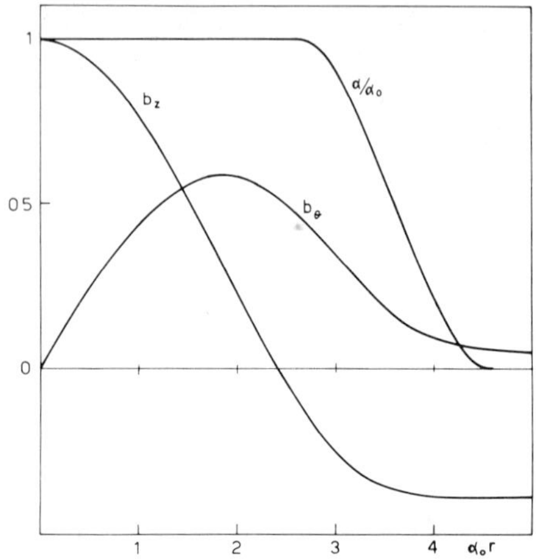

FIGURE 4.4 Profiles of cylindrical magnetic fields for an equilibrium [(4.5) with $\alpha_0 r_0 = 2.6$ and $\alpha_0 \delta = 2$] containing an axial force-free filament and an external potential-field region (Chiuderi *et al.*, 1980).

components, for a case in which $\alpha_0 r_0 > 2.4$, the first zero of $J_0(\alpha r)$. This equilibrium model is very similar to that observed in the *reversed-field pinch* (Christiansen and Roberts, 1978). For larger values of the parameters, the B_θ field can also reverse direction. At radii $r > r_0 + \delta$ the field is current-free and therefore must be locally stable, since it possesses no free energy.

It may be noted that for *cylindrically symmetric* fields the force-free condition becomes

$$\frac{\mathrm{d}}{\mathrm{d}r}(B_\theta^2 + B_z^2) = -\frac{2B_\theta^2}{r}$$

and so there are many other possible fields in addition to (4.4) and (4.5). One may choose a wide variety of functional forms for $B_\theta(r)$ (or $B_z(r)$) and then use the above equation to deduce $B_z(r)$ (or $B_\theta(r)$).

Instead of imposing $B_\theta(r)$ or $B_z(r)$, one may wish to specify the amount

$$\Phi(r) \equiv \frac{LB_\theta(r)}{rB_z(r)} = \frac{\pi}{q(r)}$$

by which a field line is twisted about the loop axis in going a distance $L(=\pi R)$ from one end of the loop to the other (Hood and Priest, 1979); $q(r)$ is the "safety factor" (Bateman, 1978) used in fusion-theory treatments of toroidal plasmas of major radius R.

4.3.3 The effects of pressure gradients

Transverse (r-directed) pressure variations have been observed, if not precisely quantified, as we have shown in Figure 4.3. To reflect such effects in an equilibrium, which is poorly constrained since there are more variables ($B_z(r)$, $B_\theta(r)$, and $p(r)$ in cylindrical symmetry) than equations, requires some added assumptions. In addition, caution must always be used in the application of equilibrium results, since the reality of the model and the self-consistency of the choices made can only be demonstrated by a subsequent stability analysis, as we will show in Section 4.4.

One physically-based set of postulates for the inclusion of pressure gradients has been devised by Chiuderi *et al.* (1977). They assume that the equilibrium is nearly force-free, so that the zero-order contribution to the right-hand side of Equation (4.2) is the first term (with α constant). The pressure variation is then taken from observations (*e.g.*, Foukal, 1978), using profiles such as those shown in Figure 4.3.

With these assumptions, the scalar product of **B** with Equation (4.2) is independent of p' (the prime denotes the r derivative) and can be transformed into a Ricatti equation (*e.g.*, Davis, 1962)

$$w' + \frac{1}{r} w - \alpha w^2 = \alpha$$

for the ratio $w = J_1(\alpha r)/J_0(\alpha r)$, so that this model has the same pitch as the Lundquist field. One should thus note that a "force-free" orientation of the field lines does not necessarily require a physically force-free condition.

With the field-direction ratio established, one can then write

$$B_z(r) = B_0(r) J_0(\alpha r), \qquad B_\theta(r) = B_0(r) J_1(\alpha r) \tag{4.6}$$

for the components, where $B_0(r)$ is a common amplitude.

The spatial variation of $B_0(r)$ is then determined by the radial component (pressure balance) of Equation (4.1). After substitution from (4.2) and (4.6), one obtains a first-order differential equation for $B_0^2(r)$. For our purposes, it is better to show the integrated result

$$B_0^2(r) = B_0^2(a) + 2\mu \int_r^a \frac{p'(s)\mathrm{d}s}{J_0^2(\alpha s) + J_1^2(\alpha s)} \tag{4.7}$$

for $r < a$, where a is the edge of the pressure profile of the loop, as shown in Figure 4.3. Finally, the assumed condition that the loop is not bounded by a

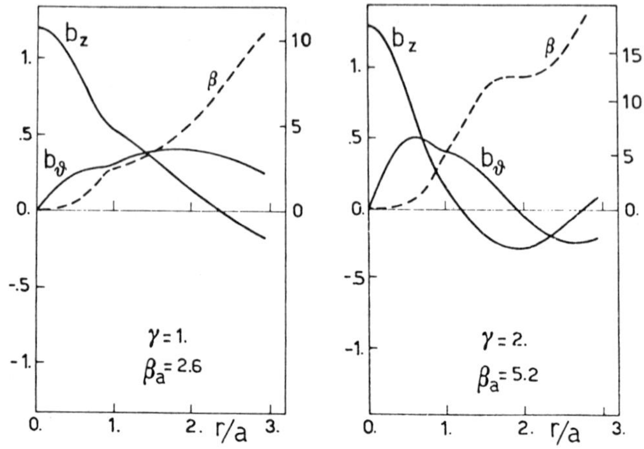

FIGURE 4.5 Profiles of pressure-modified magnetic fields, for Model II of Figure 4.3 (Chiuderi *et al.*, 1977). The value of β is given at the loop edge $r=a$.

current sheet, so that the field derivatives are required to be continuous ar $r=a$, dictates that the external value of α be the same as that inside. Thus, the field for $r>a$ is given by Equation (4.6) with $B_0(r)\equiv B_0(a)$, the uniform-pressure Lundquist field (Chiuderi *et al.*, 1977). [Equation (4.7) imposes a weak restriction on $B_0^2(a)$ such that $B_0^2(r)$ is everywhere positive. If $p'(r)<0$, then the field amplitude outside must be relatively larger.] An example of such a field profile, for a pressure variation similar to that of Model II of Figure 4.3, is shown in Figure 4.5.

A wide range of alternative, cylindrically-symmetric, analytic field expressions are possible, since (4.1) may be written

$$2\mu\frac{dp}{dr}=-\frac{d}{dr}(B_\theta^2+B_z^2)-\frac{2B_\theta^2}{r},$$

so that any two of the functions $p(r)$, $B_\theta(r)$, $B_z(r)$, may be prescribed and this equation used to deduce the third (Chiuderi *et al.*, 1977; Hood and Priest, 1979). For example, one may wish to impose $B_z(r)$ from longitudinal photospheric-field measurements and the local field twist $\Phi(r)$ from observed photospheric rotations. Two particular analytic fields have been considered for equilibrium and stability calculations by Hood and Priest (1979). The first is the *uniform-twist* (or *shear-free*) *field* (4.4), with B_θ multiplied by a constant factor $\Phi_0 b/L$ so that it can support a variable pressure, derived by integrating the force-balance equation, of

$$p=p_c+\frac{B_0^2[(b\Phi_0/L)^2-1]}{2\mu(1+r^2/b^2)^2},$$

given in terms of the external coronal value p_c at infinity. When there is no twist ($B_\theta = 0$), the pressure possesses a minimum value on the axis of $p_c - B_0^2/(2\mu)$; as the twist Φ_0 increases, the axial value rises and becomes a maximum when $\Phi_0 > L/b$. This field therefore models the effect of a uniform twist on a magnetic flux tube with a maximum field strength at $r = 0$. The second model is a *variable-twist field*, which models the effect of shear; it has a uniform axial field and a twist that decreases with r like $\Phi_0/(1 + r^2/b^2)^2$, so that

$$B_\theta = \frac{\Phi_0 r B_z}{L(1 + r^2/b^2)}$$

as before; the resulting pressure profile is

$$p = p_c + \frac{B_z^2 (b\Phi_0/L)^2}{2\mu(1 + r^2/b^2)^2},$$

which possesses a maximum at $r = 0$ that increases with Φ_0.

4.4 INFINITE–CONDUCTIVITY STABILITY

The true test of the magnetohydrodynamic self-consistency (to say nothing of the reality) of the loosely constrained equilibria described in the preceding section comes from stability considerations.

This is especially true for such tightly confined structures as coronal loops, because of the tendency of cylindrical or toroidal pinches to disrupt on the very short *MHD time-scale* (Goedbloed and Hagebeuk, 1972). This period is given by $\tau_a = a/v_a$, where $v_a = B(\mu\rho)^{-1/2}$ is the Alfvén velocity, and for typical coronal-loop parameters τ_a is of the order of seconds. However, as expressed by Foukal (1976), these narrow emissive features often exist on a "steady state" basis.

In the laboratory, z-pinches are stabilized by a combination of strong axial applied fields and close-fitting concentric conductors (Schmidt, 1979). The inapplicability of such conditions in the case of a clearly stable coronal loop presents a still incompletely answered challenge to the solar plasma theorist.

4.4.1 General dynamic considerations

The stability of a proposed equilibrium is tested by an evaluation of the dynamic behavior of a small perturbation of the plasma/field configuration (Bernstein *et al.*, 1958). Such a calculation is traditionally expressed in terms of the displacement vector $\boldsymbol{\xi}(\mathbf{r}_0, t)$ of a fluid element from its equilibrium (subscript zero) position $\mathbf{r}_0$ (Bateman, 1978; Schmidt, 1979). Various perturbed plasma quantities can then be expanded to the first order in $\boldsymbol{\xi}$ by assuming $|\boldsymbol{\xi}| \ll |\mathbf{r}_0|$.

If all of the terms in the momentum-transport equation (1.2) are treated consistently to this order, and the zero-order identity (4.1) is applied, one obtains the perturbation force equation. In the Eulerian representation this has the symbolic form (Bernstein *et al.*, 1958)

$$\rho_0 \ddot{\boldsymbol{\xi}} = \mathbf{F}(\boldsymbol{\xi}), \tag{4.8}$$

where the dot signifies time derivative and the force function

$$\begin{aligned} \mathbf{F}(\boldsymbol{\xi}) = \boldsymbol{\nabla}[\gamma p_0 \boldsymbol{\nabla} \cdot \boldsymbol{\xi} + (\boldsymbol{\xi} \cdot \boldsymbol{\nabla}) p_0] \\ + \mu^{-1}(\boldsymbol{\nabla} \times \mathbf{B}_0) \times \boldsymbol{\nabla} \times (\boldsymbol{\xi} \times \mathbf{B}_0) \\ + \mu^{-1}[\boldsymbol{\nabla} \times \boldsymbol{\nabla} \times (\boldsymbol{\xi} \times \mathbf{B}_0)] \times \mathbf{B}_0 \end{aligned} \tag{4.9}$$

is linear in $\boldsymbol{\xi}$ and otherwise depends on the equilibrium static pressure and magnetic field only.

Since $\mathbf{F}$ is independent of time, one can seek normal-mode solutions of the form $\boldsymbol{\xi}(r, t) = \boldsymbol{\xi}(\mathbf{r}) \exp(i\omega t)$, for which (4.8) becomes

$$-\omega^2 \rho_0 \boldsymbol{\xi} = \mathbf{F}(\boldsymbol{\xi}), \tag{4.8$'$}$$

wherein $(-\omega^2)$ appears as the real eigenvalue of the linear Hermitian operator $\mathbf{F}/\rho_0$ (Bernstein *et al.*, 1958).

To this equation of motion must be added boundary conditions on the flow and field vectors. These are given by the limiting forms of the linear MHD equations (Jeffrey, 1966),

$$\begin{aligned} \langle -\gamma\mu p_0 \boldsymbol{\nabla} \cdot \boldsymbol{\xi} + \mathbf{B}_0 \cdot \mathbf{B}_1 \rangle = 0, \\ \hat{\mathbf{n}}_0 \cdot \langle \boldsymbol{\xi} \rangle = 0 = \hat{\mathbf{n}}_0 \cdot \langle \mathbf{B}_1 \rangle, \\ \hat{\mathbf{n}}_0 \times \langle \mathbf{B}_1 \rangle = \mathbf{K}_1, \\ \hat{\mathbf{n}}_0 \times \langle \mathbf{E}_1 \rangle = i\omega\{(\hat{\mathbf{n}}_0 \cdot \boldsymbol{\xi})\langle \mathbf{B}_0 \rangle - (\hat{\mathbf{n}}_0 \cdot \mathbf{B}_0)\langle \boldsymbol{\xi} \rangle\}, \end{aligned} \tag{4.10}$$

where $\mathbf{B}_1$ is the Lagrangian field perturbation

$$\mathbf{B}_1 \equiv \mathbf{Q} + (\boldsymbol{\xi} \cdot \boldsymbol{\nabla})\mathbf{B}_0 \equiv \boldsymbol{\nabla} \times (\boldsymbol{\xi} \times \mathbf{B}_0) + (\boldsymbol{\xi} \cdot \boldsymbol{\nabla})\mathbf{B}_0, \tag{4.11}$$

$\langle\rangle$ denotes the "jump" discontinuity of the bracketed quantity across the equilibrium boundary with unit normal $\hat{\mathbf{n}}_0$, and $\mathbf{K}_1$ is a surface current density. These boundary conditions are written for the case of a plasma/plasma interface, applicable in the corona; different forms, for a vacuum or conducting boundary (assuming $\hat{\mathbf{n}}_0 \cdot \mathbf{B}_0 = 0$), have been given by Bernstein *et al.* (1958).

In principle, Equations (4.8′) and (4.10) may be solved for any equilibrium geometry of interest, and ω^2 evaluated to determine stability. However, this approach leads to insurmountable complexity (Goedbloed, 1971a, b, c) in all but the most idealized cases.

4.4.2 The energy principle and marginal stability

A simpler procedure is applicable, if one merely wishes to establish stability, and is not interested in the magnitudes of the eigenfrequencies ω. This method has two equivalent descriptions (Goedbloed, 1971a, b), based on the sign of the minimum perturbation potential energy (*the energy principle*) or on the conditions for having $\omega^2 = 0$ (*marginal stability*).

The energy principle (see also §6.5) results from the conservation of energy in the *first-order dynamic system* described by (4.8). This fact can be stated in the form

$$\tfrac{1}{2}\int[\rho_0\dot{\boldsymbol{\xi}}^2 - \boldsymbol{\xi}\cdot\mathbf{F}(\boldsymbol{\xi})]d^3\mathbf{r} = \delta K + \delta W = \text{constant}, \tag{4.12}$$

as shown by Bernstein *et al.* (1958). If the normal-mode eigenfunctions $\boldsymbol{\xi}_n(\mathbf{r})\exp(i\omega_n t)$ of (4.8′) form a complete set, then it is elementary to show that

$$\delta W = -\tfrac{1}{2}\int\boldsymbol{\xi}\cdot\mathbf{F}(\boldsymbol{\xi})d^3r \geqslant 0, \tag{4.13}$$

is equivalent to stability (all $\omega_n^2 \geqslant 0$) for every allowable perturbation satisfying the boundary conditions. However, the functional $\mathbf{F}$ does not usually generate a complete set of eigenfunctions, so that a more subtle proof of the necessity and sufficiency of (4.13) is required. This has been given by Laval *et al.* (1965), in a demonstration which uses only the *self-adjointness* of $\mathbf{F}$ and the requirement that the trial functions $\boldsymbol{\xi}$ satisfy proper boundary conditions.

One caution that must be observed in using the energy principle for a determination of *stability* is that the trial functions must be sufficiently general so that they can describe all allowable perturbations satisfying the boundary conditions. Any restriction on the form of the test displacements can eliminate permissible excitations which may be unstable. There is no such problem of systematic bias in the discovery of examples of instability, although restricted perturbations may not give the lowest threshold value.

The second-order energy integral can be subdivided (Bernstein *et al.*, 1958) into *v*olume

$$\delta W_v = \tfrac{1}{2}\int d^3\bar{r}[\mathbf{Q}^2/\mu + \mathbf{J}_0\cdot\boldsymbol{\xi}\times\mathbf{Q} + \gamma p_0(\boldsymbol{\nabla}\cdot\boldsymbol{\xi})^2 + (\boldsymbol{\nabla}\cdot\boldsymbol{\xi})\boldsymbol{\xi}\cdot\boldsymbol{\nabla}p_0] \quad (4.14)$$

and *s*urface

$$\delta W_s = -\tfrac{1}{2}\int d\mathbf{S}\cdot\{\boldsymbol{\xi}[\gamma p_0\boldsymbol{\nabla}\cdot\boldsymbol{\xi} + \boldsymbol{\xi}\cdot\boldsymbol{\nabla}p_0 - \mathbf{B}_0\cdot\mathbf{Q}] + \mathbf{B}_0(\boldsymbol{\xi}\cdot\mathbf{Q})\} \quad (4.15)$$

contributions, which are often useful when the boundary conditions are such that $\delta W_s = 0$. It is not clear that this is applicable in the case of the photospheric end conditions for coronal loops. An alternative, physically motivated breakdown of δW_v is given by Bateman (1978, p. 97), who describes the normal-mode contributions of the different terms.

If δW is minimized, in the sense of the calculus of variations, one can obtain a limiting non-zero form of $\boldsymbol{\xi}(\mathbf{r})$ which sets the stability boundaries as a function of the physical parameters. The requisite Euler–Lagrange equation is just the *marginal-stability* form ($\omega^2 = 0$; Goedbloed, 1971a, b) of (4.8′).

4.4.3 Cylindrical–symmetry formulation

The principal original treatment of the stability of an axial plasma/field equilibrium, the diffuse pinch, is due to Newcomb (1960), who bases his calculation on the energy principle.

Taking as his model an infinitely-long cylindrical equilibrium, with B_θ and B_z fields such as those described in Section 4.3, Newcomb expresses a general linear displacement [with the time removed, as in (4.8′)]

$$\boldsymbol{\xi}(\mathbf{r}) = \boldsymbol{\xi}(r)\exp(im\theta + ikz) \quad (4.16)$$

as a Fourier expansion in the periodic azimuthal and uniform axial directions. The latter variation is often taken to approximate the periodicity in a narrow toroidal pinch.

With this expansion, one can evaluate stability for each (m, k) excitation separately, since they are uncoupled in the absence of axial boundaries.

In order to simplify and illustrate Newcomb's minimization of the energy integral, it is useful to define the parallel displacement

$$\xi_\parallel = \boldsymbol{\xi}\cdot\mathbf{B}/B = (\xi_\theta B_\theta + \xi_z B_z)/B \quad (4.17)$$

and the perpendicular component

$$\xi_\perp = -(\boldsymbol{\xi}\times\mathbf{B})_r/B = (\xi_z B_\theta - \xi_\theta B_z)/B \quad (4.18)$$

in the r = constant magnetic surface. The quantity $B\,(\equiv B_0)$ is the magnitude of the equilibrium magnetic field, and we will henceforward drop the zero subscript. If the r component of the amplitude $\boldsymbol{\xi}(r)$ is chosen to be real, then the equation of motion (4.8′) shows that $\xi_\theta(r)$ and $\xi_z(r)$ are purely imaginary, so that $\xi_\parallel$ and $\xi_\perp$ are imaginary also.

In terms of these dependent variables, Newcomb (1960) writes the volume contribution to the energy integral (4.14), as a function of the real variables ξ_r, $\zeta_\perp \equiv -iB\xi_\perp$ and $\nabla\cdot\boldsymbol{\xi}$. The result is

$$\delta W_v = \frac{\pi}{2}\int_0^\infty r\mathrm{d}r\left\{\Lambda(\xi_r, \xi_r') + \gamma p(\nabla\cdot\boldsymbol{\xi})^2 + \left(k^2 + \frac{m^2}{r^2}\right)[\zeta_\perp - \zeta_0(\xi_r, \xi_r')]^2\right\}, \tag{4.19}$$

where Λ and ζ_0 are explicit functions of ξ_r and $\xi_r' = \mathrm{d}\xi_r/\mathrm{d}r$, (m, k) and the equilibrium parameters.

The integral is clearly minimized by the choice of *incompressible* excitations $\nabla\cdot\boldsymbol{\xi} = 0$ and by the identification $\zeta_\perp = \zeta_0\,(\xi_r, \xi_r')$. One is then left with the problem of the ξ_r extremization of the Λ contribution.

The incompressibility condition deserves some comment. It is often said that incompressible perturbations are the most unstable, but this is only true as a marginal ($\omega^2 = 0$) condition. If there is instability, compressible excitations may have higher growth rates (Goedbloed, 1971b). In addition, the compressibility term enters into the boundary integral (4.15), and is also the only place where the amplitude $\xi_\parallel$ appears.

When the remaining non-positive-definite contribution of (4.19) is integrated by parts in r, one obtains

$$\delta W_r = \frac{\pi}{2}\int_0^\infty \mathrm{d}r[f\xi_r'^2 + g\xi_r^2] \tag{4.20}$$

as the form to be minimized. The functions f and g are given by (Newcomb, 1960)

$$f = r(\mathbf{k}\cdot\mathbf{B})^2/\mu\mathbf{k}^2 \tag{4.21a}$$

and

$$g = \frac{2k^2}{\mathbf{k}^2} p' + \frac{(\mathbf{k}^2 - r^{-2})}{\mu \mathbf{k}^2} r(\mathbf{k}\cdot\mathbf{B})^2$$

$$+ \frac{2k^2}{\mu r \mathbf{k}^4}(k^2 B_z^2 - m^2 r^{-2} B_\theta^2), \tag{4.21b}$$

where $\mathbf{k}^2 \equiv k^2 + m^2 r^{-2}$ is the square of the magnitude of the *total* wavevector.

By examining the parameter dependences of various positive–definite terms in Equation (4.20), Newcomb (1960) was able to show that the infinite linear pinch is stable for all m and k, *if* it is stable for $m=0$, $k \to 0$, *and* form $m=1$, $-\infty < k < \infty$. The $m=0$ condition is easily fulfilled for equilibria satisfying $2\mu r p' \geqslant -B_z^2$ [*v.* Newcomb, Eq. (20)], which includes all force-free models. A similar argument can be used, in connection with (4.21b), to show that equilibrium profiles of the kind described by (4.6) with $p' > 0$ are more stable than the equivalent force-free case (Giachetti *et al.*, 1977).

The extremum of the energy integral (4.20) is obtained from the corresponding *Euler–Lagrange equation*

$$(f\xi_r')' - g\xi_r = 0, \tag{4.22}$$

where f and g are still as given in (4.21). Goedbloed (1971b) has shown that this is also the marginal-stability form of the r component of (4.8′) when ξ_θ and ξ_z are eliminated.

An important aspect of (4.22) is that it possesses singularities where f vanishes, or

$$\mathbf{k}\cdot\mathbf{B} = kB_z(r) + \frac{m}{r} B_\theta(r) = 0; \tag{4.23}$$

this condition specifies the so-called *singular layers* or *mode-rational surfaces*, in which the total (including radial) variation of the perturbation is directly across the magnetic field. The locations of these radii r_s depend upon (m, k) and the equilibrium field $\mathbf{B}(r)$.

The behavior of the solutions of (4.22) near the singular layers was originally studied by Suydam (1958). A clear discussion of the local asymptotic solutions

$$\xi_r = a(r - r_s)^{-n_1} + b(r - r_s)^{-n_2} \tag{4.24}$$

is given by Newcomb (1960), who derives the values

$$n_{1,2} = \tfrac{1}{2}\{1 \pm (1+\kappa)^{1/2}\}, \tag{4.25}$$

where $\kappa \propto p'$ depends on the equilibrium and its magnetic shear. We will return to this singular behavior, and its crucial influence on stability, in Sections 4.4.4.A and 4.5.2 [Equation (4.22) is also singular at $r=0$ (4.21a), but ξ_r is well-behaved there since n is real (Newcomb, 1960, p. 257)].

4.4.4 Diffuse pinch stability

There are two varieties of infinite-conductivity instability in the axial pinch. The first is *local*, involving a field-line interchange occurring in the immediate vicinity of the singular layer. The second is *global*, wherein the growing displacements range across the entire plasma. We will first develop a general formalism and then treat these in turn.

As we have stated in Section 4.4.1, an unstable solution (for ξ_r, in this cylindrical case) of Equation (4.8′) is one that satisfies the relevant boundary conditions, namely

$$r\xi_r(r) \to r^{|m|} \qquad r \to 0,$$

$$r\xi_r(r) \to 0 \qquad r \to \infty, \tag{4.26}$$

for $\omega^2 < 0$. The problem of singularities does not arise for unstable modes, since the $\omega^2 < 0$ form equivalent to (4.22) has $f \propto \omega^2 - (\mathbf{k}\cdot\mathbf{B})^2/(\mu\rho_0)$ (Gloedbloed, 1971b; Eq. (6)), which has no zeros for the real $\mathbf{k}$ determined by the boundary conditions.

The difficulty of performing this "in-principle" calculation is considerable, which led Newcomb (1960) to develop a simpler stability/instability criterion for the diffuse axial pinch, based on the *energy principle*. For the purposes of the present discussion, however, we will describe a formally equivalent treatment, based on *marginal stability* (Goedbloed, 1971, a, b), which gains some clarity from a connection with the theory of differential eigenvalue equations.

The marginal-stability criterion is based on the fact that the original ($\omega^2 \neq 0$) form of Equation (4.22) [Goedbloed (1971b), Eq. (6)] has the property that both $f(r;\, \omega^2)$ (as cited above) and $g(r;\, \omega^2)$ [which reduce to (4.21a, b) in the marginal-stability limit] are monotonic functions of the quantity ω^2. This then allows the use of Sturm's fundamental theorem (Ince, 1956), which states that the zeros of the solutions $\xi_r(r;\, \omega^2)$ of (4.18) increase in separation as (real) ω^2 *decreases*.

This property allows us to concentrate on the considerably simpler marginal-stability ($\omega^2 = 0$) problem. This is accomplished, in principle, by integrating the differential equation radially from one boundary condition until a zero (or the next boundary) is reached. If the node is inside the boundary, then ω^2 must be decreased from zero to move the node of the solution out until the required far boundary condition is satisfied. This circumstance indicates, for the parameters used in the test calculation, that one has an allowed, but undesirable, unstable ($\omega^2 < 0$) solution. The resulting operational requirement for stability is, then, the absence of interior zeros of the solutions of the marginal equation (4.22).

A remaining complication in the reduced $\omega^2 = 0$ equation is the presence of the singular points (4.23) mentioned above. Newcomb (1960) first properly treated the behavior and influence of these layers. He showed that they have one possible localized destabilizing effect, and a more important global decoupling effect which improves overall stability.

4.4.4.*A* *Local interchange stability* The first point concerns the behavior of the solutions for ξ_r near the singularities. These have been delineated in Equations (4.24) and (4.25). Since the Sturm-theorem condition for stability is the absence of an interior node of ξ_r, we must therefore require that ξ_r not oscillate locally, which is equivalent to $\kappa > -1$ in (4.25). When written in primary variables this limitation becomes

$$p'(r) \geqslant -\frac{rB_z^2(r)}{8\mu}\left\{\left(\ln\frac{rB_z}{B_\theta}\right)'\right\}^2, \tag{4.27}$$

evaluated at r_s. Equation (4.27) is a statement of the (cylindrical-geometry form of the) *Suydam* (1958) *criterion* for stability against localized interchanges. One must, therefore, require the equilibrium structure $p(r)$, $\mathbf{B}(r)$ to be Suydam-stable at all of the singular layers $r = r_s(k, m)$ given by Equation (4.23). This requirement can provide difficulty for pressure-modified *shear-free* fields, such as those derived from (4.4).

4.4.4.*B* *Global displacement stability* The most dangerous unstable perturbations of an MHD equilibrium are the large-scale ones. They must be investigated by a global integration of (4.22), along with a proper treatment of the singularities. This brings us to a consideration of the second important result of Newcomb's study of the effects of these surfaces. He carefully details their property of decoupling the plasma perturbations on their inner and outer sides. He shows that the lower sign choice in (4.25) is forbidden, since this solution has a closer zero [by Sturm's separation theorem (Ince, 1956)] than does the one corresponding to the upper sign choice, which Newcomb calls the "small" solution. He then shows that the small solution at a singular point is equivalent to a zero of the normal solution at a non-singular point. This circumstance causes the singular layer to have some of the effects of a thin rigid boundary at r_s, thereby dividing the plasma into annular "independent intervals", as far as ideal MHD perturbations are concerned. One is thus able to evaluate stability separately in each singular zone by requiring, as before, that the marginal-stability solutions fitting the boundary conditions (including the asymptotic small-solution form, where applicable) have no interior vanishing points.

The application of Newcomb's global-stability criterion, as expressed in this completely equivalent marginal-stability form, requires the wide-ranging integration of Equation (4.22). We will detail the required steps in the force-free example to be given in Section 4.4.6.

4.4.5 Toroidal effects

An important simplification in the theoretical treatment of the stability of semi-toroidal coronal loops comes from the ability to use the cylindrical approximation, as we have done in Sections 4.3 and 4.4. Although this idealization seems reasonable for large-aspect-ratio ($R/a \equiv$ major toroidal-curvature radius/minor local loop radius $\gg 1$) tori, it should be confronted more directly.

A number of papers (in the laboratory context) have devoted themselves to this topic, most of them considering the modifications of the energy integral as the aspect ratio decreases. Since *tokamaks* (which have $B_z \gg B_0$ and $\beta \ll 1$) typically have lower values of R/a than do *reversed-field pinches* (which have $B_\theta \sim B_z$ and $\beta \sim 1$ and may be more closely related to coronal loops, as explained in Section 4.3), many of the calculations have been restricted to the low-shear conditions (B_z, j_z slowly varying) applicable to the former magnetic configuration.

The lowest-order (in $\varepsilon \equiv a/R$) toroidal-geometry effect applies to the structure of the equilibrium fields, and causes the magnetic-flux surfaces (in which the field lines lie) to be displaced inward as their "radii" from the magnetic axis increase (Shafranov, 1963). [This axis at $R = R_0$ is the curve along which $\mathbf{B} = B_\varphi \hat{\mathbf{e}}_\varphi \sim B_z \hat{\mathbf{e}}_z$ is entirely in the toroidal φ-direction, corresponding to the z-direction in the cylindrical approximation.] However, the flux surfaces remain nearly circular even as their centers are displaced, and this is the basis for the expansion of δW in increasing powers of ε and the resulting evaluation of toroidal effects on stability.

The most comprehensible treatment of this problem is that by Ware and Haas (1966), who describe the coordinate system and the equilibrium (for small $\beta \sim \varepsilon^2$) and expand δW to fourth order in ε. Unfortunately, for our present purposes, they make an approximation in the lowest non-vanishing order ($\sim \varepsilon^2$ for tokamak conditions),

$$\delta W_0 \propto \sum_m \int r dr (\mathbf{k} \cdot \mathbf{B})^2 [(m^2 - 1)\xi_{r,m}^2 + r^2 (\xi'_{r,m})^2], \tag{4.28}$$

which dictates that this (cylindrical-equivalent) contribution be negligible. This is synonymous with a choice of low shear (**k** · **B** small and nearly constant, leading to Suydam-type local instabilities in higher order) *or* $m=1$, although the latter global mode is explicitly excluded (by assumption) in the next order of approximation.

The approach of Ware and Haas does, however, properly treat local instabilities and produces an approximate form [their Eq. (41)] for the Mercier (1960) toroidal modification of the cylindrical interchange-stability criterion. When their tokamak-motivated assumption $B_\phi(r) \sim B_z(r) = constant$ is removed, one obtains

$$\left[1-\left(\frac{rB_z}{RB_\theta}\right)^2\right]\frac{\mathrm{d}p}{\mathrm{d}r} > -\frac{rB_z^2(r)}{8\mu}\left\{\left(\ln\frac{rB_z}{B_\theta}\right)'\right\}^2 \tag{4.29}$$

(Shafranov and Yurchenko, 1968) as the lowest order ($\sim\varepsilon^2$ except for tokamaks) correction to the Suydam (1958) condition (4.27). Equation (4.29) shows that a negative value of the left-hand bracketed term $[1-q^2]$ can stabilize a negative pressure gradient. This bracket also contains the only appearance of the *Kruskal-Shafranov* condition (Bateman, 1978) relevant to the coronal-loop problem; normally this topological limitation is applied to a *complete* toroid, where it dictates that, for global stability, $q(r)>1$ so that the helical field lines of the torus do not close on themselves. The Kruskal-Shafranov condition is customarily used to specify the stability of the most dangerous kink mode of a complete toroidal pinch with a free boundary, as in a tokamak (Bateman, 1978). For a diffuse plasma with a conducting boundary, one can have stability even with $q(r)<1$ (Figure 4.8).

A complementary evaluation of global toroidal corrections is that given by Vahala (1975). He allows a wider range of β, small departures from circular flux surfaces, and a finite magnetic shear. His major result is that "[toroidal] stability (to leading order [in ε]) is always governed by that of the [equivalent] straight cylindrical screw pinch." The only relevant qualification to this result is that it is inapplicable to the $m=1$ mode, which is marginally stable in this order, so that δW vanishes, as shown by Robinson (1971).

This last limitation was apparently removed by Bussac *et al.* (1975) in a very brief paper on the internal-kink instability. They expand δW to the first non-zero order beyond that given in (4.28), which vanishes for $m=1$ in the cylindrical approximation. They find the next contribution δW_2 to be

$$k^2R_0^2\delta W_2 = (k^2R_0^2-1)\delta W_{2c} + \delta W_{2t}, \tag{4.30}$$

where R_0 is the major radius of the magnetic axis and δW_{2c} is the normal second-order contribution which determines $m=1$ stability in a cylindrical pinch [$R_0 \to \infty$, as in Eq. (21) of Robinson (1971)]. The purely toroidal contribution δW_{2t} is described as being *smaller* than δW_{2c}, *positive* (stabilizing) for $dp/dr = 0$, and a monotonically decreasing function of a radial average of the quantity $(-r^2\ dp/dr)$. They cite several simplified tokamak motivated sample calculations which show δW_{2t} becoming negative for large enough positive values of $\langle -r^2\ dp/dr \rangle$.

An important aspect of the result of (4.30) is that the toroidal-kink corrections only become significant for $kR_0 \simeq 1$, which particularly selects the fundamental mode $\lambda/2 = \pi R_0$ of a *half*-torus. [Similar computational results for tokamak conditions have also been exhibited, without any details, by Wesson and Sykes (1975).] The specific $kR_0 = 1$ resonance condition would seem to indicate that a larger or smaller toroidal sector (as is common in the coronal-loop context) would be dominated by the cylindrical contributions.

To summarize, there is a second-order toroidal correction to the local interchange-stability condition (4.30) for coronal-loop-like pinches. However, the results of Vahala (1975) and Bussac *et al.* (1975) indicate that *the global MHD stability of a toroidal pinch is determined primarily by the behavior of the equivalent cylindrical pinch.* The closest exception to this rule applies to the fundamental axial resonance $kR_0 = 1$ of the important $m = 1$ kink mode, where the primary toroidal correction is weakly stabilizing for small β. These analytical predictions are also verified by comparison of the results of toroidal and cylindrical computations for the stability of the reversed-field pinch (D. A. Baker, private communication), the closest laboratory analogue (having $B_z \sim B_\theta$ and $\varepsilon < 1$) to the coronal-loop configuration.

These facts have led to the general use of cylindrical-symmetry calculations to investigate the stability of reversed-field pinches (Baker *et al.*, 1977; Christiansen and Roberts, 1978) and coronal loops (Anzer, 1968; Raadu, 1972; Van Hoven *et al.*, 1977; Hood and Priest, 1979). We will now describe some of these results.

4.4.6 The stability of force-free fields

The original calculation of the stability of the *cylindrical, constant-α, force-free field* (4.3) was performed by Voslamber and Callebaut (1962). They used Newcomb's (1960) methodology and were able to show that all $m \neq \pm 1$ modes were completely stable. The $m = \pm 1$ kink case was then found to be unstable for $|k/\alpha| < 0.272$, except that these perturbations could be eliminated by the placement of a rigid cylindrical conductor at $r = b < 3.18/\alpha$. A representation of the Newcomb stability diagram for the Lundquist field (in the absence of a conductor) is given in Figure 4.6 (Giachetti *et al.*, 1977), which shows the relative locations of the zeros r_{oi} and the singularities r_{si} of the kink excitation in each (i) independent interval. There is kink instability for $r_{0i} < r_{si}$, which occurs

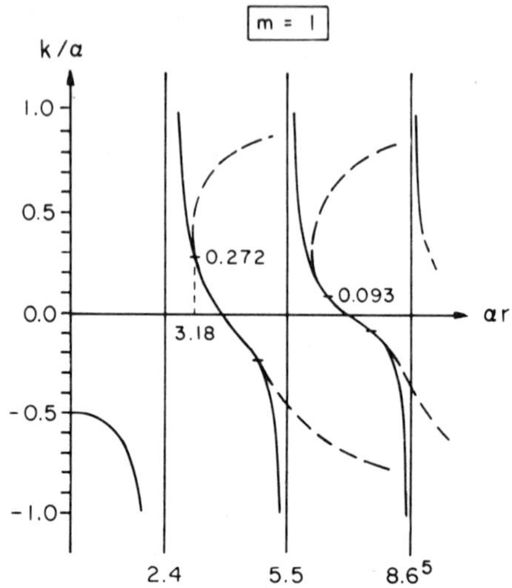

FIGURE 4.6 The locations of the marginal-equation singularities (solid) and solution zeros (dashed) of the cylindrical, constant-α, force-free field (4.3). Stability in each zone is obtained for values of k/α outside the horizontal marks (Giachetti *et al.*, 1977).

at the long wavelengths for which k is between the horizontal hash marks.

Growth-rates of the kink instability in the first and second zones of the constant-α field, with conducting boundaries in the second and third zones, have been calculated by Goedbloed and Hagebeuk (1972). They obtained results, in the incompressible approximation, which showed the growth-time in the Lundquist field to be of order ten times the radial Alfvén transit time. A similar force-free equilibrium, the paramagnetic model, which does not exhibit field reversals, was tested for stability by Whiteman (1965) and shown to have somewhat poorer characteristics in this regard.

A lengthy general discussion of the stability of cylindrical force-free fields has been given by Goedbloed (1971b, c). In particular, the latter paper treats the peculiarities of the constant-pitch force-free field (4.4), which has received a lot of solar-physics attention since its original proposal by Gold and Hoyle (1960). Goedbloed (1971c) points out that the lack of field shear means that $\mathbf{k}\cdot\mathbf{B}=F$ is constant over the field profile. He then shows that new quasi-interchange instabilities arise, for excitations satisfying $F\approx 0$ and $\xi_{\parallel}\approx 0$ (4.17), which cannot be found in the usual marginal-stability analysis.

The first application of MHD stability criteria to a force-free, coronal-loop problem was made by Anzer (1968). Although his formulation appears at first sight to be internally inconsistent, since $\nabla\cdot\boldsymbol{\xi}=0$ is assumed while an incompressible form for $\xi_r\hat{\mathbf{e}}_r+\xi_{\perp}\hat{\mathbf{e}}_{\perp}$ [see (4.18)] is used, it is actually completely general since the (unstated) value of $\xi_{\parallel}$ (4.17) can be obtained from

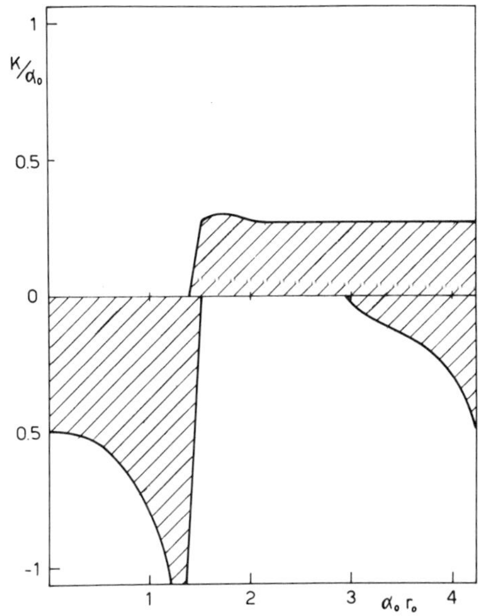

FIGURE 4.7 Regions of instability (hatched) in the parameter space of the force-free field of Figure 4.4 with $\alpha_0\delta=2$, taken from a series of determinations similar to that of Figure 4.6 (Chiuderi *et al.*, 1980).

$\nabla\cdot\xi=0$. (This omission has, however, propagated through some following papers.) Anzer showed that all (infinite-length) force-free fields are unstable for large enough axial wavelengths, and then performed corroborating Newcomb-style calculations on three specific analytic models.

A recent investigation of the stability of force-free, current filaments imbedded in external, stable (since they have no free energy), potential fields has been made by Chiuderi *et al.* (1980). The model used is that shown in Figure 4.4, with the variation of $\alpha(r)$ given in Equation (4.5). The kink-unstable regions of parameter space are shown in Figure 4.7 for a current-transition width of $\delta=2/\alpha_0$. This representation is the result of a number of different calculations of the kind shown in Figure 4.6, and gives the location of the positive and negative wavenumber stability limits as a function of the current channel width r_0. The best stability conditions (narrowest unstable-$|k|$ span) are obtained in a range around the reversal point of the B_z field at $\alpha r=2.4$, a result which agrees with that also found for the reversed-field pinch (Christiansen and Roberts, 1978).

4.4.7 Pressure–gradient effects

Actual coronal loops have observable transverse (radial) pressure gradients, as described in Section 4.3.3, which must be taken into account, at some level of

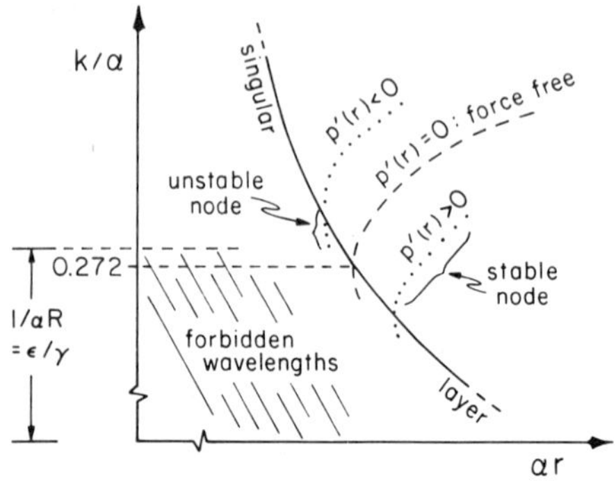

FIGURE 4.8 The effects of pressure gradients and finite length on the stability boundary of Figure 4.6. See the text below and on p. 247 for details.

accuracy, in a stability calculation. Since this situation is common in the laboratory, considerable theoretical attention has been devoted to pressure effects. Most of the relevant efforts have been applied to non-tokamak configurations (such as the reversed-field pinch), which tend to have higher values of β.

A qualitative appreciation of the effects of non-zero pressure gradients can be obtained from Figure 4.8, which shows a portion of the first independent interval of Figure 4.6. For finite-pressure fields of the class of Equation (4.6), the singularities do not move for $p' \neq 0$. Thus, one need only consider the effect on stability of the motion of the zeros of the marginal-stability equation (4.22) as g (4.21b) varies. For $p' > 0$, g increases and the node of ξ_r moves outward, so that the kink mode becomes stable over a wider range of k (and *vice-versa*). This agrees with the energy-principle comparison-theorem stated in the paragraph below Equation (4.21).

An early specific treatment of finite pressure-gradient effects was that by Whitman (1965), who supplemented the paramagnetic force-free field with a pressure profile ($p' < 0$) which marginally satisfied Suydam's criterion (4.27). The equilibrium relation (4.1) was then used to calculate a new value for B_θ or B_z. The stability of these modified fields was calculated using Newcomb's method, and the stability performance was shown to be degraded.

Robinson (1971) was the first to separate formally the p' contribution to the energy integral δW, which he pointed out was of order k^2r^2 in smallness except for the special case of the kink mode [$\delta W_0 \rightarrow 0$, as in (4.28)]; he also found some simple sufficient stability conditions on the pressure gradient. He then used these considerations to devise a number of finite-pressure equilibria (about which no details are given) and to test their stability. Once again, finite internal pressure has a deleterious effect.

Van Hoven *et al.* (1977), Giachetti *et al.* (1977) and Hood and Priest (1979),

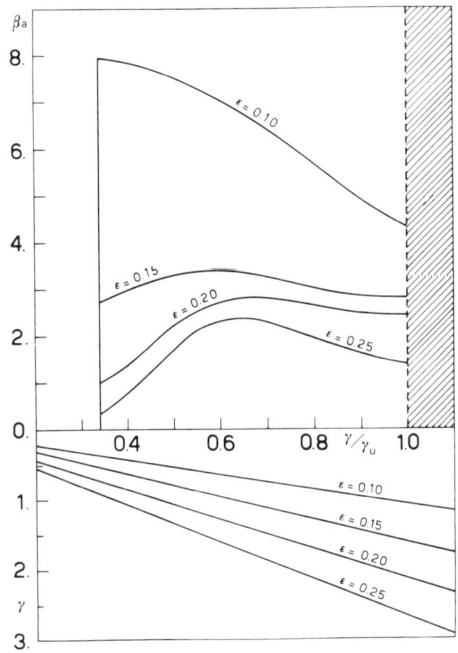

FIGURE 4.9 Stability diagram (top) for the pressure-modified force-free field of Figure 4.5 (Giachetti *et al.*, 1977). Stability is obtained above and to the left of the curves for different values of $\varepsilon = a/R$.

have considered finite-pressure profiles in the coronal-loop context. The first two papers used force-free-derived fields of the kind described in Equations (4.6) and (4.7), and assessed the kink stability of a cool-loop profile such as that shown in Model II of Figure 4.3, which has a positive gradient extending from the loop axis to the higher pressure external corona.

The primary stabilizing effect of this pressure profile comes from its positive contribution to δW_r as shown in Equations (4.20) and (4.21b) However, its influence also results from the reduction of the band of unstable wavenumbers shown in Figure 4.8, so that it depends upon the finite length of the coronal loop, whose effect will be treated in the following section (4.4.8). The sum of these effects is detailed in the normalized presentation of Figure 4.9, which indicates stability for values of $\varepsilon = a/R$ and $\beta_a = \beta(a)$ [where a is the loop edge] above and to the left of the curves. For small enough values of $\gamma/\gamma_u = 0.093\, \alpha R$ (where $\gamma \equiv \alpha a =$ pressure-scale/magnetic-scale), the loop is too short to include any unstable (long) axial wavelengths, even without assistance from the pressure profile. In the next range of $\gamma/\gamma_u > 0.342$, the pressure gradient is effective, more so when the profile is wider (ε larger for fixed R) since p' is weighted by $k^2 r^2 \sim r^2/R^2$ in (4.21b). For $\gamma/\gamma_u > 1$, the second zone of Figure 4.6

is unstable without any chance of being affected by the relatively narrower pressure gradient.

The other side of the coin of pressure effects is that negative gradients, as in the X-ray loop of Model I of Figure 4.3, generally make stability more difficult to attain. At the least, a wider range of wavelengths will be unstable as a result of the considerations illustrated in Figure 4.8.

One apparent counterexample to this trend is the result which is quoted, without details, in Christiansen and Roberts (1978). They describe the "inflation" of the Lundquist equilibrium by finite pressure in the reversed-field pinch, and show [in their Figure 3(b)] that the conducting boundary can be moved outwards by a factor of $\leqslant 1.35$ while supporting a value of β on the axis as large as 0.27. How this would translate into stable wavenumber limits is not clear, since they do not give the modified field components.

4.4.8 Finite–length stabilization: line-tying

One substantive difference between a coronal loop and a toroidal laboratory pinch is the presence of end effects. The magnetic field of the loop must be imbedded in the high-inertia ($\beta \gg 1$) photospheric boundary. This indicates that, in some sense, the ends of a coronal loop are not completely free to move, at least on the much faster Alfvén time-scale of the solar atmosphere ($\beta \ll 1$). There have been several attempts at modelling this effect, none of them completely satisfactory; a more precise analysis is certainly required.

In laboratory devices, such end constraints are usually applied by rigid conducting plates (assumed to be ideal), so that the imbedded field lines are unable to move. This effect is called *line-tying* and is believed, on general grounds, to provide a stabilizing influence on a plasma (such as a *pinch*) located between end-plates (at $z=0, L$). The basis for this conception is the fact that end-plates place an additional constraint on the allowed excitations of the plasma and thus must restrict the unstable function space, especially for the long-wavelength modes common to axial pinches.

An early discussion of line-tying is that due to Kadomtsev (1966), who uses the end condition $\boldsymbol{\xi}=0$ at the conductor. He treats a half-torus by the energy principle and, with the estimate $\partial/\partial z \sim \pi/L$, obtains a stabilizing contribution from the term $Q^2 \sim \pi^2 B_z^2 \xi_r^2/L^2$ of (4.14), which results in a simple sufficient condition for small enough L. In another example, applied to the surface-layer pinch (with $p'=0$, and an annular sheet current), he is able to write the dispersion relation in terms of k and the fields at the surface ($r=a$). By applying the requirement that the allowable wavelengths must satisfy $\lambda/2 \leqslant L$, or $k \geqslant \pi/L$, he states a kink-stability condition that is coincidentally the same as the Kruskal–Shafranov limit $q(a) > 1$. He states that this is a necessary condition, but this is only true for the simple equilibrium he has chosen, in which the energy principle is directly integrable. In fact, this result is equivalent to the sufficient condition $g>0$ in (4.20), for the case $m=1$ [*v.* Eq. (9) of Robinson (1971)], when the identification $k\to\pi/L$ is made.

Solov'ėv (1971) repeated Kadomtsev's first energy-principle argument (except for the additional assumption of incompressibility) in proper toroidal geometry and obtained a sufficient condition, to second order in $kr\sim\pi r/L$, which is again a restricted form of $g>0$ [*v.* the last bracketed term of Eq. (2) of Robinson (1971)]. He also treats (without details) a simple shear-free pinch ($j_z=0$) using the energy principle and, by *assuming* perturbations of a form equivalent to

$$\xi_r(\mathbf{r})=\xi_r(r)\sin(m\theta-kz)\sin hz$$

$$\xi_\perp(\mathbf{r})=\xi_\perp(r)\cos(m\theta-kz)\sin hz, \tag{4.31}$$

where $h\equiv n\pi/L$ and $\xi_\parallel$ (4.17) is unspecified, he solves the Euler–Lagrange equation and finds the condition that the pinch is stable for long wavelengths $k^2aL<6$.

The first realization that end conditions would be an important stabilizing influence on coronal loops was due to Raadu (1972). Basing his calculation on Anzer's (1968) *in*compressible force-free formulation, Raadu used perturbations of the form (4.31), which are compressible, so that his results are only applicable for force-free fields ($\beta<1$) where the third term of (4.14) is negligible. Within this approximation the energy integral is then minimized, producing $h=\pi/L$ so that ξ_r and $\xi_\perp$ vanish at the ends. This procedure provides an initial result

$$\delta W=\delta W_\infty+(\pi^3/\mu L^2)\int[B_z^2\xi_r^2+B^2\xi_\perp^2]r\,dr,$$

where δW_∞ is the incompressible force-free value (per unit length) of (4.14) for $L=\infty$, which is an improvement over Kadomtsev's estimate. Raadu then further minimized δW with respect to $\xi_\perp$, as Newcomb did, and obtains an Euler–Lagrange equation in $\xi_r(r)$; a property of the equation which simplifies its analysis considerably is that there are no singularities ($f=0$) beyond the origin and so the complication of having singular layers does not arise (§4.4.3). Raadu analysed the *constant-pitch* field (4.4) and showed that the most unstable kink mode has $k=0.6\,\tilde{\mu}$, and that it is suppressed in loops shorter than $6.1/\tilde{\mu}$. [Hood and Priest (1979) repeated Raadu's calculation more accurately and found the limit $L<5.2/\tilde{\mu}$; their stability diagram is reproduced in Figure 4.10a. It should be noted that this is only a *sufficient* condition for instability, since there

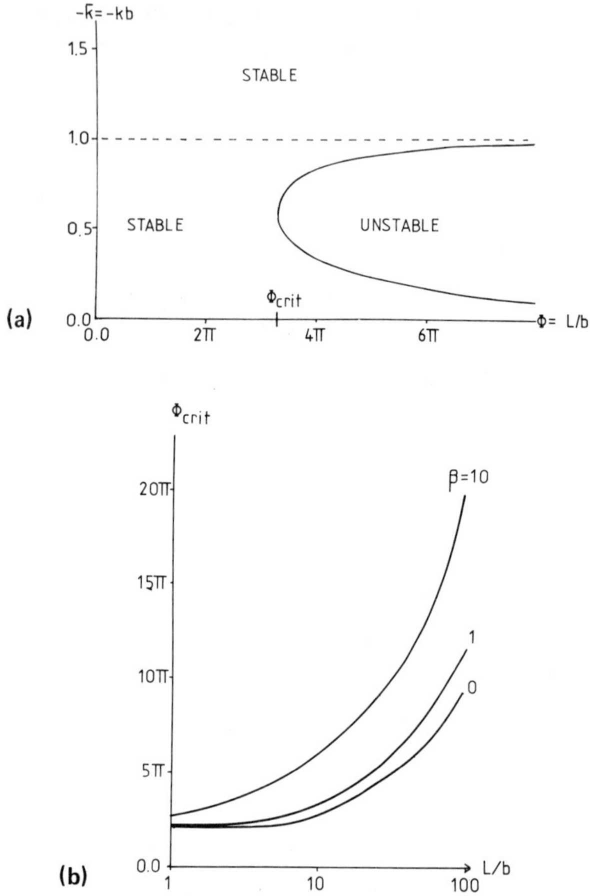

FIGURE 4.10 (a) Critical conditions for the onset of kink instability in a line-tied coronal loop of length L and width $2b$. The field is force-free, of uniform twist Φ, and the axial wavenumber is k. The critical twist for instability is $\Phi_{crit} \approx 3.3\pi$. (From Hood and Priest, 1979b.) (b) Stability diagram for a modified form of the field of (4.4), in which $B_z = constant$, so that a negative radial pressure gradient is obtained from (4.2). The twist parameter $\Phi = LB_\theta(b)/(bB_z)$ is evaluated at the magnetic radius $r=b$ of the field. (Hood and Priest, 1979b.)

may exist more general perturbations than (4.31) which are unstable at a lower threshold.]

The next attempt at finite-length stabilization of a coronal-loop model was that of Van Hoven *et al.* (1977) who used the general approach of Kadomtsev (1966). That is, a Newcomb-style calculation was performed, using eigenfunctions of the kind described in (4.16). The argument was then made that axial wavelengths λ longer than $2L = 2\pi R$ would not fit within the loop, so that $kR \geqslant 1$ was a requirement dictated by the finite length. Thus, if R were small enough, the otherwise unstable range of long wavelengths would be forbidden, as indicated in Figure 4.8. For the Lundquist field, this limitation takes the normalized form

$$|k/\alpha| \geqslant 1/(\alpha R) > 0.272,$$

resulting from the stability-boundary diagram of Figure 4.6. For non-zero pressure gradients, this limit varies with the pressure drop, as has already been discussed in Section 4.4.7, in connection with Figure 4.9. It is now clear that this method offers only a rough approximation to the true effects of end-stabilization (and provides only a sufficient condition for stability). The primary reason is the fact that the Euler–Lagrange equation (4.22) for a general infinite pinch with eigenfunctions (4.16) is not an even function of k, as may be seen from the $\mathbf{k}\cdot\mathbf{B}$ term of (4.21b). Thus $\xi_r(r;\, k)$ and $\xi_r(r;\, -k)$ are different functions of r and cannot "resonate" to form a standing wave at all radii. In addition, ξ_r and $\xi_\perp$ are in phase quadrature [as stated below (4.18)] so they cannot both vanish at the same z position.

A recent effort on the end-effect problem is that of Hood and Priest (1979), who use the formalism of Raadu, with the addition of finite pressure. They choose a perturbation form equivalent to (4.31), with the addition of $\xi_\parallel$ so that all components of $\boldsymbol{\xi}$ vanish at the loop ends, and show that $\nabla\cdot\boldsymbol{\xi}=0$ is no longer a minimizing condition, in contrast to the infinite-pinch result of Newcomb. To make the variational problem solvable, they then assume $\xi_\parallel \equiv 0$ and apply their method to the kink stability of a number of simple analytic magnetostatic fields, many with $\beta=0$. For example, the pressure-modified shear-free field (§4.3.3), with aspect ratio $L/b=10$ and $p_c = B_0^2/(2\mu)$, is found to be unstable when $L < 12\pi/\tilde{\mu}$. One result for finite β is shown in Figure 4.10b, which applies to a field similar to that of (4.4) except that $B_z = constant$ so that a negative pressure gradient may be supported. One then finds stability below the curves for different values of β in the external corona. They choose to display their results in terms of a local property [the field "twist" Φ, akin to $\tilde{\mu}L$ of (4.4) in the shear-free case] instead of the more transparent loop length or some other global parameter. This choice causes both of the coordinate scales to be functions of the value of β on the axis, so that the stability limits depend on the radial pressure drop in a way which cannot be immediately determined. Beyond these difficulties of interpretation, it is not clear that $\xi_\parallel=0$ is a reasonable assumption for a coronal-loop excitation which may involve flows to the ends. It should be noted that, because restrictive forms (4.31) for the perturbations have been chosen, the stability conditions derived are only sufficient (*v.* §4.4.2); in other words, the loop may become unstable at a lower threshold. To model line-tying more accurately and adopt a more general perturbation demands a much more sophisticated approach and a solution of the partial differential equations. For the relevance of the above stability analyses to large two-ribbon flares, the reader is referred to §6.5.3.

At the present time, it can only be said that an accurate dynamic configuration for the photospheric boundary conditions has not yet been devised, nor has a complete formulation of the resulting MHD-stability determination been found. This lack of theoretical completeness also applies to the treatment of resistive instabilities to be described in Section 4.5.

4.4.9 Limitations on ideal MHD instabilities

We have, until now, concentrated our attention on infinite-conductivity instabilities because of their short growth-times, of order τ_a. These mechanisms are particularly important in the early stages of loop formation when the magnetic shear $(rB_z/B_\theta)' = Rq'$ is small. As the field gradients increase, however, the possible growth-rate of the fastest (kink) mode decreases as $[q'(r_s)]^{-1} \propto [|\mathbf{k}\cdot\mathbf{B}|']^{-1}$ (Rosenbluth *et al.*, 1973) so that such excitations become less significant.

There is, in addition, another aspect of the behavior of ideal-MHD instabilities which causes us to look beyond the infinite-conductivity limit. That is the fact that their growth appears to be self-limiting at a relatively low level, because of the frozen-in-field restriction on the allowable magnetic distortion. The kink mode has been shown to saturate at a (rigid) displacement $\xi_r/r_s \sim (r_s/R_0)^2$, and the more stable $m \geqslant 2$ modes at r_s/R_0 (Rosenbluth *et al.*, 1973). Although these calculations involve many approximations, they seem to indicate that a loop-flare model based on an ideal-MHD instability would be severely restricted in its energy-output potential. For this reason, it seems clear that $\sigma = \infty$ stability has its best application as a constraint on the quiescent preflare state (Van Hoven *et al.*, 1980).

Finally, as the magnetic shear in a coronal loop increases, with growing axial current flow, there is a transition to the dominance of resistive instabilities. Such mechanisms involve field reconnection (Chapter 3), and have growth-rates which vary as a positive fractional power of $q'(r_s)$ (Coppi *et al.*, 1976) and the potential for significant nonlinear magnetic-energy release (Van Hoven and Cross, 1973). We will now turn to the theory of these finite-conductivity processes.

4.5 RESISTIVE MAGNETIC INSTABILITY

It is generally accepted that the energy of a solar flare is stored in stressed coronal magnetic fields (Švestka, 1976; Van Hoven *et al.*, 1980) and that the release mechanism involves the *reconnection* of these fields into a configuration which allows their rapid relaxation to a current-free, or potential-field, state (Van Hoven, 1976).

The only way that the magnetic topology of a stressed supra-potential field can be changed is through *resistive* processes, which defeat the usual, rigid, frozen-in-field constraint of ideal MHD. There are two further requirements for a plasma mechanism to be able to explain the available flare observations. The reconnection must be *anomalously rapid* compared to the simple resistive decay (at $\tau_r \sim \mu\sigma a^2$) of measured coronal structures of scale a, and the process must continue until the majority of the stored energy is liberated (Van Hoven, 1979).

The two known mechanisms for magnetic field reconnection are *resistive tearing* (Van Hoven, 1976) and *rapid current-sheet merging* (Priest, 1976). The former is a true temporally-growing instability, which can occur in any sheared magnetic configuration. It was first conceived by Furth, Killeen, and Rosenbluth (1963), and will be the focus of discussion in this section. The latter is often treated as a stationary process which occurs between mutually evolving (inflowing) and interacting magnetic structures. It was devised by Petschek (1964) and may evolve nonlinearly from a resistive instability (Van Hoven, 1976), as described in Chapter 3.

Each of these field-reconnection mechanisms, and their variations (Coppi and Friedland, 1971; Heyvaerts *et al.*, 1977), transforms magnetic energy into plasma heating and particle acceleration, and can produce the observed effects of a solar flare (Švestka, 1976; Van Hoven, 1980).

4.5.1 Magnetic reconnection

Resistive field reconnection involves a localized break-down in the frozen-in-field constraint (Ferraro and Plumpton, 1966) of infinite-conductivity MHD. Thus, it is primarily determined by the second right-hand term in the relevant form of Ohm's law (Equation (1.7))

$$\mathbf{E} = -\mathbf{v} \times \mathbf{B} + \mathbf{j}/\sigma, \tag{4.32}$$

where σ is the effective conductivity. This anomalous behavior occurs in the vicinity of the singular (4.23) or stagnation ($\mathbf{v}_\perp = 0$) layers of large-scale magnetic configurations. The location of these surfaces depends upon both the field structure and the boundary conditions. In these layers, the ideal-MHD equations are modified by the influence of the resistive term in (4.32). These dissipative effects can be due to Coulomb collisions (Spitzer, 1962), to turbulent scattering (Smith and Priest, 1972) or to kinetic effects (Drake, 1978). Finally, the anomalously fast rate of resistive diffusion, which is necessary to explain flare-like magnetic reconnection, requires that the basic process must occur in an *extremely* narrow region bracketing the singular layer. This circumstance allows the use of a planar approximation for the resistive layer, with the over-all external geometry being treated as a boundary or input condition (Coppi *et al.*, 1966, 1976; Vasyliunas, 1975).

To set the general stage for a discussion of the mechanism of resistive reconnection, we use the MHD field equations and Ohm's law (4.32) to eliminate **E** and **j**. We then obtain a variant of Faraday's law

$$\begin{aligned}\frac{\partial \mathbf{B}}{\partial t} &= (\mu\sigma)^{-1}\nabla^2\mathbf{B} + \boldsymbol{\nabla}\times\mathbf{v}\times\mathbf{B} \\ &\sim B/\tau_r + B/\tau_a \\ &= \mathbf{B}/(\mu\sigma a^2) + \mathbf{B}/(a/v_a),\end{aligned} \tag{4.33}$$

where the conductivity σ has been assumed to be uniform, a is a characteristic spatial scale, and the Alfvén speed v_a has been taken as an estimate of the fluid velocity. The relative size of the two right-hand terms in Equation (4.33) is measured by the magnetic Reynolds number

$$S \equiv \tau_r/\tau_a \sim \begin{cases} 10^{-2}\ aBT^{3/2}\ n^{-1/2}\ \text{(cgs)}, \\ 10^{7}\ aBT^{3/2}\ n^{-1/2}\ \text{(MKS)}, \end{cases} \tag{4.33$'$}$$

given here both in MKS and in the cgs units normally used by observers, for the case of Coulomb resistivity and a hydrogen plasma.

We may now see that, when S is large, the hydromagnetic (Alfvén) time-scale (τ_a) is dominant and, when S is small, resistive diffusion (at τ_r) prevails. Generally speaking, S is very large ($>10^8$) in coronal preflare situations, but this may no longer be true in special regions where a and v_a are small.

4.5.2 The resistive tearing mechanism

Since its discovery (Furth *et al.*, 1963) and the original demonstration of its significant nonlinear energy-output potential (Van Hoven and Cross, 1973), magnetic tearing has been the subject of comprehensive analysis. It has interest in connection with the loss of confinement of laboratory fusion-test devices (Van Goeler *et al.*, 1974; Bateman, 1978) and the dynamics and energetics of solar flares (Coppi, 1975; Van Hoven, 1979). The treatment to be given here must, of necessity, be brief but will be supplemented with appropriate references to available clear expositions.

To begin a discussion of resistive tearing, we must specify the equilibrium magnetic field. So that we may make the closest connection with coronal conditions, and be able to ignore presure-gradient-driven modes, we will restrict our attention to uniform-density *force-free fields* (or $\beta \ll 1$). We will consider both a *planar*, *constant-α* example (Cross and Van Hoven, 1971)

$$\mathbf{B}_0 = B_0[\hat{\mathbf{x}} \sin(\pi y/a) - \hat{\mathbf{z}} \cos(\pi y/a)], \tag{4.34}$$

with a field shear scale $a = \pi/\alpha$ (like a spiral staircase) in the y-direction, and the *cylindrical case* of Equation (4.3). These equilibria simulate the magnetic structure of a current sheet (*v.* Chapter 3) or of a coronal loop, respectively. The resistive-tearing stability behavior of these fields is quite similar, as has been made clear by Goedbloed and Dagazian (1971), and their dominant nonlinear saturation is also comparable, as we shall show in Section 4.5.3.*B*.

We then treat small perturbations of these equilibria, such as $\mathbf{B} = \mathbf{B}_0 + \mathbf{B}_1$ where $|\mathbf{B}_1| \ll |\mathbf{B}_0|$, and expand the MHD equations (with $\sigma \neq \infty$) to first order in small quantities. The solutions to these *linearized* equations may then be Fourier analyzed in the planar form

$$\mathbf{B}_1 = \mathbf{B}_1(y) \exp(\gamma t + ikx), \tag{4.35}$$

where γ ($= i\omega$ of Section 4.4.1) is the growth rate, and the selection of a particular wavelike variation along x (and the ignoring of any z-dependence) results from a choice of the location of boundaries at $x = \pm L/2 = \pm \pi/2k$ [*v.* the discussion in Van Hoven (1979)]. The cylindrical analog of this spatial variation is given in (4.16), leading to the simple correspondence $kx \to k(z + m\theta/k)$, where $z + (m/k)\theta$ is a helical coordinate.

These perturbation symmetries and boundary conditions then fix the location of the singular layers of the resulting ideal-MHD behavior. For the *planar* case,

$$\mathbf{k} \cdot \mathbf{B}_0 = kB_{0x} = kB_0 \sin \pi y/a = 0 \tag{4.36}$$

places the "singularities" at $y_{sn} = na$. The *cylindrical* equivalent is given by (4.23), where the fields of (4.3) are used. [As we shall show, the addition of resistivity eliminates the appearance of truly *singular* behavior.]

The next step in the analysis is the statement of the linearized dynamic equations, from Faraday's law (4.33) and momentum-transport (1.2). [We shall only write the planar forms since they will mainly be applied in the vicinity of the narrow singular layer.] Making the large-S approximation (Furth *et al.*, 1963) allows us to ignore compressibility and Ohmic heating, and the force-free assumption then eliminates the pressure perturbation. Finally, we disregard the very weakly coupled z-components of the velocity and field perturbations, and obtain the system of dynamic equations (Van Hoven, 1979)

$$\gamma B_{1y} = ikB_{0x}v_{1y} + (\mu\sigma)^{-1}\left(\frac{\mathrm{d}^2}{\mathrm{d}y^2} - k^2\right)B_{1y}, \tag{4.37}$$

$$\gamma\mu\rho_0 v_{1y} = ikB_{0x}B_{1y} - \frac{\mathrm{d}}{\mathrm{d}y}\left[\frac{i}{k}B_{0x}\frac{\mathrm{d}B_{1y}}{\mathrm{d}y}\right], \tag{4.38}$$

or (Cross and Van Hoven, 1971)

$$\gamma\mu\rho_0\left[\frac{\mathrm{d}^2}{\mathrm{d}y^2} - k^2\right]v_{1y} = ikB_{0x}\left[\frac{\mathrm{d}^2}{\mathrm{d}y^2} - k^2 + \alpha^2\right]B_{1y}. \tag{4.38$'$}$$

Incompressibility and $\mathbf{\nabla} \cdot \mathbf{B}_1 = 0$ have been used to eliminate v_{1x} and B_{1x}, and the latter form is the z-component of the curl of the linearized force equation. If B_{1y} is eliminated from this system, a fourth-order equation in $v_{1y} = \gamma\xi$ results which is nonsingular when $\gamma = 0$. If $\sigma = \infty$ is assumed, then one obtains the usual second-order equation of ideal MHD, equivalent to (4.8′).

The coupled differential relations (4.37) and (4.38′), expressed here in physical variables, are the equivalent of the normalized tearing-mode equations of Furth *et al.* (1963, Equations (13) and (17)). We may see the basis for this resistive instability if we use (4.37) to eliminate $\mathrm{d}^2B_{1y}/\mathrm{d}y^2$ from (4.38). We obtain

$$\gamma\rho_0 v_{1y} = -\sigma B_{0x}^2 v_{1y} - i\sigma\frac{\gamma}{k}B_{0x}B_{1y} - \frac{i}{\mu k}\frac{\mathrm{d}B_{0x}}{\mathrm{d}y}\frac{\mathrm{d}B_{1y}}{\mathrm{d}y} \tag{4.39}$$

as the y-component of the momentum-transport equation. In ideal MHD, when $\sigma \to \infty$, the first term on the right is the dominant one and it strongly prevents the relative motion of the fluid and the field. However, as first realized by Furth *et al.* (1963), this stabilizing term loses its strong influence when $\mathbf{k} \cdot \mathbf{B}_0$ $[= kB_{0x}(y)$ in this choice of symmetry$]$ vanishes for finite k. In this circumstance, the second right-hand term becomes the dominant one $[= (\gamma\sigma/k)|B_{0x}B_{1y}/v_{1y}|v_{1y}]$ and it provides a destabilizing contribution, particularly for small k or long wavelengths.

The character of the unstable perturbation is such that a reconnecting transverse field B_{1y} is spontaneously generated (4.37), as shown in the (vertically expanded) field plot in Figure 4.11. This component has a large second spatial derivative near $y = 0$. The dominant term in (4.37) then becomes

$$\gamma B_{1y} \sim (\mu\sigma)^{-1}\frac{\mathrm{d}^2B_{1y}}{\mathrm{d}y^2}, \tag{4.40}$$

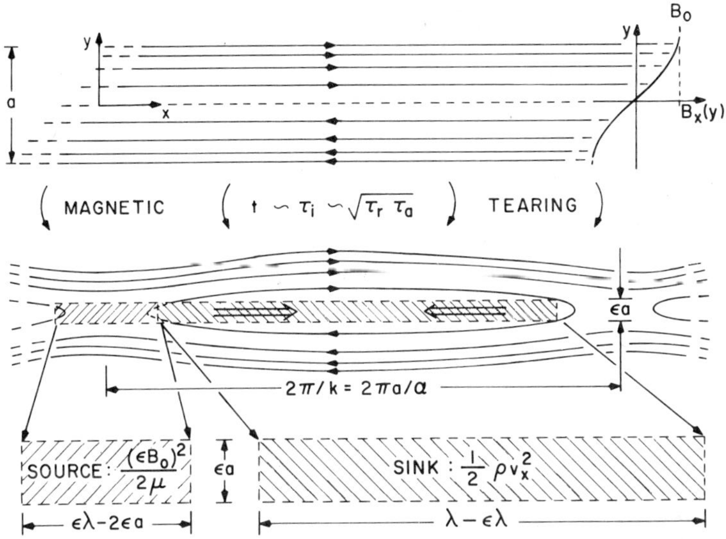

FIGURE 4.11 An expanded view of the development of the magnetic field (solid lines) in resistive tearing (Van Hoven, 1979). The equilibrium is shown at the top, and the reconnected field at the bottom, after one *e*-folding time, including an indication of the plasma flow (horizontal stippled arrows). The energy source and sink are described in the text on pp. 262–3.

where $d^2/dy^2 \sim 1/(\varepsilon^2 a^2) \gg k^2 = 4\pi^2/\lambda^2$ is large and *positive* inside the magnetic tearing layer $\varepsilon a \ll a$ shown in the figure. Equation (4.40) indicates, in passing, that the growth of the reconnecting field component depends on the spontaneous generation of a very narrow region εa where *resistive* diffusion $[\gamma\tau_{r\varepsilon} \equiv \gamma\mu\sigma\varepsilon^2 a^2 \sim 1$, from (4.33) and (4.40)] is the dominant process. Once again, *it is the thinness of the tearing layer which allows the use of a planar approximation* for the inner region (Coppi *et al.*, 1966, 1976; Drake, 1978).

An overall characterization of the results of complete calculations of the linear behavior of the magnetic-tearing instability (Furth *et al.*, 1963; Cross and Van Hoven, 1971; Coppi *et al.*, 1976) indicates the following properties. The *fastest modes generate a very thin diffusion layer* $\varepsilon a \ll a$ (where a is the field-shear scale), they have *long wavelengths* $ka \ll 1$, and the *growth rate* γ satisfies

$$\tau_a^{-1} \gg S^{\nu-1}\tau_a^{-1} \sim \gamma \sim (\varepsilon^2\tau_r)^{-1} \sim S^{\nu}\tau_r^{-1} \gg \tau_r^{-1}, \tag{4.41}$$

where $\nu \sim 1/2$. Thus γ is *intermediate between the resistive and Alfvén rates* as defined in (4.33).

Outside the tearing layer, where d^2/dy^2 and k^2 are relatively small in (4.37), we have effectively infinite-conductivity behavior. We shall now discuss the dynamic solutions in these two regions in more detail, and then match them at the common boundary (Pritchett *et al.*, 1980).

4.5.2.*A* *The external ideal-MHD layer* Beyond the tearing region, we may take $\sigma \approx \infty$ so that we have the planar frozen-in-field condition $\gamma B_{1y} = ikB_{0x}v_{1y}$ from (4.37). If this form is then used in (4.38′), we obtain

$$\frac{d}{dy}[\gamma^2 + k^2 v_a^2(y)]\frac{dv_{1y}}{dy} = k^2[\gamma^2 + k^2 v_a^2(y)]v_{1y}, \tag{4.42}$$

where $v_a^2(y) \equiv B_{0x}^2(y)/(\mu\rho_0)$. This is the ideal form [equivalent to (4.8′) with $v_{1y} = \gamma\xi$] of the planar MHD force equation (Goedbloed, 1971a, Equation (17)). Tearing modes are slow by ideal-MHD standards (4.41) so that $\gamma \ll kv_a(y)$, except near the singularities $y = na$, and the external solution can be treated as a marginal-stability problem, with $\gamma \approx 0$. In addition, these unstable excitations exhibit long wavelengths, so that one can solve the external equation (4.42) as an expansion in orders of the small quantity $(ka)^2$, $v_{1y} = v_{10} + v_{11} + \ldots$ where $v_{11}/v_{10} \sim (ka)^2 \ll 1$ (Rosenbluth *et al.*, 1973).

The proper, zero-order, exterior, velocity function, which matches the Suydam "small" solution [the upper sign in the planar equivalent of (4.25), with $\kappa \propto p' = 0$ and $\xi = v_{1y}/\gamma$] at the singularities is given by

$$v_{10}(y) = \begin{cases} v_{10} & (n-1)a < y < na, \\ -v_{10} & na < y < (n+1)a, \end{cases} \tag{4.43}$$

indicating alternating rigid motions between the singular layers $y_{sn} = na$ of (4.34), where n is integral.

The next-order solution uses (4.43) on the right-hand side of (4.42), which is integrated to obtain

$$\frac{dv_{11}(y)}{dy} = \frac{k^2 v_{10}}{v_a^2(y)} \int_{(n-1/2)a}^{y} v_a^2(y)dy, \tag{4.44}$$

for $y < na$. From the other side, one obtains the same result with integration limits $(y, (n+\frac{1}{2})a)$, so that the logarithmic derivatives

$$\frac{d}{dy}\ln v_{1y} \approx v'_{11}/v_{10} \to \left\{\frac{k^2}{v_a'^2} \int_{(n-1/2)a}^{na} v_a^2(y)dy\right\}\frac{1}{(y-na)^2}$$

$$\equiv -\frac{\lambda_h}{2\pi}(y-na)^{-2} > 0 \tag{4.45}$$

match on either side as $y \to y_{sn} = na$. We have expanded $v_a(y) \approx v_a'(na)(y - na)$ near the singular layer and denoted dv/dy by v'. The parameter λ_h has been introduced to correspond to the usage of Coppi *et al.* (1976). [The result in (4.45) is the analog, in the case of these fast-growing resistive modes, of the *finite* Δ' parameter in the original tearing-mode papers (Furth *et al.*, 1963; 1973).]

A result similar to that of (4.45) applies to the planar *double-tearing mode* with an external conducting boundary (Schnack and Killeen, 1978; Pritchett *et al.*, 1980), with the substitutions $a \to 2y_s$ and $n \to 1/2$ so that there are only *two* singular layers at $y = \pm y_s$, and $v_{10}(y) = 0$ for $|y| > y_s$ in (4.43). These changes reproduce (4.45) with the elision of the factor of 2 in the definition of λ_h, which is still negative.

Finally, when one considers the $m = 1$ *mode in cylindrical geometry*, one again reproduces a $v_{1r} = \gamma\xi_r$ analog to (4.45) for the exterior solution, as shown by Rosenbluth *et al.* (1973). Equation (4.42) with $\gamma \approx 0$ is replaced by (4.22), and $k^2 - r^{-2} = k^2 \ll \alpha^2$ appears in (4.21b), where α is the force-free parameter of (4.3). This abbreviated kink-mode form of g is denoted g_1, and is of order k^2r^2. Thus, (4.22) may be expanded in the same fashion as (4.42) and one obtains

$$\frac{\mathrm{d}}{\mathrm{d}r} \ln v_{1r} = (\ln \xi_r)' \to \left\{ \frac{\mu}{r_s^3 (\mathbf{k} \cdot \mathbf{B})_s'^2} \int_0^{r_s} g_1(r) \mathrm{d}r \right\} \frac{1}{(r - r_s)^2}$$

$$\equiv -\frac{\lambda_h}{\pi} (r - r_s)^{-2}, \tag{4.46}$$

where $(\mathbf{k} \cdot \mathbf{B})_s'$ is the radial derivative of (4.23), evaluated at the zero (singular point) $r = r_s$ of this equation. One difference from (4.45) in this last equation is that g_1 can be negative for $q(r) = krB_z/B_\theta < 1$, which is equivalent to $0 < r < r_s$ for long wavelengths. This circumstance corresponds to unstable $\sigma = \infty$ kink excitations which have ideal growth rates

$$\gamma_1 \approx \frac{|(\mathbf{k} \cdot \mathbf{B})_s'|}{(\mu\rho_0)^{1/2}} \lambda_h \tag{4.47}$$

when $\lambda_h > 0$ (Rosenbluth *et al.*, 1973; Bateman, 1978).

4.5.2.*B* *The internal resistive layer* In the near vicinity $|y - y_s| < \varepsilon a$ of the "singularity" at $y = y_s$ (or $r = r_s$), where $\mathbf{k} \cdot \mathbf{B} = 0$, we may treat the dynamic equations in a planar approximation, as given in (4.37) and (4.38). For applications inside this narrow layer, we use $\mathrm{d}^2/\mathrm{d}y^2 \gg k^2$, α^2 to obtain

$$\gamma B_{1y} \simeq ikB'_{sx}(y-y_s)v_{1y} + (\mu\sigma)^{-1}B''_{1y}, \tag{4.48}$$

$$\gamma\mu\rho_0 v''_{1y} \simeq ikB'_{sx}(y-y_s)B''_{1y}, \tag{4.49}$$

where $B_{0x}(y) \simeq B'_{sx}(y-y_s)$ [*e.g.* (4.36)] has been linearly expanded around the local singular layer at $y_s = na$. We can see that the right-hand resistive term $B''_{1y}/(\mu\sigma) = a^2 B''_{1y}/\tau_r \sim B_{1y}/\varepsilon^2\tau_r$ in (4.48) becomes important as $y \to y_s$.

Equations (4.48) and (4.49) are equivalent to Equations (5) and (6) in the normalized variables of Coppi *et al.* (1976), who proceed to derive a second-order differential equation in the even function

$$\chi(y) = (y-y_s)B'_{1y} - B_{1y} = a(\mu\rho_0)^{1/2}v'_{1y} + \chi_\infty, \tag{4.50}$$

which can be solved by an expansion in Laguerre polynominals. The logarithmic derivative of v_{1y} at the outer edge $(y-y_s \to \infty)$ of the interior layer is then written in the form

$$(\ln v_{1y})' \simeq -\chi_\infty\left[2(y-y_s)^2 \int_0^\infty \chi' y^{-1} \mathrm{d}y\right]^{-1} \tag{4.51}$$

in terms of the calculated function $\chi(y)$ and the constant χ_∞ of (4.50). Since $\chi(y)$ only depends upon the odd-parity (in y) part of $v_{1y}(y)$, the dynamic resistive behavior of the planar force-free field (Cross and Van Hoven, 1971), the double-tearing layer (Schnack and Killeen, 1978) and the internal kink (Coppi *et al.*, 1976) are effectively the same.

4.5.2.C *Tearing-mode growth-rates* The growth-rates of these resistive tearing instabilities are determined by equating (4.51) with (4.45) or (4.46). This is commonly expressed in terms of normalized parameters reflecting the external symmetry. For planar problems, including the *force-free* (Cross and Van Hoven, 1971) and *double-tearing* (Schnack and Killeen, 1978) modes, one defines

$$\lambda_h = (ka^2|B'_{sx}|S/B_0)^{1/3}\lambda_h/a, \tag{4.52}$$

where B_0 is the field amplitude, as in (4.34), which is used to define the overall Alfvén period τ_a of (4.33). For the *resistive-kink mode* (Coppi *et al.*, 1976), one uses the alternative form

$$\lambda_h = (r_s^2|(\mathbf{k}\cdot\mathbf{B})'_s|S/B_0)^{1/3}\lambda_h/r_s, \tag{4.53}$$

where $\mathbf{k}\cdot\mathbf{B}$ is taken from (4.23), with the fields given by (4.3), and the value of S is specified by using r_s as the spatial scale and B_0 as the field strength in (4.33′).

The resistive-tearing growth-rate itself is now usually expressed in terms of the characteristic ideal-MHD time of the particular geometry. [This choice differs from that in the early papers (Furth *et al.*, 1963), which used the resistive time as a reference.] In planar geometry one has

$$\lambda/a \equiv \gamma\tau_a B_0(ka^2|B'_{sx}|)^{-1} \tag{4.54}$$

and in cylindrical geometry

$$\lambda/r_s \equiv \gamma\tau_a B_0(r_s^2|(\mathbf{k}\cdot\mathbf{B})'_s|)^{-1}, \tag{4.55}$$

evaluated at the singular layer.

If a new variable λ is defined for each of these cases by the use of (4.52) and (4.53), then the universal dispersion relation becomes (Coppi *et al.*, 1976)

$$\lambda^{5/4}\frac{\Gamma[(\lambda^{3/2}-1)/4]}{\Gamma[(\lambda^{3/2}+5)/4]} = \frac{8}{\lambda_h}, \tag{4.56}$$

which specifies γ in terms of the physical parameters contained in λ_h (4.45, 4.46) and in the normalization (4.52, 4.53). For an equilibrium which is stable (or almost) in ideal MHD, and for long wavelengths, λ_h is small [$\sim k^2a^2$ as in (4.45)]. In this important limit, $\lambda \approx 1$ and the tearing-mode growth rates are given by

$$\gamma\tau_a \approx [a^2k|B'_{sx}|/B_0]^{2/3}S^{-1/3} \tag{4.57}$$

in the two planar cases and

$$\gamma\tau_a \approx [r_s^2|(\mathbf{k}\cdot\mathbf{B})'_s|/B_0]^{2/3}S^{-1/3} \tag{4.58}$$

in the cylindrical case. These results indicate that, when $S \gg 1$, magnetic tearing is much slower than ideal-MHD growth and also much faster than global resistive diffusion, since $\gamma\tau_r \approx [\ ]^{2/3}S^{2/3} \gg 1$.

For *planar current-sheet* examples there is a further S-variation hidden in the parameter ka since, for reasonably distant boundaries [$L/a \gg 1$, below (4.35)], a wide range of k values may be unstable (Cross and Van Hoven, 1971; Pritchett *et al.*, 1980). In this case, there is a fastest growing wavenumber k, whose value depends on S. For the constant-α, force-free field of (4.34), for example, the maximum value satisfies $k_m a = \pi S^{-0.14}$ (Van Hoven, 1976). When this result is inserted in (4.57), one finds $\gamma\tau_a \propto S^{-0.42}$, a result reproducing that of the complete calculation of Cross and Van Hoven (1971; Van Hoven, 1976).

When one considers *cylindrical tearing*, the resonant value of $k \sim (\mathbf{k} \cdot \mathbf{B})/B_0$ is less free to vary and $\gamma\tau_a \sim S^{-0.33}$ remains more accurate. This is especially true in the case of externally-supported tokamak-like fields, which have $B_z(r) \approx B_0 \ll B_\theta(r)$ and $j_z(r) \approx$ *constant*, so that they are far from force-free. For such a field structure, $(\mathbf{k} \cdot \mathbf{B})' \approx B_\theta q'/r_s \sim B_\theta/r_s^2$, where q is defined below (4.29), and the bracketed term in (4.58) is nearly independent of S. The tearing width for this model is given by $\varepsilon a \sim a\ S^{-0.33} \ll a$ (Coppi *et al.*, 1976), so that the planar approximation is *a posteriori* justified in this very narrow annular region.

For each of the cases treated in (4.57) and (4.58), namely the planar force-free (periodic) and double-layer tearing models, and the $m=1$ resistive internal kink, the distinguishing feature is the fact that the internal (singular-layer) dynamics are characterized by the relation

$$\gamma B_{1y} \sim (\mu\sigma)^{-1} B_{1y}'' \sim B_{1y}/\varepsilon^2 \tau_r \tag{4.59}$$

$$\sim S^{2/3} B_{1y}/\tau_r \sim S^{-1/3} B_{1y}/\tau_a,$$

in a *localized* variant of (4.33). This intermediate growth-rate $(\tau_r^{-1} \ll \gamma \ll \tau_a^{-1})$ has an essential requirement that B_{1y} *strongly varies* for $|y| < \varepsilon a[\psi \equiv B_{1y}/B_0$ is "non-constant" in the common plasma-physics terminology]. When B_{1y} is weakly variable, as it is for other kinds of tearing excitations (including $m=2$ modes in cylindrical geometry), the instability is slowly growing and its energy output quickly saturates (Rutherford, 1973).

This point can be illustrated from the tearing-mode dispersion relation (4.56). Another asymptotic limit holds when $\lambda \ll 1$, or (Pritchett *et al.*, 1980)

$$k \Delta y_s \sim ka \gg S^{-1/7}, \tag{4.60}$$

where Δy_s is the separation of two different singular, or tearing, layers. If (4.60) holds, the resistive-tearing growth-rate becomes $\gamma\tau_a \propto S^{-3/5}$, a much smaller value called the "single-tearing-mode" value which applies to an isolated (or $m \neq 1$) singular layer. Resistive tearing excitations of this kind satisfy $(\mu\sigma)^{-1} B_{1y}'' \ll B_{1y}/(\varepsilon^2 \tau_r)$, so that $B_{1y} \sim$ *constant* near the singular layer and the unstable excitations quickly diffuse inward from the outside. This circumstance leads to wider magnetic "islands" with higher inertia, and vastly slows the rate and ultimate magnitude of the magnetic energy release (Rutherford, 1973).

The linear tearing-mode morphology and growth-rate results of this section have been verified by a number of time-dependent resistive calculations applied to relevant planar and cylindrical magnetic-field geometries (Van Hoven and Cross, 1973; Waddell *et al.*, 1976; Schnack and Killeen, 1978, 1979). The major aim of these temporal computations, most of them motivated by laboratory applications, has been to delineate the nonlinear field and plasma dynamics of this instability, which we will discuss from a different perspective in Section 4.5.4.

4.5.2.*D* *Tearing-mode stability conditions* As in the ideal-MHD case, considerable analysis has been devoted to the conditions under which a cylindrical/toroidal plasma can be stable against resistive modes.

The major requirement for magnetic tearing is the existence of a "mode-rational surface" or *singular layer*, where $\mathbf{k}\cdot\mathbf{B}=0$ as in (4.23) and (4.36). There can, in principle, be many such layers, since (4.23) can be restated in the form $q(r)=rB_z/B_\theta=m/n$, if the axial wavelength k is specified to be $2L/n$ where L is the plasma-column length (or half the major circumference in the toroidal case). However, as we have shown in the previous section, only the resistive internal kink ($m=1$) has significant growth. It also exhibits the widest ranging displacement, since the other m modes have a structure similar to those of localized interchanges.

If the field profile exhibits a singular layer, the existence of instability then depends upon the radial and axial boundary conditions in a way which is similar to ideal-MHD behavior. In an approximate treatment, which is valid for large S, the resistive properties of the region around the singular layer can be replaced by vacuum-reconnection behavior (Coppi *et al.*, 1966). The stability boundary (resistive marginal condition) is then determined by matching the derivatives B'_{1r}, derived from an infinite-conductivity calculation in the $r<r_s$ and $r>r_s$ "external" layers (Furth *et al.*, 1963). This Newcomb-like method and its relationship to infinite-conductivity stability is most clearly explained by Goedbloed and Dagazian (1971). An alternative energy-principle connection was made by Furth *et al.* (1973), who then performed a series of tokamak-related stability evaulations.

An early determination of this type was that by Gibson and Whiteman (1968), who realized that the marginal ($\gamma=0$) equation could be solved analytically for the constant-α force-free case of (4.3). The solutions for B_{1r} are then just those of Voslamber and Callebaut (1962), which are continuous through the singularities. Gibson and Whiteman expressed their results in terms of the location of a rigid conducting wall which produces resistive stability of the Lundquist field. Recent calculations in this line have been performed by Robinson (1978) and Chiuderi *et al.* (1980). Robinson's work is most strongly motivated by the laboratory *reverse-field pinch* with a conducting boundary, and he has concentrated on variations of the Lundquist field (4.3) which move the stability boundary as far out in radius as possible. One difficulty with this approach is that the most unstable axial wavelengths appear to be moved to smaller values, which would make end-effect stabilization of the kind discussed in Section 4.4.8 more difficult. Chiuderi *et al.*

(1980) have investigated the effect of an external current-free zone on the tearing stability of an axial force-free region. They have had some success in establishing resistive stability for relatively narrow current channels, which have their singularities in the external potential-field region.

However, such a singularity need not exist for the crucial fastest-growing fundamental mode $k=\pi/L$ of a finite-length plasma. This circumstance can provide an axial *current threshold* for the tearing instability, as we may understand by the following argument. For this mode, (4.23) assumes the form

$$0=\mathbf{k}\cdot\mathbf{B}=\frac{\pi}{L}B_z(r)\pm\frac{1}{r}B_\theta(r), \tag{4.61}$$

where $m=\pm 1$ has been chosen. If $B_z(r)$ is assumed to be given, for example by the global, preflare, small-α, force-free, coronal field of Section 4.2.3, then consider what happens as an axial current begins to flow in the loop. Assuming the shear-inducing current density j_z to be uniform for small r, for simplicity, we can then rewrite the *necessary* resonance condition (4.61) for the fundamental resistive tearing mode in the form

$$\frac{1}{r}B_\theta(r)\sim\tfrac{1}{2}\mu j_z=\pm\frac{\pi}{L}B_z(r),$$

which cannot be satisfied until $|j_z|$ exceeds the finite-length threshold $j_{tz}=(2\pi/\mu L)B_z$. Since $B_z(r)$ usually decreases away from the axis in cylindrical geometry, the threshold is (in general) crossed at a non-zero radius r_s. Because the stored energy increases as j_z grows, one can approach the condition for instability start at a level in which there is significant free energy. The exact form of this threshold will depend, however, on a more complete understanding of the loop end effects, as previously discussed in Section 4.4.8.

4.5.3 Magnetic tearing energetics

In the solar flare context, a paramount question goes beyond that of stability to a concern about the energy release, given the existence of instability. A description of magnetic-tearing based on the energetics aspect is also natural for the flare application, since it can lead to a prediction of the observable effects of the instability and a clearer understanding of the potential of the mechanism itself. For these reasons we will include here brief treatments of the process of energy transformation, and the amount and forms of energy release in *fast* magnetic tearing of the variety described in the preceding section (Van Hoven, 1979).

4.5.3.*A* *Energy transformation in the instability* We begin by a consideration of the spatial characteristics of the equilibrium and perturbed magnetic fields, as shown qualitatively in Figure 4.11. At the top is shown the background x-component field of (4.34), oriented in the x–y coordinate system at the left, with its amplitude variation shown schematically at the right. The B_{0x}-component, parallel to **k**, as in (4.36), is the main source of energy for the instability; B_{0z}, directed into the page here, is substantially unaffected.

In the lower half of the figure is shown one wavelength ($\lambda = 2\pi/k$) of the reconnected magnetic field developed by resistive tearing. The normalized wavenumber parameter $\alpha \equiv ka$ is customarily quoted in the tearing-mode literature (Furth *et al.*, 1963), but will not be used in the following so as to eliminate confusion with the force-free-field parameter of Section 4.3. Also labelled is the width of the tearing layer εa. The elongation ratio is actually much more exaggerated than can be represented in detail on a single-page figure, since $\lambda/a \equiv 2\pi/ka \gg 1$ and $\varepsilon a/a \ll 1$ cannot be simultaneously shown.

As a first result, we examine the influence of this elongated tearing structure on the velocity and field components. Incompressibility and the fact that $\partial/\partial z = 0$ indicate

$$0 = \boldsymbol{\nabla} \cdot \mathbf{v}_1 = ikv_{1x} + v'_{1y} \sim ikv_{1x} + v_{1y}/(\varepsilon a), \tag{4.62a}$$

where the prime hereafter denotes $\partial/\partial y$, which has the variation scale εa near the tearing layer. Thus, the comparison

$$v_{1x} \sim v_{1y}/(ka\varepsilon) \gg v_{1y} \tag{4.62b}$$

may be made. Since the magnetic field is also divergence-free, the components of $\mathbf{B}_1$ are similarly related.

Next, consider the *dynamics* of the fluid near the boundaries of the fast tearing layer. *Inside*, where the field is reconnecting, the behavior is substantially resistive, as we have shown in (4.59), so that

$$\gamma\tau_r \sim 1/\varepsilon^2, \tag{4.63}$$

where ε is as-yet undetermined. Thus, we have resistive diffusion on the width-scale εa and a resulting interior diffusion velocity

$$v_y \sim \varepsilon a/\tau_t = \gamma\varepsilon a, \tag{4.64}$$

where the *t*earing growth-time τ_t has been introduced.

Just *outside* the reconnection layer, Ohm's law takes the alternative form $\mathbf{E}_1 \sim -\mathbf{v}_1 \times \mathbf{B}_0$, dominated by the fluid motion (Furth *et al.*, 1963). Using

Faraday's law (4.37) we obtain

$$\gamma B_{1y} \sim ikB_{0x}(\varepsilon a)v_{1y},$$

evaluated at $y = \varepsilon a$. This result, along with the $\mathbf{B}_1$-equivalent of (4.62), gives

$$v_{1y} \sim \gamma \varepsilon a[B_{1x}/B_{0x}(\varepsilon a)] \sim \gamma \varepsilon a, \tag{4.65}$$

since $B_{1x} \sim B_{0x}(\varepsilon a) \sim \varepsilon B_0$ for substantial reconnection at the edge of the central layer.

Thus, we have a boundary condition at $y \sim \varepsilon a$ of continuity of the fluid inflow velocity, determined by different mechanisms inside and out, as specified by Equations (4.64) and (4.65). The external velocity (4.65) describes the (fast) rate at which magnetic energy is *continually* brought into the tearing layer to be annihilated.

We are now able to examine the *energetics* of the instability. The source which drives resistive tearing is the magnetic energy which disappears in the reconnection region, as shown in the lower left of Figure 4.11. The density is of order

$$W_m \sim B_{0x}^2(\varepsilon a)/(2\mu) \sim \varepsilon^2 B_0^2/(2\mu). \tag{4.66}$$

and it vanishes in the volume (area × 1 m.), scaled from the proportions of the whole region, also sketched in Figure 4.11. Thus, at the state of development shown, the change in magnetic energy is

$$\Delta W_m \sim W_m[\varepsilon a][\varepsilon \lambda - 2\varepsilon a] \sim \frac{(\varepsilon B_0)^2}{\mu}\frac{\pi}{k}\varepsilon^2 a\left[1 - \frac{ka}{\pi}\right], \tag{4.67}$$

where the $0(ka)$ correction (which reflects the field energy intruding in the ends of the source volume) is not essential, but brings out some of the wavenumber dependence.

This magnetic energy is initially transformed into fluid energy by the action of the tearing instability. If we ignore heating at high values of the magnetic Reynolds number (Furth *et al.*, 1963, Appendix A), the output energy is entirely kinetic with density

$$W_k = \tfrac{1}{2}\rho_0 v_1^2 \sim \tfrac{1}{2}\rho_0 v_{1x}^2, \tag{4.68}$$

using the approximation of (4.62). Taking account of the volume in which this energy appears, as shown in Figure 4.11, we obtain

$$\Delta W_k \sim W_k[\varepsilon a]\lambda[1-\varepsilon] \sim \rho_0 v_{1x}^2 \frac{\pi}{k} \varepsilon a \sim \rho_0 \frac{v_{1y}^2}{(ka\varepsilon)^2} \frac{\pi}{k} \varepsilon a,$$

$$\sim \rho_0 \frac{(\gamma a\varepsilon)^2}{(ka\varepsilon)^2} \frac{\pi}{k} \varepsilon a = \pi \rho_0 \gamma^2 \varepsilon a / k^3, \tag{4.69}$$

where $\varepsilon \sim 0(k^2 a^2)$ has been dropped in the first step and Equations (4.62) and (4.65) have been used.

Equating (4.67) and (4.69), one obtains a second expression for the normalized growth-rate,

$$(\gamma \tau_r)^2 = k^2 a^2 S^2 \varepsilon^3 (1 - ka/\pi), \tag{4.70}$$

in terms of ε and the parameters of the problem. $S \equiv \tau_r/\tau_a$ is the global magnetic Reynolds number (4.33′).

Finally, the elimination of ε from (4.63) and (4.70) yields the growth-rate

$$\gamma \tau_r = [k^2 a^2 (1 - ka/\pi)]^{2/7} S^{4/7}, \tag{4.71}$$

in the form found in the full calculations of Cross and Van Hoven (1971). The instability disappears for perturbations not satisfying $0 < ka < \pi$ $[2a < \lambda < \infty]$ and the maximum growth-rate is

$$\gamma \sim \tau_a^{-4/7} \tau_r^{-3/7}, \tag{4.72}$$

a variation for the constant-shear, force-free equilibrium field of the result $\gamma \sim (\tau_a^2 \tau_r)^{-1/3}$, of Equation (4.59) (see Cross and Van Hoven, 1976, p. 1594).

Thus, we have shown that fast linear tearing-mode growth, and its rate of rise, can be explained by a balance between continuing *magnetic-energy input* and *kinetic-energy output*.

4.5.3.*B* *Magnetic energy release: nonlinear development* The first *nonlinear* calculations of resistive magnetic tearing were made by Van Hoven and Cross (1973), who treated the fast-and-far growing class of modes described in Section 4.5.2, and then by Rutherford (1973), who concentrated on the slower $B_{1y} \sim$ *constant* instabilities described at the end of Section 4.5.2.C. Van Hoven and Cross were particularly interested in the solar flare problem, so they emphasized the amount and forms of the energy release. Since that time, other computations of nonlinear fast-tearing have been made by Waddell *et al.* (1976) and Schnack and Killeen (1978, 1979), who have stressed the developing field and flow patterns but have given some attention to energy output.

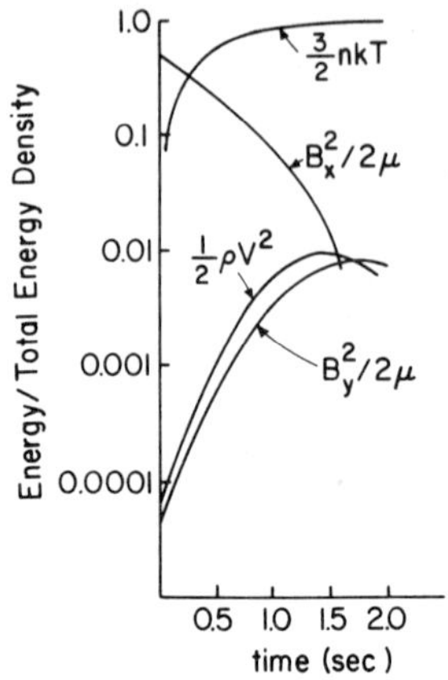

FIGURE 4.12 The energy budget of the nonlinear tearing mode of Figure 4.11, at a relatively low value of $S = 10^2$ (Van Hoven and Cross, 1973).

In order to understand the predictions of these calculations, we will describe some of the results of Van Hoven and Cross (1973; Van Hoven, 1976). Figure 4.12 shows an illustrative example of the energy budget of the planar, force-free, magnetic-tearing instability. This particular case applies to a relatively low value of the magnetic Reynolds number ($S = 10^2$), so that the instability grows only about ten times as fast as the equilibrium field resistively decays. Since the unstable perturbations begin at a very low (noise) level, most of the stored energy (including $B_z^2/(2\mu)$, which is not shown) goes into Ohmic heating under these overly-dissipative conditions.

However, the plasma kinetic energy and the transverse, reconnection component of the field energy also rise at the initial linear-instability rate given by the analog of (4.71). Eventually, when the energy release reaches the level of the energy store, the instability saturates. In this low-S example, the output is throttled by meeting the quickly-decaying background-field level. One should notice the characteristic approximate equilibration of kinetic energy and small-scale field energy at saturation.

A series of such plots, moving to higher magnetic Reynolds numbers, provides the results shown in Figure 4.13, which shows the levels of the peaks of the energy-output curves, or the energy-release efficiency. We see that larger S provides higher energy release, while maintaining the kinetic/magnetic balance mentioned previously.

There is a practical limit to how far one can increase S in such simulations, since the smallest relevant spatial scale εa and fastest time scale τ_a both decrease rapidly as shown in (4.59). One approximate way of circumventing this *computational* difficulty, as far as estimating saturated energy-release levels, is to artificially remove the limitation of the decaying equilibrium field

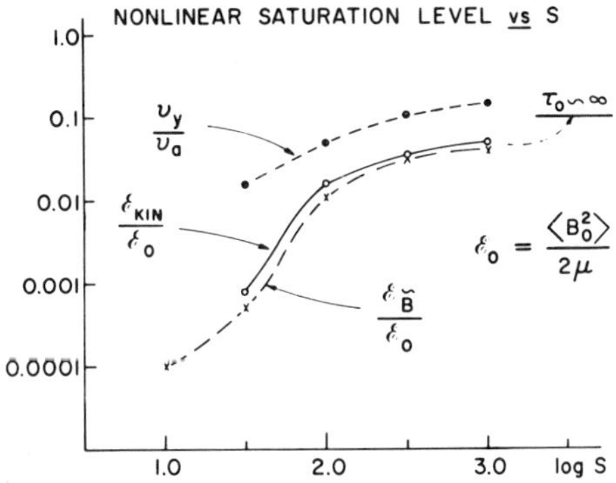

FIGURE 4.13 The energy-release efficiency of the planar, force-free, tearing instability (Van Hoven, 1976). The kinetic and magnetic levels are the peaks of a series of plots of the kind shown in Figure 4.12.

$\mathbf{B}_0 \exp(-t/\tau_{r0})$ which has the effect shown in the exaggerated case of Figure 4.12. This can be done in the computations by a procedure which effectively sets $\tau_{r0} = \infty$ (Van Hoven and Cross, 1973). This results in the increase in efficiency, equivalent to a much larger value of S, shown at the right in Figure 4.13.

The results of this solar-flare, resistive-instability, simulation shows that 20–30% of the initial stored magnetic energy can be released in 3–4 instability growth-times, given in this planar example by (4.72).

Another set of planar nonlinear calculations is that due to Schnack and Killeen (1978), who treat the *double-tearing* mode. They limit their results to low values of $S = 10^2$ and do not give the complete energy budget. The one result quoted is taken *before* the kinetic-energy peak and shows an efficiency somewhat less than that indicated in Figure 4.13.

The first nonlinear, *cylindrical-geometry*, $m = 1$ tearing-mode computations were performed by Waddell *et al.* (1976). These results apply to the Tokamak case ($B_z \gg B_\theta$) and concentrate upon the detailed magnetic-field and plasma flow patterns. They do show that the energy continues to grow for many linear-instability growth times (4.58), but do not give the relative energy-output efficiency.

A more pertinent analog to the conditions leading to a coronal-loop flare is the *reverse-field pinch*, which has an equilibrium close to force-free (4.3). A flare in this model would arise from the $m = 1$ resistive internal-kink instability. Nonlinear calculations of this process have recently appeared (Schnack and Killeen, 1979). Figure 4.14 shows two views of the helical magnetic-flux contours, at a time (in terms of τ_a) during the linear-growth phase and at a time near saturation, for $S = 10^2$. *Flux surfaces* are related to magnetic-field "lines" in that the *radial* flux-function gradient is equal to $B_\parallel \equiv \mathbf{k} \cdot \mathbf{B}/k$, where $\mathbf{B}$ is the

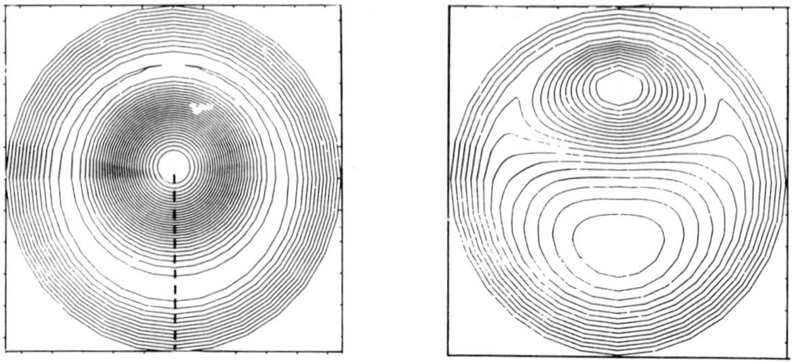

FIGURE 4.14 The resistive-kink instability of the force-free field of (4.3). The magnetic surfaces (flux contours of $\mathbf{k} \cdot \mathbf{B}/|\mathbf{k}|$ and B_r) are shown at an early linear and late nonlinear stage of magnetic tearing (Schnack and Killeen, 1979).

total field developing from the equilibrium of (4.3) and (4.23), and the *azimuthal* gradient gives B_r. The actual direction of the field changes as $\mathbf{k}(r)$ varies with radius; at the singular (open) layer r_s of Figure 4.14(a), the parallel-field-free plane (annulus) is at about 45° from the pinch cross-section for the parameters chosen ($kr_s \sim 0.96$). The vertical dashed line in the figure signifies the boundary of the next "sheet" of the helical flux surface. One can see the decrease of the parallel flux density ($\sim 1/2$ of the total magnetic energy) as the tearing instability progresses, but part of this is due to resistive diffusion (as in Figure 4.12) at this low value of S.

The extent of the available energy output is shown in Figure 4.15, plotted *vs.* $\tau_h = \tau_a$. Except for a time-lag between reconnection magnetic energy and kinetic energy, these results agree with those of Figure 4.12 and 4.13. The growth continues at the *fast* linear rate (4.58) until near to saturation; the limiting level for both forms of energy release is about ten percent, agreeing with the prediction of Figure 4.13 at $S = 10^2$.

We see here an essential agreement between the nonlinear, resistive-tearing energy-output results of the planar force-free model of Van Hoven and Cross (1973) and the cylindrical force-free resistive-kink model of Schnack and Killeen (1979). When this is added to the analogous linear dynamics of these mechanisms, in Section 4.5.2.C, one can see an apparent unity between the behavior of current-sheet and coronal-loop reconnection instabilities. Such fast-growing resistive tearing mechanisms provide the required energy release (by a reasonable extrapolation to coronal values of S) in a short enough time, using only magnetic energy as a source (Van Hoven, 1976). These facts make magnetic tearing a prime candidate for the solar flare instability.

The only other nonlinear reconnection-instability process which might offer some competition to the single-mode fast-tearing mechanism, described

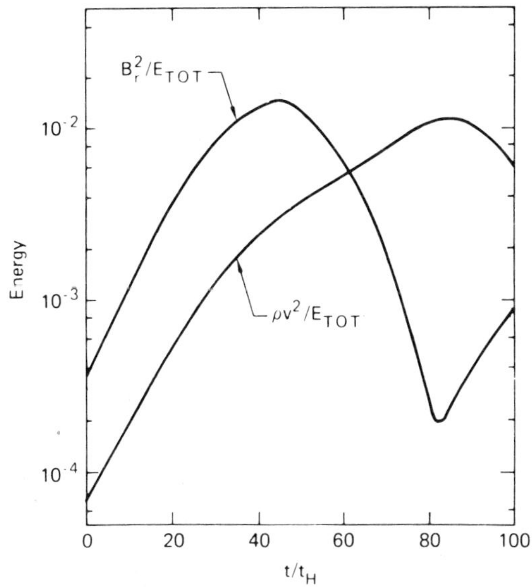

FIGURE 4.15 The energy outputs at $S=10^2$ corresponding to Figure 4.14 (Schnack and Killeen, 1979).

above, has lately been proposed by Waddell *et al.* (1979). This scenario involves the nonlinear *mode-coupling* interaction of unstable tearing modes, with *different helicities* (values of $q(r_m)=m/n$). The present calculations assume tokamak (non-force-free) conditions, the linear stability of the $m=1$ mode, and closely-spaced singular layers r_{mn}. The results show that the stimulated nonlinear growth of the fundamental resistive kink ($m=1, n=1$) is still the fastest under these conditions. The relevance of these calculations to a coronal-loop-like field has not, at the present time, been established. However, the multiply-unstable excitations and resulting chaotic magnetic structure found would seem to disagree with the clean threshold and regular global manifestations observed in flares.

4.5.3.*C* *Observable energy outputs* We have described the calculations which detail the *nonlinear* development of the resistive-tearing instability and the resultant energy release (Van Hoven and Cross, 1973). At high magnetic Reynolds numbers, there is a rough equilibrium between average fluid kinetic energy $\langle \frac{1}{2}\rho v^2 \rangle$ and transverse magnetic field energy $\langle B_y^2/(2\mu) \rangle$, each saturating at about 15% of the initial energy density (Van Hoven, 1976) in a time less than $5\tau_r$. We now wish to detail some observable aspects of these predictions (Van Hoven, 1979).

The preponderance of the fluid kinetic energy is generated by Fermi acceleration in the x-direction, away from the source region of Figure 4.11, as indicated by the arrows in the interior of the tearing layer. We can estimate the ratio of the x-outflow to the y-energy of (spontaneously induced) external inflow by using the volume (area) proportions of the figure and the incompressibility approximation of (4.62), and obtain

$$\frac{\frac{1}{2}\rho_0 v_x^2 \Delta V_i}{\frac{1}{2}\rho_0 v_y^2 \Delta V_e} \sim \frac{v_x^2 \varepsilon a}{v_y^2 a} \sim \frac{1}{k^2 a^2 \varepsilon} \sim S^{0.57}, \tag{4.73}$$

where the ΔVs are *i*nterior and *e*xterior volumes, and the values of ka and $\varepsilon(S > 10^3)$ are taken from Van Hoven (1976). The maximum values of v_x generated by the instability (Van Hoven and Cross, 1973) are of the order of the external Alfvén speed $v_a = B_0/(\mu\rho_0)^{1/2}$, so that the energy given to the bulk of the particles of species s is $2m_s v_a^2$.

The largest electric fields generated by resistive magnetic tearing occur in the reconnection layer ($y=0$). Since the instability is truly time-varying ($\partial \mathbf{B}/\partial t = 0$ in Faraday's law, in contrast to the situation in the Petschek mechanism), the electric field has a spatial structure in both x and y. The tearing-layer electric field also has a variable direction with respect to the magnetic field.

The E-field strength is greatest at $x \sim \pm\lambda/4$ in Figure 4.11, where it is almost entirely across the magnetic field. At this point, the value of the field may be estimated by the expression

$$E_z^{\perp} \sim v_x B_y \sim (2W_k/\rho_0)^{1/2}(2\mu W_m)^{1/2} \sim 2(\mu/\rho_0)^{1/2} W_{k,m}, \tag{4.74a}$$

where W_k and W_m are the kinetic and magnetic energy densities as used in (4.66) and (4.68), except they are now to be evaluated in the nonlinear stage of the development of the instability, where they have come into approximate mutual equilibrium, as shown in Figure 4.13 and used in the last step. If we now employ the further nonlinear result that these energy densities saturate (for large S-values) at $\sim 15\%$ of the initial stored energy, we may further develop (4.74a) to find

$$E_z^{\perp}[V/m] \sim 0.3(\mu/\rho_0)^{1/2}(B_0^2/(2\mu)),$$

$$= 0.15\, v_a B_0 \sim \begin{cases} 10^{4.5}\, B_0^2 n_0^{-1/2} \text{ (cgs)}, \\ 10^{15.5}\, B_0^2 n_0^{-1/2} \text{ (MKS)}, \end{cases} \tag{4.74b}$$

where B_0 and n_0 are the pre-existing magnetic field strength and number density. This estimate agrees with the values found in the full nonlinear calculation (Van Hoven, 1975). For solar-flare conditions, in which the magnetic field is 0.03 T (300 G) and the number density is 10^{15} m^{-3} (10^9 cm^{-3}), the predicted electric field is 10^5 V/m, a value similar to that inferred from Stark-effect measurements by Davis (1974).

The magnetic-field-aligned electric field is, on the other hand, largest in a narrow peak in the vicinity of the reconnection point at the origin in Figure 4.11. Taking account of the half-widths in the two directions, this (parallel) E_z-field may be estimated to affect about $0.25\varepsilon \sim 10^{-1.1}S^{-0.3}$ of the total initial plasma volume. In addition, particles are continually swept into this focus at a velocity approaching $0.1\ v_a$ as shown in Figure 4.13. The value of the electric field at this point is given by the flux-transfer rate

$$E_z^{\parallel}(0, 0) = \frac{1}{2}\cdot\frac{\partial}{\partial t}\left[\int_0^a B_x(0, y)\mathrm{d}y - \int_0^{\lambda/2} B_y(x, 0)\mathrm{d}x\right], \tag{4.75a}$$

where the symmetry of the configuration has been used to eliminate other electric-field contributions. In our calculations, this field can be estimated by the current that it drives at the origin

$$E_z^{\parallel} \sim j_z/\sigma \sim (\mu\sigma)^{-1}\partial B_x/\partial y,$$

$$\sim B_x/(\mu\sigma\varepsilon a) > aB_0/(\varepsilon\tau_r), \tag{4.75b}$$

where the last two steps use the approximation $\partial/\partial y \to (\varepsilon a)^{-1}$, as in (4.62a), and the fact, apparent in our nonlinear numerical results, that the condition $B_x > B_0$ holds outside of the reconnection layer as the instability saturates. Equation (4.75b) duplicates the parallel field found in these computations ($S \leqslant 10^3$) when the predicted value of ε (Van Hoven, 1976) is used.

These considerations of the observable energy outputs of resistive tearing have been derived with reference to a *planar* magnetic-field reversal, of the kind postulated in large-scale current-sheet or helmet-streamer flare models (Sturrock, 1968; Švestka, 1976). However, the electric fields, particularly, are concentrated in and determined by relative variations in the narrow tearing layer in a way that allows one to ignore asymmetries (in y) of the external field-shear environment (Drake, 1978). Thus, these results have considerable generality and can be applied to more complex, and realistic, sheared magnetic fields as described below.

4.5.3.*D* *Loop-flare morphology* Resistive magnetic tearing occurs in almost all physical situations in which a stressed magnetic field of sufficient scale (in one dimension) is assembled, as in the quasi-planar solar-flare configurations shown by Švestka (1976). In these cases, the geometry of the unstable electric and velocity fields are relatively simple, as we have described in the previous sections.

For a coronal-loop flare, the spatial structure, with its varying magnetic

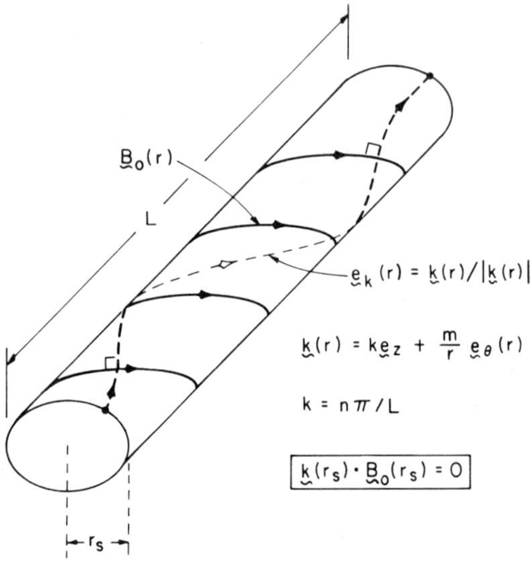

FIGURE 4.16 A schematic depiction of the fundamental ($m=1$, $k=\pi/L$), global, spatial-resonance perturbation (dashed line) and its local resonance (box) with the magnetic field (solid line) at the singular radius r_s (Van Hoven, 1979).

pitch (*v.* Section 4.3), is more complex and deserves some discussion. The primary simplifying element, which enables one to make the qualitative transition from plane geometry to the curvilinear magnetic configuration of a coronal loop, is the narrowness of the resistive-tearing layer, as we have mentioned before.

We may, therefore, concentrate our attention on the singular layer r_s, described by (4.23), and its immediate vicinity. The equilibrium-field and allowed-perturbation structures in this layer are shown schematically in Figure 4.16. The fundamental excitation ($m=1$, $n=1$) of this finite-length cylinder is depicted by the dashed-line trajectory of a fixed point on the wavefront, which makes one cycle in azimuth in going one-half wavelength along the loop axis. This (spatial-resonance) condition is satisfied at all radii for this choice of m and n, but the wavefronts and the magnetic field mutually resonate [by satisfying $\mathbf{k}\cdot\mathbf{B}=0$] only at discrete radii, of which r_s is one example.

The tearing-mode excitations near this layer are those shown in Figure 4.14(a). However, it should be made clear that this is a strange view, which is not directly across the axis of the cylindrical pinch.

Rather, the view shown is locally in the plane of the wavevector; in the case of Figure 4.16 this is actually helically twisted along the dashed line. For the fundamental excitation ($m=1$, $k=\pi/L$) shown in Figure 4.16 and Equation (4.61), Figure 4.14 actually depicts the full length of the unstable mode, which would range between the rigid ends of the coronal loop, represented schematically by the dashed line. In this curvilinear "cross-section", it is also true (as in Figure 4.11) that the largest component of the equilibrium magnetic field is perpendicular to the plane of the figure. [In the case of a tokamak, for which $B_z \gg B_\theta$ (and $k \ll 1/r_s$), the situation is somewhat different and this cross-section would be aligned nearly across the axis of the torus (Waddell *et al.*, 1976).]

Altogether, then, the view of Figure 4.14 is equivalent to that previously shown in Figure 4.11, but it is distorted by the curvilinear coordinates and inside/outside displacement asymmetry in this geometry. Because of the actual extreme narrowness of the tearing layer, we may apply most of the planar results of Sections 4.5.3.*B* and 4.5.3.*C*, including those specifying the electric fields (Drake, 1978), to the present field configuration.

In particular, the magnetic energy which drives the instability is that due to the transverse field in the plane of this figure. [In the tokamak example, it is the poloidal field which provides the energy source for magnetic tearing. If this is applied to the solar case of a force-free coronal loop (Van Hoven *et al.*, 1977), extending from one of the elemental chromospheric flux tubes of Stenflo (1975), the poloidal field alone (as a lower limit) can provide the 10^{22} J (10^{29} ergs) needed for a small loop flare (A. S. Krieger, private communication).]

The majority of the fluid which is drawn into the singular layer by the instability is accelerated down the wavevector, as described in connection with Equation (4.73). This energetic outflow would thus move, from the center to the ends, along the dashed line of Figure 4.16.

The minority of high-energy particles, on the other hand, are accelerated by the growing magnetic-field-aligned electric fields of the instability (Drake and Lee, 1977), as given by (4.75), with orientation perpendicular to the plane of Figure 4.14. These particles are subsequently channeled down the magnetic field, shown as a solid line in Figure 4.16. They would then also appear at the ends of the cylinder (or loop), albeit with different timing since the acceleration mechanism is different.

Finally, though we have not considered the effects of primary instability heating in our computations, due to the difficulty of following a time-varying Coulomb resistivity, any such heating would clearly be concentrated in the center of the reconnection layer at the top of Figure 4.14. In the solar coronal loop configuration, this would occur at the apex, which is commonly observed as the hottest region in a loop flare.

4.6 ATTEMPTS AT A LOOP-FLARE MODEL

Although a number of papers have been written which claim to describe "models" of coronal-loop flares, almost no progress has been made on the comprehensive calculations necessary to define the plasma/field dynamics and energy release of such configurations. Considerably more effort has been devoted to problems of the post-flare energy flux and dissipation, given the localized deposition of a certain amount of stored energy.

The flare models which are presently available can best be described as phenomenological, both in terms of the level of their description of coronal conditions, and in terms of their reliance on analogies with the behavior of laboratory plasma devices.

One effort in this direction is that put forward by Spicer (1977), who depends heavily on a qualitative treatment of the Skylab XUV and X-ray observations of coronal loops and flares. An attempt is made to describe briefly some twenty-odd plasma-physics mechanisms which could conceivably produce these effects. Beyond MHD topics, these cover electrostatic turbulence, the parametric oscillating two-stream instability, superheating instability, drift-wave instability, thermal convection, and shock waves. Some additional supporting material, gathered from the plasma-physics literature, is contained in an earlier unpublished report (Spicer, 1976).

In the context of our present interest, namely flare magnetohydrodynamics, Spicer devotes some attention to ideal-kink and resistive-tearing instabilities, for conditions patterned after the tokamak model. Kinks are treated by using the finite-length formalism of Solov'ev (1971; *v.* Section 4.4.8), which assumes a shear-free field ($\mathrm{d}j_z/\mathrm{d}r=0$) with a conducting boundary. The weak sufficient stability condition $q(r)\gg 1$ [equivalent to Equation (4.2) of Spicer] is the major result quoted, but this would require a *very* short loop [($\alpha_0 R\ll 2$), which implies $L\ll 10^4$ km if $\alpha_0^{-1}\sim 3000$ km] if the coronal field were nearly force-free, as described in Section 4.3 and used as an example by Spicer.

The "resistive kink" of a sheared field is then described, using the ($m\neq 1$) single-tearing-mode formalism of Furth *et al.* (1963), as described at the end of Section 4.5.2.C. Since these modes grow relatively slowly and saturate early, some attention is given to two mechanisms for further nonlinear development. Taking first a small-S situation, in which short-wavelength tearing excitations are assumed to be initially dominant, an argument is made (Finn and Kaw, 1976) that magnetic islands will sporadically coalesce to longer wavelengths. In fact, for coronal large-S conditions, the longest allowable wavelength is usually the fastest growing one (*v.* Section 3.5.2.C).

A second suggestion concerns the "overlapping of resonances" (Finn, 1975), in which it is assumed that magnetic islands from different, simultaneously unstable, tearing layers nonlinearly interact and accelerate the loss of confinement. The model taken as an illustration (Spicer, 1976) shows both of the singular layers to have the same helicity, as in the double-tearing mode. The conditions under which this interaction takes place are given by the reverse of (4.60), or $ka\sim a/R\ll S^{-1/7}$, where a is the transverse magnetic-shear scale; this limitation would require very long coronal loops for resonant overlap ($L \gg 10^{4.6}$ km for $a \sim 3000$ km and $S = 10^9$). The situation is improved if different helicities interact, as described at the end of Section 4.5.3.*B*, with results which are not yet completely clear. The only available multiple-helicity computations (Waddell *et al.*, 1979) apply to tokamak conditions, and indicate a compound growth rate 2–3 times larger than the single-tearing rate. On the other hand the $m = 1$ resistive kink (assumed stable in these fusion-motivated calculations) would grow at a rate 15–20 times faster at the chosen values of S (*v*. Section 4.5.2.C). One conclusion to be drawn from such comparisons is that the $m = 1$ tearing mode, by itself, can grow sufficiently quickly and extensively to explain a loop flare (*v*. Section 4.5.3.*B*), without the need to assume island coalescence or resonant overlap.

Another loop-flare model is that of Colgate (1978), which is principally concerned with energy transport. The plasma-dynamics discussion is based mainly on the results of some early *z*-pinch experiments (Colgate, 1958). The phenomenology of this transient discharge, under unstable conditions in which the magnetic fields become disordered, is described and analogies drawn to the loop-flare mechanism. Since the complete results of the experiment are not readily available (Birdsall *et al.*, 1962 conference proceedings), it is difficult to evaluate the relationship of this pinch device to the considerations of this chapter. From the facts given, in connection with Figure 3(c) of Colgate (1978), one can estimate that $B_\theta \sim B_z$ and $q \sim 1$; the resulting unstable excitation exhibits a substantial $m = 1$ kink perturbation. As the author states, "there is yet no complete formal theory of the laboratory observations".

Coronal-loop plasma dynamics, including resistive effects, is a difficult topic on which some progress has been made, as we have tried to show in this chapter. It is equally clear that more complete and accurate calculations, patterned on the laboratory-fusion example, but based on *coronal* conditions, as derived from better diagnostic observations, are necessary before a loop-flare model can be attained.

Acknowledgments

The writing of this chapter was supported in part by the Atmospheric Sciences Section of the National Science Foundation, and by the Solar Physics Branch of the National Aeronautics and Space Administration.

References

Alfvén, H. and Fälthammar, C. G., *Cosmical Electrodynamics*, Oxford (1963).
Anzer, U., *Solar Phys.* **3**, 298, and **4**, 101 (1968).
Baker, D. A., Burkhardt, L. C., DiMarco, J. N., Haberstich, A., Hagenson, R. L., Howell, R. B., Karr, H. J. and Schofield, A. E., *Plasma Phys. and Contr. Nucl. Fusion Res.* **1**, 419 (1977).

Bateman, G., *MHD Instabilities*, M.I.T., Cambridge, Mass. (1978).
Bernstein, I. B., Frieman, E. A., Kruskal, M. D. and Kulsrud, R. M., *Proc. Roy. Soc.* **A244**, 17 (1958).
Bussac, M. N., Pellat, R., Edery, D. and Soule, J. L., *Phys. Rev. Letts.* **35**, 1638 (1975).
Chiuderi, C., Giachetti, R. and Van Hoven, G., *Solar Phys.* **54**, 107 (1977).
Chiuderi, C., Einaudi, G., Ma, S. S. and Van Hoven, G., *J. Plasma Phys.* (to appear, 1980).
Christiansen, J. P. and Roberts, K. V., *Nucl. Fusion* **18**, 181 (1978).
Colgate, S. A., *Peaceful Uses of Atomic Energy* **32**, 129, United Nations, Geneva (1958).
Colgate, S. A., *Astrophys. J.* **221**, 1068 (1978).
Coppi, B., *Astrophys. J.* **195**, 545 (1975).
Coppi, B. and Friedland, A. B., *Astrophys. J.* **169**, 379 (1971).
Coppi, B., Greene, J. M. and Johnson, J. L., *Nucl. Fusion* **6**, 101 (1966).
Coppi, B., Galvao, R., Pellat, R., Rosenbluth, M. N. and Rutherford, P. H., *Sov. J. Plasma Phys.* **2**, 533 (1976).
Cross, M. A. and Van Hoven, G., *Phys. Rev.* **A4**, 2347 (1971).
Cross, M. A. and Van Hoven, G., *Phys. Fluids* **19**, 1591 (1976).
Davis, H. T., *Introduction to Nonlinear Differential and Integral Equations*, pp. 57–58, Dover, New York (1962).
Davis, W. D. in *Flare-Related Magnetic Field Dynamics*, High Altitude Observatory, Boulder, Colo. (1974).
Drake, J. F., *Phys. Fluids* **21**, 1777 (1978).
Drake, J. F. and Lee, Y. C., *Phys. Fluids* **20**, 1341 (1977).
Ferraro, V. C. A. and Plumpton, C., *An Introduction to Magneto-fluid Mechanics*, Oxford (1966).
Finn, J. M., *Nucl. Fusion* **15**, 845 (1975).
Finn, J. M. and Kaw, P. K., *Phys. Fluids* **20**, 72 (1977).
Foukal, P. V., *Solar Phys.* **43**, 327 (1975).
Foukal, P. V., *Astrophys. J.* **210**, 575 (1976).
Foukal, P. V., *Astrophys. J.* **223**, 1046 (1978).
Furth, H. P., Killeen, J. and Rosenbluth, M. N., *Phys. Fluids* **6**, 459 (1963).
Furth, H. P., Rutherford, P. H. and Selberg, H., *Phys. Fluids* **16**, 1054 (1973).
Gabriel, A. H. and Jordan, C., *Mon. Not. R. Astron. Soc.* **173**, 397 (1975).
Giachetti, R., Van Hoven, G. and Chiuderi, C., *Solar Phys.* **55**, 371 (1977).
Gibson, R. D. and Whiteman, K. J., *Plasma Phys.* **10**, 1101 (1968).
Goedbloed, J. P., *Physica* **53**, 412 (1971a).
Goedbloed, J. P., *Physica* **53**, 501 (1971b).
Goedbloed, J. P., *Physica* **53**, 535 (1971c).
Goedbloed, J. P. and Dagazian, R. Y., *Phys. Rev.* **A4**, 1554 (1971).
Goedbloed, J. P. and Hagebeuk, E. J., *Phys. Fluids* **15**, 1090 (1972).
Gold, T. and Hoyle, F., *Mon. Not. R. Astron. Soc.* **120**, 89 (1960).
Heyvaerts, J., Priest, E. R. and Rust, D. M., *Astrophys. J.* **216**, 123 (1977).
Hood, A. W. and Priest, E. R., *Astron. Astrophys.* **77**, 233 (1979a).
Hood, A. W. and Priest, E. R., *Solar Phys.* **64**, 303 (1979b).
Ince, E. L., *Ordinary Differential Equations*, pp. 223–226, Dover, New York (1956).
Jeffrey, A., *Magnetohydrodynamics*, Oliver and Boyd, London (1966).
Jette, A. D. and Sreenivasan, S. R., *Phys. Fluids* **12**, 2544 (1969).
Kadomtsev, B. B., *Reviews of Plasma Physics* **2**, 153 (1966).
Kahler, S. W., Krieger, A. S. and Vaiana, G. S., *Astrophys. J.* **199**, L57 (1975).
Krieger, A. S., "X-Ray Observations of Solar Structural Features," in *Proceedings of the OSO-8 Workshop*, Lab. for Atmos. and Space Phys., Univ. of Colorado, Boulder (1977).
Krieger, A. S., *Solar Phys.* **56**, 107 (1978).
Laval, G., Mercier, C. and Pellat, R., *Nucl. Fusion* **5**, 156 (1965).
Levine, R. H., *Solar Phys.* **44**, 365 (1975).
Levine, R. H., *Solar Phys.* **46**, 159 (1976).
Lundquist, S., *Phys. Rev.* **83**, 307 (1951).
Mercier, C., *Nucl. Fusion* **1**, 47 (1960).
Nakagawa, Y., in *Flare-Related Magnetic Field Dynamics*, High Altitude Observatory, Boulder, Colo. (1974).

Newcomb, W. A., *Ann. Phys.* **10**, 232 (1960).
Petschek, H. E., AAS-NASA *Symposium on the Physics of Solar Flares*, p. 425 (1964).
Priest, E. R., *Solar Phys.* **47**, 41 (1976).
Pritchett, P. L., Lee, Y. C. and Drake, J. F., *Phys. Fluids* **23**, 1368 (1980).
Raadu, M. A., *Solar Phys.* **22**, 425 (1972).
Robinson, D. C., *Plasma Phys.* **13**, 439 (1971).
Robinson, D. C., *Nucl. Fusion* **18**, 939 (1978).
Rosenbluth, M. N., Dagazian, R. Y. and Rutherford, P. H., *Phys. Fluids* **16**, 1894 (1973).
Rutherford, P. H., *Phys. Fluids* **16**, 1903 (1973).
Schmidt, G., *The Physics of High Temperature Plasmas*, Academic, New York (1979).
Schnack, D. D. and Killeen, J., in *Theoretical and Computational Plasma Physics*, Intl. At. Energy Agency, Vienna (1978).
Schnack, D. D. and Killeen, J., *Nucl. Fusion* **19**, 877 (1979).
Shafranov, V. D., *Nucl. Fusion* **3**, 183 (1963).
Shafranov, V. D. and Yurchenko, E. I., *Sov. Phys. JETP* **26**, 682 (1968).
Smith, D. F., *Coronal Disturbances, IAU Symp.* **57**, 253 (1974).
Smith, D. F. and Priest, E. R., *Astrophys. J.* **176**, 487 (1972).
Solov'ev, L. S., *Sov. At. Energy* **30**, 14 (1971).
Spicer, D. S., Ph.D. Dissertation, Univ. of Maryland (distributed as Naval Research Lab. Rept. 8036 (1976)).
Spicer, D. S., *Solar Phys.* **53**, 305 (1977).
Spitzer, L., *Physics of Fully Ionized Gases*, p. 139, Wiley Interscience, New York (1962).
Stenflo, J. O., *Solar Phys.* **42**, 79 (1975).
Sturrock, P. A., *IAU Symp.* **35**, 471 (1968).
Suydam, B. R., *Proc. of the Second Internat. Conf. on the Peaceful Uses of Atomic Energy* **31**, pp. 157–159 (1958).
Švestka, Z., *Solar Flares*, Reidel, Dordrecht, Holland (1976).
Taylor, J. B., *Plasma Phys. and Contr. Nucl. Fusion Res.* (Proc. 5th Int. Conf., Tokyo, 1974), Intl. At. Energy Agency, Vienna (1975).
Vahala, G., *Phys. Fluids* **18**, 1034 (1975).
Vaiana, G. S., Reidy, W., Zehnpfennig, G., Van Speybroeck, L. and Giacconi, R., *Science* **161**, 564 (1968).
Van Hoven, G., *Bull. Am. Astron. Soc.* **7**, 398 (1975).
Van Hoven, G., *Solar Phys.* **49**, 95 (1976).
Van Hoven, G., *Astrophys. J.* **232**, 572 (1979).
Van Hoven, G., *Highlights of Astronomy* **5** (1980).
Van Hoven, G. and Cross, M. A., *Phys. Rev.* **A7**, 1347 (1973).
Van Hoven, G., Chiuderi, C. and Giachetti, R., *Astrophys. J.* **213**, 869 (1977).
Van Hoven, G. *et al.*, "The Preflare State," in *Solar Flares*, Univ. of Colorado, Boulder, Colo. (1980).
Vasyliunas, V. M., *Rev. Geophys. Space Phys.* **13**, 303 (1975).
Van Goeler, S., Stodiek, W. and Sauthoff, N., *Phys. Rev. Letts.* **33**, 1201 (1974).
Vorpahl, J. A., Gibson, E. G., Landecker, P. B., McKenzie, D. L. and Underwood, J. H., *Solar Phys.* **45**, 199 (1975).
Voslamber, D. and Callebaut, D. K., *Phys. Rev.* **128**, 2016 (1962).
Waddell, B. V., Rosenbluth, M. N., Monticello, D. A. and White, R. B., *Nucl. Fusion* **16**, 528 (1976).
Waddell, B. V., Carreras, B., Hicks, H. R. and Holmes, J. A., *Phys. Fluids* **22**, 896 (1979).
Ware, A. A. and Haas, F. A., *Phys. Fluids* **9**, 956 (1966).
Wesson, J. A. and Sykes, A., *Plasma Phys. and Contr. Nucl. Fusion Res.*, (Proc. 5th Int. Conf., Tokyo, 1974), Intl. At. Energy Agency, Vienna (1975).
Whiteman, K. J., *Plasma Phys.* **7**, 293 (1965).
Woltjer, L., *Proc. Nat. Acad. Sci.* **44**, 489 (1958).

CHAPTER 5

Simple-loop flares: thermal evolution

I. J. D. CRAIG

Department of Astronomy, University of Glasgow†

5.1 INTRODUCTION

The central problem of solar flare theory is to explain the initial explosive energy release: how can as much as 10^{25} J (10^{32} ergs) in a large flare be liberated on a time-scale of a few hundred seconds (e.g. Sweet, 1969; Švestka, 1976)? The conventional view is that the annihilation of coronal magnetic fields provides the flare energy, but many of the details of the mechanism remain to be worked out. A complete flare theory, however, must be well enough developed to predict *radiation spectra* for comparison with observational data, typically in the X-ray, EUV, optical and radio wavebands. In this respect the initial energy release may not be as critical as the secondary flare processes that govern the relatively slow redistribution of released flare energy amongst the various layers of the solar atmosphere (see review by Somov and Syrovatsky, 1976).

A prime manifestation of flare activity, and in particular of the explosive phase, is the generation of a hot ($\approx 10^7$ K) X-ray emitting plasma in the lower corona. The details of the plasma-heating mechanism are not well understood. But unlike, say, mass ejecta or non-thermal particles, for which there is direct evidence in only a minority of events (Švestka, 1976), strong plasma heating appears to be common to all flares irrespective of individual morphology. What is more, an understanding of the *evolution of the thermal plasma* is central to the prediction of flare radiation spectra. Apart from its role as a thermal X-ray emitter the hot plasma may well provide the main energy supply for the EUV and Hα flares: flare energy is conducted and convected down from the site of primary energy release and deposited in cooler, denser regions of the solar atmosphere (e.g. Sturrock, 1973; Craig and McClymont, 1976).

In this chapter we shall consider the evolution and temperature structure of

---

† Present address: Department of Mathematics, University of Waikato, Hamilton, New Zealand

the excited flare plasma in some detail. Our starting point is the observation that simple flares are organised (presumably by the magnetic field) into filamentary loop structures. Our aim is to explore, within the context of a general (one-dimensional) hydrodynamic framework, the main physical processes that govern the evolution of flare loops and to provide a critical assessment of the strengths and limitations of current loop-models. The analysis is mainly of relevance to compact (or *simple-loop*) flares (§§1.1, 2.1.4, 2.4.2), but it is also important as a basis for understanding the more complex behaviour of loops in *two-ribbon* flares (Chapter 7). Although the discussion is slanted towards the theoretical aspects of flare evolution we shall constantly strive to emphasise the need for interplay between theoretical loop modelling and constructive data analysis.

A run-down of the observational background is given in §5.2. Various interpretative difficulties are mentioned but we make no attempt to provide a definitive assessment of the data. In §5.3 we present the basic hydrodynamic equations and introduce boundary conditions for the problem. Section 5.4 deals with quasi-static models of active-region and flare plasmas and the question of loop stability. A discussion of analytic conductive models of the flare evolution is given in §5.5, and the problem of loop superposition is briefly outlined. Finally in §5.6 we review numerical flare modelling and give a detailed interpretation of some new hydrodynamic calculations.

5.2 OBSERVATIONAL BACKGROUND

5.2.1 High resolution X-ray and EUV images

High resolution X-ray images of the quiet and active solar atmosphere obtained using satellite and rocket-borne instrumentation have shown unambiguously that the corona is composed of a complex network of individual loop-like structures. In consequence the old picture of a spherically symmetric essentially featureless corona has rapidly fallen from grace, to be replaced by an *inhomogeneous, highly-structured "atmosphere"* that is intimately related to, if not defined by, the three-dimensional magnetic field topology of the active region. Where loops appear to be absent, presumably in regions of open rather than closed magnetic field, low pressure regions — "coronal holes" — of relatively low emissivity occur.

Generally speaking, there appears to be a correlation between the brightness and the geometrical extent of the loop. Small, low-lying, arches of enhanced temperature and density ($T > 3 \times 10^6$ K, $n_e > 10^{15}$ m^{-3} (10^9 cm^{-3})) form the nucleus of the active-region assembly (length $\ll 1'$ of arc). These are overlaid by longer, broader, but more diffuse structures, while still longer arches may interconnect individual active regions. Although there is a greater tendency for small, magnetically strong loops to flare, it appears that all loops can be excited, if not directly by the initial magnetic energy release, then indirectly by some secondary flare disturbance. Thus Rust and Webb (1977) have observed flare-associated coronal shocks to propagate along magnetically weak arches that link individual active regions, while Švestka *et al.* (1977) have discussed the possibility that "slow" loop brightenings (time-scale of a few hours) can be stimulated by neighbouring flares.

Coronal radiation can be divided into two main classes depending on whether fluctuations in emission occur faster or slower than the energy loss time-scale of the plasma. In this sense quiescent active-region loops are *quasi-static*: although fluctuations from day to day occur, they appear relatively constant over a time-scale of several hours, which is appreciably greater than the conductive or radiative time-scale in the corona (10^3–10^4 s; §5.3.2). In fact to support the loop against the energy losses some form of stationary coronal heating must be present. *Flare loops*, on the other hand, are probably *dynamic* in the fullest sense of the word. Not only can the emission vary in a relatively short time-scale ($\ll 10^3$ s) but the likelihood of mass-flows powered by the explosive energy release makes a gas-dynamic description of the flare plasma inevitable.

Finally, we should mention that many fine-scale coronal features ("bright points" for instance) lie beyond the current spatial resolution of the observations ($\sim 1''$ of arc). It appears that the task of building a detector that combines sufficient spatial, spectral and temporal resolution for detailed flare study is as difficult as solving the flare problem itself — and more expensive!

5.2.2 Spectral data

Information on the thermal structure of active-region loops is derived mainly from X-ray and EUV spectral data in optically-thin collisionally-excited emissions. The interpretation of such data has been the subject of several recent reviews (e.g. Švestka, 1976; Vaiana and Rosner, 1978). It is generally accepted that spectral analysis can yield information on the distribution over temperature of material within the source volume, but the question of how informative the data really are in this respect is rather controversial (Craig and Brown, 1976a, b; Gabriel, 1977; Parkinson, 1977).

As is well known the spectrum from an optically-thin thermal source may be written

$$\varphi(\varepsilon) = \iiint_V F(\varepsilon,\ T(\mathbf{r}))n_e^2(\mathbf{r})\mathrm{d}^3\mathbf{r}, \tag{5.1}$$

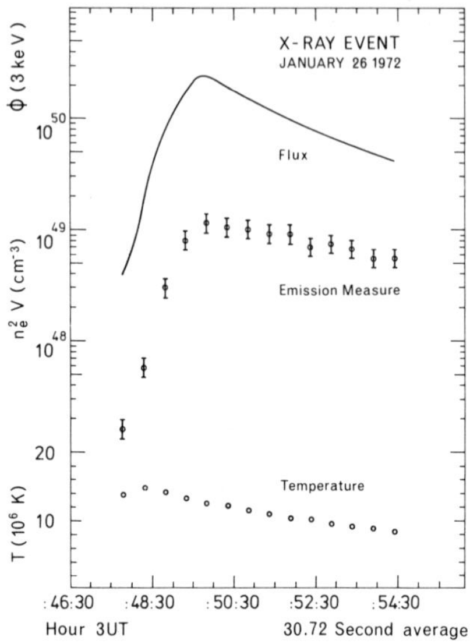

FIGURE 5.1 Evolution of the temperature and emission for a typical soft X-ray burst (based on the data of Datlowe *et al.*, 1974). The peak temperature is achieved relatively early in the event while the emission measure is still rising.

where $n_e(\mathbf{r})$ and $T(\mathbf{r})$ are the electron density and temperature at position **r** in the source volume V, ε is the photon energy and $F(\varepsilon, T)$ is the (known) theoretical spectral distribution for an isothermal source element (e.g. Tucker and Koren, 1971). It follows that for an *isothermal* source the spectrum is determined by just two parameters, the temperature T and the emission measure $\int n_e^2 \mathrm{d}V$. However, in the general multi-thermal case, an *emission measure differential in temperature* must be introduced (see below).

An important property of isothermal models of the soft X-ray flare is the time delay between the maximum temperature (typically $10\text{–}30 \times 10^6$ K) and the maximum emission measure, $10^{53}\text{–}10^{56}$ m^{-3} ($10^{47}\text{–}10^{50}$ cm^{-3}), of the burst. The temperature peaks early in the event — while the emission measure is still rising — and decays relatively slowly on a time-scale of typically $\sim 10^3$ s (e.g. Kahler *et al.*, 1970; Milkay *et al.*, 1971). The large emission measure at the burst maximum may imply that relatively dense transition zone or chromospheric material has been flare-heated to X-ray emitting temperatures (c.f. Brown, 1973).

Figure 5.1 is based on the data of Datlowe *et al.* (1974) and shows the evolution of a fairly typical soft X-ray flare. We see that the temperature peaks at least two minutes before the emission measure (and therefore possibly the density). Evidently the decay of the burst is due to a slow fall-off in both the emission measure and temperature.

Of course, the flare volume is *not really at a single temperature*: this lack of isothermality is betrayed by the fact that observations in different wavelength bands often yield different "flare temperatures". To derive a general multi-temperature description Equation (5.1) is rewritten

$$\varphi(\varepsilon) = \int_T F(\varepsilon, T)\xi(T)\mathrm{d}T, \tag{5.2}$$

where the *differential emission measure* $\xi(T)$ is formally defined by

$$\xi(T) = \sum_{i=1}^{N} \left(\iint_{S_T} n_e^2(r)|\nabla T|^{-1}\mathrm{d}S \right)_i, \tag{5.3}$$

the summation being taken over the N disjoint surfaces of constant temperature S_T in the source (Craig and Brown, 1976). This function ξ contains the totality of information on the thermal structure carried by the observations. Unfortunately, ξ is weighted away from regions of steep temperature gradient, generally the regions of prime physical interest.

As Craig and Brown (1976a, b) have emphasised, the problem of deriving ξ from typical data is mathematically ill-posed; this implies that small measurement errors in the spectrum $\varphi(\varepsilon_1)$ can give rise to arbitrarily large errors in ξ. In physical terms the data are intrinsically insensitive to thermal source structure, so that a wide range of emission-measure functions are capable of reproducing the observations. The crux of the problem is that the integral transform (5.2) smoothes out the high frequency components in ξ rendering them virtually indiscernable in the data function. Hence the data can never provide significant information on the detailed structure (viz. the high frequencies) of the source function. The severity of this effect in practice depends not only on the amplitude of the data noise but, more seriously, on the width of the kernel function $F(\varepsilon, T)$ compared to the temperature span of the emitting volume. For X-ray emission lines the kernel width $F(\varepsilon, T)$ is often comparable to the span of the coronal source, hence the instability is so severe that only a rudimentary two or three-parameter description of ξ can be derived (Craig and Brown, 1976a). For EUV lines the situation is improved slightly due to the relatively wide temperature span of material in the transition zone (say 5×10^4–10^6 K). But in either case the only sound way to tackle the data reduction is to invoke *a-priori* information to stabilise the inversion problem (see Craig, 1977, 1979).

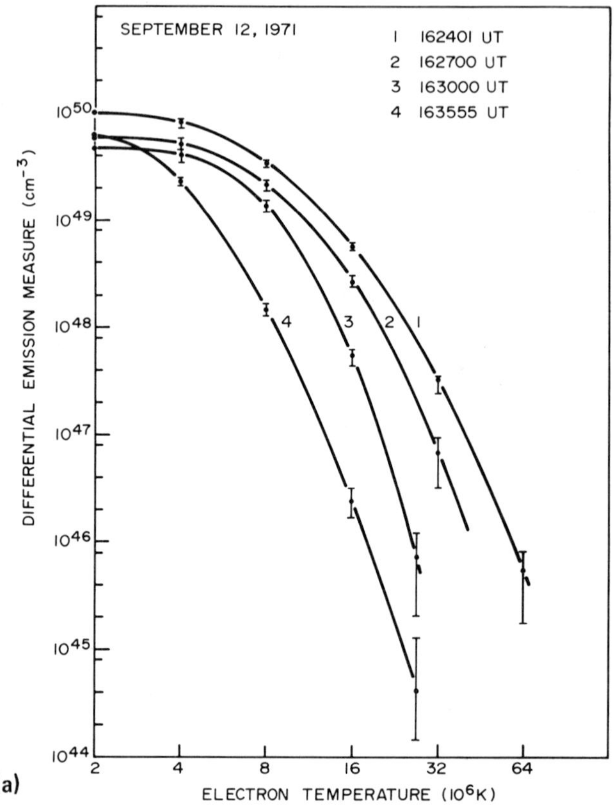

FIGURE 5.2 (a) Profiles of the differential emission measure ($\xi(T) \times T$) at four instants during the decay of a small flare (taken from Dere *et al.*, 1974).

As yet no completely convincing, rigorous analysis of spectral data has been performed.

Aside from the inversion problem there remains the difficulty of relating ξ to the physics of the solar atmosphere. Generally speaking this can only be achieved by constructing detailed theoretical models based on the energy balance of the source plasma. We note that relatively crude data on ξ can often place useful restrictions on the range of acceptable theoretical models (e.g. Brown and Smith, 1980).

Returning to the observations, we note that EUV data generally indicate that ξ follows a weakly *increasing* function of temperature (10^5–10^6 K), both for active-region and flare plasmas (e.g. Jordan, 1976). Data in the soft X-ray differential emission measure, however, are rather sparse and often ambiguous. There seems to be a tendency for ξ to *decrease* with increasing

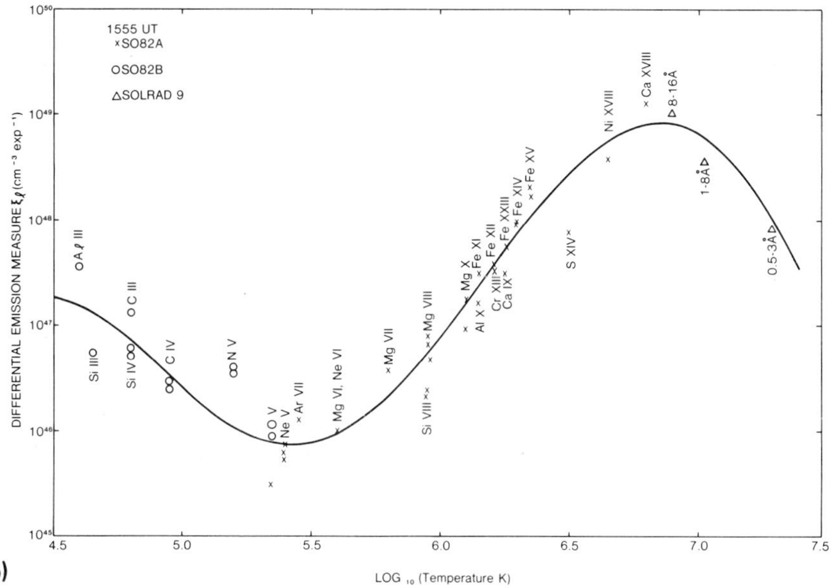

(b)

FIGURE 5.2 (b) The X-ray and EUV differential emission measure for a small flare (taken from Dere and Cook, 1978).

temperature in the corona (e.g. Chambe, 1971), but this trend is by no means universal (see Underwood and McKenzie, 1977).

Figure 5.2a is taken from Dere *et al.* (1974) and shows the evolution of the differential emission measure for a typical subflare. (Note that the curves shown here define the product $T \times \xi(T)$ not $\xi(T)$.) The cooling of the flare plasma is obviously reflected by a fall-off in the amplitude of ξ at high temperatures.

The differential emission pressure profile of Figure 5.2b was derived by Dere and Cook (1978) under the implicit assumption that ξ is smooth as well as positive–definite (here $T \times \xi(T)$ is plotted over a much wider temperature range than in Figure 5.2a). Evidently, ξ retains the same general shape, characterised by a peak near 10^7 K and a minimum near 3×10^5 K, throughout the flare evolution (c.f. Underwood *et al.*, 1978). It is not clear however, why the differential emission measure for Figure 5.2b is (in contrast to Figure 5.2a) an *increasing* function of temperature in the coronal region $10^6 < T < 10^7$ K. Although loop superposition (§5.5.6) can "explain away" this effect, it may well be that individual differences, both in the form and evolution of ξ, are expected in view of the great diversity of flares (see Sweet, 1969).

In summary, we should point out that, although the general definition of ξ is complicated, for the case of a *symmetric loop structure* of *unit cross-sectional area* it reduces to

$$\xi(T)=2n_e^2\left|\frac{dT}{dz}\right|^{-1}. \tag{5.4}$$

We shall employ this simplified form throughout the following discussion (c.f. §5.5.6).

5.2.3 Velocities and densities

There are two time-honoured ways to estimate the *electron density* in coronal flares. In the first method a density is derived by combining measurement of the emission measure with an educated guess as to the size of the emitting coronal volume; a second approach is to adjust the electron density until there is good agreement between the observed flare decay-time and the conductive or radiative time-scale of the plasma. Typical estimates generally yield average densities in the range $10^{16}-10^{17}$ m^{-3} (10^{10}–10^{11} cm^{-3}) (Švestka, 1976).

A more accurate method, at least in theory, can be devised by employing density-sensitive spectral lines (e.g. Gabriel and Jordan, 1969; Mason *et al.*, 1979), but teething problems have often been encountered in the past (see Švestka, 1976; Brown and Smith, 1980). Recent results, however, which indicate densities from $10^{16}-5\times10^{17}$ m^{-3} ($10^{10}-5\times10^{11}$ cm^{-3}) in the range 10^6–10^7 K seem more reliable (Dere *et al.*, 1979). What is certain is that densities can vary significantly throughout the flare evolution as well as from event to event (e.g. Underwood *et al.*, 1978).

Again, it is difficult to be definitive in discussing the *velocity profiles* in flare plasmas. Detailed study of broadened EUV (transition-zone) lines has yielded enigmatic relations between temperature, density and fluid velocity (e.g. Canfield *et al.*, 1980). Fluid speeds often exceed 10^5 m s^{-1} and seem to be associated with locally-enhanced plasma densities (Doschek and Feldman, 1980). Shock-wave phenomena may well be implied (see §§5.6.2 and 5.6.3).

5.3 THE HYDRODYNAMIC EQUATIONS

5.3.1 Basic assumptions

In order to provide a useful theoretical framework for the subsequent discussion we make the following simplifying assumptions.

First, the source plasma, in all respects apart from its radiative losses, is taken to be a *fully-ionised* electron-proton gas ($n_e = n_p$) in *thermal equilibrium*. The plasma pressure is then

$$p = 2n_e k_B T = 2\rho R T, \tag{5.5}$$

while the thermal energy density is simply

$$E = 3n_e k_B T = 3\rho R T, \tag{5.6}$$

Here k_B is Boltzmann's constant, R is the universal gas constant and ρ is the mass density; $\rho = m_p n_e$ where m_p is the mass of the proton.

The next assumption concerns the geometry of the flaring region. We take the plasma to be confined within a *rigid* magnetic flux tube of *uniform cross-section*. The local magnetic pressure, therefore, must always dominate the gas pressure, that is

$$B^2/(2\mu) \gg 2n_e k_B T, \tag{5.7}$$

where B/μ is the magnetic field strength. Whether this condition is satisfied for all plasmas in the lower corona is not clear. Certainly, the typical flare parameters $T = 10^7$ K, $n_e = 10^{16}$–10^{17} m^{-3} (10^{10}–10^{11} cm^{-3}) require magnetic field strengths larger than only 2–6×10^{-3} T (20–60 G) for (5.7) to be satisfied. Nonetheless, the current observational uncertainties in the coronal n_e and B values indicate that the inequality is far from universally established.

The assumption that the flux tube possesses a uniform cross-section is not so serious. It is simply an idealisation that is very convenient for analytic purposes but which can be easily relaxed in detailed numerical calculations (§5.4.4).

The equations describing the dynamics of the flaring plasma are (see §1.5) the continuity equation

$$\frac{\mathrm{D}}{\mathrm{D}t}\rho = -\rho\frac{\partial v}{\partial z}, \tag{5.8}$$

the momentum equation

$$\rho\frac{\mathrm{D}v}{\mathrm{D}t} = -\frac{\partial p}{\partial z} + \rho g_{\parallel}, \tag{5.9}$$

and the energy equation

$$\frac{\rho}{(\gamma-1)}\frac{\mathrm{D}}{\mathrm{D}t}(p/\rho)=\frac{\partial}{\partial z}\left(\kappa\frac{\partial T}{\partial z}\right)-p\frac{\partial v}{\partial z}-n_e^2Q(T)+H, \tag{5.10}$$

where v is the fluid velocity, $g_{\|}$ is the component of gravitational acceleration along the loop and $\mathrm{D}/\mathrm{D}t$ is the usual mobile operator

$$\frac{\mathrm{D}}{\mathrm{D}t}=\frac{\partial}{\partial t}+v\frac{\partial}{\partial z}.$$

The independent variables are the time t and the space coordinate, z, measured parallel to the confining magnetic field.

The energy equation written in the form (5.10) describes the rate of change of internal energy per unit mass due to inputs by the heating source H, energy transport by conduction and losses by radiation and by plasma expansion. For conduction along the field lines the classical coefficient is given by Spitzer (1962) as

$$\kappa=\kappa_0 T^{5/2}, \tag{5.11}$$

where κ_0 depends very weakly on the local temperature and density, $\kappa_0 \simeq 10^{-11}$ W m^{-1} (10^{-6} cgs units). Since heat flow across the field lines is probably negligible in practice (the gyrosynchrotron radius being several orders of magnitude smaller than the particle mean-free-path) each flux tube can be regarded as thermally insulated from its neighbours (e.g. Culhane *et al.*, 1970).

The function $Q(T)$ defines the *optically-thin* losses of the thermal plasma: it has been computed by many authors (Cox and Tucker, 1969; Tucker and Koren, 1971; McWhirter *et al.*, 1975; Raymond *et al.*, 1976). An important feature of Q is its fall-off (roughly as T^{-1}) with increasing temperature in the region from $\simeq 10^5$–10^7 K. This gives rise to the possibility of thermal instability (see e.g. Field, 1965).

The heating term H is the sum total of all energy inputs to the plasma. For most flare models H can be divided into two components; namely, a quiescent heating that supports the initial atmosphere against the radiative losses, and a rapid time-dependent heating that drives the initial flare energisation. Neither of these components is particularly well understood at present.

Finally, we should emphasise that our basic hydrodynamic system is not applicable to plasmas in the lower chromosphere (say $T<2$–3×10^4 K). The main inaccuracy arises because the radiative losses can no longer be described in the optically-thin approximation. Unfortunately, the chromospheric energy balance is so poorly understood that, even for quiescent plasmas, no self-consistent theoretical description is yet available.

5.3.2 Dimensional analysis

Rudimentary insight into the influence of the various terms in the energy equation may be gained through dimensional analysis. Consider firstly, the static processes of conduction and radiation. If l denotes the scale-length of the hot plasma then the conductive time-scale is simply

$$\tau_{\text{cond}} = \frac{3n_e k_B}{\kappa_0} l^2/T^{5/2} \simeq 4 \times 10^{-12} n_e l^2/T^{5/2} \text{ s.} \tag{5.12}$$

We see that for the typical flare parameters, viz $T = 10^7$ K, $n_e = 10^{16}$–10^{17} m^{-3} (10^{10}–10^{11} cm^{-3}) and $l = 10^{7.5}$ m (e.g. Moore and Datlowe, 1975), $\tau_{\text{cond}} = 10^2$–$10^3$ s.

Similarly, when Q is approximated by $Q \simeq 10^{-29}$ T^{-1} W m^3 (10^{-16} T^{-1} cgs units) in the range 10^5–10^7 K (e.g. Korchak, 1971; Landini and Fossi, 1975) the radiative time-scale is given by

$$\tau_{\text{rad}} = \frac{3k_B T}{n_e Q(T)} \simeq 4 \times 10^6 \text{ T}^2/n_e \text{ s.} \tag{5.13}$$

Hence, taking the same parameters as before yields $\tau_{\text{rad}} = 4 \times 10^3$–$4 \times 10^4$ s. We observe that, while conduction is generally faster than radiation at dissipating the flare energy, in the high density limit both time-scales can be closely identified with the flare decay-time of $\approx 10^3$ s. The tentative conclusion therefore, in agreement with previous studies (e.g. Culhane *et al.*, 1970; Korchak, 1971; Craig *et al.*, 1973), is that *conduction predominates in the initial flare cooling*, but radiation probably assumes a comparable role in the late decay phase by virtue of the reduced flare temperature.

But what is the role of *mass motions* in the flare cooling process? To assess the convective influence we note, following Craig and McClymont (1976), that the characteristic pressure gradient in the flare source is simply

$$\left|\frac{\partial p}{\partial z}\right| \simeq \frac{p}{l} \simeq 3 \times 10^{-23} \, n_e T/l \text{ N m}^{-3}. \tag{5.14}$$

Moreover, the pressure gradient will persist for roughly the hydrodynamic time-scale of the hot plasma, viz. l/c, where c is the sound speed given by

$$c = \left(\frac{\gamma p}{\rho}\right)^{1/2} = 170\ \mathrm{T}^{1/2}\ \mathrm{m\ s}^{-1}. \tag{5.15}$$

It follows that material will be accelerated down the pressure gradient, in accordance with the momentum equation, to the typical velocity

$$v = p/(\rho c) \simeq 100\ \mathrm{T}^{1/2},\ \mathrm{m\ s}^{-1}. \tag{5.16}$$

Hence $v \simeq 3 \times 10^5$ m s^{-1}, a value which is generally supersonic with respect to material in the undisturbed atmosphere (say at $\simeq 2 \times 10^6$ K). In fact, from (5.16) we expect the Mach number, that is, the ratio of the shock speed to the sound speed in the undisturbed medium, to depend mainly on the characteristic temperature of the hot plasma, specifically $M \sim T^{1/2}$. This trend seems to be borne out by the detailed study of McClymont (1980). Also, we note that the work done by pressure becomes comparable to the internal energy of the plasma after a time

$$\tau_{\mathrm{mass}} = \frac{l}{v} = 10^{-2}\ l/\mathrm{T}^{1/2}\ \mathrm{s}. \tag{5.17}$$

Hence τ_{mass}, being of order 10^2 s, is directly comparable with the typical conduction time-scale. The implication, therefore, is that *dynamic effects are likely to exert a profound influence* on the evolution of the hot flare plasma.

5.3.3 Boundary and initial conditions

In order to solve the hydrodynamic equations, for the temperature, density and velocity profiles, suitable boundary and initial conditions must be incorporated into the problem. For the closed magnetic structure, the boundary conditions at the *apex* or summit ($z=0$) of the loop may be derived from the natural idealisation of loop symmetry

$$T'(z=0) = n_e'(z=0) = v(z=0) = 0, \tag{5.18}$$

where the primes denote differentiation with respect to z. Though exceptions to the rule are possible, these conditions generally lead to temperature maxima at the origin (e.g. Craig and McClymont, 1976; Antiochos and Sturrock, 1976; 1978).

The boundary conditions at the *footpoints* of the loop, however, represent something of a problem. A common strategy is to assume that the temperature gradient and the fluid velocity vanish at some point deep in the chromosphere,

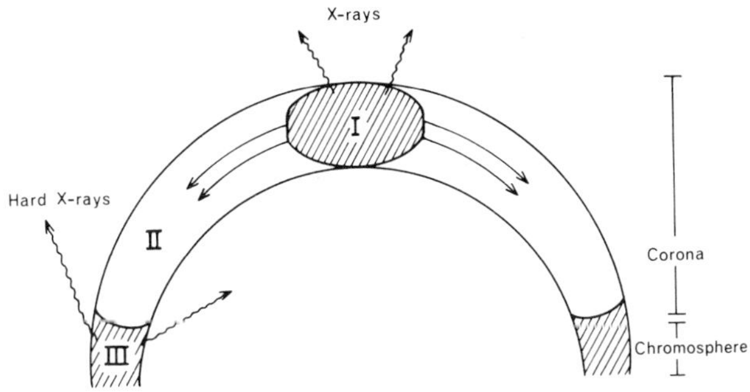

FIGURE 5.3 Schematic model of the impulsive phase of a solar flare (based on Vlahos and Papadopoulos, 1979). The loop comprises three distinct regions: I a hot truncated Maxwellian confined by acoustic turbulence; II a region of electron beams (with reverse current); III precipitation in the chromosphere.

the region below which is unaffected by the flare disturbance; alternatively, the deep chromosphere is sometimes represented by a rigid barrier of fixed temperature (see e.g. Kostyuk and Pikelner, 1975; Somov *et al.*, 1977). Of course, in the absence of a chromospheric flare model, the applicability of such conditions can always be questioned (§§5.4.2, 5.4.4 and 5.6.2).

Initial conditions for the flare loop are determined by the thermal structure of the quiescent active-region plasma. Static energy balance is generally assumed (§§5.4.2 and 5.4.4). However, complications arise when only a certain phase of the flare evolution is modelled. It is by no means clear, for instance, that static conditions can be applied at the onset of the flare decay (c.f. Antiochos and Krall, 1978; §§5.5.2 and 5.5.3).

5.3.4 Flare heating by electron collisions

Although resistive heating by coronal currents is often mentioned (e.g. Tucker, 1973; Somov *et al.*, 1977), the only impulsive flare heating mechanism which has been vigorously examined to date invokes the *collisional energy loss* of non-thermal particles accelerated during the initial flare instability. It is generally assumed that electrons accelerated high in the corona provide the bulk of the energy deposition to the underlying plasma (Brown, 1971; Cheng, 1971; Strauss and Papagiannis, 1971; Syrovatsky and Shmeleva, 1972) but proton heating has also been examined (Jefferies and Orrall, 1965). Here we concentrate only on the conventional *electron-heating model.*

Figure 5.3 is based on the sketch of Vlahos and Papadopoulos (1979) and gives some indication of what might happen during the impulsive phase of the flare. Here, in contrast to the conventional view, the hard X-ray burst is *primarily thermal* in origin (Brown *et al.*, 1978). The hard X-ray source (Region

I) is formed when the ambient *electron gas* of the high corona is impulsively heated by some (unspecified) primary flare mechanism. The initial free expansion of the hot electrons into the cool surroundings drives a reverse-current (from Region II) which is unstable to the generation of ion-acoustic waves (Brown *et al.*, 1978). The resulting ion-acoustic turbulence "bottles up" all except an escaping high energy tail of the electron gas, while maintaining the "trapped" electron distribution close to Maxwellian ($T_e \simeq 10^8$ K). Precipitation of the escaping electron beam into the chromosphere (Region III) may then give rise to non-thermal hard X-ray ($\geqslant 80$ keV), EUV and optical flashes.

The ramifications of the above model have not yet been adequately investigated and it is more conventional to regard the hard X-ray burst ($E > 10$ keV) as entirely non-thermal in origin. Traditionally the flux spectrum of accelerated electrons (typically $E \geqslant 10$ keV) is presumed to approximate a power law, viz.

$$F_e(E, t) = \frac{\delta - 2}{2E_c} f(t)\theta(E - E_c)(E/E_c)^{-\delta}, \tag{5.19}$$

where $f(t)$ is the energy flux into the loop and θ is a unit step function that accounts for the (idealised) low-energy cut-off at E_c; typically $E_c \simeq 10$–20 keV and $\delta \simeq 4$ (Kane and Anderson, 1970; Hoyng *et al.*, 1975). The heating due to the collisional loss of the beam in the ambient plasma is given by (e.g. Brown, 1973)

$$H(\bar{\zeta}, t) = \frac{\delta - 2}{12} \frac{n_p}{\bar{\zeta}_c} f(t) \left(\frac{\bar{\zeta}}{\bar{\zeta}_c}\right)^{-\delta/2} B_x(\delta/2, 1/3). \tag{5.20}$$

Here $\bar{\zeta}$ is the column density measured from the injection point at the origin,

$$\bar{\zeta}(z) = \int_0^z n_p(z')\mathrm{d}z', \quad (n_p \simeq n_e); \tag{5.21}$$

B_x is the incomplete β-function with $x = \min(1, \bar{\zeta}/\bar{\zeta}_c)$, $\bar{\zeta}_c = E_c^2/3K_c$ and $K_c = 3.64 \times 10^{-22}$ keV^2 cm^2. Equation (5.20) is based on the analysis of Brown (1972): it applies to small pitch-angles at injection but takes into account scattering by protons to first order (c.f. Emslie, 1978).

We observe that, while the energy deposition generally falls-off as a power law in column density, close to the origin H rises from zero to achieve a sharp maximum at $\bar{\zeta}=\bar{\zeta}_c$. This behaviour is clearly an artifact of the idealised (and poorly determined) low-energy spectral cut-off (e.g. Hoyng *et al.*, 1976). Moreover, in practice, the distribution will be further smeared out due to the finite size of the injection region. For this reason McClymont and Craig (1980) smooth H near the origin, keeping the total injection rate fixed. On the other hand, Equation (5.20) has been adopted without modification by many authors (e.g. Kostyuk and Pikelner, 1975; Kostyuk, 1976).

The observation that many flares do not possess measurable non-thermal components (Švestka, 1976) represents strong evidence against the general applicability of the collisional heating model (c.f. the detector threshold effect of Brown and Smith, 1980). Yet, non-thermal emissions, when they are observed during the flash phase, appear to reflect the rate of energy input to the thermal plasma (see Neupert, 1968; Datlowe *et al.*, 1974). And the close synchronisation between Hα flare kernels and hard X-ray and EUV bursts (e.g. Vorpahl, 1972) is most readily explained in terms of descending particle streams (§5.4.3).

But the theoretical difficulties of the collisional heating model should not be overlooked. Aside from the problem of finding a flare mechanism which is efficient enough energetically to account for the initial electron acceleration (see Brown and Smith, 1980), there remains the difficulty of dealing self-consistently with the induced *reverse-current* of slow electrons that is required to maintain charge neutrality in the target plasma (see e.g. Knight and Sturrock, 1977; Hoyng *et al.*, 1976). Evidently, the joule-heating of the reverse current may sometimes exceed the collisional beam deposition (see also Emslie, 1980).

Finally, we mention that, since the beam heats the electrons of the ambient plasma rather than the protons, a *two-temperature description* of the atmosphere may be required. In particular, a single temperature model is only valid when the heating time-scale is significantly larger than the electron-proton energy-exchange time

$$\tau_{ep}=1.2\times 10^7\ T_e^{3/2}/n_e \text{ sec.} \tag{5.22}$$

The fact that τ_{ep} may exceed 30 s ($T_e=10^7$ K, $n_e=10^{16}$ m^{-3}) has motivated two-temperature formulations by Somov *et al.* (1977) and by McClymont and Craig (1980).

5.3.5 Numerical and analytic flare modelling

Due to the complexity and non-linearity of the hydrodynamic equations we must generally resort to a numerical simulation of the flare evolution. Several points should be remembered in deriving and interpreting numerical results. First, the numerical solution is generally facilitated by working in the *Lagrangian variable* $\bar{\zeta}$ (Equation 5.21) rather than in the usual space variable z: $d\bar{\zeta} = n_e dz$ or more conventionally $d\zeta = \rho dz$ (e.g. Zeldovich and Raiser, 1967). The transformation from (z, t) to $(\bar{\zeta}, t)$ is very straightforward:

$$\frac{D}{Dt} \rightarrow \frac{\partial}{\partial t}; \quad \frac{\partial}{\partial z} \rightarrow n_e \frac{\partial}{\partial \bar{\zeta}}. \tag{5.23}$$

To avoid severe restrictions on the time-step imposed by the parabolic nature of the heat conduction equation an *implicit* rather than an explicit finite difference scheme must be developed. In addition, since shock waves can be generated, it is expedient to augment the momentum equation by a *pseudo-viscous term* to spread the discontinuity over several grid points (Craig and McClymont, 1976; Somov *et al.*, 1977). Finally, the numerical scheme should include an automatically *adjustable time-step* to take into account the possibility of rapidly forming radiative condensations —failure to do this probably renders unsound the most complete flare analysis to date (Kostyuk and Pikelner, 1975; §5.6.2).

Analytic flare models are usually based on the assumption of *constant gas density* or *uniform pressure*. Since radiation is also neglected (§5.5), these models are valid only for highly localised regions of the flare plasma. The constant-density approximation is applicable, moreover, only for times which are short compared to the dynamical time-scale of the plasma, that is the *first minute* or so of the flare evolution. Similarly, while uniform pressure is a reasonable approximation for quiescent plasmas (the gravitational scale-height is very long in the corona and transition region) it is only likely to be valid for flares during the *late decay* when dynamic effects have died down. The analytic approach is important, however, because it readily yields *insight into the basic physics* of the problem. In the absence of analytic calculations to check and interpret numerical results, flare simulations — explain the observations though they may — rapidly degenerate into indigestable numbers churned out by the computer.

5.4 QUASI-STATIC MODELS OF THE SOLAR ATMOSPHERE

5.4.1 Introduction

Since the flare, by its very nature, is a highly transient phenomenon, it is obviously unrealistic to expect any quasi-static model to provide a complete description of the hot flare plasma. But the quasi-static approach, because of its inherent simplicity, is useful in isolating the influence of individual energy-transfer mechanisms on the loop evolution, especially conduction and radiation. Also, the relatively simple analytic formulation of the problem made possible under the assumption of uniform pressure allows any difficulties associated with boundary conditions on the plasma to be clearly brought out.

5.4.2 Conductive and radiative energy balance (Shmeleva and Syrovatsky)

One of the first attempts to describe the structure of quiescent active region plasmas was provided by Moore and Fung (1972). In their approach the coronal regions are not explicitly modelled; the corona is simply regarded as a source of conductive flux which supports the radiative losses of the low-temperature ($T < 10^6$ K) material. By solving numerically the hydrostatic equation

$$\frac{\partial p}{\partial z} = -\rho g_{\parallel}, \tag{5.24}$$

and the energy equation

$$\frac{\partial}{\partial z}\left(\kappa \frac{\partial T}{\partial z}\right) = n_e^2 Q(T), \tag{5.25}$$

Moore and Fung were able to take into account relatively minor effects that occur through incomplete ionisation of the source plasma, for instance, conduction by neutral hydrogen in the chromosphere (below $\sim 2 \times 10^4$ K). For a fixed value of the coronal heat flux, they find that the lower boundary temperature T_b of their model, that is, the point of vanishing conductive flux, depends rather sensitively on the assumed coronal density (c.f. Shmeleva and Syrovatsky, 1973; see below). In particular, if this density is too low, the radiative losses are not sufficient to exhaust the downward heat flux and no point of vanishing temperature gradient is achieved. However, for reasonable values of the base temperature (say $T_b \simeq 10^4$ K) Moore and Fung were able to reproduce the classic transition region structure: thus a narrow region ($\approx 10^5$ m) of steep temperature gradient, but almost uniform pressure, forms an interface between the near-isothermal coronal and chromospheric plasmas.

A slightly more general analysis of the conductive-radiative energy balance, of relevance to both quiescent and flaring plasmas, was given by Shmeleva and Syrovatsky (1973). Although *gravitational effects are neglected*, the central assumption follows Moore and Fung, in that the thermal structure is supported by heat flux that emanates from a hot source plasma overlying the region of interest. In the case of quiescent active-region plasmas it is suggested that the source heating proceeds by waves generated in the solar convection zone (the conventional hypothesis e.g. Kuperus, 1969), whereas, for flares, the heating is attributed to the collisional energy loss of non-thermal electrons that are accelerated during the primary energy release. Either way, the *energy input is assumed to be negligible below a certain critical temperature*, T_0, that forms the upper boundary to their model.

Shmeleva and Syrovatsky express (5.25) in terms of the Lagrangian variable $\bar{\zeta}$ and obtain a semi-analytic solution for $T(\bar{\zeta})$ subject to the boundary conditions (c.f. §5.3.3)

$$\left.\begin{aligned} F_0 &= \kappa \left.\frac{dT}{d\bar{\xi}}\right|_{\bar{\xi}=\xi_0}, \\ T &\to 10^4 \text{ K as } \bar{\xi} \to \infty, \end{aligned}\right\} \tag{5.26}$$

F_0 being the conductive flux at the upper boundary: note that since an "external" heat source of magnitude $n_e^2 Q(10^4 \text{ K})$ is assumed to balance the radiative losses at the temperature minimum, both the conductive flux and its divergence must vanish simultaneously at this point.

Two limiting cases of the energy balance are considered: first, a rapid flare heating in which the particle density has no time to adjust ($n_e = constant$); second, a slow prolonged heating for which uniform pressure is valid. In both these cases a narrow transition layer is found to separate the hot ($>10^6$ K) and cool ($\leqslant 10^4$ K) regions of the source. This prompted Shmeleva and Syrovatsky to suggest that the EUV emitting layer in solar flares ($10^4 - 5 \times 10^5$ K) may really be quite thin — say, 10^5 m in the uniform pressure case — in contradiction to Švestka's view of a thick flare with filamentary structure (see §2.3.3 and Švestka, 1976).

The analysis of Shmeleva and Syrovatsky, useful though it is, can be criticised on several rather general grounds. Certainly, while few astronomers would question the validity of the conductive-radiative energy balance for quiescent plasmas, it is not clear to what extent solar flares follow this description. Take for example the case of rapid flare heating. It is true that the atmosphere will initially be heated at constant density, but it is unlikely that a conductive-radiative equilibrium can be established without hydrodynamic motions of the plasma. This follows from the observation that any temperature enhancement in the flare source ($T > T_0$) leads to increased conductive deposition in the region of interest without significantly increasing the radiative losses. A similar problem arises under the uniform pressure assumption: even during the flare decay when the assumption may be valid, it is not clear that the evolutionary term (viz., $\frac{3}{2}\dot{p}$) accounting for the changing thermal energy of the plasma can always be neglected.

The lower boundary condition presents a further difficulty. The assumption that the temperature gradient and its derivative must vanish at 10^4 K is obviously highly restrictive and cannot possibly be valid for all quiescent and flaring plasmas. Shmeleva and Syrovatsky do not investigate how critically the computed solution depends on this rather artificial boundary condition (see §§5.4.4 and below).

The Shmeleva and Syrovatsky flare model has also been strongly criticised by Brown (1975). His criticisms are threefold. Firstly, he maintains that the collisional energy loss from the high energy tail of accelerated electrons is likely to dominate conductive heating in the chromosphere. He argues that, even ignoring the decline in the temperature gradient around 10^4 K, the chromospheric conductive flux must be reduced to approximately 10^{-7} times its coronal value (F_0) by virtue of the temperature sensitivity of κ ($\simeq T^{5/2}$). On the other hand, the electron collisional loss falls off as $n_p\zeta^{-\delta/2}$ (Equation 5.20) and since there are roughly 10^2 more particles per m^2 in the chromosphere than in the flare source ($\sim 10^7$ K), the electron energy flux is reduced less severely, typically (taking $\delta = 4$) by a factor of 10^4. As a result, Brown contends that Shmeleva and Syrovatsky achieve conductive dominance only because they impose (via 5.26) an *artificial temperature gradient* on the chromospheric plasma (see also Švestka, 1976).

A second point of disagreement arises because Shmeleva and Syrovatsky express their final solution in the form of a "universal function of temperature". Brown points out that since this universality results from taking *optically-thin* radiative losses, it can never be appropriate to the chromosphere ($\approx 10^4$ K) which requires an optically-thick formulation (§5.4.3).

Brown's final argument is based on the relatively slow conduction speed in the chromosphere: as thermal waves generally propagate below the sound speed (Craig and McClymont, 1976; §5.5.1) it is unlikely that conduction is fast enough at transferring energy through space to account for the observed close synchronisation ($\sim$seconds) of Hα kernels and EUV and Hard X-ray bursts (e.g. Vorpahl, 1972). Since a similar objection can be raised against energy transfer by mass flow, it appears that *non-thermal electron streams provide the most natural explanation* of these phenomena.

Finally, we should mention that Somov and Syrovatsky (1976) and Švestka (1976) have attempted to unite conduction and particle-heating models of the flare chromosphere. They emphasise that, while conduction generally predominates over particle deposition, there may be certain events characterised by hard-electron spectra (and presumably low coronal densities) for which this is not the case. The resulting hybrid model, however, is rather disappointing: it lacks the attractive simplicity of earlier models while still neglecting all hydrodynamic effects (see Kostyuk and Pikelner, 1975).

5.4.3 Chromospheric energy balance between electron collisions and radiation (Brown)

In contrast to the previous conduction model, Brown (1973) postulates that the flash phase of the *chromospheric* flare is the result of a steady-state energy balance between the *radiative losses* and the *collisional heating* of non-thermal electrons. The input of the beam is superposed on the normal acoustic heating that supports the radiative losses of the quiescent atmosphere (e.g. Osterbrock, 1961). Brown argues that since the dynamical time-scale of the chromosphere (~ 50 s) is comparable to the characteristic time variation of the beam — at least for large events — the atmosphere will respond at uniform pressure to the electron bombardment. With this simplification Brown employs a semi-analytic solution to the energy equation to compute the response of a typical quiet chromosphere (Gingerich *et al.*, 1971) to the flare heating. Above a certain critical temperature the atmosphere is radiatively unstable and the quasi-steady approximation breaks down. By identifying the mass of material lying above this level ($T \approx 6 \times 10^4$ K) with the hot coronal plasma ($\approx 20 \times 10^6$ K), Brown derives a plausible estimate of the X-ray emission measure and the up-flow expansion (or "evaporation") velocity in large flares (c.f. §5.5.3).

Canfield (1974), however, has pointed out that Brown's model is unable to reproduce satisfactory Hα profiles. The discrepancy is mainly due to radiative losses, which, in the simplified optically-thin approximation, become over-estimated by at least an order of magnitude. In a revised calculation, Brown, Canfield and Robertson (1978) employ an iterative numerical scheme to incorporate radiative transfer throughout the flare atmosphere. They demonstrate that, when spatial inhomogeneities and macroturbulent velocities are included in the model, satisfactory Hα profiles can be obtained.

But the matter does not end here. In the one event for which satisfactory data are available, the energy fluxes implied under the non-thermal interpretation of the hard X-ray burst (e.g. Hoyng *et al.*, 1976) are *one order of magnitude larger* than the fluxes required for description of the Hα kernel. Although Brown *et al.* cite several possibilities (e.g. certain conductive and radiative losses from the Hα region may have been neglected), they freely admit that the discrepancy is rather difficult to explain away. It appears that only further detailed optically-thick investigation can shed light on this matter.

5.4.4 The energy balance in closed loops

To extend quasi-static models to closed loop geometries it is necessary to specify the quiescent coronal heating mechanism. Various forms of heating are possible: the dissipation of acoustic waves generated in the convection zone (e.g. Kuperus, 1969), the damping of Alfvén waves (Wentzel, 1974) and the ohmic dissipation of induced coronal currents (e.g. Tucker, 1973) have all been extensively discussed in the literature. Yet due to the theoretical difficulty of modelling non-linear wave propagation in the solar atmosphere no mechanism is generally accepted or even particularly well understood. In consequence most thermal loop models assume rather *simple empirical forms* for the coronal heating function.

The closed-loop analysis of Landini and Fossi (1975) rests on two prime physical assumptions. Firstly, they suppose that a spatially-uniform "acoustic" heating mechanism supports the global thermal structure against the optically-thin radiative losses. Secondly, in common with Shmeleva and Syrovatsky, they take the temperature gradient to vanish at some fixed level ($\approx 10^4$ K) deep in the chromosphere. Direct integration of the energy equation (from the base upwards) then yields a second point of vanishing conductive flux which may be identified with the peak temperature, T_0, at the apex of the loop. By this device, Landini and Fossi demonstrate that a relationship must exist between the loop length, l, and the coronal temperature and density (T_0, n_0) — a relationship subsequently re-emphasised in the related analysis of Rosner, Tucker and Vaiana (1978). However, while good agreement with the differential emission measure in the EUV temperature range was obtained, their model was unable to provide a satisfactory explanation of X-ray coronal observations. Landini and Fossi take this to imply that *many loops, of differing peak temperature, are present* in the active region assembly, an idea independently supported by the semi-empirical EUV modelling of Jordan (1975; 1976) (see §5.5.6).

Craig, McClymont and Underwood (1978) (see also Hood and Priest, 1979) attack the problem of coronal loop structure from an entirely different viewpoint. They maintain that, since coronal heating and chromospheric energy balance are poorly understood, it is unwise to construct models based on questionable assumptions about the energy deposition function and on vanishing conductive flux at the lower boundary. To this end, they write the static energy equation as

$$\frac{\mathrm{d}}{\mathrm{d}z}\left(\kappa \frac{\mathrm{d}T}{\mathrm{d}z}\right)=n_e^2 Q(T)-H(T), \tag{5.27}$$

the energy input being an *unknown function* of temperature rather than of position, a transformation which is always possible for monotonic temperature profiles typical of the corona–transition region structure. In particular, writing Equation (5.27) in terms of the differential emission measure and assuming *uniform pressure* yields

$$H(T)=\left(\frac{p}{2k_B}\right)^2 Q(T)T^{-2}-2\kappa_0\left(\frac{p}{2k_B}\right)^4 \frac{\mathrm{d}}{\mathrm{d}T}(T/\xi^2(T)), \tag{5.28}$$

which directly links H to the observational parameter ξ (Equation 5.4). Unfortunately, as ξ is poorly prescribed by the data (§5.2.2), this equation has limited practical value; however, it may be employed theoretically to investigate the sensitivity of the loop structure in the heating function. Due to the strong temperature dependence of the conductive and radiative terms, it turns out that ξ must generally be an *increasing* function of temperature ($10^5 \leqslant T \leqslant 10^7$ K) *almost independent of the form of H*. Exceptions to this behaviour are possible, but a necessary condition is that H follows the radiative loss function ($n_e^2 Q(T)$) over an observationally resolvable range of temperature space. The physical independence of the heating and radiative loss functions renders this possibility rather implausible.

To parameterise the loop Craig *et al.* proceed as follows: multiplying (5.27) by $(T^{7/2})'$ and integrating readily yields the conductive flux

$$F_c(T) = (2\kappa_0 \psi(T))^{1/2} \tag{5.29}$$

and the differential emission measure

$$\xi(T) = \left(\frac{p}{2k_B}\right)^2 (2\kappa_0 \psi(T)/T)^2 \tag{5.30}$$

in terms of the positive semi-definite incomplete integral

$$\psi(T') = \int_{T'}^{T_0} (H(T) - n_e^2 Q(T)) T^{5/2} \mathrm{d}T, \tag{5.31}$$

where $n_e = p/(2k_B T)$. We observe that prescription of $H(T)$, together with the central temperature and density (T_0, n_0), completely determines ψ and hence the structure of the loop. In this sense ψ is a generalisation of the universal function of temperature discussed by Shmeleva and Syrovatsky (1973) (§5.4.2).

The problem is that $H(T)$ is unknown. To get over this difficulty Craig *et al.* write

$$\int_{T_b}^{T_0} H(T) T^{5/2} \mathrm{d}T = \beta H(T_0) \int_{T_b}^{T_0} T^{5/2} \mathrm{d}T,$$

where T_b (taken as 10^5 K) is a lower boundary temperature for the model: accordingly, β is a "shape factor" associated with the heating function. With the half-length of the loop defined by (from 5.29)

$$l = \int_{T_b}^{T_0} \frac{T^{5/2}}{(2\psi(T)/\kappa_0)^{1/2}} \, dT,$$

it is possible to show that the coronal density is given by

$$n_0 = 3.8 \times 10^8 \ \beta^{1/2} [1 + (F_c(T_b)/F_0)]^{-1/2} \frac{T_0^{9/4}}{l}, \tag{5.32}$$

$F_0 = 2 \times 10^{-20} \ n_0 T_0^{5/4}$ W m^{-2} being a reference coronal flux. This expression links the three important loop parameters T_0, n_0 and l with the heating shape factor (β) and the base conductive flux $F_c(T_b)$. (A similar result has been obtained independently by Hood and Priest, 1979.)

Two limiting cases of (5.32) deserve comment. In the case where $F_c(T_b)$ is small compared with F_0, we find

$$n_0 = 3.8 \times 10^8 \ \beta^{1/2} \frac{T_0^{9/4}}{l}; \qquad F_c(T_b) \ll F_0, \tag{5.33}$$

implying that an increase in the peak temperature of the loop is compensated by an increase in the coronal density. This represents an "evaporative" solution in the sense that the chromosphere must respond dynamically to a temporary increase in the heat flux by driving mass into the loop and increasing the radiative losses ($T > T_b$) until the steady-state situation described by (5.33) is established. For the special case of a uniform heating mechanism ($\beta = 1$), Equation (5.33) yields the "parameter-free" scaling law deduced by Landini and Fossi (1975) and Rosner *et al.*, (1978): note that slight deviations in this "law" are due to differing approximations for $Q(T)$ and have no inherent physical significance (see Figure 5.4).

In the other extreme where $F_c(T_b) \gg F_0$, the chromosphere is able to absorb the enhanced coronal heat flux without significant dynamical response; the result is a simple conducting loop for which no well-defined relationship between the temperature and density can exist. For both limiting cases, however, the gross form of ξ (outside the singularity at $T = T_0$) is a weakly increasing function of temperature, $\xi(T) \sim T^{\alpha}$ where $\frac{1}{4} \leqslant \alpha \leqslant \frac{1}{2}$ for $10^5 \leqslant T \leqslant 10^7$ K. The lower value of α, which is achieved for the "flux limited" chromosphere (5.33), arises because the temperature profile has to be slightly "flatter" ($T'' < 0$, for $T \ll T_0$) in order to support the stronger coronal radiative losses.

Vesecky, Antiochos and Underwood (1979) have performed a numerical study of closed loop structures for the case of uniform energy deposition. They assume a *flux-limited* or "*thermally isolated*" chromosphere ($F_c(T_b) \simeq 0$ for T_b

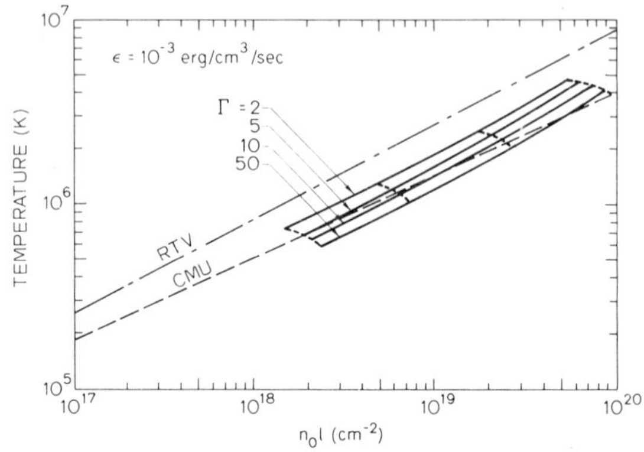

FIGURE 5.4 Variation of the loop apex temperature T_0 with the product of central density N_0 and loop length l (taken from Vesecky *et al.*, 1978). Here ε refers to the magnitude of the (uniform) energy deposition while Γ defines the magnetic compression ratio of the loop. The analytical models of Craig *et al.* (CMU) and Rosner *et al.* (RTV) are shown for comparison.

$\simeq 3 \times 10^4$ K) but take into account effects due to gravity and the variable cross-section of the loop (as do Wragg and Priest, 1980). Figure 5.4 is taken from their paper and compares these numerical results with the analytic models of Craig *et al.* (1978) and Rosner *et al.* (1978) (Γ is the ratio of apex area to base area). As can be seen, for the most part the analytic and numerical models are in fairly good agreement. Moreover, due to the relative insensitivity of the loop parameters to β (5.32), these results should be valid for most forms of heating. Conversely, we note, following Craig *et al.* (1978), that in the absence of extremely high quality data on ξ suitable for use with (5.28), the factor β probably represents the totality of information on the heating function that is derivable from observation.

Other aspects of loop behaviour have been examined in more recent studies. Taking thermally isolated coronal loops only as a special case Hood and Priest (1979) discuss the *non-equilibrium* that can occur when the pressure (or length) becomes too large or the heating too small. This analysis is extended by Wragg and Priest (1980), who include loops in hydrostatic equilibrium of *non-uniform* cross sectional area. Both of these papers relax the condition that the conductive flux vanish at 2×10^4 K, since the plateau at that level is caused by optically thick Lyman α effects and so cannot be modelled in the optically thin approximation.

Finally, we mention that the transverse structure of a loop depends essentially on the variation of the energy input function and pressure *across* the magnetic field. Hood and Priest included such transverse structure in their modelling. Also, Landini and Fossi (1975), in trying to explain X-ray data, constructed a global active region model from an assembly of individual loops by taking an empirical form for the heating, namely $H \alpha B^{\alpha}$ where α is an

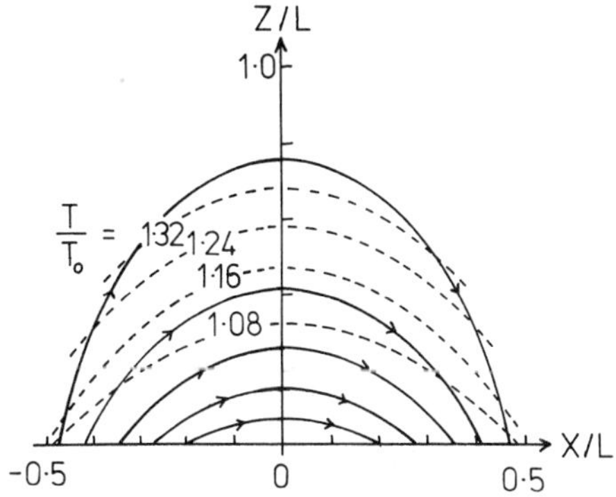

FIGURE 5.5 The isotherms (dashed) and magnetic field lines (solid) for a typical coronal arcade. Here the arcade width is 5000 km, the coronal base temperature T_0 is 10^6 K and base heating is twice the radiative loss (from Priest and Smith, 1979).

adjustable parameter of order unity. More recently, Priest and Smith (1979) have modelled an arcade of loops: Figure 5.5 shows the thermal structure of one of their models.

We return to the general theoretical problem of loop superposition in §5.5.6.

5.4.5 Stability of closed-loop solutions

Various authors have speculated that certain loop configurations may be prone to *radiative instability*. Rosner *et al.* (1978), for instance, argue that the temperature at the loop centre must be maximal lest the radiative instability be manifest. In essence, they assume that a coronal plasma is always unstable whenever it is supported purely by a stationary heating function: because the radiative losses decrease with increasing temperature in the coronal domain ($10^5 < T < 10^7$ K), it appears that any slight perturbation on the plasma is sufficient to trigger the instability. This argument, however, is incorrect. Only a plasma of sufficiently large scale-length is unstable. This can be seen from Field's isobaric criterion

$$n_0 < 2.2 \times 10^9 \frac{T_0^{9/4}}{l} \tag{5.34}$$

and follows from the fact that conduction is able to stabilise the plasma provided the wavelength of the perturbation is not too great (see Priest, 1978; Craig *et al.*, 1978; Vesecky *et al.*, 1979; Hood and Priest, 1979).

More seriously, Hearn (1975) has attempted to construct solar and stellar coronae from a "maximum flux principle" allegedly derived on thermal stability arguments (also Vaiana and Rosner, 1978). Again the analysis is unsound (Antiochos and Underwood, 1978). Since Hearn fails to provide a rigorous stability analysis it appears that his conjecture is most generously regarded as a working hypothesis, rather than a proper physical criterion. Theory aside, Hearn's models, unless propped up by various other ad-hoc assumptions, apparently fail to explain the observations (see Antiochos and Underwood, 1978).

Although the Field (1965) criterion applies only to homogeneous structures, it is instructive to compare Equation (5.34) with the general loop scaling law derived previously (5.32). No hard and fast conclusions can be reached, but we anticipate that instability is more likely to occur for high β-values and low conductive fluxes through the base ($F_c(T_b)$). This makes good sense physically since it corresponds to radiation being more closely supported by the quiescent heating, implying weaker conductive influence on the loop. Of course, to determine whether a particular configuration is unstable requires a detailed perturbation analysis.

Hood and Priest (1980) and Antiochos (1979) have recently performed simplified *global perturbation analyses* for static, closed loop structures. Models which assume vanishing conductive flux at the base (e.g. Rosner *et al.*, 1978) are always thermally unstable. However, conduction is able to quash the instability if the base conductive flux is large enough (typically 15% of its peak coronal value). According to Hood and Priest, equilibria on an upper branch of static solutions are stable, while those on a lower branch (including the thermally isolated loop) are unstable. The form of the quiescent heating is shown by Antiochos to have minor effect on the instability, essentially in agreement with our earlier argument concerning β. Antiochos goes on to speculate that because of the instability, regions of low conductive flux in the deep transition layer may develop oscillatory behaviour (periods of ~ 1 s) as material heats, expands and then cools and condenses. As yet it appears that detailed hydrodynamic models provide little support for this assertion (see §5.6.3).

A review of flows and instabilities in active-region loops is given by Priest (1980); further details on the thermal stability analyses (including one by Habbal and Rosner, 1979) can be found there.

5.4.6 Thermal non-equilibrium — a possible non-magnetic flare mechanism

The present chapter describes the thermal evolution of a loop in response to some prescribed flare heating mechanism, whose origin is deliberately left unspecified. Several magnetic sources for the heating have been proposed (§1.6.1) such as magnetic reconnection (Chapter 3), the kink instability, or a resistive instability (Chapter 4). However, it has long been speculated that a non-magnetic effect, namely thermal runaway (e.g. Kahler and Kreplin, 1970), may act as a trigger mechanism for the explosive flare heating. Whether the effect actually *causes* the flare eruption, or whether it is simply an artifact of some more fundamental flare instability, is evidently a thorny issue (see Švestka, 1976).

More recently Hood and Priest (1980a) have investigated a thermal mechanism by which loops with cool cores (which in extreme cases may show up as fibrils in Hα) may suddenly heat up to temperatures of around 10^7 K. They point out that "thermal non-equilibrium" can give rise to a sudden flaring which may be relevant to simple-loop or compact flares. Their analysis is based on equilibrium solutions of the quasi-static energy equation (5.27) and is not strongly influenced by the details of the heating. As before, the summit temperature of the loop is found to depend on the base pressure, the heating and the loop length $2L$(c.f. (5.33)); the heating H is expressed in units of the optically-thin radiative loss at $T=2\times10^4$ K and $n_e=5\times10^{14}$ m^{-3}, while the dimensionless half-length $\bar{L}$ is defined by

$$\bar{L}^2 = n_e^2 \frac{Q(T)L^2}{\kappa_0 T^{7/2}}.$$

The way in which the equilibrium summit temperature depends on the heating and loop length for a fixed base pressure is shown in Figure 5.6a. As the heating rate H slowly increases so the peak temperature increases through equilibria in the lower branch. But when the heating exceeds a critical value H_{crit}, there is no longer a neighbouring equilibrium and so the plasma rapidly heats up along the dashed line and eventually reaches a new quasi-steady equilibrium at a temperature above 10^6 K, as denoted by a dot.

This process evidently satisfies one of the basic requirements of a flare mechanism, namely it is explosive in the sense of §1.3. The new quasi-static temperature, T_f, represents the temperature of the flare and depends on the particular value of L, as shown in Figure 5.6b. In practice, of course, the transition from the low to the high temperature state depends on the plasma dynamics: the detailed hydrodynamic experiment of §5.6.3 represents a specific example of this transition.

Hood and Priest also speculate that thermal non-equilibrium in an active-region (or plage) filament may act as a *trigger* for the eruption of the filament in a *two-ribbon flare*. It may cause the plasma to heat up, expand along the field lines and fall down to the photosphere, so that the unloaded field lines ascend. As the filament slowly rises, the threshold for MHD instability can be

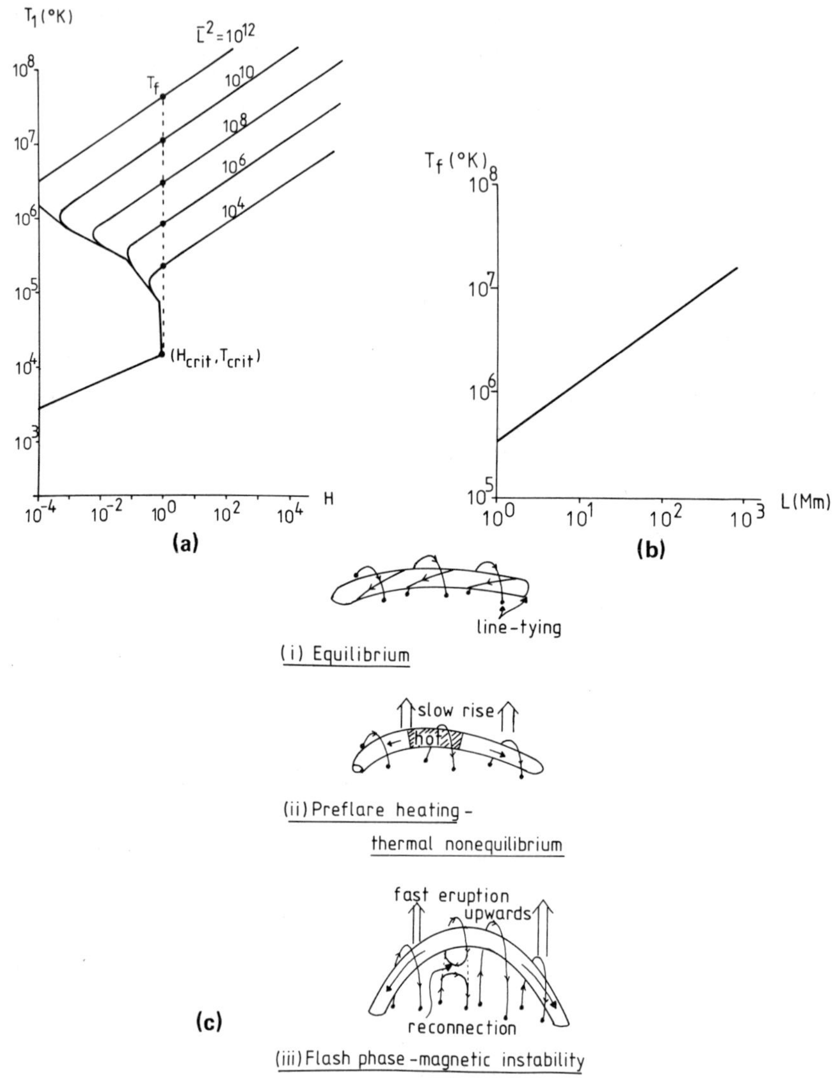

FIGURE 5.6 (a) The equilibrium summit temperature T_1 as a function of the heating parameter H, for various values of the dimensionless loop length $\bar{L}$. The heating is uniform and the base temperature 2×10^4 K. As H increases T_1 follows the cool lower solution until, at a critical heating H_{crit}, non-equilibrium sets in: the temperature rapidly heats up along the dashed line to a flaring value T_f, which depends on $\bar{L}$ and is denoted by a dot. (From Hood and Priest, 1980a.) (b) The flaring temperature T_f as a function of the loop half-length for a base temperature and density of 2×10^4 K and $5 \times 10^{17}\ \mathrm{m}^{-3}$. (From Hood and Priest, 1980a.) (c) Possible stages in the evolution of a two-ribbon flare: (i) the basic preflare state with a cool filament and the overlying sheared arcade, line-tied at its foot-points; (ii) thermal non-equilibrium sets in and causes a preflare temperature rise; material flows from the hot spot and the filament *slowly* ascends; (iii) when the filament reaches a critical height, an MHD instability causes the filament to erupt violently. (From Hood and Priest, 1980b.)

exceeded, so that the filament becomes MHD unstable and erupts much more rapidly (Hood and Priest, 1980b), as indicated in Figure 1.11 and 5.6c. The main phase of the energy release is associated with the subsequent reconnection of the magnetic arcade as the field lines close back down.

5.5 ANALYTIC CONDUCTION MODELS

5.5.1 Heating by conduction in the flash phase (Zeldovich and Raiser)

Several authors have argued that heat transfer by thermal conduction can power the low temperature, chromospheric flare during the initial explosive phase (e.g. Shmeleva and Syrovatsky, 1973; Švestka, 1973, 1976). The central idea is that the energy flux which emanates from a localised coronal flare source is transferred down the temperature gradient into the chromosphere, where it is eventually dissipated by radiation. But though conduction probably controls the thermal structure of the decaying flare plasma, its role during the flash phase is somewhat controversial (e.g. Craig and McClymont, 1976; Brown, 1973). It is not clear that conduction alone is efficient enough at transferring energy through space to account for the initial chromospheric brightening.

Various possibilities for conductive energy transfer in the flare loop are indicated in Figure 5.7. For sufficiently impulsive heating the plasma will not have time to move significantly, implying that the static conductive model of Figure 5.7(a) may be applicable to the initial phase. Later on however, it is clear that mass flows resulting from pressure imbalance will be coupled to the conductive energy redistribution. Two possible modes are illustrated in Figure 5.7: in diagram (b) both conduction and mass flow carry energy away from the coronal flare source (see §5.5.4), whereas in (c) the downward heat flux is compensated by an upward flow of relatively cool evaporated material (see §5.5.3). Although both modes are likely to be important during the initial flare decay, mass flows eventually die down as the loop approaches the quasi-steady equilibrium of Figure 5.7(d).

Returning to the question of the initial flare heating, we note that in a quiescent active-region loop virtually all the volume is occupied by material at coronal temperatures and densities (§5.4.2). Thus if the flare heating is strongly localised at the loop centre — the typical situation discussed by Strauss and Papagiannis (1971), Shmeleva and Syrovatsky (1973) and Craig and McClymont (1976) — then the thermal disturbance must propagate to the ends of the loop before the transition region and the chromosphere can be energised.

For conductive energy transfer during the initial phase the disturbance

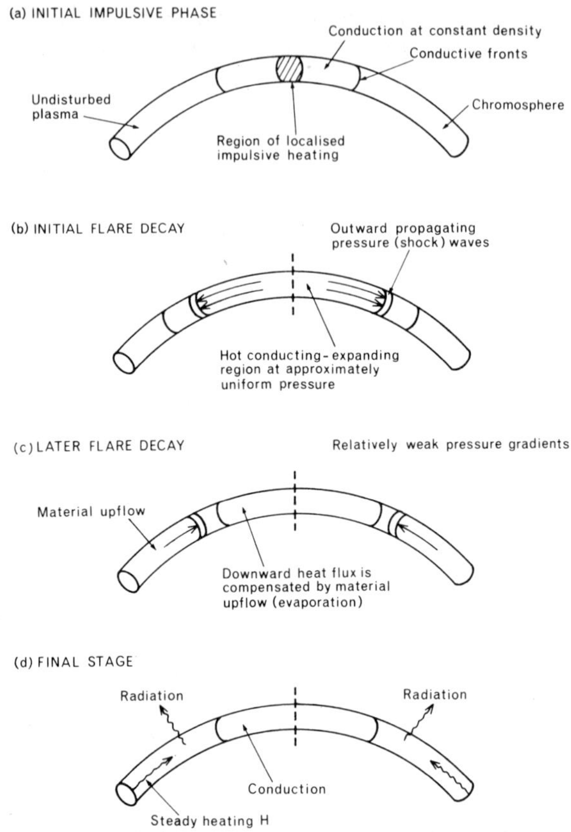

FIGURE 5.7 Possible conductive modes in the hot flare plasma. After the initial energy deposition (a) both conduction and mass flow redistribute flare energy within the loop. In (b) conduction and mass downflow act in unison to carry energy away from the flare source, whereas in (c) material upflow tends to counterbalance the conductive energy loss. The quasi-static equilibrium is illustrated in (d).

propagates according to the heat equation

$$\frac{\partial T}{\partial t} = \chi_0 \frac{\partial}{\partial z}\left(T^{5/2}\frac{\partial T}{\partial z}\right), \tag{5.35}$$

where $\chi_0 = \kappa_0/(3n_e k_B)$ is approximately constant for times shorter than the hydrodynamic time-scale of the plasma (§5.3.2). Although a general analytic solution to (5.35) cannot be obtained (due to the non-linearity of κ), in the limit of impulsive, localised heating the *self-similar analysis* of Zeldovich and Raiser (1967) is appropriate. Details of the self-similar solution are included in Figures 5.8 and 5.9. The temperature profile is given by

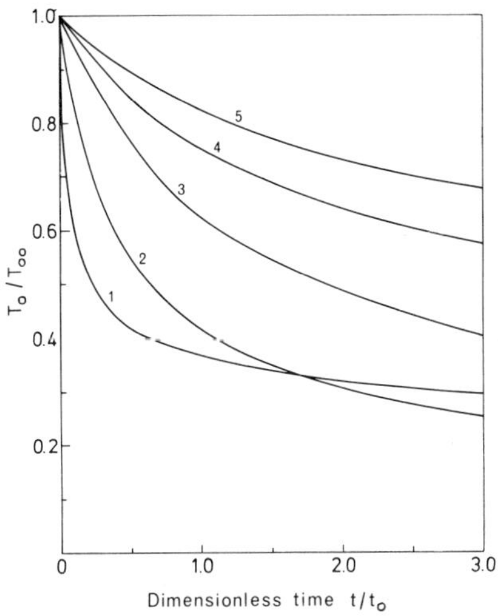

FIGURE 5.8 Cooling of the hot flare plasma for various conductive models. Curve 1 refers to the self-similar conductive solution of Zeldovich and Raiser (1967). Curve 2 applies to the conduction-expansion mode (with $\alpha = 2.5$) of Craig and McClymont (1978); also shown (curve 3) is the profile for cooling by expansion in an isothermal plasma ($\alpha = 0$). Curves 4 and 5, respectively, refer to the static and evaporative models of Antiochos and Sturruck (1978).

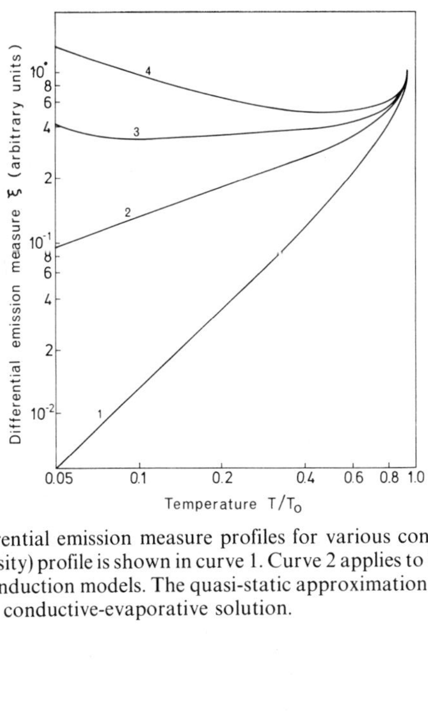

FIGURE 5.9 Differential emission measure profiles for various conductive models. The self-similar (constant-density) profile is shown in curve 1. Curve 2 applies to both static and expanding (uniform pressure) conduction models. The quasi-static approximation is shown in curve 3, while curve 4 indicates the conductive-evaporative solution.

$$T(z, t) = T_0(t)(1 - z^2/l^2(t))^{2/5} \tag{5.36}$$

and the wave propagates at constant density according to

$$l(t) \simeq (2.3 \times 10^{11}/n_e)^{2/9} Q_0^{5/9} t^{2/9}, \tag{5.37}$$

where Q_0 defines the strength of the flare energisation; formally $Q = Q_0 \delta(z)\delta(t)$. (Equation (5.37) can also be derived by elementary dimensional analysis.) Since the average temperature of the wave may be closely identified with the temperature of the near-isothermal conductive plateau that emanates from the origin ($T \simeq T_0$) it follows, by conservation of energy, that $Q_0 \simeq T_0 l$ (c.f. Zeldovich and Raiser, 1967). Hence, normalising Q_0 to correspond to a flare plasma of temperature $T = 10^7$ K and length $l \simeq 10^{7.5}$ m, we find that, for $n_e = 10^{16}$ m^{-3}, it takes approximately 10^2 seconds for the thermal wave to reach the end of the loop. A longer tube, a higher density or a weaker energisation (Q_0) increases this time-scale greatly. Hence, if conduction, or more precisely a conductive wave, is really responsible for energising the chromosphere we should expect a *measurable time-delay between the coronal and chromospheric brightenings.* The observations are not definitive on this point (c.f. Antiochos and Sturrock, 1978).

Another question which must be answered is whether or not conduction is capable of transferring the bulk of the released flare energy into the chromosphere. Certainly the local conductive flux cannot exceed the "free-streaming" upper limit (c.f. Spicer, 1977)

$$F_0 < (n_e kT)\bar{v}, \tag{5.38}$$

where $\bar{v}$ is the thermal speed of the electrons. This condition reduces to

$$n_e T^{3/2} > 2 \times 10^{19} \; F_0. \tag{5.39}$$

Hence, to maintain a typical flare heat flux of 10^6 W m^{-2} (appropriate to $T = 10^7$ K and $l \simeq 10^{7.5}$ m) a density of $n_e \geqslant 10^{18}$ m^{-3} at 10^5 K is required. Although the initial atmosphere in flare loops is highly uncertain there is clearly no reason to expect that (5.39) should be fulfilled for all events. When the condition is violated, the heat flow gets "bottled up", pressure gradients are enhanced and mass flows are more strongly accelerated (c.f. Somov, 1978). The net effect therefore is to *transfer flare energy by mass flow rather than by heat conduction* (see also Craig and McClymont, 1976; Antiochos and Sturrock, 1976).

Finally, we note that flux limitation is expected to occur only during the initial energisation of the plasma, not after the quasi-steady state has been established, which is the case discussed by Somov (1978).

5.5.2 Static conduction-cooling (Antiochos and Sturrock)

A static conductive analysis of the flare decay was recently presented by Antiochos and Sturrock (1976). Although their work attempts to assess the influence of magnetic field compression on flare cooling, our present intention is to investigate only their basic conduction model.

As Antiochos and Sturrock *neglect mass flows and gravity* the momentum and continuity equations imply

$$n_e = n_e(z), \qquad p = p(t). \tag{5.40}$$

The neglect of radiation and heat sources then leads to a simple form for the heat equation

$$\frac{3}{2}\frac{\mathrm{d}p}{\mathrm{d}t} = \frac{\partial}{\partial z}\left(\kappa \frac{\partial T}{\partial z}\right). \tag{5.41}$$

Antiochos and Sturrock replace T by $p/(2n_e k_B)$, separate variables, and integrate to yield (see Figure 5.8)

$$T_0(t) = T_{00}(1 + t/\tau)^{-2/5}, \qquad T(z, t) = T_0(1 - z^2/l^2(t))^{2/7}, \tag{5.42}$$

where $T_{00} = T_0$ $(t=0)$, τ is a free parameter (defined for fixed T_{00} by the separation constant) and

$$l^2(t) = \frac{4}{21}\left(\frac{\kappa_0}{k_B}\right)\frac{T_0^{7/2}}{n_0}\left(-\frac{\mathrm{d}T_0}{\mathrm{d}t}\right)^{-1}. \tag{5.43}$$

The differential emission measure predicted by the model is readily derived from Equation (5.42):

$$\xi(T) \sim \frac{(T/T_0)^{1/2}}{(1 - (T/T_0)^{7/2})^{1/2}}, \tag{5.44}$$

as illustrated in Figure 5.9. Outside the singularity which arises through the vanishing temperature gradient at the loop centre, ξ has the typical form expected for models of constant conductive flux, namely $\xi \sim T^{1/2}$ (see §5.4.4). However, because of the increasing volume of material per unit temperature range, the amplitude of ξ slowly increases as the loop cools. The *total* emission measure ($n_e^2 V$; $T \geqslant 0$), of course, remains constant.

In discussing the applicability of this model, Antiochos and Sturrock concede that the high heat flux ($\sim 10^6$ W m$^{-2} = 10^9$ erg cm^{-2} sec^{-1}) at the onset of cooling may result in an evaporation of chromospheric material. The possibility of an initial "explosive" evaporation, probably in response to chromospheric bombardment by a stream of high energy particles in the flash phase (Brown, 1973; Kostyuk and Pikelner, 1975; §5.4.3), is also mentioned. Indeed Antiochos and Sturrock speculate that the static model may be valid *after the explosive evaporative phase* has ended — presumably they assume that mass flows are insignificant by this stage (e.g. Strauss and Papagiannis, 1971). After further cooling, they suggest that *radiation* takes over as the dominant loss mechanism.

5.5.3 Conductive–evaporative cooling (Antiochos and Sturrock)

The *evaporative cooling* calculation of Antiochos and Sturrock (1978) represents one of the most interesting flare decay models to date. The analysis itself is deceptively simple. Motivated by the large coronal emission measures ($\sim 10^{55}$ m$^3 = 10^{49}$ cm^{-3}) during the initial flare decay, Antiochos and Sturrock postulate that the *energy loss due to heat flow away from the flare source is compensated by an upward flowing mass of "evaporated" chromospheric material.* Since *gravity is neglected*, and since *up-flow velocities are presumed small* compared to the local sound speed, the pressure distribution is expected to remain approximately *uniform* throughout the evaporative cooling. In addition, the neglect of energy sources and sinks ensures that the *pressure remains constant in time* as well as in space. Hence the energy equation is simply

$$\frac{\partial}{\partial z}\left(\kappa \frac{\partial T}{\partial z}\right) = \frac{5}{2} p \frac{\partial v}{\partial z}, \tag{5.45}$$

where $p(z, t) = constant$. With the usual assumption of loop symmetry (§5.3.3) the energy equation yields

$$v = \frac{2}{5} \frac{\kappa}{p} \frac{\partial T}{\partial z}, \tag{5.46}$$

which shows that the local velocity is proportional to the magnitude of the conductive flux. Substituting Equation (5.46) into the continuity equation and eliminating n_e from the equation of state then yields

$$\frac{5}{2}\frac{p}{T}\frac{\partial T}{\partial t}=\frac{\partial}{\partial z}\left(\kappa\frac{\partial T}{\partial z}\right)-\frac{\kappa}{T}\left(\frac{\partial T}{\partial z}\right)^2. \tag{5.47}$$

Although this equation is non-linear Antiochos and Sturrock show that for the (*assumed*) *separable solution* the loop cools according to

$$T_0(t)=T_{00}(1+t/\tau)^{-2/7} \tag{5.48}$$

(as shown in Figure 5.8) where T_{00} is the peak temperature at the onset of evaporation and τ is defined in terms of the initial temperature and density (viz. T_0 and n_{00}) together with a separation constant that effectively determines the length of the loop. Moreover, from the spatial variation of the solution it can be demonstrated that the heat flux is linearly proportional to the temperature ($T \ll T_0$). Accordingly, the gross form of the differential emission measure

$$\xi(T)\sim\frac{(T/T_0)^{-1/2}}{(1-(T/T_0)^{3/2})^{1/2}} \tag{5.49}$$

follows a weakly decreasing function of temperature $\xi \sim T^{-1/2}$ (Figure 5.9; c.f. §§5.4.4, 5.5.2 and 5.5.4). The concave thermal profile ($T''>0$) is due to the fact that the local conductive flux is depleted quite rapidly (c.f. §5.4.4) in supporting the energy losses of the expanding low-temperature plasma.

As Antiochos and Sturrock (1978) emphasise, the main attraction of the evaporative process is that the heat flux entering the chromosphere may be reduced by several orders of magnitude over the heat flux anticipated on static models. Indeed, we have already seen that some form of evaporation always occurs in active-region plasmas, for it is improbable that the chromosphere can absorb arbitrarily high heat fluxes without a significant dynamic response (§5.4.4). But whether the *steady-state "slow" evaporative process* of Antiochos and Sturrock can provide an adequate description of the flare decay is really a separate issue. Certainly, the assumption that the *pressure remains almost uniform*, to the extent that the momentum equation can be ignored, is rather questionable. In particular, as mass flows cannot die down faster than the energy loss time-scale of the plasma, it is not clear that fluid velocities (resulting from the primary flare energisation) can be neglected during the initial decay. Rather, it appears that evaporation may be coupled to the natural hydrodynamic oscillations of the loop (as discussed in detail in §5.6.3).

The presence of *radiative losses also tends to destroy pressure equilibrium* within the source. Since radiation cooling is most effective for dense, low-temperature plasmas ($\approx 10^5$ K) we anticipate that material downflow may occur due to the reduced pressure at the base of the loop. In fact, Antiochos and Krall (1978), by numerically modelling an evaporating-radiating loop, show that mass-downflow develops quite quickly, after approximately 100 seconds of cooling. Their main conclusion is that a (non-unique) combination of four such loops, which differ mainly in length and area, can explain typical soft X-ray data. It is not clear, however, to what extent these conclusions depend on their assumptions of static initial conditions at the onset of flare cooling. Obviously, the only sound way to test the applicability of the analysis is to model the complete flare evolution, from the flash phase onwards.

As already mentioned, the evaporative cooling of Antiochos and Sturrock is designed to explain the appearance of a large mass of X-ray emitting material in the corona. Yet, as Brown and Smith (1980) remark, the chromosphere is highly insensitive to the form of heating that drives the evaporation. It is misleading therefore, to believe that any heating model, whether powered by particles, conduction, or by descending shocks, can be adequately tested by the observation of enhanced coronal mass alone (§5.6.3).

5.5.4 Conduction–expansion cooling (Craig and McClymont)

Craig and McClymont (1978) have recently investigated the influence of convection on conduction cooling. The analysis is motivated by a specific numerical model (Craig and McClymont, 1976; McClymont and Craig, 1978a), which shows that, in response to flare heating, a pair of outward propagating shock-waves emerge from the region of energy injection (symmetry being assumed) and leave behind a hot ($\gg 10^7$ K) thermal plasma that cools mainly by expansion and by heat conduction. Since pressures in the hot region have been approximately equalised by the passing shocks the energy equation is simply

$$\frac{3}{2}\frac{\mathrm{d}p}{\mathrm{d}t}=\frac{\partial}{\partial z}\left(\kappa\frac{\partial T}{\partial z}\right)-\frac{5}{2}p\frac{\partial v}{\partial z}. \tag{5.50}$$

Solving (5.50) simultaneously with the momentum and continuity equations Craig and McClymont show that (Figure 5.8)

$$\begin{aligned} T_0(t) &= T_{00}[(\alpha+1)\tau^{5/2}-\alpha]^{-2/5}, \\ T(z,t) &= T_0(t)(1-z^2/l^2(t))^{2/7}. \end{aligned} \tag{5.51}$$

Here $\tau \geqslant 1$ is the dimensionless time ($\tau = t/t_0$, t_0 being the time at the onset of cooling) and l is related to α by

$$l^2(t) = \frac{4}{7} \frac{\kappa_0 T_{00}^{7/2} t_0}{p_0 \alpha} \tau^2, \tag{5.52}$$

so that α effectively defines the importance of thermal conduction. The special case $\alpha = 0$ corresponds to cooling by expansion in an isothermal plasma (see Figure 5.8).

The above equation shows very clearly how the thermal disturbance spreads out linearly with increasing time. Moreover, though the form of the differential emission measure (Figure 5.9) is given by Equation (5.44), the amplitude of ξ (for $T \ll T_0$) increases according to $[(\alpha+1) - \alpha/\tau^{5/3}]^{3/5}$. Hence the emission measure rises monotonically by a factor of $(\alpha+1)^{3/5}$ as $\tau \to \infty$ (c.f. Zaumen and Acton, 1974). In contrast, static models, because of the relatively slow propagation of the conductive front, imply only a weak increase in the effective flare volume and therefore cannot account for a significant rise in the amplitude of the differential emission measure.

The above model has certain attractive features: it represents a simple but exact solution to the complete set of hydrodynamic equations (unlike say, the evaporation model of §5.5.3) which is applicable within the framework of a more general numerical treatment (Craig and McClymont, 1976). As a model of the initial flare decay, however, its validity is rather marginal. By the time the solution is established — and this cannot be shorter than the hydrodynamic time-scale of the hot plasma — there is a danger that the uniform-pressure assumption may be violated by the interaction of shocks with the chromosphere (§5.6.2). The model, therefore, can be *valid only for the case of rapid heating* (say < 100 s) *in relatively long* ($\sim 10^8$ m) *flux tubes*. For this reason it is probably best applied to predicting X-ray spectra for relatively long-lasting coronal shock phenomena (see Rust and Webb, 1977; McClymont, 1980).

5.5.5 Synopsis of conductive flare models

Table 5.1 presents a concise summary of the analytic flare models illustrated in Figures 5.8 and 5.9. Except for the evaporative solution, *all models predict that the differential emission measure should follow a weakly increasing function of temperature* ($T < T_0$). Since the amplitude of ξ also increases as the loop cools, it is clear that the detailed evolution of the thermal structure, rather than the gross form, is required to distinguish between the various models.

In constructing Table 5.1 we have assumed that the quasi-static model of Craig *et al* (1978) may be applied to the late flare decay where the pressure is changing fairly slowly ($\dot{p}$ small). As discussed in §5.4.4, only very weak restrictions on the form of the energy deposition H need be imposed. In the

TABLE 5.1

Energy equation	Approximate range of validity	Gross form of $\xi(T)$	Reference
$3n_e k_B \frac{\partial T}{\partial t} = \frac{\partial}{\partial z}\left(\kappa \frac{\partial T}{\partial z}\right)$	Initial flare heating by conduction (n_e constant)	$\xi \sim T^{3/2}$ ($T > 5 \times 10^6$ K)	Zeldovich and Raiser (1967)
$\frac{3}{2}\frac{\mathrm{d}p}{\mathrm{d}t} = \frac{\partial}{\partial z}\left(\kappa \frac{\partial T}{\partial z}\right)$	Flare decay by conduction (p uniform)	$\xi \sim T^{1/2}$ ($T > 5 \times 10^6$ K)	Antiochos and Sturrock (1976)
$\frac{3}{2}\frac{\mathrm{d}p}{\mathrm{d}t} = \frac{\partial}{\partial z}\left(\kappa \frac{\partial T}{\partial z}\right) - \frac{5}{2} p \frac{\partial v}{\partial z}$	Initial flare decay by conduction and expansion (p uniform)	$\xi \sim T^{1/2}$ ($T > 5 \times 10^6$ K)	Craig and McClymont (1978)
$\frac{\partial}{\partial z}\left(\kappa \frac{\partial T}{\partial z}\right) = \frac{5}{2} p \frac{\partial v}{\partial z}$	Later flare decay by conduction and evaporation (p uniform and constant)	$\xi \sim T^{-1/2}$ ($T > 10^6$ K)	Antiochos and Sturrock (1978)
$\frac{\partial}{\partial z}\left(\kappa \frac{\partial T}{\partial t}\right) = n_e^2 Q(T) - H$	Final stage (p uniform)	$\xi \sim T^{\gamma}(\frac{1}{4} < \gamma < \frac{1}{2})$ ($T > 10^5$ K)	Craig, McClymont and Underwood (1978); Hood and Priest (1979)

case where the heating is absent ($H=0$ locally) we recover the solutions ($\gamma \simeq \frac{1}{4}$) obtained by Shmeleva and Syrovatsky (1973) (also Moore and Fung, 1972).

The following schematic flare model gives some indication of how a loop might evolve non-linearly through several conductive modes. In the initial heating phase we note that the temperature gradient in the transition zone must steepen in order to maintain the enhanced conductive flux; the emission-measure profile steepens also, and, in the limit of impulsive localised heating about the apex of a static loop, ξ approaches the $T^{3/2}$ dependence of the static conduction model. However, as mass flows rapidly become coupled to the conductive energy redistribution, the transition zone thickens and the differential emission measure ($T > 10^5$ K) begins to "flatten out". In fact we may anticipate a dynamic non-linear oscillation (on a time-scale of roughly l/c) between the expanding and evaporative conduction modes ($\xi \sim T^{1/2}$ and $T^{-1/2}$ respectively). Finally, as mass flows die down through radiative and conductive damping, the loop approaches a final quasi-steady equilibrium ($\xi \sim T^{\gamma}$; $\frac{1}{4} < \gamma < \frac{1}{2}$).

5.5.6 Inhomogeneities — the superposition of loops and kernels

Although analytic conductive models yield insight into the behaviour of single flare loops, it is clear that radiation from the active corona is generally emitted by an assembly of evolving loop structures (e.g. Landini and Fossi, 1975; Jordan, 1976; Craig *et al.*, 1978; Priest and Smith, 1979). In this case, the X-ray spectrum may not depend critically on the thermal structure of an individual loop or flare kernel; rather it is the superposition of all such loops that is important.

In general, the problem of loop superposition may be formulated in terms of a distribution function $G(T_0)$ that defines the instantaneous distribution of peak temperatures in the flare (c.f. Craig *et al.*, 1978) — all other loop parameters are effectively "integrated out". The total emission measure, say ξ_Σ, is then given by

$$\xi_\Sigma(T) = \int_T^{T_m} \xi(T; T_0) G(T_0) \mathrm{d} T_0, \tag{5.53}$$

where T_m is the maximum temperature in the source and $\xi(T;T_0)$ is the emission-measure function for a loop of peak temperature T_0.

By way of illustration consider the ξ-function for the conducting-expanding plasma of §5.5.4. For this case we have

$$\xi_\Sigma(T) = T^{1/2} \int_T^{T_m} \frac{G(T_0)\eta^{1/2}(T_0)}{(\eta(T_0) - \eta(T))^{1/2}} \mathrm{d}T_0, \tag{5.54}$$

where $\eta = T^{7/2}$. This expression shows how, for an assembly of conducting-expanding plasmas, the total differential emission measure departs significantly from the form ξ when the assembly fails to congregate around a single peak temperature; that is, when $G(T_0)$ departs from a delta-function form.

Since equation (5.54) is an Abel equation, it may be inverted to yield

$$G(T_0) = -\pi^{-1}\eta^{-1/2}(T_0)\frac{\mathrm{d}}{\mathrm{d}T_0} \int_{T_0}^{T_m} \frac{\eta'(T)\xi_\Sigma(T)T^{-1/2}}{(\eta(T) - \eta(T_0))^{1/2}} \mathrm{d}T, \tag{5.55}$$

which indicates that observations of ξ_Σ can lead to rudimentary information on the unknown function $G(T_0)$ (Craig *et al.*, 1978). This fact has direct relevance to the important, but as yet unexplored, question of the heating variation from loop to loop in the quiescent and flaring corona. Conversely, the above argument shows, all else being equal, that the differential emission measure of a single loop cannot be identified with ξ_Σ if the loopwise variation of the heating is significant.

In a recent paper Levine and Pie (1980) have attempted to derive information on $G(T_0)$ from X-ray data on quiescent active-region plasmas. They point out that only a *crude representation can be derived*, due mainly to the poor numerical conditioning of Equation (5.53). However, this is unlikely to be the main source of the trouble. The Abel inversion (which also arises in their case due to the conductive dominance around the peak temperature), though mathematically ill-posed, is quite well-conditioned numerically compared with most integral transform problems; in any event weak *a-priori* information stabilises the inversion (see Craig, 1979). The root cause of the trouble probably resides in the fact that ξ_Σ *itself is poorly determined by the X-ray data* (§5.2.2). Clearly, if the function to be inverted is poorly defined, we can expect no more than rudimentary information from this type of analysis.

Finally, we mention that Brown, Craig and Karpen (1980) have pointed out that a consistent thermal interpretation of the hard X-ray burst also requires the superposition of many individual "flare kernels". Their argument follows from the observation that the temperature and emission measure of the burst decline concurrently, in apparent contradiction to models which assume a gradual conductive redistribution of flare energy. Apparently, the hypothesis of kernel superposition allows an anomalous conduction model to be reconciled with the observations, at least provided the burst spectra are not too hard.

5.6 HYDRODYNAMIC MODELS OF A FLARING LOOP

5.6.1 Introduction

The importance of modelling the global loop evolution, from the initial energisation to the late flare decay, has already been emphasised (§5.5.3). The adoption of a *non-stationary heating* mechanism, however, when coupled to the *flash phase dynamics*, greatly increases the complexity of the loop behaviour. Moreover a prime difficulty with any numerical model is the incorporation of *realistic boundary conditions* into the problem (§§5.4.2, 5.4.4). To handle the flare evolution properly obviously requires a coupled *radiative transfer–hydrodynamic* scheme (e.g. Kneer and Nakagawa, 1976). Only when this aim is achieved can gas-dynamic models be confidently tested against observational data.

5.6.2 Hydrodynamic models

Probably the most comprehensive gas-dynamic analysis to date is provided by Kostyuk and Pikelner (1975). They model the response of an idealised initial atmosphere ($T > 5.3 \times 10^3$ K) to steady heating by a stream of non-thermal electrons ($F_e \sim 10^7$ W m$^{-2} = 10^{10}$ erg cm^{-2} s^{-1} for 100 s duration). As a lower boundary condition they take the fluid velocity and temperature gradient to vanish low in the chromosphere (specifically at $\bar{\zeta} = 10^{25.4}$ m^{-2}, $T = 5300$ K and $n_e = 10^{20}$ m^{-3}). In the original treatment an open-loop geometry is assumed (see §5.4.2), but more recently the model has been extended to incorporate closed-loop topologies (Kostyuk, 1976). For either case the form of the initial atmosphere ensures that the *bulk of the heating takes place in the chromosphere* rather than in the corona. The final results quantify and greatly extend the earlier computations of Brown (§5.4.3). Above a certain temperature determined by the threshold of thermal instability ($> 5 \times 10^4$ K), the *atmosphere expands upward at shock velocities* ($\sim 10^6$ m sec^{-1}) to form a hot coronal region ($T = 20 \times 10^6$ K). During the cooling phase *mass downflows occur in the transition region* which compress and heat the underlying chromospheric material (c.f. Nakagawa *et al.*, 1973). The overall velocity profile in the chromosphere is quite complex, but fluid velocities generally decrease with depth. Kostyuk and Pikelner emphasise that *collisional, conduction and compressional heating all play an important role* in energising the chromospheric plasma (cf. §5.4.3).

Surprisingly, Kostyuk and Pikelner do not publish density profiles for the flare. Apart from the obvious difficulty of comparing their results with observational data, this omission makes it hard to verify the theoretical self-consistency of the model. Somov and Syrovatsky (1976) have already pointed out that the one-second time-step which Kostyuk and Pikelner adopt in their numerical scheme may be significantly longer than the time of formation of a radiative condensation (see §5.6.3). Indeed, from the criterion that the time-step must never exceed the radiative time-scale of the plasma, it is easy to derive an upper limit to the density in their model transition region ($\approx 10^5$ K). It follows from Equation (5.13) that Kostyuk and Pikelner's analysis is *only valid for transition region densities below* 4×10^{16} m^{-3}. This upper limit on the density is suspiciously low, particularly when it is remembered that the condition must hold throughout the entire flare evolution. Similar criticisms can also be levelled against the more recent flare analysis of Nagai (1980).

McClymont and Craig (1980) have attempted to remedy the lack of density information in the Kostyuk and Pikelner analysis (see Canfield *et al.*, 1980). By estimating the fluid acceleration, then integrating up the momentum equation, they are able to estimate the local flare pressure. Evidently, the early cooling phase is characterised by a rapidly decreasing coronal temperature, but increasing soft X-ray emission measure ($T > 10^6$ K). Moreover, the emission measure is higher by an order of magnitude, and peaks considerably later (≈ 500 s) in the transition region ($10^5 < T < 10^6$ K) than in the corona. But, as Canfield *et al.* point out, this behaviour is *not consistent* with the flare described by Underwood *et al.* (1978).

The calculation of Craig and McClymont (1976) deals *only with the coronal manifestation* of the flare. They point out that significant mass-flows are a natural consequence of the rapid (≈ 100 s) flare heating and stress that models which ignore the convective influence (e.g. Strauss and Papagiannis, 1971; Landini *et al.*, 1975) are theoretically inconsistent and physically unrealistic (c.f. §§5.3.2, 5.5.2 and 5.5.3). For uniform initial conditions ($T \approx 2 \times 10^6$ K, $n = 10^{16}$–10^{17} m^{-3}) and symmetric, localised ($\simeq 10^7$ m) energy deposition, it is shown that *a sizeable fraction of the flare energy* ($\sim 50\%$) *is carried off by outward propagating pressure waves;* these develop on the hydrodynamic time-scale of the hot plasma and rapidly steepen into shocks (Mach 2–3; c.f. §5.3.2). Apart from greatly enhancing the rate of energy propagation through space, convection is found to be important in transferring energy through the moving fluid. And, in contrast to previous assumptions, (e.g. Strauss and Papagiannis, 1971; Shmeleva and Syrovatsky, 1973) the energy loss by *plasma expansion* at the temperature maximum ($T \simeq 25 \times 10^6$ K) is generally comparable to conduction cooling.

McClymont and Craig (1978) have also developed a *two-temperature (electron-proton) model* to investigate the detailed coronal response to collisional heating by non-thermal electrons (§5.3.4). Although the increased conductive efficiency — due to the hotter electron gas — tends to drive the shock considerably faster, the gross thermal evolution of the loop is not dramatically affected.

The main drawback of the McClymont–Craig analysis is the total neglect of chromospheric effects. Loosely speaking, this renders the calculation invalid for times in excess of the sound travel-time along the loop. At best, therefore, the analysis can only be *realistic for the initial two or three minutes* of the flare evolution. Assuming it may be applied to the *initial* flare decay, however, the model predicts a rapid decline in the peak temperature of the plasma accompanied by a rise in the amplitude of the differential emission measure (see Zaumen and Acton, 1974; §5.5.4).

The original McClymont–Craig analysis has now been extended to include the interaction of thermal waves in the transition region and chromosphere, and a probabilistic description (Athay 1972) of the optically-thick radiative losses. Some preliminary results are presented in the following section.

Somov and Syrovatsky (1976) have recently emphasised the role of *thermal instability* in the gas dynamics of the flare loops. They show — at least within the limited framework of Field's linear theory — that the temperature profiles derived by Shmeleva and Syrovatsky (1973) are stable against perturbations having the fastest growth rate. But the same cannot be said for models which equate, point by point, the local radiative losses with steady-state electron heating. In this case the conduction time-scale is several orders of magnitude greater than the radiative (and also the dynamical) time-scale. Consequently, Somov and Syrovatsky contend that *powerful heating by non-thermal electrons leads by thermal instability to the formation of cool, dense condensations*, which strongly modify the gas-dynamic behaviour of the flare. In their view this invalidates the Kostyuk and Pikelner model.

Somov *et al.* (1977) have extended the qualitative arguments of the previous study. A *two-temperature* description is used to model the response of the deep chromosphere ($T = 6700$ K) to dynamic beam heating ($F_e \simeq 10^8$ W m^{-2}). The initial isothermal atmosphere is in hydrostatic equilibrium; temperature gradients and fluid velocities vanish in the low chromosphere ($\zeta_{max} = 2 \times 10^{25}$ m^{-2}); and the corona is represented as a freely expanding upper boundary.

As with Brown (1973) and Kostyuk and Pikelner (1975), flare plasma above a certain temperature ($\sim 10^5$ K) rapidly expands into the corona achieving shock velocities of $\sim 10^6$ m s^{-1}. Yet, for the deeper layers, a *thermal condensation forms ahead of a downward propagating shock* (Mach ≈ 10). The shock decelerates rapidly as it encounters increasing gas densities and by the end of the flare simulation ($t = 13$ s) it has slowed up completely. Even ignoring the inaccurate radiative losses, this short simulation times makes the model very difficult to test against observational data. None the less, the analysis is quite successful in driving home the radiative-dynamical coupling in the low chromosphere.

Finally, we should mention that Bloomberg *et al.* (1977) have also pointed out that convective behaviour can dramatically affect the emission characteristics of EUV lines (c.f. David *et al.*, 1977). The calculation is based on the assumption of mono-energetic beam-deposition for a single-temperature source plasma and is restricted to the first 25 seconds of the flare evolution. Since no new convective phenomena are involved we will not present a detailed discussion here (see Brown and Smith, 1980).

5.6.3 New results

A. The dynamic corona To illustrate the detailed hydrodynamic behaviour of a thermal loop we first of all examine the dynamic formation of a complete quasi-static atmosphere. The boundary conditions we adopt are the usual ones, based on loop symmetry and vanishing fluid velocity and temperature gradient at the loop base (§5.3.3). For ease of exposition we *neglect gravitational acceleration* and, apart from a zeroth order correction, we *ignore the finite optical opacity* of the chromosphere. Computations including these effects are available however; they indicate that the present simplifications in no way invalidate our central conclusions.

The equations to be investigated, then, are the continuity equation,

$$\frac{\mathrm{D}}{\mathrm{D}t}\rho = -\rho\frac{\partial v}{\partial z},$$

the momentum equation in the absence of gravity,

$$\rho\frac{\mathrm{D}v}{\mathrm{D}t} = \frac{-\partial p}{\partial z},$$

and the energy equation

$$\frac{3}{2}\rho\frac{\mathrm{D}}{\mathrm{D}t}(p/p) = \frac{\partial}{\partial z}\left(\kappa\frac{\partial T}{\partial z}\right) - p\frac{\partial v}{\partial z} - n_e^2 Q(T) + H.$$

As the initial atmosphere is to a large extent arbitrary, we assume the following uniform initial conditions

$$T = 6.0\times 10^4\ \mathrm{K},$$
$$n_e = 1.0\times 10^9\ \mathrm{cm}^{-3}\ (=10^{15}\ \mathrm{m}^{-3}). \tag{5.56}$$

The loop is supported against the radiative losses by a quiescent energy input which is assumed *constant per unit mass*. A transient gaussian pulse of half-width σ —which is short compared to the loop length l—is then used to heat the quiescent atmosphere; we adopt the values

$$\begin{aligned} \sigma &= 5.0\times10^{8}\ \text{cm}\ (=5\times10^{6}\ \text{m}),\\ l &= 3.3\times10^{9}\ \text{cm}\ (=3.3\times10^{7}\ \text{m}). \end{aligned} \tag{5.57}$$

The pulse is sinusoidal in time, reaching a maximum amplitude of 10^{-2} ergs $\text{cm}^{-2}\ \text{s}^{-1}=(10^{-5}\ \text{W m}^{-2})$ after 100 s; explicitly

$$H=H_0 \sin\left(\frac{\pi t}{\tau}\right)\exp\left(\frac{-z^2}{2\sigma^2}\right)\theta(\tau-t), \tag{5.58}$$

with $\tau=200$ s. The heating amplitude is chosen so that the material in the centre of the loop is "lifted" into the domain of radiative instability ($T\geqslant10^5$ K). When the pulse has switched off ($t>200$ s) the atmosphere is allowed to relax dynamically towards a new quasi-steady state. Figures 5.10 to 5.13 show the detailed loop evolution.

The dynamic response proceeds as follows. During the first minute or so of the transient energy deposition the plasma is heated at approximately constant density. As soon as the central temperature exceeds 10^5 K the local radiative condensation ($T\simeq2\times10^4$ K; $n_e\simeq10^{11}\ \text{cm}^{-3}=10^{17}\ \text{m}^{-3}$); the local conduction is able to quash the instability at approximately 2×10^6 K. However, the enhanced thermal energy density in the "corona" then drives a pressure wave outwards into the cooler material. By 280 s we see from Figure 5.10a that sufficient mass has "piled up" in this region to form a cool, dense radiative condensation ($T\simeq2\times10^4$ K; $n_e\simeq10^{11}\ \text{cm}^{-3}=10^{17}\ \text{m}^{-3}$); the local density is enhanced almost two orders of magnitude beyond its initial quiescent value of $10^9\ \text{cm}^{-3}$. As the pressure wave advances into the undisturbed plasma the radiative condensation gradually develops into a loop chromosphere. Material immediately ahead of the wavefront, which at first is compressionally heated (see Figure 5.10a), begins to cool rapidly by radiation due to the subsequent density enhancement. In fact by 400 s the condensation has almost completely "expanded" into the undisturbed atmosphere.

After 500 s the loop structure is virtually fully-developed: coronal and chromospheric regions in Figure 5.10b are separated by a classical narrow transition layer ($\sim10^7$ cm thick). Chromospheric material is now flowing upwards through the transition region enhancing the coronal temperature and pressure. This is a *dynamic evaporation* of the chromosphere. Figure 5.10c shows that by 800 s the coronal pressure has built up significantly to bring about a further velocity reversal at the origin. Yet, while complex velocity

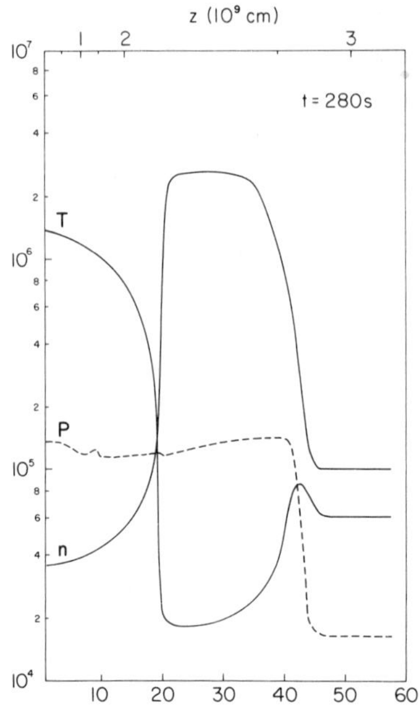

(a)

FIGURE 5.10(a) Evolution of the dynamic corona after 280 s. The initially uniform atmosphere ($T=6\times10^4$ K, $n_e=10^9$ cm^{-3}) is beginning to form a coronal-transition zone structure as the pressure wave propagates outward: the density of the newly formed chromosphere is enhanced by almost two orders of magnitude above that of the surrounding plasma. The half-length of the loop is 3.3×10^9 cm and the development, which is symmetric about the apex ($z=0$), is plotted against both Eulerian (z) and Lagrangian coordinates (ζ). Units are T(K), n_e(10^4 cm^{-3}) and pressure (10^{-6} dyne cm^{-2}).

profiles are still possible, the loop is now able to relax through relatively minor adjustments in the thermal structure.

The evolution of the velocity profile is illustrated in Figure 5.11. Although the initial trend is for plasma to be expelled from the flare source, a local "suck-back" of material due to a temporary reduction in the coronal pressure is evident after ~300 s. The chromospheric velocity has reversed completely after 500 s but coronal material is still moving downwards: from Figure 5.11a we observe that coronal speeds are generally a factor of six or seven enhanced over chromospheric values, indicating the approximate equipartition of kinetic energy. But, even as mass flows begin to die down, there is little reduction in the complexity of the velocity profile: downward flowing transition zone material in Figure 5.11b at 1020 s separates upward flowing coronal and chromospheric plasma.

Figure 5.12 illustrates the evolution of the temperature and density at the top of the loop. Quasi-static equilibrium is attained after ~2000 s, by which

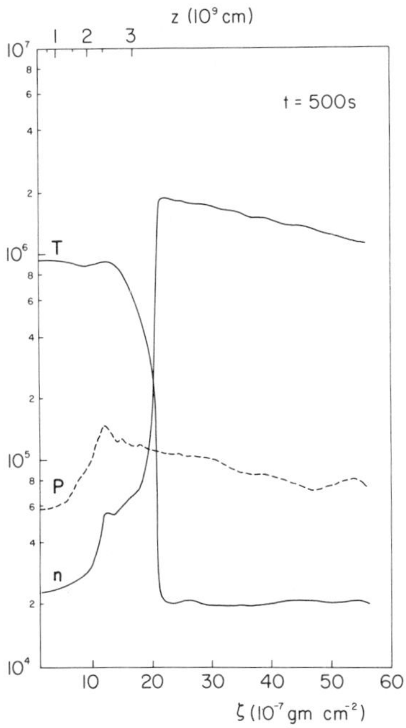

FIGURE 5.10(b) Density, temperature and pressure profiles after 500 s. Material is now flowing up through the transition layer and piling into the corona. This is the *dynamic evaporation* of chromospheric plasma. Note how chromospheric speeds are now lower than coronal and transition zone speeds (Figure 5.11a).

time the hydrodynamic oscillation of the loop is almost absent. Indeed, the final coronal parameters show excellent agreement with the T_0, n_0, l scaling law of Equation (5.32).

The thermal profile represented by Figure 5.12 is evidently characterised by several physical time-scales: the rise-time of temperature and pressure is determined mainly by the heating time-scale ($\sim 10^2$ s), while the envelope of the decay curve can be explained, quite naturally, by conductive and radiative damping; for $T \simeq 10^6$ K and $n_e \simeq 3 \times 10^8$ cm^{-3} (3×10^{14} m^{-3}) an e-folding time of $\tau_{\mathrm{cond}} \simeq \tau_{\mathrm{rad}} \simeq 1200$ s is implied. Evidently, the more rapid quasi-periodic oscillations are due to compressions and rarefactions that occur on the hydrodynamic time-scale (~ 200 s) of the coronal plasma.

A global picture of the hydrodynamic relaxation is shown in Figure 5.13. It is clear that the quasi-steady state is established once the ratio of the global kinetic to thermal energy of the plasma has decreased from $\sim 10^{-1}$ to 10^{-4}, while the complex transient behaviour at earlier times is replaced by a simple oscillation with period corresponding to the acoustic propagation time along the loop (~ 300 s).

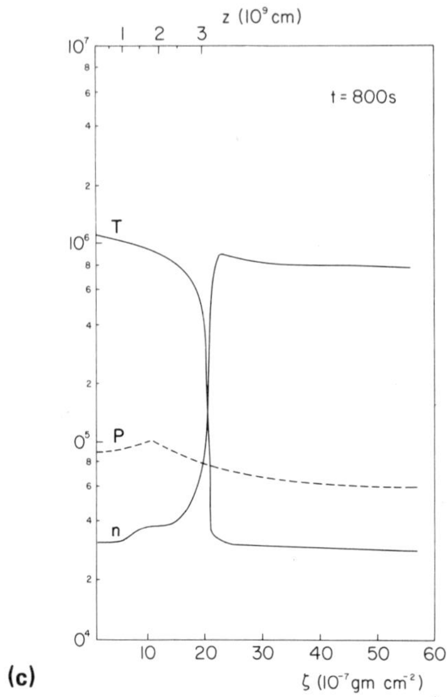

FIGURE 5.10(c) Profiles after 800 s. The pressure gradient has once again reversed and material is now flowing into the chromosphere.

It is instructive to ask to what extent the above dynamic solution approximates a truly static coronal structure. Suffice it to say that we have constructed a static solution of the same length-scale and column mass as above which approximates the dynamic model (after some 5000 s evolution) to better than 10% in the coronal and chromospheric temperature and density. As discussed below, however, it is far more difficult to maintain comparable accuracy in a typical flare heating calculation.

B. The effect of flare energisation To gain insight into the possible effects of flare heating we now investigate the response of the fully-formed quasi-static loop to a further injection of energy at 1600 s. (The dynamic heating H remains the same as before.) In this case a substantial fraction of the energy flux emerging from the coronal source is reflected back off the denser material in the transition zone and chromosphere. The net effect, as illustrated by the dotted lines in Figure 5.12, is a *dynamic "evaporation" of chromospheric material*. Thus the coronal density, which is initially diminished by motions down the pressure gradient, becomes rapidly enhanced by the initial backflow of relatively dense material through the transition zone. Likewise, the pressure achieves its maximum value after the transient heating has ceased, almost

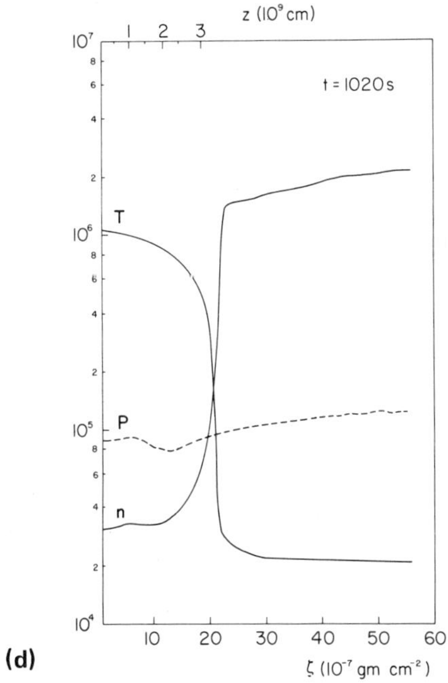

FIGURE 5.10(d) Profiles after 1020 s. Although the thermal structure is now close to its final quasi-static equilibrium the velocity profile is still quite complex (Figure 5.11b).

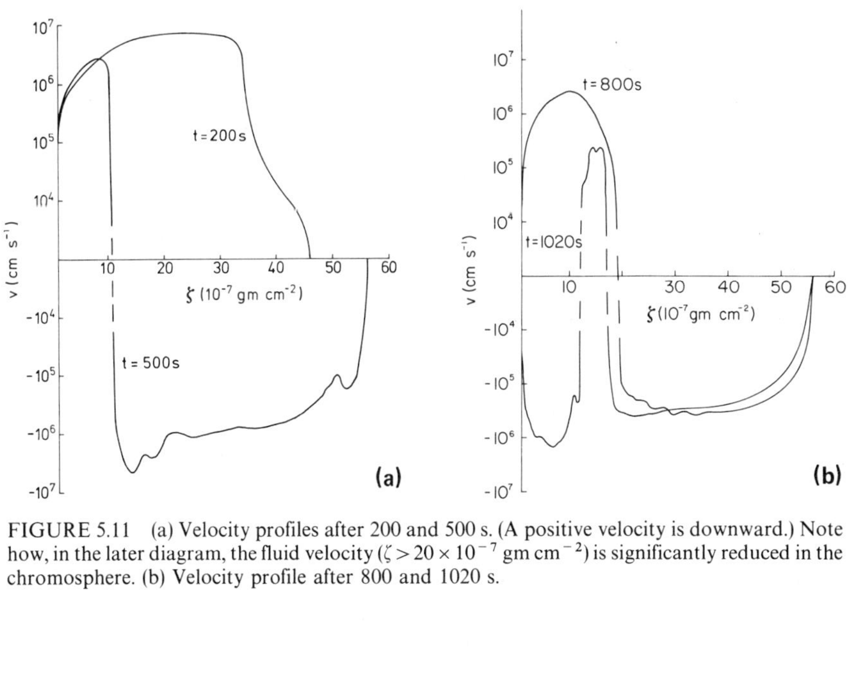

FIGURE 5.11 (a) Velocity profiles after 200 and 500 s. (A positive velocity is downward.) Note how, in the later diagram, the fluid velocity ($\zeta > 20 \times 10^{-7}$ gm cm^{-2}) is significantly reduced in the chromosphere. (b) Velocity profile after 800 and 1020 s.

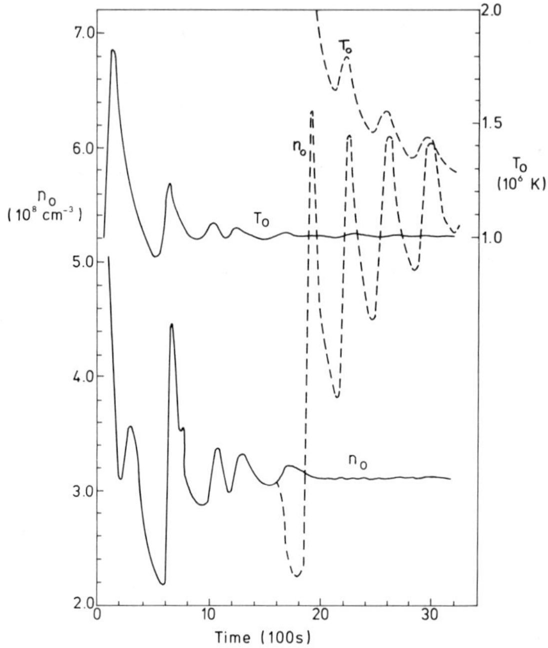

FIGURE 5.12 Relaxation of the central coronal density and temperature (n_0, T_0). Dotted lines show the response of the loop to a further injection of energy after 1600 s.

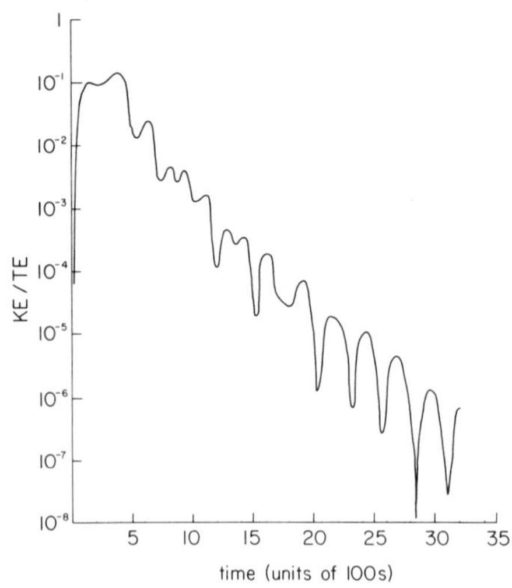

FIGURE 5.13 This shows the ratio of the kinetic to thermal energy of the loop. After 1800 s the quasi-steady equilibrium is attained and a simple hydrodynamic oscillation is set up.

concurrent with the density maximum. In contrast, the coronal temperature declines steadily throughout the initial relaxation. However, the fact that the loop settles down to a higher temperature state than before is probably spurious, a manifestation of numerical inadequacy (see below).

An important feature of this analysis is the superposition of the hydrodynamic oscillation of the loop on the much slower variation of temperature and density. Similar effects are revealed by the independent flare heating calculations of McClymont and Canfield. Figure 5.14 is based on their work and shows the response of an idealised static loop (of half-length 3×10^9 cm $= 3 \times 10^7$ m) to an impulsive injection of 3×10^9 erg cm^{-2} (3×10^6 J m^{-2}) at the apex. The calculation represents only a very small flare since the peak coronal temperature reached is below 5×10^6 K. Initial temperature and density at the top of the loop are 2×10^6 K and 10^9 cm^{-3} (10^{15} m^{-3}), respectively. In this example gravity is "switched-off" so that a density of 10^{11} cm^{-3} (10^{17} m^{-3}) is associated with the chromospheric temperature of 2×10^4 K. The deep chromosphere is represented, as in the previous calculation, by a rigid insulating wall. Also an *empirical opacity* has been introduced to reduce the chromospheric radiative losses to a realistic level. The vertical axis of Figure 5.14a is a non-linear Lagrangian (column-mass) coordinate; here height and distance refer to the initial position of a fluid element in the undisturbed atmosphere; hence the positions of features are shown relative to fluid particles rather than physical space. Velocity contours are shown at 1, 2, 5, 10, 20 and 50 km s^{-1}, positive velocities being directed away from the top of the loop (i.e. downward). Broken lines denote points of zero velocity.

Several results are immediately apparent. Firstly, the initial oscillations of the atmosphere show transient behaviour, but after about 10 minutes the oscillations settle down to a steady state. Secondly, the oscillations are heavily damped; the peak velocity decays on a time-scale of ~ 6 minutes, comparable to the oscillation period of ~ 5 minutes. The damping-time also agrees with the conduction damping time-scale τ_{cond} (Equation 5.12) for a coronal temperature of 2.5×10^6 K, which is not unrepresentative of its value during the decay ($n_e = 10^9$ cm$^{-3} = 10^{15}$ m^{-3}, and $l = 3 \times 10^9$ cm $= 3 \times 10^7$ m). Since gravity is not represented in this calculation, coincidence of the oscillation period with that of the ubiquitous 5 minute solar oscillation must be regarded as fortuitous. In fact, the fundamental acoustic resonance period of the loop for symmetric modes of oscillation is around 6–7 minutes, in reasonable agreement with the numerical results. It is interesting to note that chromospheric velocities are $\approx 1/3$ as large as those in the corona. Since the chromospheric density is 100 times larger than the coronal density, while the chromospheric height is only $\approx 1/15$ of the loop half-length, approximate equipartition of kinetic energy ($\frac{1}{2}\rho v^2 l$) between the corona and chromosphere is indicated.

Let us now examine the initial evolution in more detail. Coronal material is

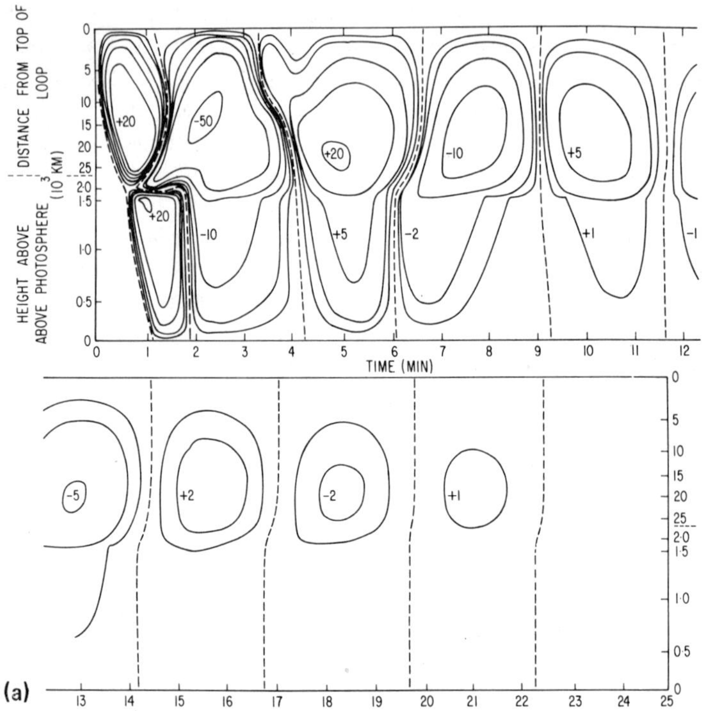

FIGURE 5.14 The dynamical response of a coronal loop to the injection of 3×10^6 Jm^{-2} (3×10^9 erg cm^{-2}) at its apex (unpublished results of McClymont and Canfield, 1979). (a) Contours of constant velocity are plotted at 1, 2, 5, 10, 20 and 50 km s^{-1}, positive velocities being directed outwards and downwards from the top of the loop. Broken lines show points of zero velocity. The vertical scale is a nonlinear Lagrangian coordinate denoting the *initial* position of the fluid elements.

driven down the legs of the loop by the "explosion" at the apex, reaching a peak velocity of ≈ 40 km s^{-1} near the end of energy injection ($t = 30$ s). After one minute this material collides with the dense chromosphere, the upper part of which is thrust downwards at 20 km s^{-1} while coronal material begins to reflect upwards again. The wedge of upward moving material is seen to grow in extent, until, at $t \simeq 1.5$ minutes, the whole corona is moving upwards. Note the material in the upper part of the loop actually reversed direction slightly earlier due to decreasing pressure at the loop apex. Slightly later, the chromosphere reverses direction *en masse*, reaching a peak upward velocity of ≈ 10 km s^{-1} after 2.5 minutes. Meanwhile, coronal plasma is beginning to slow down after reaching a peak upward velocity of 50 km s^{-1}. Around 3.5 minutes, the next velocity reversal, caused by reflection at the top of the loop, occurs. While this velocity reversal propagates down to the chromosphere, a

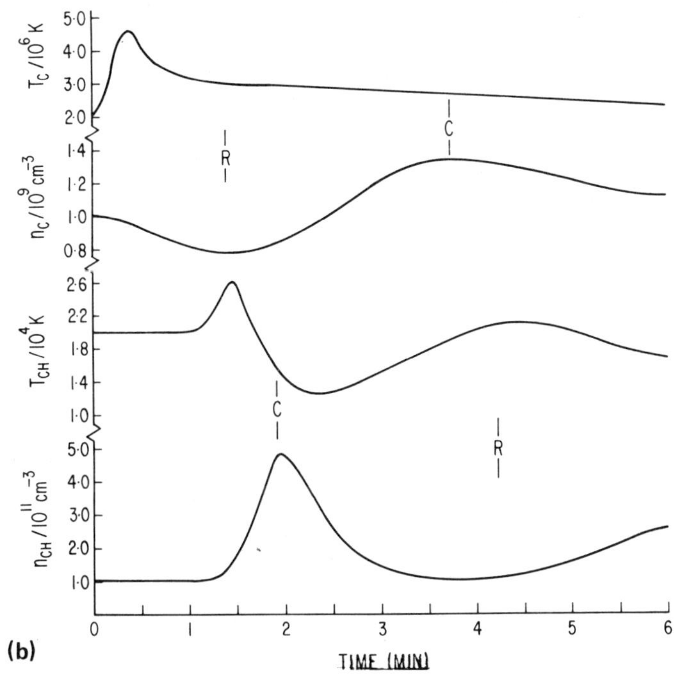

(b) Initial evolution of temperature and density at the top of the loop (T_c and n_c) and at the base of the chromosphere (T_{ch} and n_{ch}). Times of maximum compression and rarefaction at these points, corresponding to velocity reversals, are denoted by C and R, respectively.

small transient oscillation is seen at the top of the loop where material slows and almost moves upwards again for a brief period. From this time onwards, transient effects gradually die out and the oscillation appears to settle down to a fundamental eigenmode.

Figure 5.14b shows the initial evolution of temperature and density at the top of the loop (T_c and n_c) and at the base of the chromosphere (T_{ch} and n_{ch}). Times of maximum compression and rarefaction at these points, corresponding to velocity reversals, are denoted by C and R respectively on the diagram. As coronal material is driven out of the energy injection site, the density at the top of the loop falls from 10^9 cm^{-3} (10^{15} m^{-3}) to under 8×10^8 (8×10^{14} m^{-3}) at the time of maximum rarefaction, that is, when the velocity reverses at 1.5 minutes. At maximum compression (the next velocity reversal) the coronal density rises to a peak of 1.3×10^9 cm^{-3}. In the chromosphere the increase in density is more pronounced; at the peak of compression ($t\simeq2$ m) the density rises from 10^{11} cm^{-3} (10^{17} m^{-3}) to nearly 5×10^{11} cm^{-3}. At the next velocity reversal the density drops back to $\sim10^{11}$ cm^{-3} (but no further) before rising to its next peak of 2.7×10^{11} cm^{-3} at $t\simeq6$ minutes.

The variation of temperature at the top of the loop, being heavily damped by conduction, is not so spectacular. At the end of heating, a peak coronal temperature of 4.6×10^6 K is attained. The temperature then decays monotonically with a very small oscillatory component superposed. This oscillation represents evidence of compressional (quasi-adiabatic) heating; the corona cools slightly slower during compression and slightly faster during rarefaction. In the chromosphere, which is not subject to conductive damping, temperature changes are larger. Here compressional heating seems to compete with the enhancement of radiative losses at higher density. The temperature at the base of the chromosphere, initially 2×10^4 K, peaks at 2.6×10^4 K after $t = 1.5$ min and before maximum compression is reached. As the density increases further the temperature begins to fall in response to the increased radiative losses, reaching a minimum of 1.2×10^4 K at 2.3 minutes, approximately 0.4 minutes after the density peaks. The next temperature peak, 2.2×10^4 K, coincides with maximum rarefaction at $t \simeq 4.3$ minutes, suggesting that compressional heating is important only in the initial, highly transient phase. Excursions of temperature and density from their steady state values decrease along with the velocity as the atmosphere relaxes back to a static equilibrium.

C. Interpretation of results The previous numerical modelling has illustrated several points of physical interest. In the first place, we have seen that a dynamic loop of fixed mass which, apart from radiative losses, is energetically isolated from its surroundings, will relax on the conductive and radiative timescales to a quasi-steady equilibrium. Natural hydrodynamic oscillations of the loop are superposed on this relaxation. The oscillation period (first symmetric mode) is approximately

$$\tau \simeq 2l/c \simeq 10^{-4}\ l/T^{1/2}\ \text{s},$$

where l is the loop half-length (in cm), c is the coronal sound speed and T the coronal temperature. We would expect, therefore, for both active-region and flare plasmas, variations in coronal emission of approximately this period. Of course, if an assemblage of coronal flux tubes is being observed, these oscillatory effects may be smeared out and rendered indiscernible in the data (see §5.5.6). Contrary to the speculation of Antiochos (1979), we see no evidence of high frequency oscillations ($\tau \sim 1$ s) in the lower transition zone due to effects of radiative instability. However, we agree with Antiochos in that, despite thermal instability, quasi-static models should describe the gross features of quiescent coronal loops.

Of relevance to the solar flare, we have seen that, even for small events (hardly more than transient loop brightenings), dynamic effects can be of the

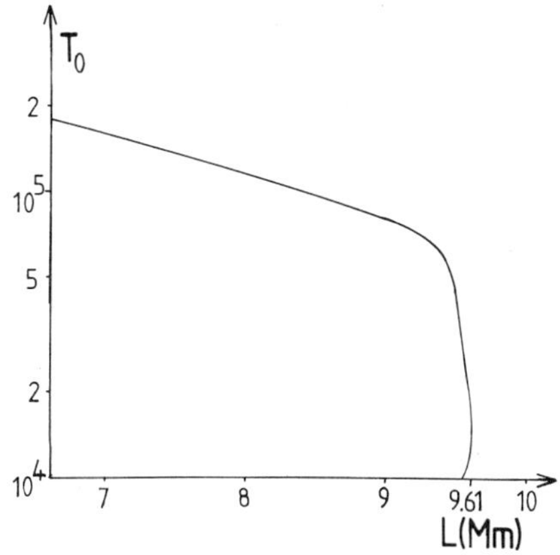

FIGURE 5.15 The base temperature T_b as a function of the half-length $L$of a static, thermally-isolated loop having a given mass and heating. In this case, the mass and heating are such that a temperature of 6×10^4 K and a density of 10^{15} m^{-3} give a uniform state. (From Wragg, 1980.)

utmost importance. In particular, the transition zone and chromosphere respond dynamically to enhanced coronal flux by driving mass into the upper atmosphere; also the ensuing hydrodynamic oscillations give rise to compressions and rarefactions that strongly influence the loop evolution.

Yet, there is one respect in which the previous "flare-analysis" may be misleading: the trend of the temperature and density profile after flare heating (Figures 5.12 and 5.14b) suggests that the atmosphere may not relax to its previous quasi-steady state; rather, it appears to relax to a new higher energy equilibrium. This implies that an infinite family of steady-state solutions is possible, all with the same stationary heating but differing in thermal energy content. On the other hand analytic arguments indicate that the quasi-static equations introduced in §5.4 admit only two solutions, the original isothermal state and one "corona-chromosphere" state. (A second isothermal state due to bremstrahling losses at $T \simeq 10^{10}$ K is also possible in principle.) It follows that this discrepancy is probably due to the poor finite-difference representation of the flare transition-zone: the relatively poor spatial resolution in this region prevents the temperature gradient from steepening as much as it should, effectively bottling up the coronal conductive flux. Thus when equilibrium is reached in the (new spurious) numerical solution (Figure 5.12) the corona is too hot and the chromosphere too cool.

The suspicion that the second solution may be spurious has been substantiated by Wragg (1980), who has determined the thermally-isolated, static solutions that are possible for a loop of constant mass and heating. The resulting base temperature T_b for one particular mass and heating is shown in Figure 5.15 as a function of the loop half-length L. It can be seen that static solutions are possible only when $L < 9610$ km and that, for most such half-lengths, only one non-uniform solution is possible; only for an extremely small range of loops $9550 < L < 9610$ km are there two solutions (with $T_b > 10^4$ K).

That an accurate description of the temperature profile in solar flares represents a particularly severe numerical problem can be seen from the condition that the finite-difference grid-spacing should not exceed the temperature scale-height of the plasma viz.,

$$\Delta z \leqslant \kappa_0 T^{7/2}/F_c(T).$$

Thus a minimum grid-spacing of 10^6 cm (10^4 m) through the flare transition zone (typical of a Lagrangian scheme with a few hundred grid points) would imply (taking $T \simeq 10^5$ K) that heat fluxes in excess of $\sim 10^6$ erg cm^{-2} s^{-1} (10^3 W m^{-2}) cannot be accurately described. In the previous section we restricted our attention to relatively weak coronal fluxes ($< 10^7$ erg cm^{-2} s^{-1}) in an attempt to minimise the non-physical effects that arise through poor numerical resolution in the flare transition zone. The same, however, cannot be said of other numerical flare studies (Kostyuk and Pikelner 1975; Somov *et al.*, 1977; Nagai, 1980). Consequently, we recommend that numerical results obtained by previous authors should be critically re-examined to determine the extent to which they model the flare transition zone in a realistic manner.

5.7 SUMMARY: THEORY AND OBSERVATION REVISITED

At present it is probably unrealistic to expect detailed agreement between observational data and theoretical loop models. Even within the confines of a one-dimensional treatment and in the absence of gravity, the loop evolution can be rather complex, the detailed thermal structure depending on a multitude of *poorly prescribed input parameters*, in particular those specifying the initial atmosphere and the quiescent and dynamic heating mechanisms (§5.6). There are certain effects, moreover, which have not yet been adequately studied. The *dynamic-radiative coupling* at high flare densities ($\simeq 10^{18}$ m^{-3} $= 10^{12}$ cm^{-3} in the corona) falls in this category, as does the influence of collective plasma effects on the initial flare energisation (§5.3.4). In addition, the *magnetic field* has been considered only to the extent that it defines the geometry of the flaring plasma: no account is taken of the magnetic influence on the energy-deposition mechanism and all first-order MHD effects have been ignored. By contrast, for the "post"-flare loop phenomenon in two-ribbon flares, the magnetic effects have so far been fairly well treated (Chapter 7), but at the expense of largely neglecting the thermal effects: in future, it should be possible to couple the magnetic effects (in the two-ribbon flare) with the physical ideas of this chapter.

As Brown and Smith (1980) have mentioned, the relationship between hydrodynamic loop models and observational data often appears obscure. The observations themselves, however, are often difficult to interpret and sometimes rather surprising. The results of Underwood *et al.* (1978), for instance, indicate a high coronal density ($\simeq 10^{18}$ m^{-3} = 10^{12} cm^{-3} at the flare peak) and a differential emission measure that increases very sharply with temperature from 10^6 to 10^7 K; yet earlier results would indicate a fall-off of the emission measure with temperature and certainly lower plasma densities (e.g. Dere *et al.*, 1974; Figure 5.2a). With such strong variations in the temperature from 10^6 to 10^7 K; yet, earlier results would indicate a fall-off of to expect a single hydrodynamic calculation, no matter how sophisticated, to explain the details of a given data set.

In marked contrast to earlier studies of the coronal flare, Underwood *et al.* conclude that their event is probably *radiatively dominated.* However, their demonstration that both static and evaporative conductive models are incompatible with the data is more easily anticipated. Since the dynamic heating is highly impulsive in this event ($\leqslant$100 s), and since all time-scales are of order 10^2 s ($l \simeq 10^7$ m), it is natural to expect complex *hydrodynamic behaviour* to develop. Hence mass flows powered by the primary energy deposition and driven further by the uneven radiative cooling (perhaps instability) are likely to play a prominent role in the initial flare decay. In this case it is clear that simple conductive models based on the assumption of uniform pressure (§5.5) can never provide an adequate description of the burst: a thorough understanding probably requires a detailed numerical simulation of the radiative-dynamic coupling in the loop.

There has been some progress, however, in our *general* understanding of both dynamic and quasi-static loops. We have seen that even a highly dynamic plasma will naturally evolve a corona-transition region structure very early in its evolution, almost independent of the form of the energy deposition (§§5.4 and 5.6). It follows that the thermal structure of a flare loop differs from the quasi-static atmosphere in two crucial respects: firstly, the temperature gradient through the flare transition-zone must initially steepen due to the temporarily enhanced heat flux emanating from the overlying coronal regions; secondly, and more importantly, the thermal profile will be altered by the *dynamic exchange of mass between the coronal and chromospheric regions* which is driven by pressure gradients resulting from the primary flare energisation in addition to the evaporative effects of thermal conduction. As the loop cools and the hydrodynamic oscillations diminish — due to conductive and radiative damping — the flare will relax towards a quasi-static equilibrium ($\xi \sim T^{\alpha}$ where $0 < \alpha < \frac{1}{2}$). This would account quite naturally, for the generally similar EUV emission measure profiles in both quiescent and flaring plasmas (Wood and Noyes, 1972; Jordan 1976; Švestka, 1976). However, the emission measure functions derivable from X-ray data *cannot in general be associated with the physics of a single loop or flare kernel* (§5.5.6): unless the data are highly resolving spatially, the differential emission measure is determined more by the instantaneous distribution of peak loop temperatures in the flare than by the thermal structure of any particular component in the source.

Acknowledgements

The author is grateful to Sandy McClymont for reading a half-completed manuscript. Any barbarisms that remain, of course, are entirely the author's responsibility. Discussions with John Brown have also been much appreciated. Ken Dere and John Vesecky kindly supplied three of the Figures at short notice.

Thanks are due also to many of the participants at the Skylab Solar Flare and Active Region Workshops and to the Science Research Council for financial support.

Finally, the author would like to thank Dick Canfield and Sandy McClymont for providing a detailed analysis of their unpublished computations.

References

Antiochos, S. K., *Ap. J.* **232**, L125 (1979).
Antiochos, S. K. and Sturrock, P. A., *Solar Phys.* **49**, 359 (1976).
Antiochos, S. K. and Sturrock, P. A., *Ap. J.* **220**, 1137 (1978).
Antiochos, S. K. and Underwood, J. H., *Astron & Astrophys.* **68**, L19 (1978).
Antiochos, S. K. and Krall, K., *SUIPR Report* 751 (1978).
Athay, R. G., *Ap. J.* **176**, 659 (1972).
Bloomburg, H. W., Davis, J. and Boris, J. P., *J.Q.S.R.T.* **18**, 237 (1977).
Brown, J. C., *Solar Phys.* **18**, 489 (1971).
Brown, J. C., *Solar Phys.* **26**, 441 (1972).
Brown, J. C., *Solar Phys.* **31**, 143 (1973).
Brown, J. C., *Solar Gamma, X- and EUV Radiation, IAU Symp.* **68**, 245 (1975).
Brown, J. C., Melrose, D. B. and Spicer, D. S., *Ap. J.* **228**, 592 (1978).
Brown, J. C., Canfield, R. C. and Robertson, M. N., *Solar Phys.* **57**, 399 (1978).
Brown, J. C. and Smith, D. F., *Rep. Prog. Phys.* **43**, 125 (1980).
Brown, J. C., Craig, I. J. D. and Karpen, J., *Solar Phys.* **67**, 143 (1980).
Canfield, R. C., *Solar Phys.* **34**, 339 (1974).
Canfield, R. C. *et al.*, Chapter 6 of *Solar Flares*; P. Sturrock (ed.) Colo. Ass. Univ. Press (1980).
Chambe, G., *Astron. & Astrophys.* **12**, 210 (1971).
Cheng, C. C., *Solar Phys.* **23**, 178 (1971).
Cox, D. P. and Tucker, W. H., *Ap. J.* **157**, 1157 (1969).
Craig, I. J. D., *Astron. & Astrophys.* **61**, 575 (1977).
Craig, I. J. D., *Atron. & Astrophys.* **79**, 121 (1979).
Craig, I. J. D., Culhane, J. L., Phillips, K. J. H. and Vesecky, J. F., in *High Energy Phenomena on the Sun* R. Ramaty and R. G. Stone (eds), p. 276 (1973).
Craig, I. J. D. and Brown, J. C., *Nature* **264**, 340 (1976a).
Craig, I. J. D. and Brown, J. C., *Astron. & Astrophys.* **49**, 239 (1976b).
Craig, I. J. D. and McClymont, A. N., *Solar Phys.* **50**, 133 (1976).
Craig, I. J. D. and McClymont, A. N., Unpublished Report (1978).
Craig, I. J. D., McClymont, A. N. and Underwood, J. H., *Astron. & Astrophys.* **70**, 1 (1978).
Crannell, C. J., Frost, K. J., Mätzler, C., Ohki, K. and Saba, J. L., *Ap. J.* **223**, 620 (1978).
Culhane, J. L., Vesecky, J. F. and Phillips, K. J. H., *Solar Phys.* **15**, 396 (1970).

Datlowe, D. W., Hudson, H. S. and Peterson, L. F., *Solar Phys.* **39**, 193 (1974).
Davis, J., Kepple, P. C. and Strickland, D. J., *J.Q.S.R.T.* **17**, 711 (1977).
Dere, K. P. and Cook, J. W., Preprint (1978).
Dere, K. P., Horan, D. M. and Kreplin, R. W., *Solar Phys.* **36**, 49 (1974).
Dere, K. P., Mason, H. E., Widing, K. G. and Bhatia, A. K., *Ap. J. Suppl.*, submitted (1979).
Doschek, G. A. and Feldman, U., in Canfield *et al.*, (1980).
Emslie, A. G., *Ap. J.* **224**, 241 (1978).
Emslie, A. G., *Ap. J.* **235**, 1085 (1980).
Field, G. B., *Ap. J.* **142**, 531 (1965).
Gabriel, A. H., *Nature* **267**, 410 (1977).
Gabriel, A. H. and Jordan, C., *M.N.R.A.S.* **145**, 241 (1969).
Gingerich, O., Noyes, R. W., Kalkofen, W. and Cuny, Y., *Solar Phys.* **18**, 347 (1971).
Habbal, S. R. and Rosner, R., *Ap. J.* **234**, 1113 (1979).
Hearn, A. G. *Astron. Astrophys.* **40**, 355 (1975).
Hood, A. W. and Priest, E. R., *Astron. & Astrophys.* **77**, 233 (1979).
Hood, A. W. and Priest, E. R., *Solar Phys.* (1980a).
Hood, A. W. and Priest, E. R., *Solar Phys.* **66**, 113 (1980b).
Hood, A. W. and Priest, E. R., *Astron. & Astrophys.* **87**, 126 (1980).
Hoyng, P., Brown, J. C. and van Beek, H. F., *Solar Phys.* **48**, 197 (1976).
Jefferies, J. T. and Orrall, F. Q., *Ap. J.* **141**, 519 (1965).
Jordan, C., *IAU Symp.* **68**, 101 (1975).
Jordan, C., *Phil. Trans. Roy. Soc.* **281**, 391 (1976).
Kahler, S. W., Meekins, J. F., Kreplin, R. W. and Bowyer, C. S., *Ap. J.* **168**, 531 (1970).
Kane, S. R. and Anderson, K. A., *Ap. J.* **162**, 1003 (1970).
Kaplan, S. A., Pikelner, S. B. and Tsytovich, V. N., *Physics Reports* (North Holland Publ. Co.) **45**, 1 (1974).
Kneer, F. and Nakagawa, Y., *Astron. & Astrophys.* **47**, 65 (1976).
Knight, J. W. and Sturrock, P. A., *Ap. J.* **218**, 306 (1977).
Korchak, A. A., *Solar Phys.* **18**, 284 (1971).
Kostyuk, N. D., *Soviet Astron.* **19**, 458 (1976).
Kostyuk, N. D. and Pikelner, S. R., *Soviet Astron.* **18**, 590 (1975).
Kuperus, M., *Space Science Reviews*, **9**, 713 (1969).
Landini, M. and Fossi, B. C., *Astron. and Astrophys.* **42**, 213 (1975).
Landini, M., Fossi, B. C. and Pallavicini, F., *Solar Phys.* **44**, 101 (1975).
Levine, R. H. and Pye, J. P., *Solar Phys.* **66**, 39 (1980).
Mason, H. E., Doschek, G. A., Feldman, U., Bhatia, A. K., *Astron. & Astrophys.* **73**, 74 (1979).
McClymont, A. N., in Rust, R. M. *et al.* (1980).
McClymont, A. N. and Craig, I. J. D., Unpublished Report (1978).
McClymont, A. N. and Craig, I. J. D., in Canfield *et al.*, (1980).
McWhirter, R. W. P., Thonemann, P. C. and Wilson, R., *Astron. & Astrophys.* **40**, 63 (1975).
Milkey, R. W., Blocker, N. K., Chambers, W. H., Fehlau, P. E., Fuller, J. C. and Kunz, W. E., *Solar Phys.* **20**, 400 (1971).
Moore, R. L. and Fung, P. C., *Solar Phys.* **23**, 79 (1972).
Moore, R. L. and Datlowe, D. W., *Solar Phys.* **43**, 189 (1975).
Nagai, F., *Solar Phys.* in press (1980).
Nakagawa, Y., Wu, S. T. and Han, S. M., *Solar Phys.* **30**, 111 (1973).
Neupert, W. M., *Ap. J.* **153**, L59 (1968).
Osterbrock, D. E., *Ap. J.* **134**, 347 (1961).
Parkinson, J., *Nature* **267**, 409 (1977).
Priest, E. R., *Solar Phys.* **38**, 57 (1978).
Priest, E. R., Chapter 9 *The Skylab Workshop on Active Regions*, F. Orrall (ed) (1980).
Priest, E. R. and Smith, E. A., *Solar Phys.* **64**, 267 (1979).
Raymond, J. C., Cox, D. P. and Smith, B. W., *Ap. J.* **204**, 290 (1976).
Rosner, R., Tucker, W. H. and Vaiana, G. S., *Ap. J.* **220**, 643 (1978).
Rust, R. M. and Webb, D. F., *Solar Phys.* **54**, 403 (1977).
Rust, R. M. *et al.*, Chapter 7 of *Solar Flares*, P. Sturrock (ed), Colo. Ass. Univ. Press (1980).

Shmeleva, O. P. and Syrovatsky, S. I., *Solar Phys.* **33**, 341 (1973).
Somov, B. V., *Solar Phys.* **60**, 315 (1978).
Somov, B. V. and Syrovatsky, S. I., *Sov. Phys. Usp.* **19**, 813 (1976).
Somov. B. V., Spektor, A. R. and Syrovatsky, S. I., *Izo. Acad. Sci. USSR, Phys. Ser.* **41**, 273 (1977).
Sturrock, P. A., in 'High Energy Phenomena on the Sun', R. Ramaty and R. G. Stone (eds), p. 3 (1973).
Sturrock, P. A. *et al.*, *Solar Flares*, Colo. Ass. Univ. Press (1980).
Şyrovatsky, S. I. and Shmeleva, O. P., *Astron. Zh.* **49**, 334 (1972).
Şvestka, Z., *Solar Phys.* **31**, 389 (1973).
Şvestka, Z., *Solar Flares*, D. Reidel, Dordrecht Holland (1976).
Švestka, Z., Krieger, A. S., Chase, R. C. and Howard, R., *Solar Phys.* **52**, 69 (1977).
Sweet, P. A., *Ann. Rev. Astron. Astrophys.* **7**, 149 (1969).
Tucker, W. H., *Ap. J.* **186**, 285 (1973).
Tucker, W. H. and Koren, M., *Ap. J.* **168**, 283 (1971).
Underwood, J. H. and McKenzie, D. L., *Solar Phys.* **53**, 417 (1977).
Underwood, J. H., Antiochos, S. K., Feldman, U. and Dere, K. P., *Ap. J.* **224**, 1017 (1978).
Vaiana, G. S. and Rosner, R., *Ann. Rev. Astron. Astrophys.* **16**, 393 (1978).
Vesecky, J. F., Antiochos, S. K. and Underwood, J. H., *Astrophys. J.* **233**, 987 (1978).
Vlahos, L. and Papadopoulos, D., *Ap. J.* **233**, 717 (1979).
Vorpahl, J. A., *Solar Phys.* **26**, 397 (1972).
Wentzel, D. G., *Solar Phys.* **39**, 129 (1974).
Wood, A. T. and Noyes, R. W., *Solar Phys.* **24**, 180 (1972).
Wragg, M., Ph.D. Thesis, St. Andrews University (1980).
Wragg, M. and Priest, E. R., *Solar Phys.* (in press) (1980).
Zaumen, W. T. and Acton, L. W., *Solar Phys.* **36**, 139 (1974).
Zeldovich, Y. B. and Raiser, Y. P., *Physics of Shock Waves and High-Temperature Hydrodynamic Phenomena*, Academic Press, N. York, and London (1967).

CHAPTER 6

Two-ribbon flares: magnetostatic equilibria

J. BIRN and K. SCHINDLER

Ruhr-Universität Bochum, Theoretische Physik IV, Universitätsstrasse 150, 4630 *Bochum* 1, *West-Germany.*

6.1 INTRODUCTION

One of the central problems of the theory of solar eruptive processes is to provide quantitative models of the transition from conditions of slow passive evolution to fast active evolution. This problem has a variety of different aspects such as the modelling of preflare states, their stability, onset criteria, identification of dynamic modes, growth rates and nonlinear development. In this chapter we shall concentrate on *passive evolution* and on some aspects of onset and stability. Although, to be specific, we will address the flare phenomenon directly, most of the theoretical work described may equally apply to other eruptive processes such as eruptive prominences. Indeed, whether an erupting prominence produces an Hα two-ribbon flare or not may depend on just the magnetic field strength (§1.1).

The recent development of the theory of preflare configurations has been comparatively successful in the following two ways:

— quantitative models are replacing earlier qualitative pictures;

— the theory in part applies not only to specific configurations but to rather large classes of quasi-static states.

The latter point is particularly important in view of the observational difficulties in obtaining three-dimensional information in sufficiently high spatial resolution, so that the details of the relevant configurations are usually not well-known.

Our description of the various existing models is by no means complete. We try to concentrate on those approaches that include quantitative information on *equilibrium sequences* and on *onset conditions.*

It may seem surprising that an analysis of equilibrium states alone is able to provide information on the onset conditions for a *dynamic* phenomenon.

However, it is perhaps appropriate to remind the reader that qualitative stability properties follow from mere inspection of the shape of the potential in many simple mechanical problems and that equilibrium thermodynamics provides information on conditions under which phase transitions occur. In these cases — as in our present case — existence and non-uniqueness of solutions play an important role.

One of the most stringent limitations of the present theory of eruptions is the spatial two-dimensionality assumed in almost all cases. Neglecting the variations of all physical quantities along one of the coordinate directions, one naturally concentrates on applications where a two-dimensional model may be relevant, at least in the sense of a crude approximation. One may, for instance, argue that this is the case for two-ribbon flares, where the variation along the ribbons seems to be less pronounced than in the two other directions. This is the reason why we address the two-ribbon flare explicitly in the title of this chapter, although the range of applications of the theory is possibly much larger.

6.2 BASIC EQUATIONS AND ASSUMPTIONS OF THE QUASI-STATIC THEORY

We will first discuss the basic MHD equations in a more general form to see which terms will be omitted later and to determine the conditions under which they can be omitted:

$$\frac{\mathrm{D}\rho}{\mathrm{D}t} \equiv \frac{\partial \rho}{\partial t} + \mathbf{v}\cdot \boldsymbol{\nabla}\rho = -\rho\boldsymbol{\nabla}\cdot\mathbf{v}, \tag{6.1}$$

$$\rho\frac{\mathrm{D}\mathbf{v}}{\mathrm{D}t} \equiv \rho\frac{\partial v}{\partial t} + \rho\mathbf{v}\cdot\boldsymbol{\nabla}\mathbf{v} = -\boldsymbol{\nabla}p + \mathbf{j}\times\mathbf{B} + \rho\mathbf{g}, \tag{6.2}$$

$$\mathbf{E} + \mathbf{v}\times\mathbf{B} = \frac{1}{\sigma}\mathbf{j} + \frac{1}{ne}\mathbf{j}\times\mathbf{B} - \frac{1}{ne}\boldsymbol{\nabla}p_e$$

$$+\frac{m_e}{ne^2}\left[\frac{\partial \mathbf{j}}{\partial t} + \boldsymbol{\nabla}\cdot(\mathbf{vj}+\mathbf{jv})\right], \tag{6.3}$$

$$\frac{\rho^\gamma}{\gamma-1}\frac{\mathrm{D}}{\mathrm{D}t}\left(\frac{p}{\rho^\gamma}\right) = \nabla\cdot(\boldsymbol{\kappa}\cdot\boldsymbol{\nabla}T) + \frac{j^2}{\sigma} - L_{\mathrm{rad}} + H_w, \tag{6.4}$$

$$p = \frac{R}{\tilde{\mu}} \rho T, \tag{6.5}$$

$$\nabla \times \mathbf{B} = \mu \mathbf{j}, \tag{6.6}$$

$$\nabla \cdot \mathbf{B} = 0, \tag{6.7}$$

$$\frac{\partial \mathbf{B}}{\partial t} = -\nabla \times \mathbf{E}. \tag{6.8}$$

SI-units are used and the notation is the same as in Chapter 1 (see Equations (1.1)–(1.12)). The assumptions which have already been made in stating these equations are:

— quasi-neutrality ($n_i \simeq n_e$) for a fully-ionized hydrogen plasma,
— non-relativistic flow ($v^2 \ll c^2$),
— pressure isotropy,
— neglect of terms of order m_e/m_i in the generalized form of Ohm's law (6.3),
— isotropic conductivity σ,
— neglect of the displacement current, consistent with $v^2 \ll c^2$ and quasi-neutrality.

The set of Equations (6.1)–(6.8) must be closed by appropriate assumptions on the radiative losses L_{rad}, e.g.

$$L_{\text{rad}} = n_e^2 Q(T), \tag{6.9}$$

with $Q(T)$ given, for instance, by Hildner (1974), and by assumptions on the wave heating term H_w, the heat conductivity tensor $\boldsymbol{\kappa}$ and the electrical conductivity σ. For non-turbulent conditions these are given approximately by (Spitzer, 1962):

$$\sigma = 10^{-3}\ \mathrm{T}^{3/2} \qquad (\mathrm{Ohm}^{-1}\ \mathrm{m}^{-1}), \tag{6.10}$$

$$\kappa_{\parallel} = 2 \times 10^{-11}\ \mathrm{T}^{5/2} \qquad (\mathrm{Watt\ m}^{-1}\ \mathrm{deg}^{-1}), \tag{6.11}$$

$$\kappa_{\perp} = 2 \times 10^{-31}\ \mathrm{n}^2\ \mathrm{T}^{-3}\ \mathrm{B}^{-2} \kappa_{\parallel} \qquad (\mathrm{Watt\ m}^{-1}\ \mathrm{deg}^{-1}), \tag{6.12}$$

where the heat conductivity tensor $\boldsymbol{\kappa}$ is represented by its components parallel and perpendicular to the magnetic field, $\kappa_{\parallel}$ and $\kappa_{\perp}$, respectively, and the temperature is given in degrees Kelvin.

To compare the relative importance of the different terms in equations (6.2)–(6.4) we adopt two sets of characteristic parameters, which might represent (A) coronal and (B) chromospheric conditions before the onset of a flare:

(A) $$n=5\times10^{15}\ \mathrm{m}^{-3},\ T=3\times10^{6}\ \mathrm{K};$$ (corona)

(B) $$n=10^{17}\ \mathrm{m}^{-3},\ T=10^{4}\ \mathrm{K};$$ (chromosphere)

with $B=10^{-2}$ T, $L=10^4$ km, $v=10$ km/s, and $g=270$ m/s^2 in both cases. For both sets of parameters, the magnetic pressure $B^2/2\mu$ dominates the other terms in Equation (6.2), such that approximately

$$\mathbf{j}\times\mathbf{B}=\mathbf{0}, \tag{6.13}$$

provided the time-scale is not as short as about 5 sec and the local density values do not reach about 10^{18} m^{-3} (10^{12} cm^{-3}) in the corona and 10^{21} m^{-3} (10^{15} cm^{-3}) in the chromosphere.

Equation (6.2) provides also an equation of motion parallel to **B**. In this equation, the characteristic length follows from a balance of the appropriate terms, whereas the adopted value of $L\approx10^4$ km corresponds predominantly to gradients perpendicular to the magnetic field. The balance of pressure and gravitational forces along the field yields scale-heights of about 100,000 km for the coronal parameters (A) and about 300 km for the chromospheric parameters (B), whereas a balance of the inertia and the gravitational forces implies a length-scale of about 400 km for both sets. Therefore, the convection term may be neglected in case (A), whereas it may be important in case (B), such that,

$$\mathbf{B}\cdot(\nabla p-\rho\mathbf{g})=0 \tag{6.14a}$$

in case (A) and

$$\mathbf{B}\cdot(\nabla p-\rho\mathbf{g}+\rho\mathbf{v}\cdot\nabla\mathbf{v})=0. \tag{6.14b}$$

The explicit time-dependence has been omitted in both equations, which is justified if the time-scale is larger than about 1 min. This is obviously the case for the adopted time-scale $L/v\simeq10^3$ sec, which means that the preflare structures may be treated as sequences of quasi-static (or quasi-steady in case (B)) equilibria which are predominantly force-free.

For both cases, (A) and (B), $\mathbf{v}\times\mathbf{B}$ is completely dominant over the terms on the right-hand side of Equation (6.3), so that the component perpendicular to reads approximately

$$\mathbf{E}_{\perp} + \mathbf{v} \times \mathbf{B} = \mathbf{0}. \tag{6.15}$$

The dominant terms of the parallel component yield

$$\mathbf{E}_{\|} = \frac{1}{\sigma}\mathbf{j}_{\|} - \frac{1}{ne}\boldsymbol{\nabla}_{\|} p_e. \tag{6.16}$$

For smooth non-turbulent conditions, $E_{\|}$ is considerably smaller than $E_{\perp}$, so that the ideal Ohm's law

$$\mathbf{E} + \mathbf{v} \times \mathbf{B} = \mathbf{0} \tag{6.17}$$

holds on a larger scale. Locally strong pressure gradients or anomalous resistivity may, however, enhance the relative importance of $E_{\|}$ (see also Section 6.6).

The dominant terms of Equation (6.4) are heat conduction parallel to **B** in case (A) and the radiation losses in case (B). In both cases, these terms must be compensated by the heating term to achieve equilibrium.

It must be emphasized that the above estimates hold only for variations on a large scale ($>$ 10,000 km) under non-turbulent conditions. Other terms may be important if anomalous resistivity and/or thin current sheets are involved (see Chapter 3 and also Section 6.6).

The estimates carried out earlier in this section suggest that the preflare stage is characterized by a sequence of quasi-equilibria evolving slowly due to changing boundary conditions. We restrict ourselves to large coronal flares, so that the convection term may be neglected. At any instant of time the equilibrium equations

$$-\boldsymbol{\nabla} p + \mathbf{j} \times \mathbf{B} + \rho \mathbf{g} = \mathbf{0}, \tag{6.18}$$

where

$$\boldsymbol{\nabla} \times \mathbf{B} = \mu \mathbf{j} \quad \text{and} \quad \boldsymbol{\nabla} \cdot \mathbf{B} = 0$$

must be fulfilled. Although the magnetic field must be force-free on a large scale, as discussed above, we have retained the pressure gradient and gravity terms because they are important for the equilibrium along the magnetic field lines and possibly for embedded current sheets and for the stability properties.

In general ρ and p are connected through the energy equation (6.4), which yields information on the temperature. If heat conduction parallel to the magnetic field plays the dominant role in the energy transport, it seems reasonable (as a first approximation) to replace Equation (6.4) by the assumption of constant temperature on the field lines. The force balance may be treated separately from the energy balance if the density is low enough (e.g. in the corona) that the $\rho\mathbf{g}$ may be neglected in (6.18).

In the following section we shall discuss some general existence properties which hold for two-dimensional solutions of (6.18).

6.3 EXISTENCE OF EQUILIBRIUM SOLUTIONS IN TWO DIMENSIONS

We adopt a coordinate system with x and z measured parallel to the surface of the Sun and y perpendicular to it. We assume that the configuration depends only weakly on z, so that it may be treated as independent of z in a first approximation. Large two-ribbon flares seem to be reasonably approximated by such a configuration. One can show (e.g. Birn *et al.*, 1978a) that the magnetic field can be represented by

$$\mathbf{B}=\nabla A(z,\ y)\times\mathbf{e}_z+B_z(A)\mathbf{e}_z, \tag{6.19}$$

where A represents the z-component of the vector potential (flux function) and must obey the following non-linear partial differential equation:

$$-\nabla^2 A=\frac{\partial}{\partial A}\{\mu P(A,\ y)+\tfrac{1}{2}B_z^2(A)\}\equiv J(A,\ y), \tag{6.20}$$

with the plasma pressure

$$P(A,\ y)=P_0(A)e^{-y/H(A)} \tag{6.21}$$

and scale-height

$$H(A)=RT(A)/(g\tilde{\mu}) \tag{6.22}$$

for a constant gravitational acceleration g and assuming a constant temperature on each field line. Since in a steady state there are no additional differential equations for $P(A, y)$ and $B_z(A)$, a mathematically obvious way of finding equilibrium solutions is to assume $P(A, y)$ and $B_z(A)$ as given functions and then to obtain $A(x, y)$ from the partial differential equation (6.20) for given boundary conditions. Some general properties of this problem will be discussed in this section, and then special sequences of solutions obtained in this way will be discussed in the following section.

Although the procedure described above is useful for constructing equilibrium solutions, it does, however, not in all cases appear to be the best way of modelling the actual time-evolution. In fact, sometimes it is *preferable to prescribe the position of footpoints* of field lines rather than the sheared magnetic field component B_z, the reasons being that, in some cases, the Sun is more likely to do this and a tangential component of the magnetic field cannot be fixed arbitrarily (Jockers, 1978; Priest and Milne, 1980). We shall return to this problem at the end of this section and more generally in Section 6.6.

6.3.1 Solutions on a finite domain for prescribed J(A,y)

We assume that the right-hand side of Equation (6.20) is a given function which changes due to changing boundary conditions. For simplicity, we assume that this change is described by a certain factor λ such that the equation now reads

$$-\nabla^2 A = \lambda f(A, y), \tag{6.23}$$

where $f(A, y)$ is a fixed function of A and y. An appropriate boundary condition is to fix the value of A on the boundary (Dirichlet boundary condition).

Depending on the form of $f(A, y)$ the following properties hold if Dirichlet boundary conditions are imposed (see e.g. Birn *et al.*, 1978a):

(i) for $\partial f/\partial A \leqslant 0$, Equation (6.23) has a unique solution for every $\lambda \geqslant 0$;

(ii) for $f \geqslant 0$, $\partial f/\partial A \geqslant 0$, $\partial^2 f/\partial A^2 > 0$ (or equivalently $f \leqslant 0$, $\partial f/\partial A \geqslant 0$, $\partial^2 f/\partial A^2 < 0$) there exists a positive number λ^* such that (6.23) has no solution for $\lambda > \lambda^*$, and at least one solution for $0 \leqslant \lambda < \lambda^*$.

A typical situation of case (ii) is shown in Figure 6.1. For any value of λ between 0 and λ^* there exist two solutions. (Sufficient conditions for the existence of at least two solutions are discussed by Birn *et al.*, 1978a.) At the critical point $\lambda = \lambda^*$ the solution curve bends back, with $\lambda = \lambda^*$ representing a catastrophe point in catastrophe theory (Thom, 1975). Such points were proposed by Birn *et al.* (1978a) to represent the onset points of eruptive phenomena such as solar flares, since no neighbouring equilibrium state could be achieved if λ was forced to exceed the critical value λ^*. Non-equilibrium may also occur in the thermal structure of a loop (§5.4.6).

It is a convenient feature of this approach that information on the dynamic behaviour may be provided from an *equilibrium* analysis alone. Nevertheless, as discussed later in this paper, this result has to be confirmed by a stability analysis.

6.3.2 Solutions on an unbounded domain for prescribed J(A,y)

Unlike the case of a finite domain, no similarly *general* mathematical statements exist at present for unbounded domains. Using a conformal

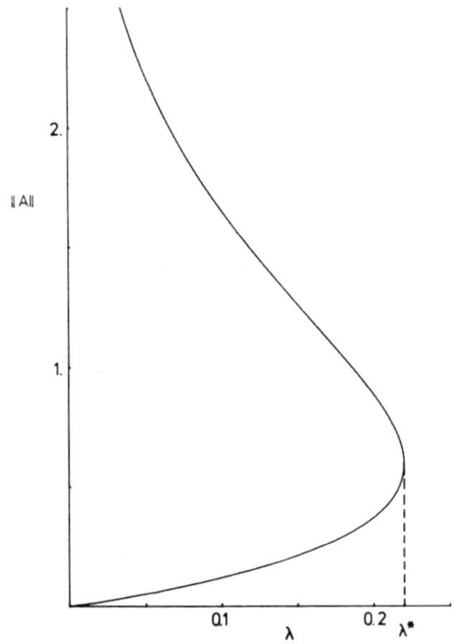

FIGURE 6.1 Curve representing solutions of (6.23) for a one-dimensional neutral sheet with $f = \exp(-2A)$ and the boundary conditions $A(\pm 1) = 0$. $\|A\|$ represents the maximum value of $|A|$ for a given solution.

mapping, Birn *et al.* (1978a) showed that, for the special case $f(A) \propto \exp(-cA)$, statements (i) and (ii) hold also for a half-plane.

Jockers (1978) studied numerically a number of cases with f being a polynomial. Varying the parameter λ, he found also in those cases critical points, beyond which no neighbouring equilibrium could be obtained.

Heyvaerts *et al.* (1979) studied a class of bounded functions f. They found that solutions of (6.23) might exist for all values of λ but that the solutions are generally not unique and that there still exist "catastrophe points" at which the solution curve bends back, as illustrated by Figure 6.2, so that there is no *neighbouring* equilibrium if λ exceeds such a critical value. They improved this result numerically by replacing the Dirichlet problem on an unbounded region by a problem on a finite domain with mixed Dirichlet and von Neumann boundary conditions (see Section 6.4).

6.3.3 Prescribed foot-points

Now let us assume $B_z \neq 0$ and consider the more physical boundary condition of prescribing the position of footpoints (e.g. by externally driven motions). For field lines with both ends at the boundary, the displacement d(A) of footpoints of the field line belonging to a certain value of A is defined by the

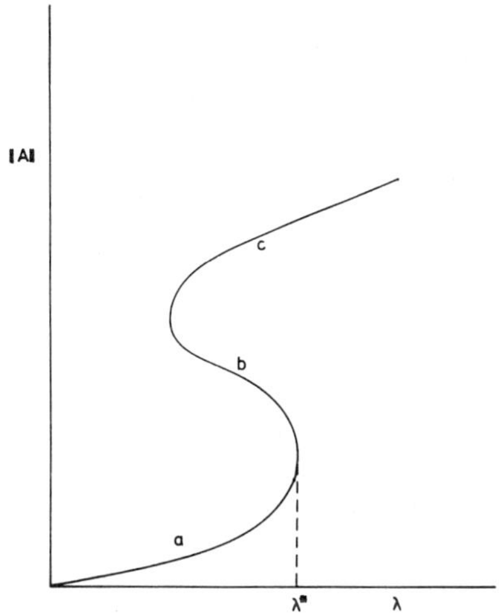

FIGURE 6.2 Qualitative sketch of a possible solution curve corresponding to Fig. 6.1 for the case of an unbounded domain with a bounded function *f*. (After Heyvaerts *et al.*, 1979.)

difference of the z-coordinate of the two end-points at the boundary; it can be obtained for a given solution by integrating dz along the field line from one end-point to the other:

$$\mathrm{d}(A) = \int\limits_{A=\mathrm{const}} \mathrm{d}z = \int\limits_{A} \frac{B_z}{|\nabla A|}\,\mathrm{d}s = B_z(A) \int\limits_{A} \frac{\mathrm{d}s}{|\nabla A|}, \tag{6.24}$$

where s is the arc-length of the projection of the field line on the x, y plane. If there are closed field lines, d(A) and thus $B_z(A)$ must be zero for the corresponding values of A. There is no restriction for field lines with one or no end at the boundary.

The equilibrium problem now consists of solving (6.20) and (6.24) with boundary conditions for A, p and the displacement d. The problem may become even more involved, if the pressure is not prescribed by the boundary condition but by the thermodynamic relations. We will return to this problem in Section 6.6. There are no general statements on the solution of (6.20) with (6.24), even for the simpler force-free case with negligible pressure. Of course, any sequence of solutions obtained with a certain sequence of functions $B_z(A)$ can be reinterpreted by assuming that the corresponding displacement of footpoints was prescribed. But in that case, two different solutions for the same $B_z(A)$ generally belong to different footpoint displacements and in general nothing is known about the number of solutions corresponding to the *same* displacement of footpoints (Jockers, 1978; Priest and Milne, 1980).

There is, however, at least one special case (Priest and Milne, 1980, see also section 6.4) showing that the non-uniqueness of the equilibrium problem is sometimes still retained if the displacement of the footpoints is prescribed. However, it is an open question whether there exist critical "catastrophe points") $\lambda=\lambda^*$ if the displacement is changed with the parameter λ.

6.4 SPECIAL SOLUTIONS

As was pointed out earlier, our main interest — besides finding single quasi-equilibrium solutions which may represent actual pre-flare configurations — is to study sequences of solutions containing critical points that may be interpreted as the onset of an unstable dynamic evolution. Such sequences have been found by several authors, both analytically and numerically. These solutions and some of their properties will now be discussed in some detail.

Barnes and Sturrock (1972) computed numerically some force-free configurations with cylindrical symmetry, representing a sunspot surrounded by a region of opposite polarity. The magnetic field was represented by Euler potentials $\alpha(r, y)$ and $\beta=\delta-\gamma(r, y)$ as

$$\mathbf{B}=\nabla\alpha\times\nabla\beta, \tag{6.25}$$

in terms of cylindrical coordinates $r=(x^2+z^2)^{1/2}$, $\delta=\arctan x/z$ and y, with y representing the height above the photosphere. (This somewhat unusual definition is chosen to be consistent with the notation in the other parts of this chapter.) The corresponding force-free equations,

$$\begin{aligned} \mathbf{j}\cdot\nabla\alpha&=0,\\ \mathbf{j}\cdot\nabla\beta&=0, \end{aligned} \tag{6.26}$$

were solved numerically by means of an iteration technique in a cylindrical region $0\leqslant r\leqslant R_M$, $0\leqslant y\leqslant y_M$. The boundary conditions were

$$r=0\text{: } \alpha=0 \text{ and } \gamma=\gamma_M/2,$$

$$r=R_M\text{: } \alpha=0 \text{ and } \partial\gamma/\partial r=0,$$

$$y=0\text{: } \alpha=(r/R)\exp(1/2-r^2/2R^2) \text{ and}$$

$$\gamma=(\gamma_M/2)\cos(\pi r/2R) \text{ for } r\leqslant 2R$$

$$\gamma = -\gamma_M/2 \text{ for } r > 2R,$$

$$y = y_M\text{: } \alpha = 0 \text{ and } \partial\gamma/\partial y = 0, \tag{6.27}$$

with $R = 14$, $y_M = 86.6$, and $R_M = 88.7$ in arbitrary units. The variation of the parameter γ_M was assumed to represent a quasi-static change of the photospheric boundary conditions. As γ_M increases from 0, the footpoints of the field lines rotate differentially around the line of symmetry (see Figure 6.4) and the field pattern expands (see Figures 6.3a and b). Barnes and Sturrock compared the energy of the twisted force-free configurations with that of an open field consisting of potential fields in two regions separated by an infinitely thin current sheet (see Figure 6.3c). The photospheric condition at $y = 0$ was the same as that for the closed configurations, as well as the symmetry condition at $r = 0$. The boundary conditions at $y = y_M$ and $r = R_M$, however, *had to be changed to allow for an open configuration* to:

$$\begin{aligned} y = y_M\text{: } \alpha &= 2(1 - (1 + r^2/y_M^2)^{-1/2}), \\ r = r_M\text{: } \alpha &= 2(1 - (1 + R_M^2/y^2)^{-1/2}). \end{aligned} \tag{6.28}$$

Barnes and Sturrock found that the energy of the twisted force-free field exceeded that of the open field if the rotation angle γ_M exceeds about 180°; they concluded that the force-free configuration should become *metastable* (§1.3) beyond this point (i.e. stable against small perturbations but unstable against sufficiently large perturbations) and that the configuration might evolve explosively towards the open field. However, *no rigorous stability analysis was done* and an actual threshold that should lie somewhere beyond the point of equal energy was not determined.

The two-dimensional problem (6.20) in cartesian geometry has been tackled by many authors. Since the large-scale coronal magnetic field must be predominantly force-free, it is the force-free case that has usually been treated. It is, however, obvious from (6.20) that any force-free field $\mathbf{B} = \nabla A \times \mathbf{e}_z + \mathbf{B}_z^*(A)$ can be replaced by a non-force-free field $\mathbf{B} = \nabla A \times \mathbf{e}_z + \mathbf{B}_z(A)$ with $B_z^2(A) < B_z^{*2}(A)$ and a pressure function $P(A) = (B_z^{*2}(A) - B_z^2(A))/(2\mu)$, provided that gravity can be neglected (which is true when the gravitational scale-height H defined by (6.22) is large compared with the typical scale of magnetic structures, as seems reasonable for typical coronal conditions). Therefore the following sets of solutions are not necessarily restricted to the force-free case.

Low and Nakagawa (1975) computed numerically solutions of (6.20) with $p = 0$. They used two functions for $B_z(A)$, namely,

$$B_z(A) = -\frac{2\varepsilon A}{(1 + 3A^2)^{1/2}}, \tag{6.29}$$

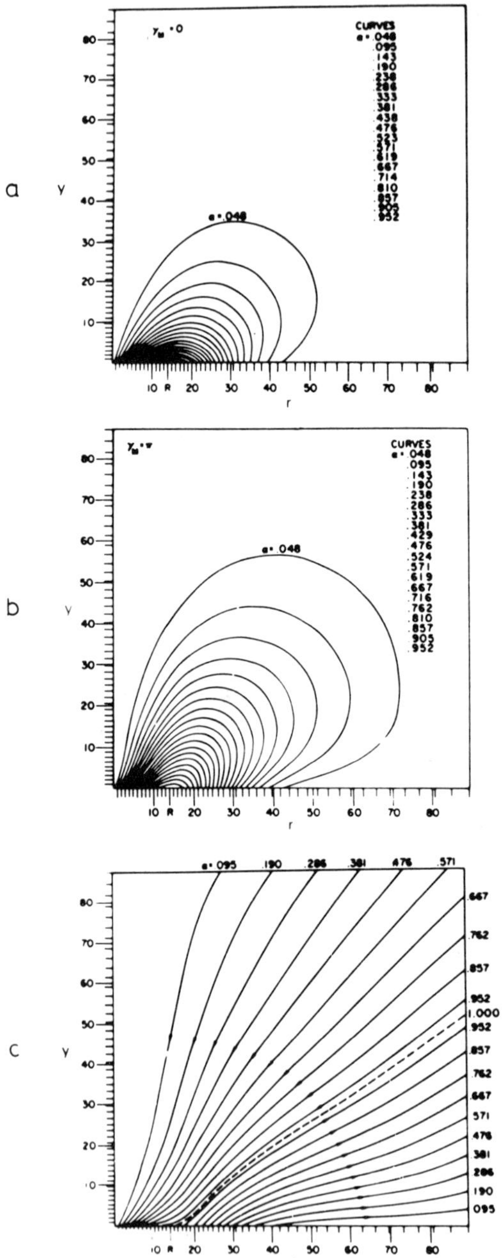

FIGURE 6.3 Axisymmetric magnetic field configurations in a vertical plane through a model sunspot, after Barnes and Sturrock (1972) with boundary conditions (6.27) and (6.28). (a) current-free case $\gamma_M = 0$, (b) force-free case $\gamma_M = \pi$, (c) open-field configuration.

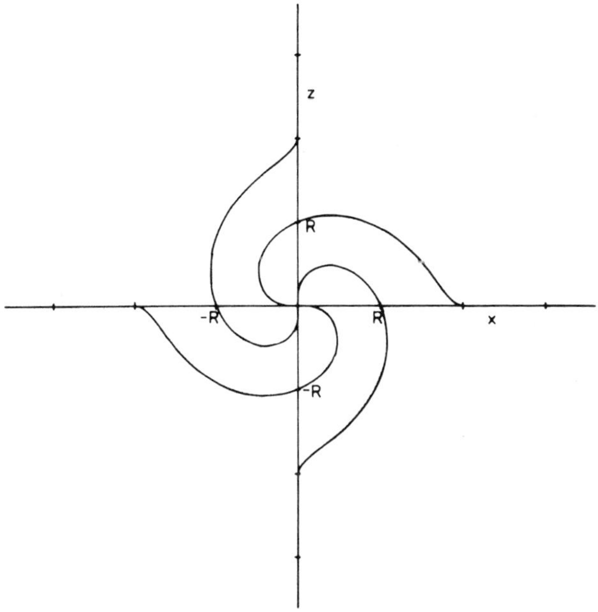

FIGURE 6.4 Rotation of foot-points of field lines in the horizontal x,z plane for $\gamma_M = \pi$, corresponding to the configuration in Figure 6.3(b).

$$B_z(A) = -\frac{\varepsilon A^2}{1 + A^2}. \tag{6.30}$$

The parameter ε, which was assumed to increase with time, equals $\lambda^{1/2}$ in Equation (6.23). Dirichlet boundary conditions were imposed:

$$A(x, 0) = \cos x \text{ for } |x| \leqslant \pi/2,$$
$$A = 0 \text{ for } |x| = \pi/2,\ 0 \leqslant y \leqslant 10, \tag{6.31}$$
$$A = 0 \text{ for } y = 10,\ |x| \leqslant \pi/2.$$

For the ε-values considered, two solutions were found in each case. One of them was ruled out, however, by requiring that all field lines connect with the photospheric boundary. The adopted solutions for (6.29) and (6.30) differed in the qualitative evolution of the field pattern and in the behaviour of the footpoint displacement. The solutions obtained with (6.29) (see Figure 6.5a) lead to field lines rising as ε increases and a footpoint displacement that increases with x (Figure 6.5b). For the solutions obtained with (6.30) the field lines descend with increasing ε (Figure 6.6a) and the footpoint displacement

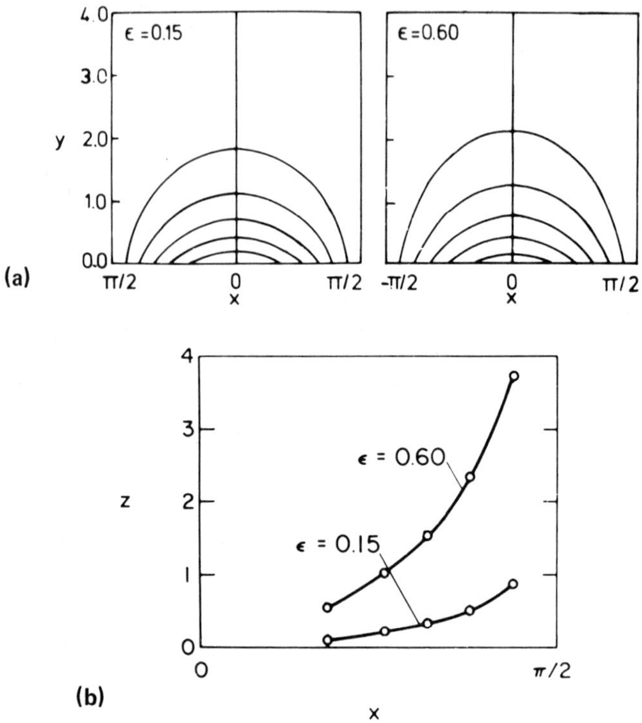

FIGURE 6.5 (a) Magnetic field lines projected in the vertical x,y plane (after Low and Nakagawa, 1975) obtained with the function (6.29) for two different values of ε. (b) The corresponding displacement of foot-points of field lines in the horizontal x,z plane.

attains a maximum at some value of x (Figure 6.6b). No attempt was made, however, to determine a threshold of existence or of stability of the solutions and relate it to the onset of an eruption.

Low (1977) derived analytic solutions of Equation (6.20) with $p=0$ and

$$B_z=\sqrt{\lambda}\mathrm{e}^{-A} \tag{6.32}$$

or

$$B_z=(K^2-\lambda\mathrm{e}^{-2A})^{1/2} \tag{6.33}$$

and the boundary condition

$$A(x, 0)=\log(1+x^2). \tag{6.34}$$

Whereas the function (6.33) yielded unique solutions, as is obvious from

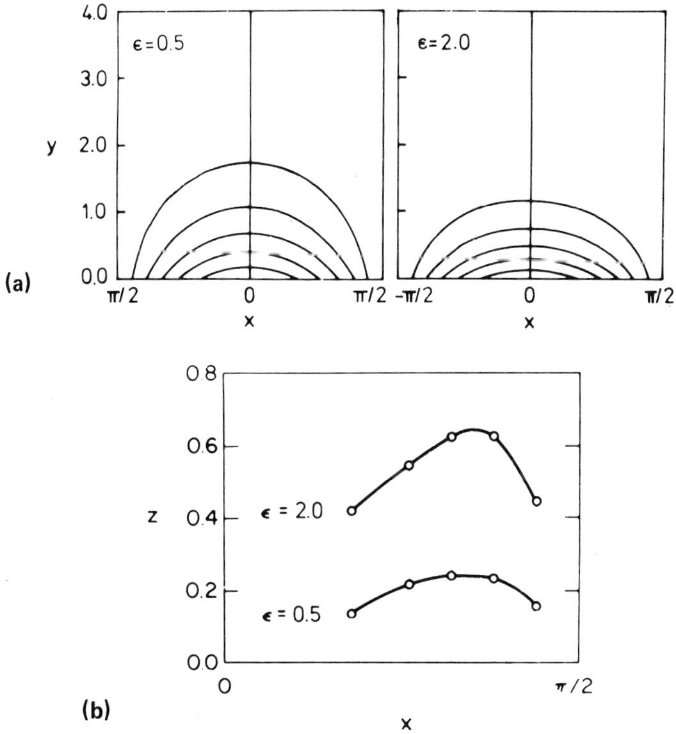

FIGURE 6.6 (a) Magnetic field lines projected in the vertical x,y plane (after Low and Nakagawa, 1975) obtained with function (6.30) for two different values of ε. (b) The corresponding displacement of foot-points of field lines in the horizontal x,z plane.

theorem (i) in Section 6.3, Low found two solutions for the function (6.32):

$$A_{\pm} = \log(1 + x^2 + y^2 - 2h_{\pm}y), \tag{6.35}$$

with

$$h_{\pm} = \pm\sqrt{1 - \lambda/4} \tag{6.36}$$

for any $\lambda < \lambda^* = 4$. The solutions consist simply of concentric helical field lines with the symmetry axis located at $y = h$ (see Figure 6.7). As λ increases from $\lambda = 0$ (current-free case) to the critical value λ^* the symmetry axis rises from $h = -1$ until it reaches the boundary ($h = 0$) at $\lambda = \lambda^*$ (Figure 6.7a, b). This solution branch corresponds to the lower part of Figure 6.1. As one follows the upper branch of Figure 6.1 from $\lambda = \lambda^*$ to smaller values of λ, the symmetry axis rises above the boundary, such that there are field lines without connection to the

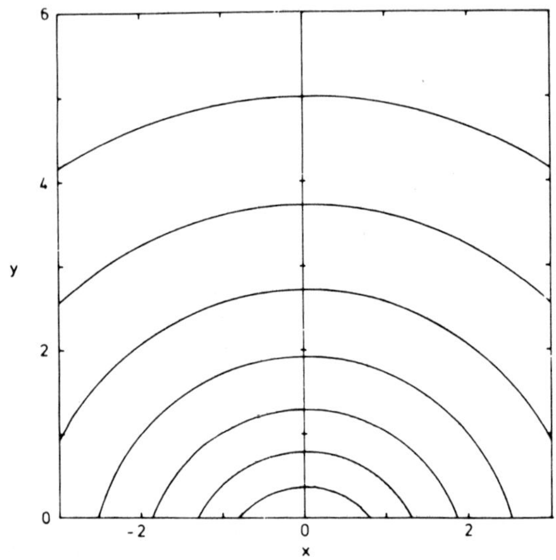

(a)

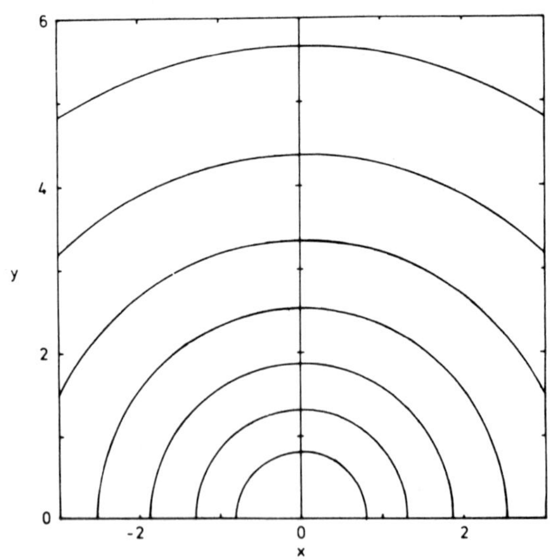

(b)

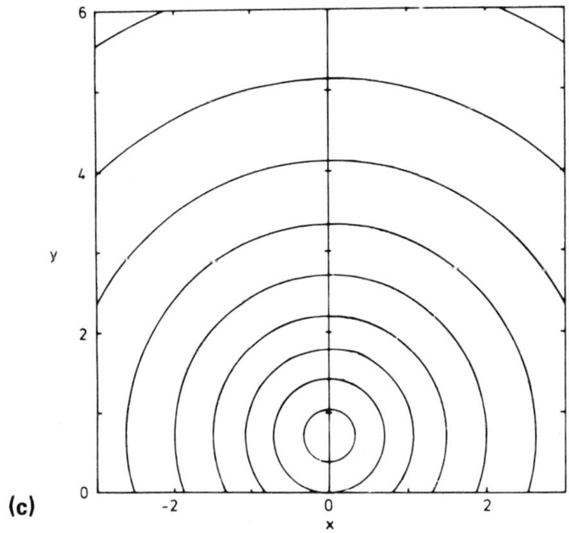

FIGURE 6.7 Magnetic field pattern in a vertical plane of the solution (6.35) (after Low, 1977) for different values of λ. (a) $\lambda=2$, $h=-\sqrt{0.5}$, (b) $\lambda=\lambda^*=4$, $h=0$, (c) $\lambda=2$, $h=+\sqrt{0.5}$.

boundary (Figure 6.7c); they are closed circles in the projection of Figure 6.7. The displacements of footpoints at the photosphere

$$d(x)=\begin{cases}\sqrt{1-h^2}\arctan(x/|h|) \text{ for } h<0,\ x>0,\\ \sqrt{1-h^2}[\pi-\arctan(x/h)] \text{ for } h>0,\ x>0,\end{cases} \tag{6.37}$$

attain a constant value at large distances from the origin (Figure 6.8). This value is the same for both solutions belonging to the same λ. It increases with λ, such that it reaches its maximum at $\lambda=\lambda^*$. The behaviour of $d(x)$ close to the origin however, is different for two solutions belonging to different branches of the solution curve Figure 6.1.

Low noted that the two solutions for the same λ are topologically different, because the ones on the upper branch of Figure 6.1 contain field lines without connection to the boundary whose projections in the x, y plane are closed circles. He related this change of topology at the critical point $\lambda=\lambda^*$ to the onset of a flare.

Low's solution was extended by Priest and Milne (1980) to a whole family of cylindrically symmetric pairs of force-free solutions:

$$\begin{aligned} A(x,y)&=\frac{(1-h^2+r^2)^{1-n}}{2(n-1)}, \qquad n\neq 1,\\ B_z^2&=\frac{1-h^2+2(1-n)r^2}{(2n-1)(1-h^2+r^2)^{2n}}, \end{aligned} \tag{6.38}$$

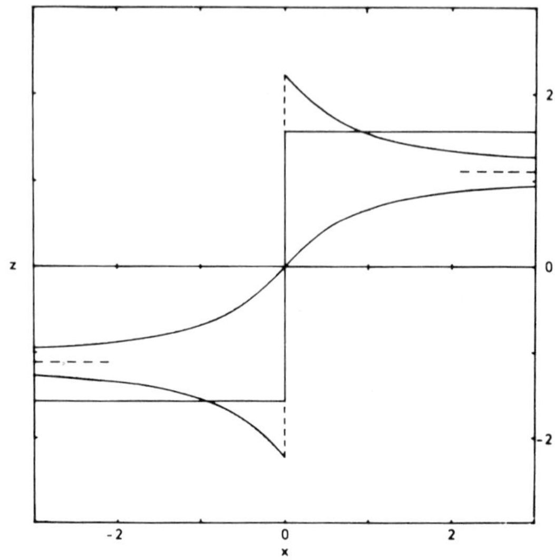

FIGURE 6.8 Displacement of foot-points of field lines in the horizontal x,z plane corresponding to the configurations in Figure 6.7.

with $r^2 = x^2 + (y-h)^2$, where the limit $n=1$ corresponds to Low's solution. If one requires that the solutions (6.38) hold also for large r and that **B** tends to zero with $r \to +\infty$, n must lie between 1/2 and 1. These configurations are similar to Low's solutions. However, for $n \neq 1$, the sequences described by the parameter h do not include a potential field as a special case.

The displacement of footpoints at the boundary is given by

$$d(x) = \left(\frac{1 - h^2 + 2(1-n)(x^2 + h^2)}{2n - 1} \right)^{1/2} \varphi_0(x), \tag{6.39}$$

where

$$\varphi_0(x) = \begin{cases} \arctan(x/|h|) \text{ for } h<0 \text{ and } x \geqslant 0, \\ \pi - \arctan(x/h) \text{ for } h>0 \text{ and } x \geqslant 0. \end{cases}$$

For large distances from the origin, $d(x)$ increases linearly with x, with the slope independent of h.

In their family of cylindrically symmetric solutions, Priest and Milne found also two sequences with *different* B_z, but the *same* footpoint displacement at the boundary. These are given by

$$B_\varphi^2=(1-cI(\Phi))\cos^2\Phi,$$

$$B_{z+}^2=\frac{c\Phi^2}{\pi(\pi-2\Phi)}, \tag{6.40}$$

$$B_{z-}^2=B_{z+}^2+c,$$

where

$$I(\Phi)=\int_0^\Phi \frac{2\varphi(\pi-\varphi)\mathrm{d}\varphi}{\pi(\pi-2\varphi)^2\cos^2\varphi}$$

and

$$\cos\Phi=1/r.$$

The solutions consist of field lines which are concentric spirals with the symmetry axis located at $y=\pm 1$ for the indices + and −, respectively. They contain the free parameter c. The displacement of footpoints for both solutions is given by

$$d(x)=\frac{2c\Phi(\pi-\Phi)}{(\pi-2\Phi)\cos\Phi\sqrt{(1-cI(\Phi))}}, \tag{6.41}$$

with

$$\tan\Phi=x.$$

$d(x)$ reaches infinity at a finite distance x_0 which is given by

$$I(\arctan x_0)=1/c. \tag{6.42}$$

B_φ and the normal magnetic field component at the boundary both become imaginary for $|x|>x_0$.

Therefore, it seems questionable whether these sequences of solutions have direct applications. Their importance, however, is to demonstrate, that non-uniqueness can sometimes still exist if the photospheric footpoints of field lines are prescribed instead of the tangential magnetic field component B_z.

Jockers (1978) calculated numerically some force-free solutions of (6.23) with the boundary condition

$$A(x, 0) = \frac{1}{1+x^2} \tag{6.43}$$

and two different kinds of function J:

$$J = \begin{cases} \lambda(n+1)A^n & \text{for} A \geqslant 0, \\ 0 & \text{for } A < 0, \end{cases} \tag{6.44}$$

and

$$J = \begin{cases} \lambda(n+1)A^n(a-A)^2 & \text{for } 0 \leqslant A \leqslant 1, \\ 0 & \text{elsewhere}, \end{cases} \tag{6.45}$$

and different values of n (especially 8, 12, 16).

The reason for choosing the function (6.45) is to inhibit the development of field lines without connection to the boundary. Jockers found that in *both* cases a maximum value λ^* exists such that there are two solutions for $0 < \lambda < \lambda^*$ and no solution for $\lambda > \lambda^*$. For the function (6.44) he found two types of topologically different solutions (see Figure 6.9a). The point at which the topology changes, however, does not coincide with the critical point $\lambda = \lambda^*$, at which B_z becomes maximal. Figure 6.9b shows the footpoint displacement for configurations close to the critical case $\lambda = \lambda^*$.

A similar result was obtained by Birn *et al.* (1978a) for a sequence of analytic solutions. Conformal mapping was used to construct solutions on a half-plane from known solutions on different domains with the function

$$J(A) = -\lambda e^{-2A}. \tag{6.46}$$

Mapping Bennett's (1934) solutions on a circle onto the half-plane lead again to Low's (1977) solution, which we have discussed above. From Harris' (1962) solutions on a strip $|y| < a$, Birn *et al.* found new pairs of solutions with the boundary condition

$$A(x, 0) = \log |x|. \tag{6.47}$$

The two solutions are given by

$$A_{\pm} = \log (r \cosh(2b_{\pm}\varphi/\pi)/\cosh b_{\pm}) \tag{6.48}$$

for cylindrical coordinates $r = (x^2 + y^2)^{1/2}$ and $\varphi = \arctan(x/y)$ and using a suitable normalization of the original expressions given by Birn *et al.* The two values of b follow from

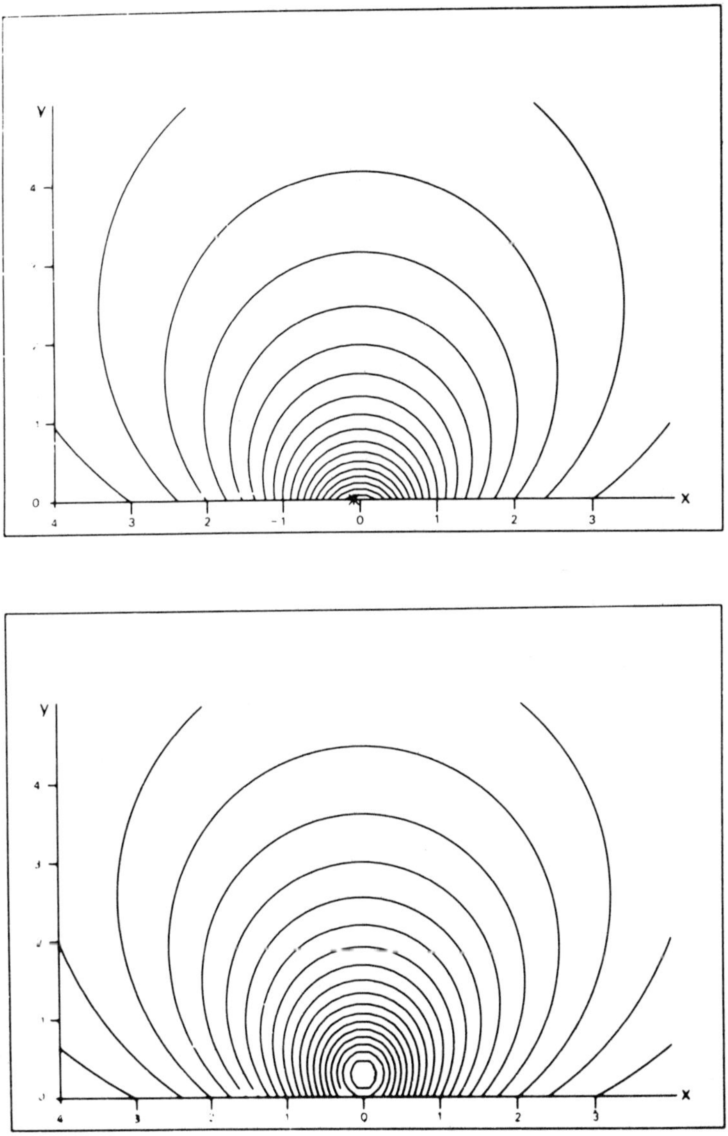

(a)

FIGURE 6.9 (a) Magnetic field line projections in the vertical x,y plane for two solutions obtained with the function (6.44) and the values $n=8$ and $\lambda=0.3$. (After Jockers, 1978 and Zwingmann, 1979.)

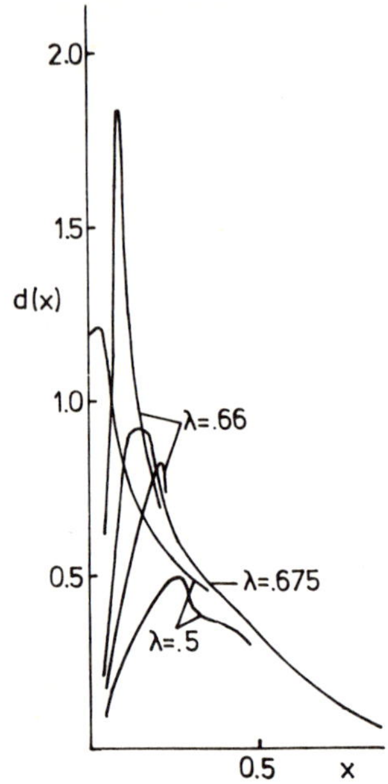

FIGURE 6.9 (b) The corresponding foot-point displacement in the horizontal x,z plane for $n=16$, and different values of λ. (After Jockers, 1978.)

$$b_{\pm}/\cosh b_{\pm} = \frac{\pi}{2}\sqrt{\lambda} \tag{6.49}$$

for $\lambda < \lambda^* \simeq 0.44$. There is no solution for $\lambda > \lambda^*$.

The two solutions for any $\lambda < \lambda^*$ show no topological difference (see Figure 6.10); they become, however, singular at the origin.

The curve connecting the displaced footpoints of the force-free solutions that initially lie on the x-axis are straight lines through the origin:

$$d(x) = x \sinh b. \tag{6.50}$$

As one follows the solution curve similar to Figure 1.6 from the lower branch to the upper one, the slope of $d(x)$ increases monotonically with no singular behaviour at the critical point $\lambda = \lambda^*$.

The critical point $\lambda = \lambda^*$ was interpreted by Birn *et al.* as the onset point of a dynamical evolution, just as Low (1977) did for his sequence of solutions. Birn *et al.* emphasize, however, the importance of the lack of equilibrium (if the parameter λ could be forced to exceed λ^*) rather than a change of topology.

By studying solutions that are separable in the cylindrical coordinates r and φ, Priest and Milne (1980) found a class of solutions that represent a

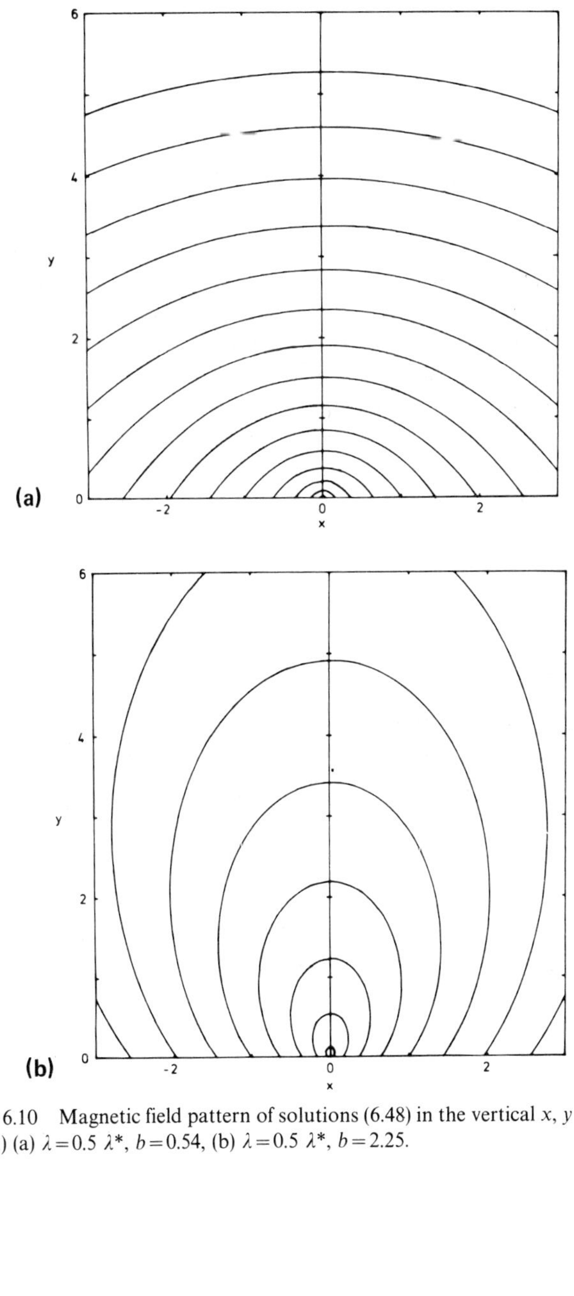

FIGURE 6.10 Magnetic field pattern of solutions (6.48) in the vertical x, y plane. (After Birn *et al.*, 1978a.) (a) $\lambda=0.5\ \lambda^*$, $b=0.54$, (b) $\lambda=0.5\ \lambda^*$, $b=2.25$.

generalization of the solution given by Birn *et al.* (1978a). An example can be given analytically by

$$A = r^{1/2}(1+\mu\cos\varphi)^{1/2},$$
$$J(A) = -\lambda/(4A^3), \tag{6.51}$$

where

$$\mu = \pm\sqrt{1-\lambda}, \qquad \lambda \leqslant 1.$$

Figure 6.11 shows the field lines of two solutions for $\lambda = 0.5$ projected in the vertical x, y plane. The solutions show similar characteristic properties to that given by Birn *et al.* The footpoint displacement $d(x)$ is a linear function of x for all solutions of this class and there is *no singular behaviour* at the critical point $\lambda = \lambda^*$ where $J(A)$ becomes maximal. This fact led Priest and Milne to the conclusion that the onset of a flare should be connected with the instability and non-uniqueness of equilibrium solutions rather than the non-existence of neighbouring solutions (which is connected with catastrophe points).

Heyvaerts *et al.* (1979) investigated the problem (6.20) for bounded functions of the form

$$J(A) = \lambda C A^{\Delta},\ \Delta > 3 \text{ for } 0 < A < A_0,$$
$$J(A) = 0 \text{ for } A \geqslant A_0.$$

The main general results about the existence of multiple solutions and "catastrophe points" have already been discussed in Section 6.3.2. A special sequence of solutions was obtained numerically on a finite domain with von Neumann boundary conditions at the "open" boundaries and Dirichlet conditions at the photospheric boundary:

$$A(x,0) = \begin{cases} 0.2 & \text{for } |x| \leqslant 1.3, \\ 1.5 - |x| & \text{for } 1.3 < |x| < 1.5, \\ 0 & \text{for } |x| \geqslant 1.5, \end{cases}$$
$$\frac{\partial}{\partial x} A(x, y) = 0 \text{ for } |x| = 6, \tag{6.52}$$
$$\frac{\partial}{\partial y} A(x, y) = 0 \text{ for } y = 3,$$

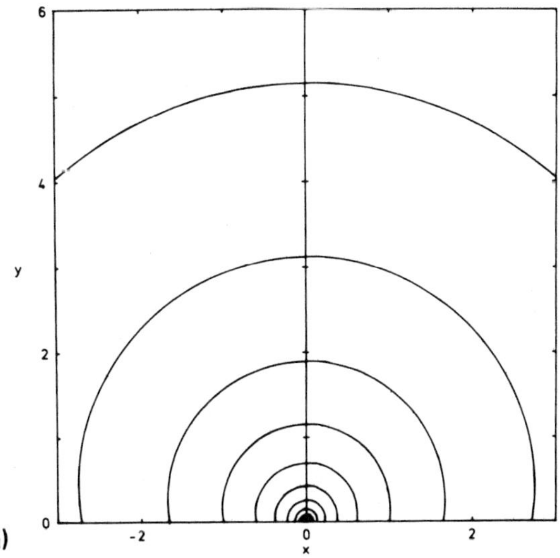

(a)

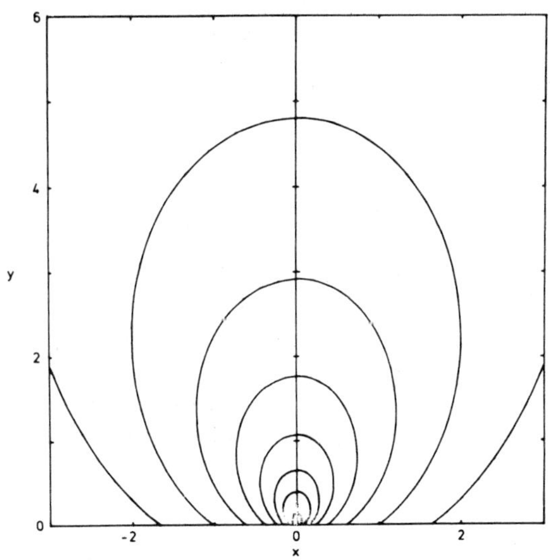

(b)

FIGURE 6.11 Magnetic field pattern of solutions (6.51) in the x,y plane. (After Priest and Milne, 1980.) (a) $\lambda = 0.5$, $\mu = -\sqrt{0.5}$, (b) $\lambda = 0.5$, $\mu = +\sqrt{0.5}$.

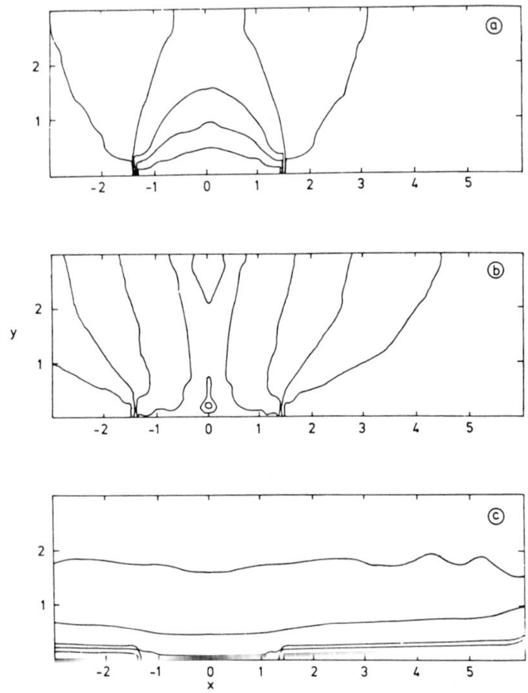

FIGURE 6.12 Magnetic field pattern in the x,y plane for three solutions obtained with (6.53), corresponding to a solution curve similar to Figure 6.2. (a) Lower branch, (b) medium branch, (c) upper branch. (After Heyvaerts *et al.* (1979).)

with

$$J(A)=\lambda A^4(1-A)^2. \tag{6.53}$$

Figure 6.12 shows three topologically different solutions which correspond to different branches of the "solution curve" similar to Figure 6.2.

Heyvaerts *et al.* took the same point of view as Birn *et al.* (1978a): i.e. they interpreted the critical "catastrophe" points $\lambda=\lambda^*$, where the solution curves like Figures 6.1 or 6.2 bend back, as possible points for the onset of dynamic evolution. They proposed that this evolution might lead to an open field solutions like Figure 6.12c.

6.5 STABILITY

6.5.1 General two-dimensional analysis

The results of the previous sections support the view that the onset of a solar flare (or of other types of eruptive process) may occur when a quasi-statically evolving configuration reaches the point $\lambda = \lambda^*$ (Figure 6.1). This concept, however, is applicable only if it is possible for the system to reach this point in a stable way; i.e. if all systems corresponding to the associated equilibrium branch are stable against arbitrary perturbations. Confining our discussion to linear stability and to two dimensions, it is comparatively easy to show that, under a large variety of circumstances, the equilibrium branch on which $|A - A_v|_{\max}$ assumes its smallest values (lower branch in Figure 6.1) is stable. The restriction to two dimensions means that not only the equilibrium quantities but also the perturbations are independent of z ($k=0$) for modes of the form $\boldsymbol{\xi}(x, y)\exp(ikz)$). A brief discussion of three-dimensional modes will be given at the end of this section.

Before going into specific dynamical models it is convenient to introduce and briefly discuss the following eigenvalue problem of Schrödinger type. Consider

$$-\Delta U - \frac{\partial J(A_0)}{\partial A_0} U = \bar{\mu} U, \tag{6.54}$$

with $U=0$ on the boundary and $\bar{\mu}$ denoting the eigenvalue. Obviously, this eigenvalue problem is closely related with our equilibrium problem (6.23), since $\bar{\mu}=0$ corresponds to an infinitesimal perturbation of the flux function $A = A_0 + U$ (neighbouring equilibrium).

In the mathematical literature (e.g. Amann, 1976) it is shown that, if conditions (ii) of Section 6.3.1 hold, no negative eigenvalue $\bar{\mu}$ exists, if A_0 lies on the lower branch (Figure 6.1). Since it is beyond the scope of the present survey to give a rigorous proof of this result, we confine ourselves to the following plausible argument: for $\lambda = 0$ (i.e. for $J(A_0) = 0$), (6.54) reduces to a linear equation such as appears typically in wave propagation problems for homogenous media, implying that the eigenvalues $\bar{\mu}$ are positive. On the other hand, at $\lambda = \lambda^*$ two solutions become arbitrarily close, so that we expect there to be an eigenvalue $\bar{\mu} = 0$ for $\lambda = \lambda^*$. The reason for the lowering of the minimum of the eigenvalue $\bar{\mu}$ as λ varies from 0 to λ^* is that the (negative) Schrödinger potential $(-\partial J(A_0)/\partial A_0)$ becomes more and more pronounced.

Since it is possible to show that there is no vanishing eigenvalue on the lower branch for $0 \leqslant \lambda < \lambda^*$ (as follows e.g. from the results of Amann, 1976), by continuity one expects $\bar{\mu} > 0$ for $0 \leqslant \lambda < \lambda^*$.

The non-existence of a negative eigenvalue $\bar{\mu}$ is equivalent to the statement that the minimum of the variational functional

$$\delta F = \frac{1}{2\mu} \int [(\nabla U)^2 - \frac{\partial J}{\partial A_0} U^2] \mathrm{d}^2\mathbf{r} \tag{6.55}$$

is non-negative. In fact, if we normalize $\int U^2 \mathrm{d}^2\mathbf{r} = \text{const}$, the minimum of δF coincides with the smallest eigenvalue $\bar{\mu}$. (Equation (6.54) is the Euler–Lagrange equation associated with (6.55).)

Let us now return to our stability problem. We will base our analysis on the fact that for a variety of dynamical plasma models one can derive variational principles that allow a direct comparison with (6.55).

Considering ideal MHD in the absence of gravity forces with two-dimensional perturbations (ignoring variations and plasma displacements along the z-direction), we find that we can cast the variational principle of ideal MHD (Bernstein *et al.*, 1958) in the following form (see also §4.4.2):

$$\delta W = \delta F + \delta Q, \tag{6.56}$$

with

$$\delta Q = \tfrac{1}{2} \int \gamma p (\boldsymbol{\nabla} \cdot \boldsymbol{\xi})^2 \mathrm{d}^3\mathbf{r} \geqslant 0,$$

where $\boldsymbol{\xi}$ denotes the displacement of the plasma volume element and γ the ratio of specific heats. Since δQ is non-negative, we immediately conclude that *any equilibrium situated on the lower branch* (where $\delta F \geqslant 0$) *is stable* against two-dimensional ideal-MHD-perturbations.

One could argue that for solar applications ideal MHD may not be appropriate. Therefore, it is important to repeat the same stability analysis for more refined models also. This was done for the case of Vlasov theory and for several models of resistive MHD (e.g. resistivity constant along magnetic field lines), where, however, in both theories again the perturbations were restricted to two dimensions (Schindler *et al.*, 1980). In these cases the corresponding variational principles assume a form similar to that of ideal MHD, (6.54); although the term δQ is different for different dynamical models, δQ is non-negative in all cases considered. This means that the lower branch in Figure 6.1 is stable under a rather large variety of circumstances. Our results support the concept that *the critical point $\lambda = \lambda^*$ may be reached by quasi-static evolution* along the lower equilibrium branch. Note, however, that for the upper branch(es) even in two dimensions a similarly general stability result is not available. Depending on the dynamical model used, the upper branch(es) may be in part (or entirely) unstable.

Since the functional δF plays a central role in stability questions it is of interest to ask whether δF is integrable, in the sense that there is a functional F such that δF is its second variation. Evidently, the functional F has the form

$$F = \int \left[\frac{(\boldsymbol{\nabla} A)^2}{2\mu} - \Pi(A) \right] \mathrm{d}^2\mathbf{r}, \tag{6.57}$$

with

$$\Pi(A) = P(A) + B_z^2(A)/(2\mu). \tag{6.58}$$

This result applies to a restricted class of perturbations ($k=0$) only. Any final stability discussion must include also the case $k \neq 0$ (see discussion below).

For $B_z = 0$ the functional F coincides with a functional introduced by Grad (1964) for ideal MHD, namely

$$F = \int \left[\frac{B^2}{2\mu} - p \right] \mathrm{d}^3\mathbf{r}, \tag{6.59}$$

which was shown by Grad to provide sufficient stability results under three-dimensional isobaric variation. (Note that (6.57) keeps the generalized pressure Π fixed on field lines automatically.)

It is also possible to show that the modified equilibrium theory described in Section 6.3.3 (footpoints fixed rather than B_z) would directly lead to (6.59) rather than to (6.57). The latter property demonstrates that the proper choice of the expression that assumes a minimum for a stable equilibrium configuration depends on the boundary conditions. A rather simple case is realized if the footpoints of the field lines are fixed and if the configuration is force-free. In that case (6.59) applies with $p \to 0$, so that the appropriate functional simply reduces to the magnetic energy. (Note, however, that due to the boundary conditions in general the minimizing $\mathbf{B}$ will *not* coincide with the vacuum field.) This confirms that the point of view intuitively adopted by Barnes and Sturrock (1972) is essentially correct.

6.5.2 Three-dimensional analysis

We now turn to a brief discussion of some aspects of a full three-dimensional stability analysis, where both a plasma displacement and a variation along the z-axis are admitted. We restrict the considerations, however, to the case of ideal MHD, neglecting an external gravity force for the sake of simplicity.

Inserting the two-dimensional equilibrium (6.20) with (6.19) into the full MHD variational principle we obtain

$$\delta W = \frac{1}{2\mu} \int \mathrm{d}^3\mathbf{r} \Big\{ |\nabla_p A_1|^2 - \frac{\partial J}{\partial A_0} |A_1|^2 + |\mathbf{B}_p \cdot \nabla \xi_z - B_z \nabla \cdot \boldsymbol{\xi}_p|^2$$

$$+ \gamma \mu p |\nabla \cdot \boldsymbol{\xi}|^2 + k^2 |B_z \boldsymbol{\xi}_p - \xi_z \mathbf{B}_p|^2$$

$$+ik\{[-\nabla_p A_1 \times \mathbf{e}_z\cdot(B_z\boldsymbol{\xi}_p^* - \xi_z^*\mathbf{B}_p) - \xi_z J A_1^*]$$

$$\left. -ik[-\nabla_p A_1^* \times \mathbf{e}_z\cdot(B_z\boldsymbol{\xi}_p - \xi_z\mathbf{B}_p) - \xi_z^* J A_1 + J B_z \boldsymbol{\xi}\times\boldsymbol{\xi}^*\cdot\mathbf{e}_z]\}\right\}, \tag{6.60}$$

where $\mathbf{B}_p = \nabla A_0 \times \mathbf{e}_z$, $\boldsymbol{\xi}_p$, and $\nabla_p A_1$ denote projections of the vectors $\mathbf{B}_0$, $\boldsymbol{\xi}$, and ∇A_1 into the x, y plane and an asterisk marks a conjugate complex quantity. The perturbation A_1 is connected with $\boldsymbol{\xi}$ by

$$A_1 = -\boldsymbol{\xi}\cdot\nabla A_0 = -\boldsymbol{\xi}_p\cdot\nabla A_0. \tag{6.61}$$

Here we have chosen trial functions with a z-dependence of the form $\sim e^{ikz}$. Obviously, a straightforward comparison theorem between δW and δF is available only in special cases, such as the two-dimensional case ($k=0$) as discussed above, or when B_z and ξ_z vanish. The same applies to perturbations with all three components of $\boldsymbol{\xi}$ in phase ($\boldsymbol{\xi}$ real), as was tacitly assumed by Birn *et al.* (1978b) dealing with $k \neq 0$ cases (Hood, 1979). For the general case one can, however, proceed in a slightly different way, rewriting (6.60) in the form

$$\delta W = \frac{1}{2\mu}\int d^3\mathbf{r}\left\{\left|\frac{\partial A_1}{\partial s}\right|^2 + \left(\nabla A_0\cdot\nabla\frac{J}{B_p^2} - 2\frac{J^2}{B_p^2} - \frac{\partial J}{\partial A_0}\right)|A_1|^2 + \right.$$

$$+k^2 B_z^2|\xi_\perp|^2 + \gamma\mu p|\nabla\cdot\boldsymbol{\xi}|^2 + |\mathbf{B}_p\cdot\nabla\xi_z - B_z\nabla\cdot\boldsymbol{\xi}_p|^2 +$$

$$\left. +|(\nabla A_0\cdot\nabla A_1 + JA_1)/B_p - ik(\xi_z B_p - \xi_\parallel B_z)|^2\right\} \tag{6.62}$$

Here $\boldsymbol{\xi}_\parallel$ and $\boldsymbol{\xi}_\perp$ are the components of $\boldsymbol{\xi}_p$ parallel and perpendicular to $\mathbf{B}_p$, respectively; $\boldsymbol{\xi}_\perp$ is related to A_1 by

$$A_1 = -\boldsymbol{\xi}_\perp\cdot\nabla A_0 = -B_p\xi_\perp; \tag{6.63}$$

$\partial/\partial s$ represents the derivative in the direction of $\mathbf{B}_p$, such that

$$\mathbf{B}_p\cdot\nabla = B_p\frac{\partial}{\partial s}.$$

Obviously, it is sufficient for stability if the minimum of the functional

$$\delta\tilde{F} = \tfrac{1}{2}\int\left[\left|\frac{\partial A_1}{\partial s}\right|^2 + \left(\nabla A_0\cdot\nabla\frac{J}{B_p^2} - 2\frac{J^2}{B_p^2} - \frac{\partial J}{\partial A_0}\right)|A_1|^2\right]d^3\mathbf{r} \tag{6.64}$$

is positive. Note that the "Schrödinger potential"

$$V = \nabla A_0 \cdot \nabla \frac{J}{B_p^2} - 2\frac{J^2}{B_p^2} - \frac{\partial J}{\partial A_0}$$

is no longer simply given by $-\partial J/\partial A_0$ but has a more complicated structure.

It is instructive to apply (6.64) to simple configurations, such as Low's solution (6.35). Minimizing $\delta\tilde{F}$ in this case leads to the following eigenvalue problem:

$$-\frac{\partial^2 A_1}{\partial s^2} + V(r)A_1 = \Lambda A_1, \tag{6.65}$$

with

$$V = -\frac{\lambda}{r^2(\lambda/4 + r^2)},$$

where $r^2 = x^2 + (y-h)^2$ and $s = r\varphi$ is the arc-length along the circular $\mathbf{B}_p$-lines. Equation (6.65) is solved by

$$\begin{aligned} A_1 &= \hat{A}_1(r)\sin K(r)s, \qquad 0 \leqslant s \leqslant 2r\varphi_0, \\ K &= m\pi/(2r\varphi_0), \qquad m = 1, 2, 3, \ldots, \\ \varphi_0 &= \pi/2 + \arcsin(h/r), \end{aligned} \tag{6.66}$$

$$\Lambda = V + K^2. \tag{6.67}$$

At first sight it might seem unacceptable that the eigenvalue Λ depends on r. Note, however, that (6.65) does not involve derivatives perpendicular to the direction of the $\mathbf{B}_p$-lines, so that one can include modes that are localized on single $\mathbf{B}_p$-lines; the corresponding eigenvalues then are $\Lambda(r)$, where r denotes the field line chosen. It also follows that for a suitable normalization the lowest value of (6.64) coincides with the lowest value assumed by $\Lambda(r)$.

The result (6.67) implies that Low's *lower branch solutions* (Figure 6.1) *are stable* for $\lambda \leqslant 0.84\lambda^*$. Note that this result is independent of whether $J(A)$ is derived from a particle pressure p or from a magnetic pressure $B_z^2/(2\mu)$ or from a suitable combination of both.

Since we disregarded several positive terms in δW we can expect that the actual stability boundary will be associated with a value of λ which is larger than 0.84 λ^*. To demonstrate this point we may choose $B_z=0$ and take into account the $\gamma(\nabla\cdot\xi)^2$ in δW. Minimizing δW we find that the entire lower branch $\lambda\leqslant\lambda^*$ is stable for $\gamma\geqslant\frac{4}{3}$. More general cases seem to require further detailed analysis and possibly a considerable amount of numerical work.

6.5.3 The instability of pre-flare magnetic arcades

(This section is authored by the editor.)

The magnetic configuration near an active-region (or plage) filament before it erupts to give a two-ribbon flare is not yet known in any detail. In particular, it is not known which of the following two structures best represents the configuration:

I) a simple sheared arcade, with a topology similar to that of Figure 6.7a and with field lines crossing the filament;

II) a large flux tube contained within an arcade (Figure 6.13a), with a topology in a plane transverse to the arcade similar to that of Figure 6.7c and with the filament located along the flux tube.

Hood and Priest (1980) have considered the stability of configurations of both types using the energy principle of Bernstein *et al.* (1958) for force-free fields. The change in potential energy due to a perturbation $\boldsymbol{\xi}$ is

$$\delta W=\int\{|\nabla\times(\boldsymbol{\xi}\times\mathbf{B}_0)|^2-[\boldsymbol{\xi}\times(\nabla\times\mathbf{B}_0)]\cdot[\nabla\times(\boldsymbol{\xi}\times\mathbf{B}_0)]\}\mathrm{d}^3\mathbf{r}.$$

For a cylindrically symmetric basic state

$$\mathbf{B}_0=(0,\ B_\theta(r),\ B_z(r)),$$

a perturbation of the form

$$\boldsymbol{\xi}=\{\xi_r(r)\cos(m\theta+kz)\hat{\mathbf{r}}-\xi_0(r)(B_z/B)\sin(m\theta+kz)\boldsymbol{\theta}$$
$$+\xi_0(r)(B_\theta/B)\sin(m\theta+kz)\hat{\mathbf{z}}\}\cos h\theta \qquad (6.68)$$

is considered, where m is arbitrary,

$$h=2\pi i/L, \qquad i=1,\ 2,\ \ldots$$

$$h=(2n+1), \qquad n=0,\ 1,\ 2,\ \ldots$$

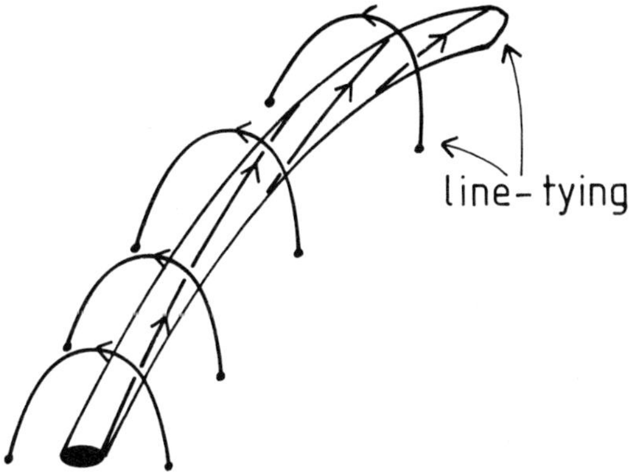

(a)

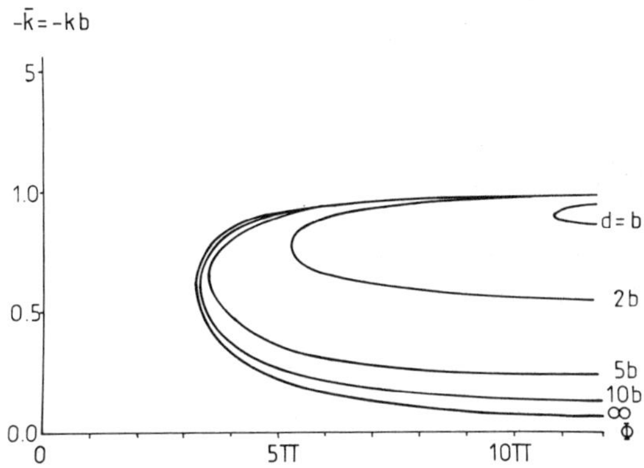

(b)

FIGURE 6.13 (a) A possible preflare configuration consisting of a weakly twisted flux tube anchored at its ends and located within a magnetic arcade. The plage filament is assumed to be located along the flux tube. (b) Sufficient conditions for the amount of twist $\Phi \equiv 2L/b$ required to produce instability of a flux tube embedded in an arcade. The flux tube has length $2L$and its axis is situated at a height d above the photosphere. For Φ greater than some critical value (i.e. to the right of each curve), the equilibrium is unstable for a range of wavenumbers k (From Hood and Priest, 1980.)

This perturbation is perpendicular to **B** and has been chosen to vanish at $\theta = \pm\frac{1}{2}\pi$, so as to simulate the effect on the arcade of *line-tying* in the dense photosphere. It should be noted that a full stability analysis to all possible three-dimensional perturbations has not yet been undertaken, since it requires the solution of partial differential equations. The present analysis provides necessary and sufficient stability conditions only to the above class of perturbations; in other words, if an instability is found it certainly exists but it may commence at a lower threshold.

Minimizing first with respect to ξ_0 gives

$$\xi_0 = \frac{r\{(krB_\theta - mB_z)\xi_r' - (krB_\theta + mB_z)\xi_r/r\}}{B(m^2 + h^2 + k^2r^2)},$$

where a dash denotes differentiation with respect to r. This reduces δW to an integral over r alone, namely,

$$\delta W = \tfrac{1}{2}L\int_0^\infty (F\xi_r'^2 + 2H\xi_r\xi_r' + G\xi_r^2)\mathrm{d}r,$$

where F, G, H are functions of r that depend on $B_\theta(r)$, $B_z(r)$ and the parameters k, m, h. Minimizing next with respect to r leads to the Euler–Lagrange equation

$$(F\xi_r')' = (G - H')\xi_r \tag{6.69}$$

and the stability of the basic state may then be decided by solving for $\xi_r(r)$ and determining its zeros: if ξ_r has no zeros the equilibrium is stable, but if it has at least one zero instability ensues.

Hood and Priest investigated first the stability of a sheared arcade of type I above, but, for a wide range of equilibria and (line-tied) perturbations, they were unable to find instability. This and the results of §6.5.2. strongly suggests (but does not rigorously prove) that *equilibria on the lower branch are stable.* It appears that line-tying is able to prevent instability of this type of field altogether. They considered, for example, the cylindrically symmetric arcade

$$B_\theta = \frac{B_0(r/b)}{1 + r^2/b^2},$$

$$B_z = B_0\left(a^2 + \frac{1}{(1 + r^2/b^2)^2}\right)^{1/2},$$

with its axis in the photosphere, and also Low's field (6.35) with its axis below the photosphere. In addition, they tested the lower branch of Birn *et al.*'s field (6.48) and found it stable to a wide range of line-tied perturbations, including two-dimensional ones in the x-y plane, in agreement with §6.5.1. Finally, the more general fields of Priest and Milne (1980), such as (6.51), were considered: the lower branch was found to be stable to line-tied perturbations of the form

$$\boldsymbol{\xi} = (\tilde{\boldsymbol{\xi}}(r, \theta) + \hat{\mathbf{z}}\tilde{\xi}_z(r, \theta))\mathrm{e}^{ikz},$$

with the real and imaginary parts of $\tilde{\boldsymbol{\xi}}$ and $\tilde{\boldsymbol{\xi}}_z$ in phase. Furthermore, some of the upper branch solutions are cylindrically symmetric and are stable to all perturbations of the form (6.68): thus it is not safe to assume that a field on the upper branch will necessarily be linearly unstable and erupt to reform as a field on the lower branch.

Preflare configurations of type (II), which have the magnetic field directed predominantly along the filament, have also been tested for stability. The fact that motions are frequently observed along the length of a filament suggests that the magnetic field is directed along the filament and forms a (possibly twisted) flux tube that is embedded within an external arcade (Figure 6.13a). All the field lines are assumed to be line-tied. The stability of a twisted flux tube without the external arcade was investigated by Hood and Priest (1979). They (and others) demonstrated that the dominant stabilizing mechanism for coronal loops is *photospheric line-tying* and found that a flux tube becomes kink unstable when it is twisted too much (§4.4.8). This was suggested as the *basic cause for the eruption of a plage filament* at the start of a two-ribbon flare (see also Sakurai, 1976). Their analysis was extended, as follows, by Hood and Priest (1980) to derive sufficient conditions for instability of a more realistic preflare configuration, containing the extra stabilizing effect of an overlying arcade.

The configuration consists of a cylindrical structure of length $2L$, line-tied at its ends and surrounded by a magnetic arcade; a vertical section through the structure has a similar topology to that of Figure 6.7c and the weak curvature normal to the plane of the Figure is neglected. The axis is located at distance d above the photosphere and the form adopted for the magnetic field is simply the cylindrically symmetrical field of uniform twist, namely

$$B_\theta = \frac{B_0(r/b)}{1+r^2/b^2}, \quad B_z = \frac{B_0}{1+r^2/b^2}, \quad \text{for } r \leqslant d, \tag{6.70}$$

with the distance r measured from the flux tube axis. Sufficient conditions for instability are obtained by integrating the Euler–Lagrange equation (6.69). The critical twist Φ that is sufficient for instability depends on the axial wavenumber k, the height d of the axis above the photosphere and the constant b (which appears in (6.70) and determines the rate at which the magnetic field falls off with distance). The results are plotted in Figure 6.13b. For Φ greater than a critical value Φ_{crit}, there exists a range of values of k for which the tube is unstable. To the right of each curve the equilibrium is definitely unstable, but, because the conditions are only sufficient, we cannot state that the region to the left of each curve is stable. However, the region to the left of the curve $d=\infty$ is certainly stable to perturbations of the form (6.68). This curve gives us a lower bound on the critical twist for these perturbations, whereas the curve for a given d supplies an upper bound: thus, for example, when $d=3b$, the configuration is stable to perturbations (6.68) when $\Phi<3.3\pi$ and has certainly become unstable by the time $\Phi>4.2\pi$. Moreover, a stronger result for stability to all kink ($m=1$) perturbations can be proved, namely that when $\Phi\leqslant 2\pi$ the flux tube is certainly stable.

In conclusion, simple sheared arcades of type (I) appear so far to be stable, whereas configurations of type (II) can become unstable if the length, twist or filament height are too great. Two possible ways for this to happen are as follows. In the first place, if the length or twist of the flux tube is increased while its height d remains fixed, the effect of line-tying is gradually reduced until instability ultimately ensues; for example, when $d=5b$, $\Phi_{\text{crit}}=3.6\pi$. On the other hand, if the height increases, while the length and twist remain fixed, the stabilizing effect of the overlying arcade is gradually reduced until again the structure becomes unstable; for example, with $\Phi=5\pi$ we find $d_{\text{crit}}=2.2b$. A means of making the height increase and so triggering the flare is suggested in §5.4.6. However, in general a filament may rise as it is twisted.

6.6 SLOW PASSIVE EVOLUTION

In the previous sections, special examples and general existence and stability properties for sequences of two-dimensional quasi-static equilibria have been considered. They were described by a certain parameter λ, that was assumed to vary with time. A (slow) time variation, however, involves (small) bulk velocities and electric fields. A discussion of slowly evolving configurations must thus include the equations for these quantities. In this section we will do this for the general three-dimensional MHD problem. Since we are dealing with slow time variations, we retain only those terms that are of first order in a suitable parameter measuring the rate of change. It is obvious from the structure of the MHD equations that the static equilibrium conditions (6.18) remain unchanged. For the purpose of this section, we will write these conditions in a different form, expressing the magnetic field **B** by Euler potentials α and β as

$$\mathbf{B}=\nabla\alpha\times\nabla\beta$$

and the gravitational field by the gravitational potential ψ:

$$\mathbf{g}=-\nabla\psi.$$

If the pressure is expressed as a function of these potentials $p=P(\alpha, \beta, \psi, t)$, the following three equations can easily be derived:

$$\frac{\partial P}{\partial \alpha} = \mathbf{j} \cdot \nabla \beta, \tag{6.71}$$

$$\frac{\partial P}{\partial \beta} = -\mathbf{j} \cdot \nabla \alpha, \tag{6.72}$$

$$\frac{\partial P}{\partial \psi} = -\rho. \tag{6.73}$$

(The only restriction on the validity of these equations is that the gravitational potential must not be constant on field lines.) In the case of force-free fields, (6.71) and (6.72) simplify to

$$\mathbf{j} \cdot \nabla \beta = 0, \tag{6.71'}$$

$$\mathbf{j} \cdot \nabla \alpha = 0. \tag{6.72'}$$

To these equations, the first order equations (6.1), (6.17) and (6.8) must be added:

$$\frac{\partial \rho}{\partial t} = -\nabla \cdot (\rho \mathbf{v}), \tag{6.1}$$

$$\mathbf{E} + \mathbf{v} \times \mathbf{B} = 0, \tag{6.17}$$

$$\frac{\partial \mathbf{B}}{\partial t} = -\nabla \times \mathbf{E}. \tag{6.8}$$

These equations must be completed by an energy equation, which is complicated and may be of zeroth or first order, depending on which terms are dominant. A reasonable simplification for certain conditions (see also Section 6.2) seems to be to prescribe the temperature as a function of α, β and ψ

$$\frac{\tilde{\mu}}{R} \frac{P}{\rho} = T(\alpha, \beta, \psi, t).$$

Simplified cases of interest are conditions where T is constant on field lines (i.e. $T(\alpha, \beta, t)$) or even constant throughout. Assuming $T(\alpha, \beta, \psi, t)$ as given, (6.73) can be integrated, leading to

$$P = P_0(\alpha, \beta, t)\exp\left\{-(\tilde{\mu}/R)\int_{\substack{\psi_0 \\ \alpha,\,\beta=\text{const}}}^{\psi}(1/T)\mathrm{d}\psi\right\}, \tag{6.74}$$

where $P_0(\alpha, \beta, t)$ remains arbitrary. If T is assumed constant on field lines and g is constant, (6.74) assumes a form corresponding to (6.21), namely

$$P = P_0(\alpha, \beta, t)\exp(-y/H(\alpha, \beta, t)), \tag{6.75}$$

where

$$H = RT/(g\tilde{\mu}).$$

A solution of the (slow) time-dependent problem thus requires the solution of the static equations (6.71) to (6.73) at any instant of time with boundary conditions for α, β, P and T. Prescribing α and β at the boundary means prescribing the footpoints of field lines. The "natural" boundary conditions for the three-dimensional problem are thus more closely related to the actual physics than is the two-dimensional problem (6.20). This three-dimensional problem, however, has not yet been solved in a general way.

Once the solution for α and β and p is known for a sequence of time-varying boundary conditions, **E** and **v** can be calculated uniquely from (6.1), (6.17), and (6.8), in the way described by Schindler and Birn (1978) for slow convection in the earth's magnetosphere, provided that all field lines cross the boundary and the boundary conditions are formulated in the appropriate way. What "appropriate" means will become clear from the discussion of the method of solution given further below. At this point, we note that the pressure can be prescribed only at one point on each field line, as is obvious from (6.74). This means that p cannot be prescribed over the whole boundary if there are field lines which intersect the boundary more than once, as is the case for most pre-flare structures discussed in the literature.

We suppose now that the pressure function $P(\alpha, \beta, \psi, t)$ and the corresponding sequence of solutions $\alpha(\mathbf{r}, t)$, $\beta(\mathbf{r}, t)$ are known. Equation (6.8) is satisfied identically if the electric field is expressed as

$$\mathbf{E} = \frac{\partial\beta}{\partial t}\nabla\alpha - \frac{\partial\alpha}{\partial t}\nabla\beta - \nabla\Phi, \tag{6.76}$$

in terms of an electrostatic potential Φ, which must be constant on field lines if the ideal Ohm's law (6.13) is to hold. In an appropriate gauge, Φ may be set identically zero, and then (6.13) yields

$$\frac{\mathrm{D}\alpha}{\mathrm{D}t}=\frac{\partial\alpha}{\partial t}+\mathbf{v}\cdot\boldsymbol{\nabla}\alpha=0 \tag{6.77}$$

$$\frac{\mathrm{D}\beta}{\mathrm{D}t}=\partial\beta\partial t+\mathbf{v}\cdot\boldsymbol{\nabla}\beta=0, \tag{6.78}$$

which means that the values of α and β are "frozen" in the plasma. Since $\alpha(\mathbf{r}, t)$ and $\beta(\mathbf{r}, t)$ are known, the electric field follows explicitly from (6.76). Note that the representation of **B** by Euler potentials α and β is not unique, as was pointed out by Grad (1978). Therefore, the velocity components following from (6.77) and (6.78) would in general not correspond to the real motions. If, however, the boundary conditions on α and β are in accordance with "frozen-in" motion, which is consistent with $\Phi=0$, the non-uniqueness is removed and (6.77) and (6.78) yields the real fluid velocity components.

Alternatively, the velocity components perpendicular to **B** can be calculated from (6.17) once the electric field is known from (6.76):

$$\mathbf{v}_{\perp}=\frac{1}{B^2}\mathbf{E}\times\mathbf{B}. \tag{6.79}$$

The parallel component can then be calculated from (6.1) by integrating along the field lines

$$\frac{\rho v_{\|}}{B}=\left.\frac{\rho v_{\|}}{B}\right|_{s_0}-\underset{\alpha,\,\beta=\text{const}}{\int_{s_0}^{s}}\left(\frac{\partial\rho}{\partial t}+\boldsymbol{\nabla}\cdot(\rho\mathbf{v}_{\perp})\right)\frac{\mathrm{d}s}{B}, \tag{6.80}$$

where the density ρ follows from the pressure p either through (6.73) or through the temperature function.

Although this derivation seems to indicate that the (ideal MHD) electric field can be calculated from any arbitrary time-dependent magnetic field, this is not the case, as was discussed by Syrovatsky (1978). If a field configuration contains closed field lines, a suitable choice of $\partial\mathbf{B}/\partial t$ might produce a non-vanishing $\oint \mathbf{E}\cdot\mathbf{ds}$ on such a field line, which must be connected with a non-vanishing $\mathbf{E}\cdot\mathbf{B}$. Such a parallel electric field would produce an infinite current density on this closed loop unless finite resistivity is imposed. Obviously, parallel electric fields would also arise if two footpoints of a field line were not at the same electric potential. An arbitrary change of the boundary conditions for α, β, p and Φ must therefore lead to a situation where the "smooth" equilibrium conditions (6.71) to (6.73) with (6.1) (6.17), (6.8) are violated, at least locally. From the estimates in Section 6.2 it is apparent that this implies very short time-scales (i.e. MHD waves, if the equilibrium is stable) or very short length-scales (e.g. thin current sheets (Chapter 3) or electrostatic double-layers).

A smooth solution thus requires that the boundary conditions be consistent with a "frozen-in" condition at the boundary.

Equation (6.80) yields another restriction on the possible choice of boundary conditions, since $v_{\parallel}$ can be prescribed only at one point on each field line. For closed field lines, this implies that ρ (and thus p) cannot be chosen arbitrary on these field lines, but must be determined from the conservation laws, as has been described by Grad *et al.* (1975) for adiabatically evolving two-dimensional configurations.

For field lines with both ends at the boundary, the problem is similar to the case of closed field lines if $v_{\parallel}$ is prescribed (e.g. to be zero) at both ends. An alternative procedure for getting "smooth" solutions is to prescribe p at the boundary and to assume that the boundary is "open" (i.e. will subtract or supply the necessary amount of plasma at the "desired" speed). In this case $v_{\parallel}$ can be prescribed only at one point on each field line; e.g. if symmetry is imposed, $v_{\parallel}=0$ at the plane of symmetry.

Fixing p and $v_{\parallel}$ at the boundary, however, and assuming slow time-dependence implies the existence of thin layers where one or more of the assumptions of ideal MHD, quasi-neutrality, and low plasma speed must be violated (Birn, 1980).

6.7 DISCUSSION

The general equilibrium properties and the special examples discussed in Sections 6.3 and 6.4 have shown that, under a large variety of circumstances, the quasi-static evolution in the preflare stage might lead to critical points, at which neighbouring equilibrium solutions cease to exist. It has been shown that such points can be reached from potential fields by sequences of equilibria that are stable in two dimensions according to ideal MHD theory, resistive MHD theory and collision-free Vlasov theory; and, at least in some cases, it could be shown that they are also stable against three-dimensional ideal MHD perturbations.

The critical points are interpreted by several authors as the onset of a dynamic (eruptive) evolution. The advantage of this interpretation is that it provides a quantitative criterion for the onset of the dynamic phase. A further advantage is that the critical points can be determined from equilibrium theory alone. The character of the instability, however, (e.g. its topology and whether it is resistive or ideal MHD) cannot be deduced from the equilibrium theory.

The main draw-back of the theory is its restriction to two dimensions. Fortunately, the big two-ribbon flare provides an application for which the neglect of the dependence on the space coordinate along the ribbons seems to be reasonable. It remains, however, an open question, whether critical "catastrophe" points exist also in the general three-dimensional problem.

It is important to note that, in the framework of the theory described in this chapter, the "catastrophe" concept is not the only way to envisage the onset of a solar flare. A possible alternative is the triggering of the large-scale instability by a small-scale instability, possibly via a thremal instability (see §3.5 and Heyvaerts *et al.*, 1977). In this case the pre-flare evolution might proceed on a solution branch which is stable in ideal MHD, but unstable in resistive MHD, the ordinary resistivity, however, being too small to give a sufficiently high growth-rate. The effect of the small-scale instability would be to increase suddenly the resistivity and thereby the growth-rate of the resistive instability. In this case the onset point cannot be determined from the large-scale equilibrium theory, but from a microscopic theory or a detailed discussion of the energy balance, which is beyond the scope of this chapter. In cases where the plasma is sufficiently collision-free, transitions from stability to instability on a suitable branch can also arise from particle inertia effects.

Acknowledgements

We acknowledge important comments on the three-dimensional stability problem by A. Hood and L. Janicke.

This work was supported by the Deutsche Forschungsgemeinschaft through the Sonderforschungsbereich Plasmaphysik Bochum/Jülich.

REFERENCES

Amann, H., *SIAM Review* **18**, 620 (1976).
Barnes, C. W. and Sturrock, P. A., *Astrophys. J.* **174**, 659 (1972).
Bennett, W. H., *Phys. Rev.* **45**, 90 (1934).
Bernstein, I. B., Friemann, E. A., Kruskal, M. D. and Kulsrud, R. M., *Proc. Roy. Soc.* **A244**, 17 (1958).
Birn, J., to be published (1980).
Birn, J., Goldstein, H. and Schindler, K., *Solar Phys.* **57**, 81 (1978a).
Birn, J., Goldstein, H. and Schindler, K., in *Pleins Feux sur la Physique Solaire*, Proc. Sec. Europ. Conf. on Solar Physics, ed. by CNRS, Toulouse, p. 237 (1978b).
Grad, H., *Phys. Fluids* **7**, 1283 (1964).
Grad, H., preprint (1978).
Grad, H., Hu, P. N. and Stevens, D. C., *Proc. Nat. Acad. Sci. U.S.A.* **72**, 3789 (1975).
Harris, E. G., *Nuovo Cim.* **23**, 115 (1962).
Heyvaerts, J., Priest, E. R. and Rust, D. M., *Astrophys. J.* **216**, 123 (1977).

Heyvaerts, J., Lasry, J. M., Schatzman, M. and Witomsky, G., preprint (1979).
Hildner, E., *Solar Phys.* **35**, 123 (1974).
Hood, A. W., private communication (1979).
Hood, A. W. and Priest, E. R., *Solar Phys.* **64**, 303 (1979).
Hood, A. W. and Priest, E. R., *Solar Phys.* **66**, 113 (1980).
Jockers, K., *Solar Phys.* **56**, 37 (1978).
Low, B. C., *Astrophys. J.* **212**, 234 (1977).
Low, B. C. and Nakagawa, Y., *Astrophys. J.* **199**, 237 (1975).
Priest, E. R. and Milne, A. M., *Solar Phys.* **65**, 315 (1980).
Sakurai, T., *Pub. Astron. Soc. Japan* **28**, 117 (1976).
Schindler, K. and Birn, J. *Physics Reports* **47**, 109 (1978).
Schindler, K., Birn, J. and Janicke, L., to be published (1980).
Spitzer, L., *Physics of Fully Ionized Gases*, Interscience Publ., John Wiley, New York–London (1962).
Syrovatsky, S. I., *Astrophys. Space Sci.* **56**, 3 (1978).
Thom, R., *Structural Stability and Morphogenesis*, Benjamin, Reading–Massachusetts (translated from the French edition) (1975).
Zwingmann, W., private communication (1979).

CHAPTER 7

Two-ribbon flares: (post)-flare loops

G. W. PNFUMAN†

Max-Planck-Institute for Physics and Astrophysics, Karl-Schwarzschild-Strasse 1, 8046 *Garching bei München, Fed. Rep. Germany*

7.1 INTRODUCTION

Probably the most spectacular and persistent features associated with the classic large two-ribbon flares are the so-called flare loops or "post-flare" loops. These loop systems are observed to rise upward slowly into the corona during the main phase of the flare. The velocity of ascent decreases with height from about 10–20 km s^{-1} at the beginning to a fraction of a km s^{-1} when they are finally observed in soft X-rays at great heights. As will be stressed again later, the system does not appear to consist of single loops rising upward but, rather, of newly formed or activated stationary loops appearing at successively higher levels. The entire event is relatively long-lived, lasting for some 10–20 hours after the flash phase of the flare.

The term "post-flare" as applied to these loop systems is, unfortunately, quite misleading, since it implies that the phenomenon occurs after the flare and is somehow a byproduct of the flare process. This designation probably originated historically, because the loops were first clearly seen on the limb in Hα only after the system had risen quite high in the corona; that is, unless the flare occurred exactly on the limb, the loops lower down would not be clearly seen against the disc. By this time, the chromospheric flare was well in progress. Now, however, as a result of recent EUV and soft X-ray observations taken aboard Skylab and the detailed comparison of these observations with ground-based Hα photographs taken simultaneously, a much more complete and unified picture of the solar flare is emerging. These comprehensive observations, spanning a wave-length scale and with spatial resolution never before achievable, now strongly indicate that the "post-flare" loops are a major aspect of the flare. The Hα chromospheric flare ribbons have now been clearly identified as the footpoints of the flare loops and, therefore, occupy only

† On leave from the High Altitude Observatory, National Center for Atmospheric Research, Boulder, Colorado, USA.

a small fraction of the total flare volume. Furthermore, the observations suggest that the chromospheric flare does not produce the flare loops — rather *the flare loops produce the chromospheric flare.* Hence, our main concern here will be with the question of what produces the flare loops. The answer to this question is still a subject of some dispute. However, the one most consistent with the observations and our understanding of solar MHD is that the *loops are formed via magnetic reconnection* as an open magnetic field configuration evolves to a closed configuration. The energy for the flare is derived from the magnetic field via Joulean dissipation in the vicinity of the reconnection site.

All these arguments will be amplified as we proceed. But first, a review of the pertinent observations on flare loops is in order.

7.2 OBSERVATIONS

Before proceeding further, we must first distinguish between two apparently different types of flares (§1.1, §2.1.4, §2.4.2). One is the *two-ribbon flare*, which generally occurs along a neutral line in the photospheric magnetic field and is seen in Hα on the disc as two bright ribbons expanding outward from the neutral line. These are the largest and most energetic of all flares and are the ones that are accompanied by the rising loop systems in the corona. The majority of flares, however, are of a different nature — much smaller in the scale and also 10 to 100 times less energetic than the two-ribbon flares. These are the *simple-loop flares*, sub-flares or "compact" flares. They appear as small bright loop structures, usually away from the main neutral lines of the active region; the loops brighten in place and there is no apparent increasing separation of the footpoints and no expansion of the system upward into the corona. In this chapter, we will exclude these subflares from our discussion, since they do not produce rising flare-loop systems and have been considered in Chapters 4 and 5. Therefore, the reader should keep in mind that the word "flares" as used here refers only to the two-ribbon type. Of this type, the most reliable and extensive observations were taken during the Skylab period in 1973. Two large flares, those of July 29 and September 7, exhibited classic flare-loop systems which were singled out for concentrated study at the Solar Flare Workshop held in Boulder, Colorado in 1976–1977 (cf. Moore *et al.*, 1980). Our discussion will lean heavily on the results of that study.

7.2.1 Morphology

A classic morphological description of flare loops as seen in Hα is given by Bruzek (1964a), based upon a detailed study of a particular active region which was present on the Sun in July, 1961. The region contained a rather complex spot group and produced four large flares. The most conspicuous features of the flares were the two long flare ribbons which expanded laterally away from a neutral line in the magnetic field of the active region. When the separation of the ribbons became obvious, loops connecting the flare ribbons became visible. An upward expansion of these loops was apparent, with the expansion speeds being about 5–10 km s^{-1}. The footpoints of the loops were rooted in the flare ribbons. However, as pointed out by Bruzek, there was not much real expansion of the loops. Rather the whole system enlarged as new blobs or knots of material condensed out of the corona at progressively higher levels. The individual loops did not move very much. The tops of the loops were visible in the yellow coronal line and had extremely broad Hα wings (Jefferies and Orrall, 1963a), suggesting a high-energy phenomenon. The overall evolution of the loops and their relation to the flare ribbons appeared to Bruzek as a continuation of the flare upward into the corona. Figure 7.1 shows a typical flare-loop system as seen on the limb in Hα, whereas, in Figure 7.2, we see the loops as projected against the disc. The overall evolution of the loop system, i.e., the slow rise upward into the corona, is clearly shown in the sequence of Hα limb photographs of Figure 7.3.

It is quite clear from a wide variety of observations (cf. Bruzek, 1964a, b; Neupert *et al.*, 1974; Kahler *et al.*, 1975; Cheng and Widing, 1975; Vorphal *et al.*, 1975; Pallavicini *et al.*, 1975; Rust and Webb, 1977; MacCombie and Rust, 1978) that the basic structure involved in the post-flare loop phenomenon is a system or arcade of arches which successively appear at greater heights in the corona. On the limb they appear bright, with the very brightest parts occurring at the tops of the loops where the most intense soft X-ray emission is observed (cf. Rust, 1976; Kahler, 1976; Pallavicini *et al.*, 1975; Vorpahl *et al.*, 1977; Nolte *et al.*, 1979; Petrasso *et al.*, 1979; MacCombie and Rust, 1979). They generally appear after filament eruptions, beginning as a small bright mound in the location where the filament was previously situated. As the mound grows both in height and laterally, distinct loop-like structures then become apparent. On the disc, they appear as dark fibril-like structures spanning the magnetic neutral line and filament cavity and terminating in the bright flare ribbons.

The deduced heights of the loops as a function of time depend strongly upon what type of emission is observed. The lowest loops are seen in Hα. These rise to heights considerably less than 100,000 km. The highest loops are observed in soft X-rays at heights of even greater than 100,000 km (Nolte *et al.*, 1978; MacCombie and Rust, 1978). Above the Hα loops but below the X-ray loops are seen loops at intermediate temperatures such as in the EUV lines (Withbroe, 1978). One particularly good example is the loop system following the class 3b flare of July 29, 1973 in which good coverage was obtained in Hα, EUV, and thermal X-rays (from Skylab). The complete flare was observed in the Hα line at the McMath-Hulbert Observatory and with high spatial resolution ($\sim$1 arc sec) by the Big Bear Solar Observatory (see Figure 4). High resolution ($\sim$2 arc sec)

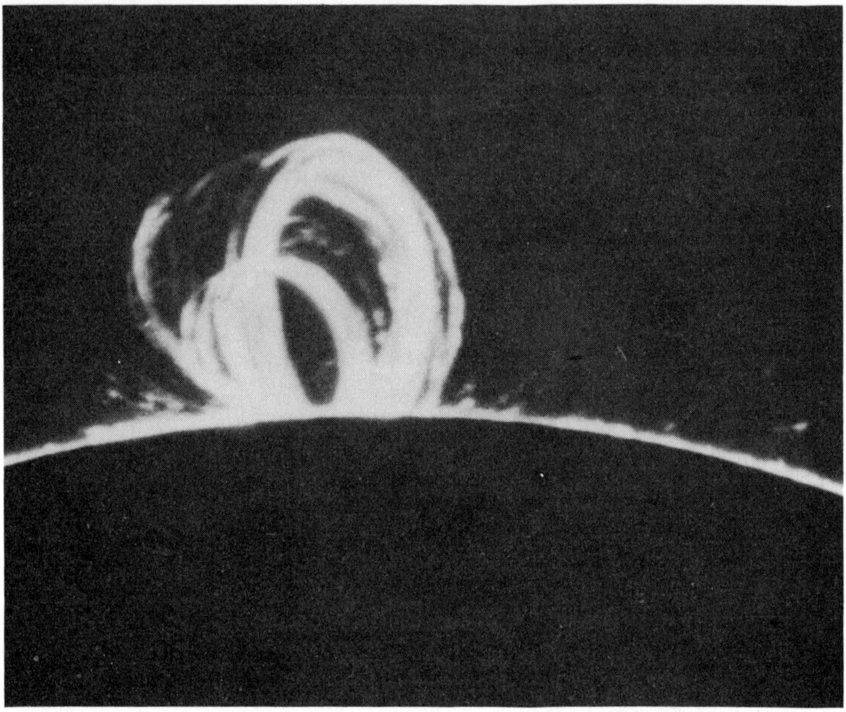

FIGURE 7.1 Limb photograph of a typical flare-loop system taken in Hα using a 3 Å wide band filter. (Courtesy, Association of Universities for Research in Astronomy, Inc., Sacramento Peak Observatory.)

X-ray observations were obtained with the AS&E S-O54 telescope aboard Skylab, an example of which is shown in Figure 5.

The flare began with the eruption of a large quiescent prominence which became activated at 13:02 (Moore *et al.*, 1979). The Hα flare began at 13:12 and reached maximum between 13:27 and 13:31. Since the Skylab satellite was in the earth's shadow when the flare began, the X-ray coverage began somewhat later. Figure 6, taken from Moore *et al.* (1978), shows the deduced heights of the classic flare loop system that accompanied this event. The cold Hα loops were first discernable at about 14:00 at a height of slightly less than 20,000 km*. The system grew to about 60,000 km in 3 hours when, apparently due to their low density, they faded from view. The X-ray observations which began just about when the loops disappeared in Hα showed, however, that the event was by no means over. They appeared at 80,000 km, about 20,000 km above the cold loops and rose slowly to over 100,000 km during a period of about 8 hours. Earlier coverage was obtained for the 7 September 1973 flare, showing that *the X-ray loops system actually originates quite early in the flare and outlasts the optical event*

FIGURE 7.2 An Hα photograph of flare-loops seen projected against the disc. This picture was taken at Hα + 0.8 Å with a narrow band filter (0.25 Å) and so only the dark plasma moving with the proper Doppler velocity is apparent. (Courtesy, Association of Universities for Research in Astronomy, Inc., Sacramento Peak Observatory.)

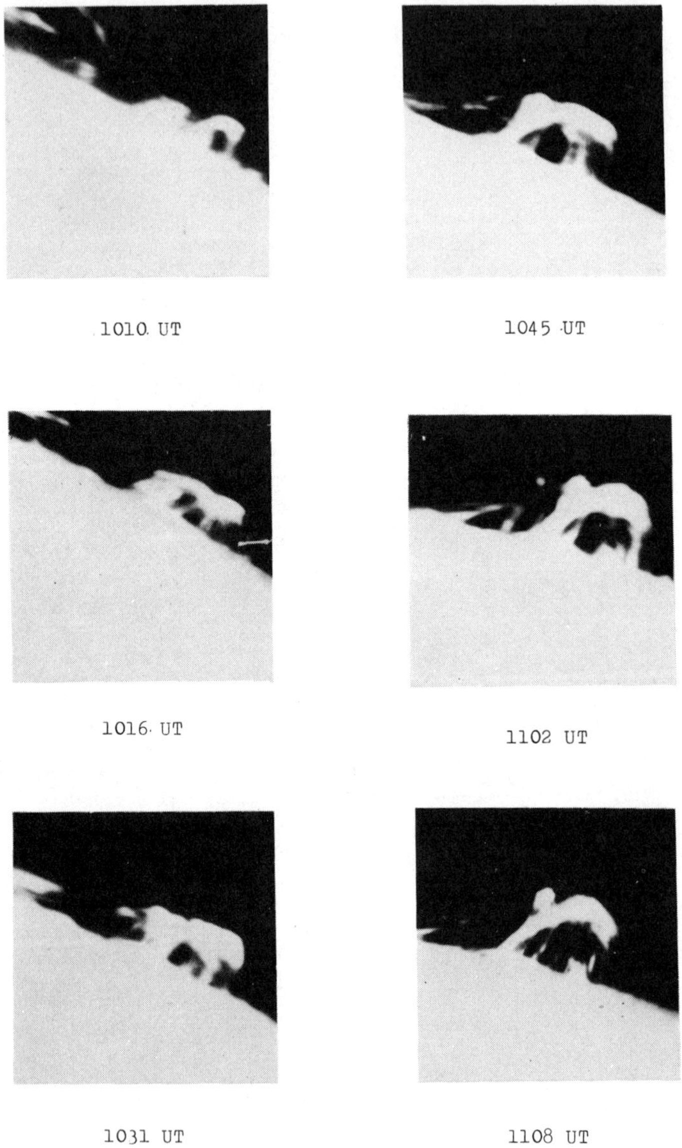

FIGURE 7.3 A sequence of Hα photographs showing the growth of a typical flare-loop system — this one being for the flare of 12 September, 1966. (Courtesy, W. Mattig, Kiepenheuer-Institut für Sonnenphysik.)

1129 UT

1154 UT

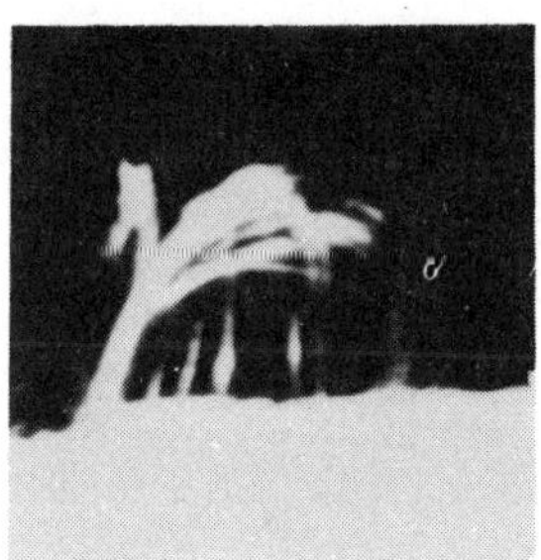

1230 UT

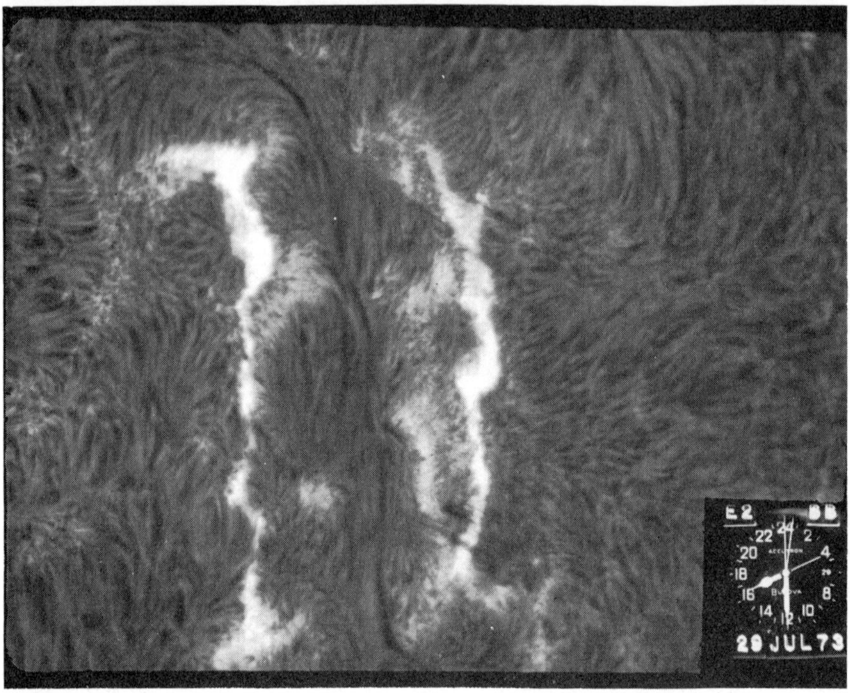

FIGURE 7.4 Flare ribbons, in Hα, for the large class 3b flare of 29 July 1973. (Courtesy, Big Bear Solar Observatory.)

by many hours. The reader is referred to MacCombie and Rust (1979) for a discussion on the growth of X-ray loop systems for other flares observed by Skylab.

The velocity of growth of flare loops is also dependent upon the emission line observed. In Hα, the system initially rises with speeds in the range of 10–20 km s^{-1} (Bruzek, 1964a; Roy, 1972; Martin, 1980). As it rises, the speed steadily decreases in an exponential manner to just a few km per second when the loops later disappear. The X-ray loops also show a similar velocity versus time profile but the speeds are considerably lower, being in the range 0–1 km s^{-1} (MacCombie and Rust, 1979; Nolte *et al.*, 1979). Švestka *et al.* (1978) has found that the best fitting curve for the velocity–time profile observed in the 29 July flare is given by the formula

$$v = 34.7 \exp(-t^{0.46}),$$

where v is in km s^{-1} and t in units of 20 min beginning at 13:20. The observed speed versus time for the combined observations of the 29 July event along with Švestka's fit is shown in Figure 7.7. One very important feature of this figure should be pointed out. Note that, in the velocity of growth curve, *the Hα*

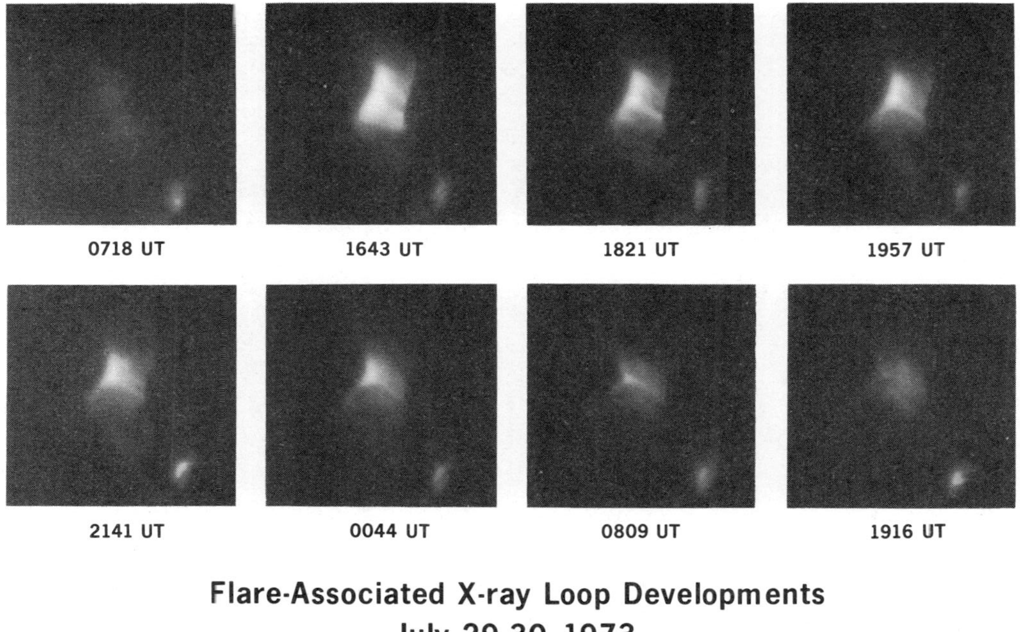

FIGURE 7.5 Flare loops as seen in soft X-rays for the 29 July 1973 flare. These loops lie well above the corresponding Hα loops and rise to considerable heights in the corona. (Courtesy, Solar Physics Group, American Science and Engineering.)

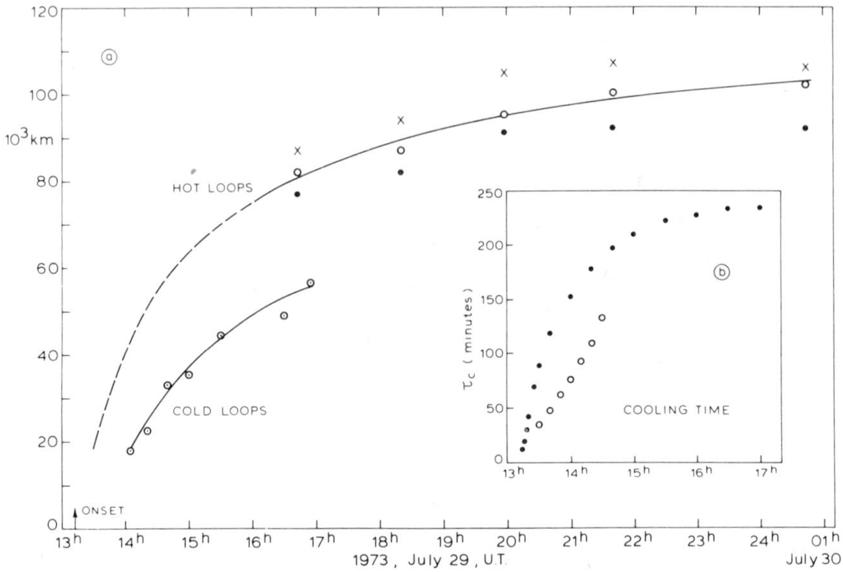

FIGURE 7.6 Heights versus time of the cool (Hα) and hot (X-ray) flare loops during the 29 July 1973 flare. The arrow denotes the Hα flare onset time. The symbols ○, ⊙, ●, and × refer to various geometrical assumptions used to deduce the loop heights and are described in Moore *et al.* (1980). These heights have recently been revised upward somewhat (Švestka, 1979) and should be considered as lower limits. The box in the lower right-hand corner shows the cooling time, τ_r, of the individual loops derived from the loop growth (○), and ribbon separation (●). (From Moore *et al.*, 1980.)

and X-ray observations tend to fall on the same general curve. Although for this flare, the observations do not overlap, we might speculate that the observations might follow the curve at all heights. If this is true, it means that the velocity of ascent of the system is independent of the physical properties of the individual loops (e.g., temperature) and depends only upon the height in the corona at which the system is observed. This is a strong argument in favor of the viewpoint that we are observing a rising system of discretely-formed stationary loops, i.e., there is no actual material motion upward, but a sequence of discrete events. We will touch again upon this point later.

The overall evolution of the loop system seems to follow very closely the evolution of the flare ribbons. Note from Figure 7.7 that the velocity of separation of the ribbons and the rise-velocity of the loop system at a given time are similar — as they should be if, indeed, the flare ribbons are the loci of the footpoints of the loops. The physical connection between the Hα loops and Hα chromospheric ribbons is well-known (Bruzek, 1964a, b). Moreover, more recent observations indicate that the Hα loops are rooted on the inside edges of the ribbons (Roy, 1972; Rust and Bar, 1973; Beckers, 1977), whereas the X-ray

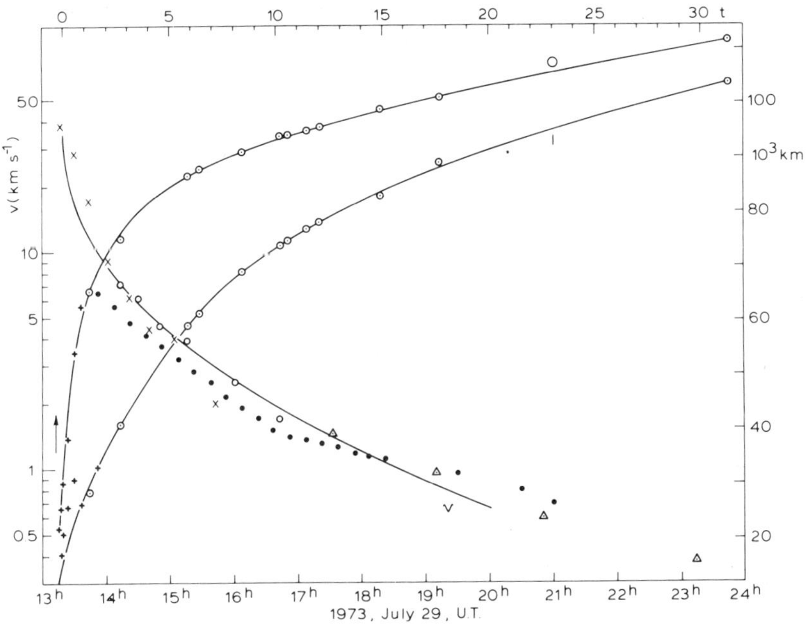

FIGURE 7.7 Velocities (left scale) of loop growth (○ — cool loops; △ — hot loops) and ribbon separation (×, ●) for the 29 July 1973 flare. The solid curve is the empirical fit obtained by Švestka *et al.* (1978) as discussed in the text. Also shown in the figure is the ribbon separation distance (⊙ — right scale) as measured from the outer (O) and inner (I) edges. (From Moore *et al.*, 1980.)

loops originate in the middle and outer portions of the ribbons (Moore *et al.*, 1980). This is then consistent with the concept that the hot loops are the highest with the cool loops nested beneath.

One very conspicuous feature of the flare loop systems is the *downward motion* seen in the legs. As soon as a loop is seen, Hα material is observed to be streaming downward at velocities comparable with the free-fall speed. Kleczek (1964) has estimated that the total mass delivered to the chromosphere by this downflow is about 10^{12}–10^{13} kg. This is comparable with the total mass of the corona. Hence, it is difficult to explain the origin of this material as condensing out of coronal material (Jefferies and Orrall, 1963a, b, 1965; Kleczek, 1963, 1964). Also, considering the high magnetic Reynolds numbers appropriate for coronal material, how can the material be transported across field lines into the loops? Mass could be brought up from below, but upward motions are not observed. However, due to the low temporal resolution of the X-ray experiments aboard Skylab, nothing can be said about motions of any kind in the hot X-ray loops. Perhaps the upward transport of material from below is taking place here.

To summarize the morphological description of a typical post-flare loop event, we may say this. Usually, the first sign of an upcoming two-ribbon flare is *the activation of a filament* about one hour before the flare (Martin and Ramsey, 1972). This activation is manifested as a darkening which becomes most apparent about 10–20 minutes before the flare. Then a brightening and break-up occur about 2–3 minutes before. The flare itself begins as *two bright ribbons* of chromospheric material lying parallel to the neutral line where the prominence was. Very quickly, the two ribbons begin to expand away from the neutral line. At this time loop systems are seen moving upward into the corona. The system consists of discretely formed stationary loops at progressively higher levels. The *footpoints of the loops are rooted in the flare ribbons*, which move laterally outward from the neutral line as the loop system rises (see Figure 7.8). As shown by observations of the same system in different emission lines, the whole structure shows a definite temperature stratification, with the *coolest loops being nested within progressively hotter loops* as one proceeds upward. The hottest loops are the most newly formed loops at the top of the system. The velocity of ascent of the process is similar to the speed at which the flare ribbons spread apart, with the general characteristic being that the speed decreases with increasing height and time. The entire event has a duration of about 10 hours.

7.2.2 Physical properties

The physical structure of loop prominence systems is extremely complex since we are dealing with individual loops in different stages of evolution. Thus the spatial and temporal distribution of any given physical property are intimately linked. Nevertheless, the structure of the whole system can be understood quite nicely by combining two concepts. (1) The temporal variation of the physical properties in a single flux tube, once it is created, depends upon the characteristic of that flux tube only and is independent of variations in flux tubes around it. (2) The composite structure of the whole system evolves in time as the individual loops evolve but with each loop being in a different stage of evolution, beginning with the youngest at the top of the system and the oldest at the bottom.

The hottest part of the loop prominence is at the top where temperatures of 10 million degrees or greater are observed in X-rays (Hudson and Ohki, 1972; Cheng and Widing, 1975; Widing and Dere, 1977; Vorpahl *et al.*, 1977; MacCombie and Rust, 1979; Petrasso *et al.*, 1979; Moore *et al.*, 1980) and the EUV (Neupert *et al.*, 1974; Withbroe, 1978). The Hα loops at the bottom are the coolest with temperatures of only about 2×10^4 K (Jefferies and Orrall, 1964). Thus, there exists a strong vertical temperature gradient across the loops which, presumably, can be maintained by the inhibition of thermal

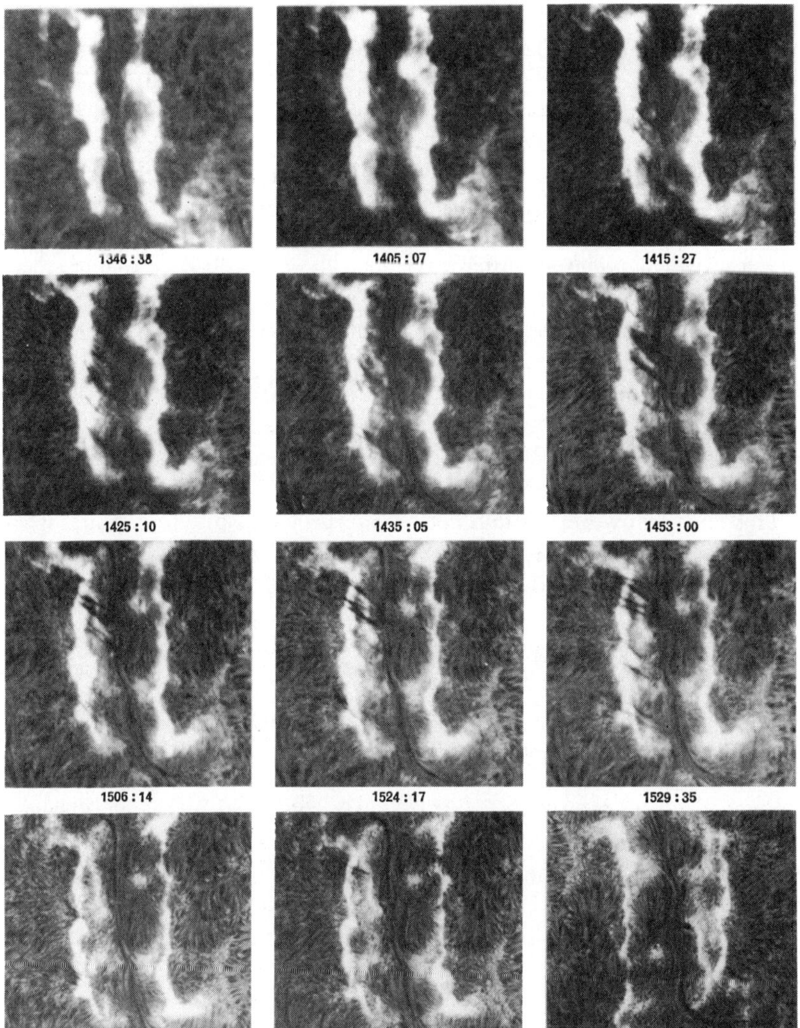

FIGURE 7.8 An Hα time sequence photograph of the flare ribbons of the 29 July 1973 flare. Note the obvious increasing separation of the ribbons as the flare progresses. (Courtesy, Big Bear Solar Observatory.)

conduction across field lines. The electron density at the apex of the loop system lies in the range 10^{16}–10^{17} m^{-3} (Hudson and Ohki, 1972; Widing and Cheng, 1974; Withbroe, 1978; Švestka *et al.*, 1978; MacCombie and Rust, 1979; Moore *et al.*, 1980).

The recent Skylab observations and the detailed analysis of these observations by authors such as Withbroe (1978), Švestka *et al.* (1978), MacCombie and Rust (1979), Nolte *et al.* (1979) and Moore *et al.* (1980) now give us some idea about how the physical properties of the flare plasma vary along a given loop from top to bottom as well as how the properties of the top of the system vary with time. The results of these studies are more-or-less in agreement and indicate that the *temperature is highest at the loop top* with the difference in temperature between top and bottom being a few times 10^6 K. The electron density, however, is essentially uniform. Hence, the gas pressure is higher at the loop top than at the bottom resulting in a *downward* pressure gradient. This could pose a problem for models of loop prominences involving mass injection from below.

The variation with time of the temperature and density at the top of the loop system for specific X-ray events is given by Švestka *et al.* (1978) and MacCombie and Rust (1979). Both the temperature and density decay quite slowly with time. Hence the thermal energy density, $3n_e k_b T$, also decays with time. This decay, however, should *not* be interpreted as due to cooling of an individual loop that is rising in time. Rather, the system consists of individual stationary loops formed at different times and at different temperatures. One does not observe the cooling of an individual loop (which is quite fast, as indicated in the inset of Figure 7.6). Rather, we are seeing *the rate at which energy is being added to the most newly formed loop.*

Little is known about the magnetic field strength in loop prominences since direct measurements are not presently available. One can infer the field strengths, however, using some rather crude physical considerations. Since the flare plasma in loop prominences is obviously contained by magnetic fields, one can infer a lower limit to the field strength by merely requiring that the magnetic pressure exceed the gas pressure, i.e., $B^2/(2\mu) > 2n_e k_b T$ (Moore *et al.*, 1980). Taking $T = 10^7$ K and $n_e = 10^{17}$ m^{-3}, we find $B \sim 10^{-2}$ tesla (100 gauss) at the top of the system. Another way of estimating the field is based upon the consideration that, if indeed the field is dominating the plasma, then one might expect it to be nearly potential. In fact, potential field calculations have met with reasonable success in describing the geometrical structure of post-flare loops (Correll and Roberts, 1958; Rust and Roy, 1971; Roy, 1972; Rust and Bar, 1973). For example, Rust and Bar (1973) find field strengths of 5–8 $\times 10^{-3}$ tesla (50–80 gauss) in loops 40,000 km above the photosphere at times of 3–4 hours after flare onset. These results are consistent with the previous calculations based upon the magnetic pressure argument.

7.3 LOOP PROMINENCES AS A RECONNECTION PROCESS

Since the loop prominence system is situated relatively high in the corona and since it possesses a relatively dense top of material at coronal temperature, it is quite natural that the earliest theories to explain the phenomenon evoked the condensation of material out of the corona. Hence, the well-known term "sporadic coronal condensation". Among those earliest models were those of Kleczek (1957, 1958), Lüst and Zirin (1960), and Olson and Lykoudis (1967). These theories all meet with the difficulty that one loop prominence system would drain the entire corona in about a day. The *enormous mass downflows observed seem to require that the prominence material be brought up from below.* Energetic particles emanating from the flare region or below and stored in the corona where they thermalize have been proposed by Kleczek (1964), Jefferies and Orrall (1964, 1965) and Newkirk (1973). This would explain why upward motions are not observed since the energies would be too high. However, to contain the required number of particles at high energies would require an inordinately strong magnetic field. De (1973) has proposed that loop prominences are paths of electrical current in the corona. This model, however, does not address the question of the origin of the mass and also seems to be in disagreement with conclusion of the last section that the field lines in loop prominences seem to be potential.

Unfortunately, none of the above concepts adequately deal with several important observed properties of post-flare loops — namely, the intimate *association of the loops with the flare ribbons*, the fact that the *loops are formed discretely* in situ, and the *overall rising of the loop system* upward into the corona. In the remainder of this section we will attempt to show that all these properties can be logically explained as the consequence of the *magnetic reconnection of lines of force which were pushed open by the original eruptive process.* We envisage the active region (in the simplest sense) to contain a closed magnetic arcade or helmet bridging the neutral line, with a filament containing its own field structure lying underneath and along the neutral line. Hence, the filament and helmet are perpendicular. For some reason (see Chapter 6), the magnetic field configuration associated with the filament erupts, forcing open the overlying field. After the filament leaves, the open field collapses back to the original closed configuration via magnetic reconnection. This process produces the flare-loops along with the two flare ribbons in the chromosphere. The rise of the loop system is explained by the rising neutral point as the reconnection proceeds upwards. This also naturally explains the outward expansion of the flare ribbons, as well as the discreteness and stationarity of the loops. Actually, a smooth continuous reconnection process would not produce discrete loops. Evidently, if reconnection is actually taking place, it is occurring in a pulse-like manner.

The concept that two-ribbon flares could be explained as the formation of a closed magnetic field configuration from an open one by reconnection was first introduced by Carmichael (1964). He proposed that the energy for the flare was derived from the solar wind which drags field lines of the plage outward into an open configuration, thus setting up the stage for the subsequent collapse. He estimates that the solar wind does work on the plage magnetic fields at the rate of about 10^{23} J/day (10^{30} erg/day) and that this energy could be released occasionally to produce flares. One problem with this suggestion, however, is that it places the flare quite high in the corona where the solar wind dominates. This is not consistent with the observation that the flare loops seem to emerge from very low levels.

Somewhat similar ideas were put forward by Bruzek (1969) and Schmidt (1969). Bruzek suggested that, initially, there exists a filament in a stable state supported on closed magnetic field lines, as in the Kippenhahn–Schlüter (1957) model. Due to energy accumulation in the closed field by shearing, twisting, or perhaps turbulence, the field eventually expands lifting the filament. This produces a quasi-open field configuration in which field reconnection can take place via the Petschek mechanism (Petschek, 1964). The dense tops of the flare loops are produced by the downward compression of material by the reconnected field lines.

The flare process as being manifested by the reconnection of open field lines was also suggested by Sturrock (1968, 1972); however, the dynamical process is somewhat different (§1.4.3). He proposed that the flare occurs in a pre-existing *helmet streamer* type configuration consisting of closed field lines with open field lines above. The flare occurs at the neutral point between the closed and open regions via reconnection. As a result of this, high energy particles are produced at the reconnection site which propagate downward on the closed field lines to produce the flare ribbons and also escape outward to produce Type III bursts. Although this theory has many attractive features, it also has some drawbacks. In helmet streamers, the neutral point is quite high in the corona. If this is where the flare occurs, it is questionable whether there is enough magnetic energy available at this height to account for the flare energy. Also, the theory does not account for the rising of the flare loop system. It seems more plausible that the flare occurs during the *creation* of a helmet streamer configuration rather than in a pre-existing one. This feature is present in the model of Barnes and Sturrock (1972) in which a closed force-free field configuration is presumed to expand outward when its energy exceeds that of an open field corresponding to the same boundary conditions (§6.4). Once the field has been opened, the flare can then be produced by reconnection as the open field relaxes back to a potential field which is at an even lower energy state. More recent work on the way the preflare configuration can become unstable and erupt outwards is described in Chapter 6.

Hirayama (1974), expanding on Bruzek's model, proposed that the material seen falling out of the prominence system was originally brought upward from the chromosphere by evaporation. He assumed that downward heat flow

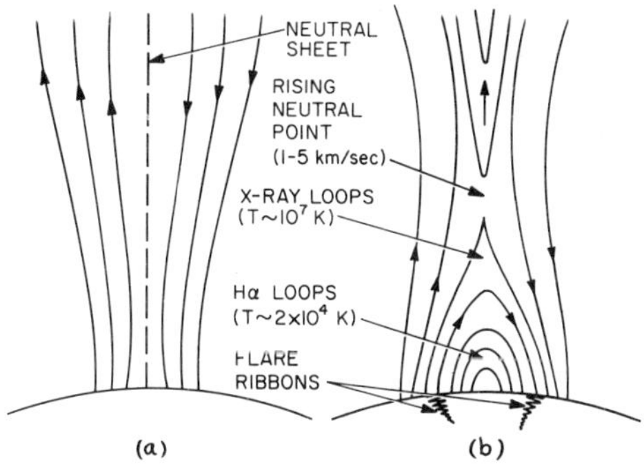

FIGURE 7.9 Schematic of flare loop evolution. (a) Open field configuration immediately after the eruption, and (b) the field topology at some arbitrary time during the flare. Here, the open field is reconnected to form closed loops with the hottest loops at the top where the magnetic energy is being released, and the cool (Hα) loops lower down. These cool loops were formerly hot but have lost most of their thermal energy content due to radiation and thermal conduction. The chromospheric flare ribbons are presumably produced by energy travelling downward (via either thermal conduction or energetic particles) from the reconnecting region.

generated at the reconnection site reaches the chromosphere, produces the flare ribbons, and evaporates chromospheric material upward to the loop tops. The role of evaporation in the solar flare mechanism has been stressed by several authors (Sturrock, 1973; Hudson, 1973; Hirayama, 1974; Hirayama and Endler, 1975; Antiochos and Sturrock, 1978; Withbroe, 1978) and will be discussed again in Section 7.5. A summary of the effect of evaporation in loops is also presented in Sections 5.5 and 5.6.

The overall dynamics of the loop prominence phnomenon envisaged as a reconstruction process has been examined in some detail by Kopp and Pneuman (1976). Here it is assumed that prior to the flare the magnetic field configuration consists of closed loops bridging the neutral line of the active region. Due to some eruptive process, the field is pushed outward into an "open" configuration with a neutral sheet separating the regions of opposing polarity. This sets up lateral inward stresses in the magnetic field which cause it to collapse toward the neutral sheet and reconnect. This process begins at the coronal base and proceeds upward. At any given instant of time the configuration will appear as in Figure 7.9b with open field lines transforming to closed ones at a rising neutral point. Assuming no changes in the boundary conditions occur meanwhile, the neutral point will continue to rise until the original helmet structure is restored.

During the reconnection process, the geometry of any given flux tube is altered drastically as it moves from an open radial configuration to a closed one. For example, the outflow velocity is significantly increased by the rapidly increasing divergence of the tube which occurs during the process. This increasing divergence reduces the gas pressure at higher levels producing an increase in the outward pressure gradient. Another important effect takes place immediately after a flux tube closes. When this happens, all the material below the neutral point is captured in the tube. A shock is immediately generated at the top of the closed region which propagates downward towards the solar surface. Material continues to flow into the tube until the shock reaches the coronal base signaling the flow to stop. This shock can raise the temperature of the gas to 3–4 × 10^6 K. As we shall see later, however, Joulean dissipation at the reconnection site is probably a more effective heating mechanism. Subsequent radiative cooling of this hot gas eventually produces the Hα prominence while new hotter loops as seen in X-rays are being produced above.

7.3.1 Response of the plasma to magnetic field line motion

In order to understand this process in more detail, let us write the basic equations applicable to coronal flow on moving flux tubes. The component of the momentum equation along the field, of course, does not contain the $\mathbf{j} \times \mathbf{B}$ force and may be written in the form

$$\hat{\mathbf{e}}_s \cdot \left[\frac{\partial \mathbf{v}}{\partial t} + (\mathbf{v} \cdot \boldsymbol{\nabla})\mathbf{v} \right] = -\frac{1}{\rho}\frac{\partial p}{\partial s} - \frac{GM_\odot}{r^2}\hat{\mathbf{e}}_r \cdot \hat{\mathbf{e}}_s, \tag{7.1}$$

where $\mathbf{v}$ is the velocity vector, ρ the density, p the gas pressure, G the gravitation constant, $M_\odot$ the solar mass, and r the radial distance from the solar center. $\hat{\mathbf{e}}_r$ and $\hat{\mathbf{e}}_s$ are unit vectors in the radial direction and along the field respectively, and the partial derivative with respect to s refers to differentiation along the field, i.e. $\hat{\mathbf{e}}_s \cdot \boldsymbol{\nabla}$. The relevant physics of the process we are concerned with here can be demonstrated by considering a two-dimensional configuration. In this case, we can write $\mathbf{v}$ in the form

$$\mathbf{v} = v_s\hat{\mathbf{e}}_s + v_n\hat{\mathbf{e}}_n, \tag{7.2}$$

where v_s and v_n are components of the velocity parallel to and perpendicular to the field. $\hat{\mathbf{e}}_n$ is a unit vector normal to the field.

In substituting Equation (7.2) into Equation (7.1), we must keep in mind that the unit vectors $\hat{\mathbf{e}}_s$ and $\hat{\mathbf{e}}_n$ are functions of both space and time. Therefore, we will need to evaluate their derivatives. These are given by

$$\frac{\partial \hat{\mathbf{e}}_s}{\partial t} = \hat{\mathbf{e}}_n \frac{\partial \alpha}{\partial t}, \qquad \frac{\partial \hat{\mathbf{e}}_n}{\partial t} = -\hat{\mathbf{e}}_s \frac{\partial \alpha}{\partial t},$$

$$\frac{\partial \hat{\mathbf{e}}_s}{\partial s} = \hat{\mathbf{e}}_n/R, \qquad \frac{\partial \hat{\mathbf{e}}_n}{\partial s} = -\hat{\mathbf{e}}_s/R,$$

$$\frac{\partial \hat{\mathbf{e}}_s}{\partial n} = \hat{\mathbf{e}}_n/R_n, \qquad \frac{\partial \hat{\mathbf{e}}_n}{\partial n} \doteq -\hat{\mathbf{e}}_s/R_n,$$

where R is the radius of curvature of the field line and R_n the radius of curvature of the field line normal. α is the angle between the field line and the radius vector. Now, Equation (7.1) becomes

$$\frac{\partial v_s}{\partial t} + v_s \frac{\partial v_s}{\partial s} + v_n \frac{\partial v_s}{\partial n} = -\frac{1}{\rho}\frac{\partial p}{\partial s} - \frac{GM_\odot}{r^2}\cos\alpha + v_n \frac{\partial \alpha}{\partial t} + \frac{v_n v_s}{R} + \frac{v_n^2}{R}, \tag{7.3}$$

where the partial derivatives $\partial/\partial t$ and $\partial/\partial s$ refer to changes seen by a fixed observer. However, it will be more convenient to reformulate the equation in terms of derivatives in a frame of reference fixed on a given field line, i.e., derivatives as seen by an observer moving with the field line normal to itself. Denoting these derivatives as $\mathrm{D}/\mathrm{D}t$ and $\mathrm{D}/\mathrm{D}s$, we can easily show

$$\frac{\partial v_s}{\partial t} = \frac{\mathrm{D}v_s}{\mathrm{D}t} - v_n \frac{\partial v_s}{\partial n},$$

$$\frac{\partial v_s}{\partial s} = \frac{\mathrm{D}v_s}{\mathrm{D}s},$$

$$\frac{\partial \alpha}{\partial t} = \frac{\mathrm{D}\alpha}{\mathrm{D}t} - v_n \frac{\partial \alpha}{\partial n}.$$

Then, using the geometrical relation

$$\frac{1}{R_n} = \frac{\partial \alpha}{\partial n} + \frac{\cos\alpha}{r},$$

we finally obtain

$$\frac{\mathrm{D}v_s}{\mathrm{D}t} + v_s \frac{\mathrm{D}v_s}{\mathrm{D}s} = -\frac{1}{\rho}\frac{\mathrm{D}p}{\mathrm{D}s} - \frac{GM_\odot}{r^2}\cos\alpha + v_n \frac{\mathrm{D}\alpha}{\mathrm{D}t} + \frac{v_n^2}{r}\cos\alpha + \frac{v_n v_s}{R} \tag{7.4}$$

In a similar manner, we write the continuity equation in the form

$$\frac{\mathrm{D}}{\mathrm{D}t}(\rho A)+\frac{\mathrm{D}}{\mathrm{D}s}(\rho v_s A)-\frac{\rho v_n A}{R}=0, \tag{7.5}$$

where A is the cross-sectional area of the flux tube. The last term in Equation (7.5) represents the stretching of an infinitesimal length of a curved flux tube as it moves normal to itself.

In place of an energy equation we will assume, for simplicity, that the gas is isothermal. Also, using the perfect gas law, we can relate the gas pressure to the density by the relation

$$p=\rho c^2,$$

where $c^2=2k_B T/m_p$ is the isothermal sound speed, k_B is Boltzmann's constant, and m_p the proton mass.

In applying Equations (7.4) and (7.5), we will adopt the following philosophy. We will choose an analytic model for a time-dependent reconnecting magnetic field geometry. This specifies the parameters v_n, α, R, and A as functions of space and time. Then Equations (7.4) and (7.5) are solved for the two unknowns v_s and p. Since the system is hyperbolic in time, we can employ the method of characteristics for the formal solution.

The characteristic equations corresponding to Equations (7.4) and (7.5) can be written in the form

$$\left.\frac{\mathrm{d}\xi}{\mathrm{d}t}\right|_{C^+}=-v_s\frac{\partial \ln A}{\partial s}-\frac{\partial \ln A}{\partial t}-\frac{GM_\odot}{r^2 c}\cos\alpha+\frac{v_n}{c}\frac{\partial\alpha}{\partial t}+\frac{v_n^2}{cr}\cos\alpha, \tag{7.6}$$

$$\left.\frac{\mathrm{d}\eta}{\mathrm{d}t}\right|_{C^+}=-v_s\frac{\partial \ln A}{\partial s}-\frac{\ln A}{\partial t}+\frac{GM_\odot}{r^2 c}\cos\alpha+\frac{v_n}{c}\frac{\partial\alpha}{\partial t}-\frac{v_n^2}{cr}\cos\alpha, \tag{7.7}$$

where $\xi=\ln p+v_s/c$ and $\eta=\ln p-v_s/c$. The derivatives are taken along the positive (C^+) and negative (C^-) characteristics defined in s, t space by the expressions

$$\begin{aligned}\frac{\mathrm{d}s}{\mathrm{d}t}&=v_s+c \text{ along } C^+\\ \frac{\mathrm{d}s}{\mathrm{d}t}&=v_s-c \text{ along } C^-.\end{aligned} \tag{7.8}$$

To obtain a unique solution of Equations (7.6) and (7.7), the initial dynamical state $p(s)$ and $v_s(s)$ must be prescribed at $t=0$. In addition, one boundary condition must be specified at the base of a given flux tube for all times. That these are the necessary and sufficient conditions can be derived from the theory of characteristics, a point discussed in more detail by Kopp and Pneuman (1976).

A convenient analytical model of a reconnecting magnetic field geometry containing the pertinent physical features can be constructed by superposing a solar-centered dipole and a uniform field parallel to the dipole axis. A suitable choice of the relative field strengths enables us to make this field become radial at any desired radial distance from the solar center, r_1. For $r > r_1$, the field is set radial out to infinity. $r_1(t)$ will be chosen to be an increasing function of time to simulate the rising neutral point during the reconnection process. The radial and transverse components of this field are given by

$$B_r = B_0 \left\{ \frac{1 + 2\left[\frac{r_1(t)}{r}\right]^3}{1 + 2\left[\frac{r_1(t)}{r_0}\right]^3} \right\} \sin\theta,$$

$$B_\theta = -B_0 \left\{ \frac{1 - \left[\frac{r_1(t)}{r}\right]^3}{1 + 2\left[\frac{r_1(r)}{r_0}\right]^3} \right\} \cos\theta, \tag{7.10}$$

for $r_0 \leqslant r \leqslant r_1(t)$, where r_0 is some reference level (chosen here to be the coronal base) and B_0 is the field strength there. At greater heights, the field is just a radial extension of the field given by Equation (7.10), i.e.,

$$B_r = \left\{ \frac{3B_0\left[\frac{r_1(t)}{r}\right]^2}{1 + 2\left[\frac{r_1(t)}{r_0}\right]^3} \right\} \sin\theta, \tag{7.11}$$

for $r \geqslant r_1(t)$. For $r_1 = r_0$, the field is purely radial and open, but as r_1 increases with time, closed loops are formed below. The field is potential for $r < r_1$ and non-potential for $r > r_1$. At any given instant of time the field configuration appears as in Figure 7.10.

From Equations (7.10) and (7.11) we can now derive the relevant quantities v_n, α, R, and A. In terms of the non-dimensional coordinates $x = r/r_0$ and $x_1(t) = r_1(t)/r_0$, these are

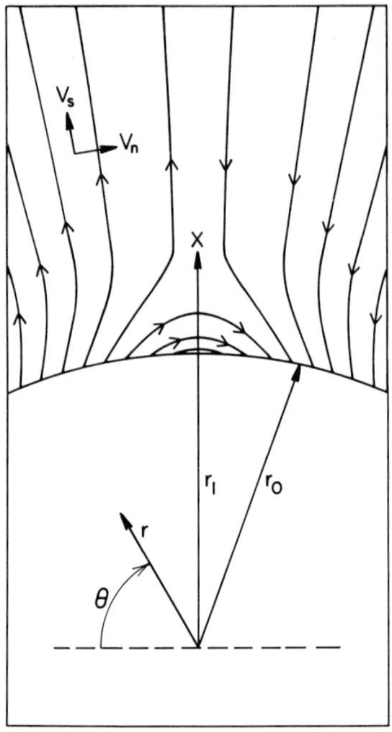

FIGURE 7.10 Field configuration at some arbitrary time during the reconnection process as calculated from Eqs. (7.10) and (7.11). Here r and θ are spherical coordinates with origin at a distance r_0 beneath the solar surface which is the center of the dipole. r_1 gives the location of the neutral point, v_s is the flow speed along the field, and v_n is the velocity of the field line normal to itself.

$$v_n(x, t)=\left(\frac{3x_1^2x}{1+2x_1^3}\right)\left(\frac{x^3-1}{x_1^3-x^3}\right)\frac{\mathrm{D}x_1}{\mathrm{D}t}r_0\sin\alpha, \tag{7.12}$$

$$\alpha(x, t)=\tan^{-1}\left[\frac{x^3-x_1^3}{x^3+2x_1^3}\operatorname{ctn}\theta\right], \tag{7.13}$$

$$R(x, t)=\left\{\frac{\cos^2\alpha\sin\alpha}{x^3+2x_1^3}\left[\left(\frac{9x_1^3x^2}{x^3-x_1^3}\right)-\left(\frac{x^2-x_1^3}{x}\right)(1+\operatorname{ctn}^2\theta)\right]+\frac{\sin\alpha}{x}\right\}r_0, \tag{7.14}$$

$$A(x, t)=A_0\left\{\frac{(1+2x_1^3)^2}{\left[1+2\left(\frac{x_1}{x}\right)^3\right]^2\sin^2\theta+\left[1-\left(\frac{x_1}{x}\right)^3\right]^2\cos^2\theta}\right\}^{1/2}, \tag{7.15}$$

for $x \leqslant x_1(t)$, and

$$v_n(x, t) = \frac{x_1^3 - 1}{1 + 2x_1^3}\left(\frac{x}{x_1}\right)\frac{\mathrm{D}x_1}{\mathrm{D}t} r_0 \operatorname{ctn}\theta, \tag{7.16}$$

$$\alpha(x, t) = 0 \tag{7.17}$$

$$R(x, t) = \infty \tag{7.18}$$

$$A(x, t) = A_0\left(\frac{x}{x_1}\right)^2 \frac{1 + 2x_1^3}{3 \sin\theta} \tag{7.19}$$

for $x \geqslant x_1(t)$.

We now must specify the upward velocity of the neutral point located at $x = x_1(t)$. As discussed previously, the observations indicate that the low-lying loop systems (as seen in Hα) rise at a higher velocity than the higher X-ray loops. This falls in line with our physical reasoning that the reconnection process is a relaxation phenomenon, where opened field lines are collapsing back to the original equilibrium configuration — at which time, the neutral point would come to rest at some finite height in the corona. Hence, we want to choose a functional form for $x_1(t)$ in which $\mathrm{D}x_1/\mathrm{D}t$ is a steadily decreasing function of time and eventually goes to zero. A suitable form having these properties is

$$x_1(t) = \bar{x}_1 - (\bar{x}_1 - 1)\mathrm{e}^{-\omega t}, \tag{7.20}$$

where ω^{-1} is the time-scale of the reconnection process and $\bar{x}_1$ is the equilibrium value of x_1, which is approached asymptotically as $t \to \infty$. Here we choose $r_0 = R_\odot$ and $r_1 = 1.5\, R_\odot$, giving $\bar{x}_1 = 1.5$. We also take $\omega = 5.7 \times 10^{-5}$ s^{-1} which gives $\mathrm{d}r_1/\mathrm{d}t = 20$ km s^{-1}.

In order to integrate Equations (7.6) and (7.7) along individual flux tubes as the geometry evolves in time, we must specify v_s and p at $t = 0$ everywhere and also fix one boundary condition at the base for all times. Since the flux tubes are all initially open and radial, the most logical choice for v_s and p at $t = 0$ would be the distributions given by Parker (1958) for an isothermal solar wind in a radial geometry. For the base boundary condition we will require that the pressure, p, remain fixed at the base for all time.

Having now specified all that is needed to solve the equations, let us now consider the time-variation of the velocity distribution along a *single flux tube* as it participates in the reconnection process. This is shown in Figure 7.11. The curves, proceeding from bottom to top, show the velocity as a function of radial distance along the particular flux tube at progressively later times. This

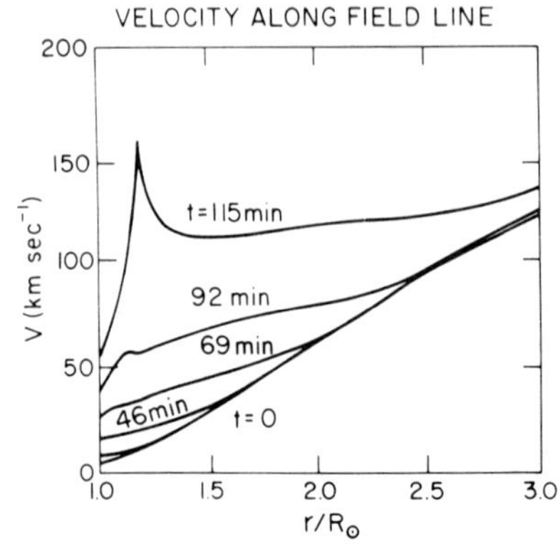

FIGURE 7.11 Velocity versus radial distance along a typical flux tube shown in Figure 7.10 at different times during the reconnection process (Kopp and Pneuman, 1976). In this case, the coronal temperature and base electron density were taken to be 1.5×10^6 K and 10^{16} m^{-3}. The bottom curve is the Parker (1958) isothermal solution for radial flow and represents the initial velocity profile when the flux tube is radial. Proceeding upward, the successive curves show how the velocity distribution changes as the tube swings over towards the neutral line. We see that the outflow is significantly enhanced during the process and reaches a maximum when the tube reaches the neutral point (here, at $t = 115$ min.). For this example, the field line closes when the neutral point has risen to a height of 0.2 $R_\odot$ above the solar surface.

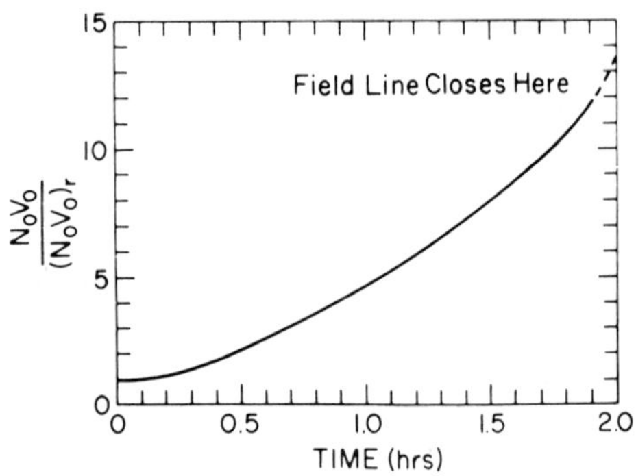

FIGURE 7.12 Particle flux at the coronal base, normalized to that for radial flow (denoted by the subscript 0), as a function of time for the flux tube described in Figure 7.11. Here, when the tube closes, the flux has been enhanced over that of the normal solar wind by a factor of about 12.

tube closes when the neutral point reaches 1.2 $R_\odot$ which occurs at $t = 115$ min. The distribution begins at $t = 0$ with the Parker (1958) solution but, as the tube swings laterally toward the neutral line, we see that the speed increases. This increase starts lower down but later spreads outward, yielding velocities of over 100 km s^{-1} quite low in the corona. Notice the spike in the distribution which develops near the location where the field line closes. This is produced by the large outward forces (centrifugal and pressure) which develop in this

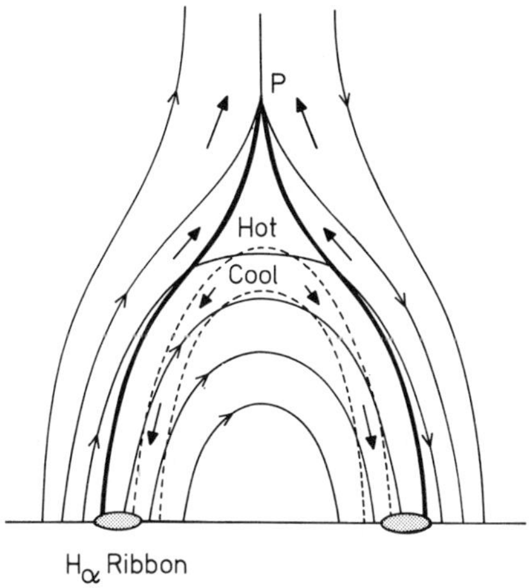

FIGURE 7.13 Configuration of the expanding slow MHD shock from Heyvaerts *et al.*, (1977). The thin lines represent magnetic field lines whereas the thick line denotes the shock. Heavy-headed arrows indicate the direction of plasma motion.

region. An important consequence of this velocity increase is the resulting increase in mass flow into the tube. This is demonstrated in Figure 7.12, where the ratio of the particle flux to its initial value is plotted as a function of time. Note that the flux is enhanced by over a factor of ten just before closure occurs.

7.3.2 The shock wave

When the tube first closes, a shock is immediately set up at the top due to the instantaneous stopping of the flow. This shock propagates down the tube to the coronal base and eventually stops the inflow. However, until the shock reaches the bottom, mass continues to flow into the tube. When considering all the flux tubes together, this phenomenon can be viewed as two large-scale slow magnetohydrodynamic shocks expanding laterally outward as the neutral point rises (Heyvaerts *et al.*, 1977). This process is depicted pictorially in Figure 7.13.

The shock mach number, M, is given by

$$M-\frac{1}{M}=\frac{\gamma+1}{2}\left(\frac{v_+-v_-}{c}\right),$$

where v_+ is the speed of the upward moving unshocked material and v_- the velocity behind the shock. In this application $v_-=0$ and v_+ as seen from Figure 7.11 is about the sound speed for a temperature of 1.5×10^6 K. This gives, for $\gamma = 5/3$, $M = 1.9$. The density and temperature ratios are then given by the Rankine–Hugoniot relations, i.e.,

$$\frac{n_+}{n_-} = \frac{(\gamma+1)M^2}{(\gamma-1)M^2+2} = 2.2,$$

$$\frac{T_+}{T_-} = 1 + \frac{2(\gamma-1)}{(\gamma+1)^2}\frac{\gamma M^2+1}{M^2}(M^2-1) = 1.8.$$

Hence, if the unshocked material is at normal coronal temperatures (1.5–2 $\times 10^6$ K), the hot material behind the shock will be at $3\text{–}4 \times 10^6$ K. These temperatures are *lower* than those observed in the hot X-ray loops discussed in the previous section. This means that, in order to explain the unusually high energy densities seen in X-rays, an additional heating mechanism must be incorporated. This mechanism is, in all probability, Joulean dissipation at the reconnection site and will be dealt with in more detail later (Section 7.6).

Let us now extend our results for a single flux tube to the prominence system as a whole. This is shown schematically in Figure 7.9b. The hottest loops are at the top and are the most recently formed, being heated both by shocks and Ohmic dissipation. These, presumably, are those seen in soft X-rays. These loops subsequently cool by radiation and thermal conduction and are eventually seen in Hα when they have reached temperatures of the order of 10^4 K. By this time, new X-ray loops are being formed above. Thus, to understand the evolution of the entire loop system, one should envisage each reconnected loop as being stationary in space but passing through different temperature regimes beginning with temperatures of the order of 10^7 K and cooling down as time increases. Thus, the time history of the apex of the prominence system does not reflect the temporal behavior of any single flux tube but, rather, how energy is being added to the reconnection region as the neutral point rises. The Hα post-flare loops, then, represent a relatively late stage in the process.

With this picture, the role of the flare ribbons is easy to understand. They merely represent the chromospheric intersections of the reconnecting field lines and outline those regions of the chromosphere which are being heated from above. The increasing separation of the ribbons is seen, not as material motion, but as an expanding front of excitation produced by the formation of progressively higher and broader loops in the corona. This concept is given considerable observational support in that the highest X-ray loops are rooted on the middle to outside of the ribbons (Moore *et al.*, 1980) while the Hα loops are anchored on the inside (Roy, 1972; Rust and Bar, 1973; Beckers, 1977). The ribbons are heated either by high-energy particles penetrating downwards from the neutral point (Sturrock, 1968; Kopp and Martin, 1980), by downward heat conduction (Hirayama, 1974; Antiochos and Sturrock, 1978; Withbroe, 1978), or perhaps by impact of the large-scale downflow (Hyder, 1967a, b). It is clear, however, that the chromospheric ribbons occupy only a small portion of the total flare volume.

7.4 MAGNETIC FIELD DYNAMICS

7.4.1 Role of the filament

In the analysis of Section 7.3, we were concerned mainly with the motion and heating of the coronal material as it participated in the reconnection process. However, the evolution and motion of the magnetic field was specified so that the problem was purely kinematic. In this section, we wish to touch upon some of the MHD aspects of the flare-loop phenomenon. Clearly, the magnetic structure of the active region dominates the entire flare event. Hence, it is of importance to clarify its role more explicitly.

First of all, what process opens the field in the first place? One possibility, as suggested by Carmichael (1964), is that the higher weaker parts of emerging new flux are pulled outward by the solar wind thus setting the stage for reconnection of some of the opened field lines. There are some problems with this suggestion, however. As mentioned earlier, this process puts the flare high in the corona in contradiction to the observation of flare loops at low levels. Moreover, if flare-associated coronal transients are an indication of field-line opening, the observation that they are apparently magnetically dominated (Dulk *et al.*, 1976; Gergely *et al.*, 1979) would seem to negate the supposition that the solar wind provides the driving force.

What is this role of the filament? Normally, the eruption of a preflare filament precedes two-ribbon flares. In fact, Hirayama (1974) makes a stronger statement based upon a study of published papers and Hα observations: "Limb flares do not exist if surges, rising prominences and post-flare loops are excluded." 297 Importance 1 or larger flares were studied by Martin and Ramsey (1972) who concluded that 53% of them were preceded by rising filament activity. It is not known what percentage of these flares were of the two-ribbon type. Typically, there is a marked darkening of the prominence about one hour before the flare itself, becoming most apparent about 10 min. before. Then, about 2 minutes before the flare, expansion, break-up, and brightening of the filament takes place. These general patterns are shown quite clearly in Figure 7.14. Roy and Tang (1975) studied six flares with very high resolution cinematography in Hα. They noted soft X-ray enhancements associated with preflare filament activity which they interpreted as being due to a compression of the overlying corona by the rising filament. This preflare role of the filament in specific events is discussed by many authors (cf. Neupert

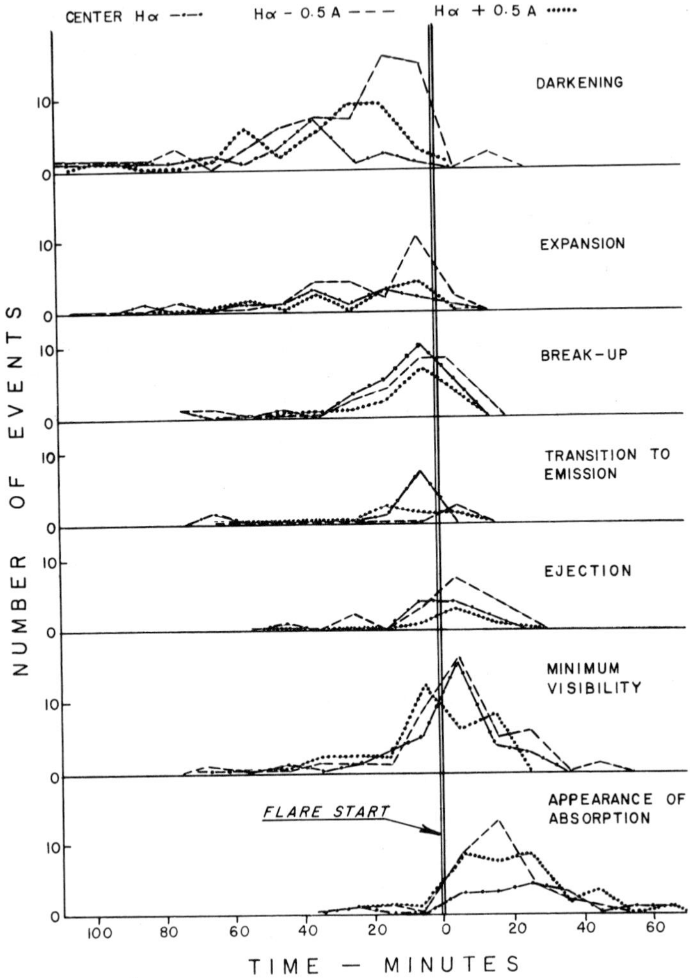

FIGURE 7.14 Distribution in time of observed changes in filaments relative to the starting time of flares (from Martin and Ramsey, 1972).

et al., 1974; Foukal, 1970; Zirin and Tanaka, 1973; Pallavicini *et al.*, 1975; Thomas, 1975; Widing, 1975) and the reader is referred to Rust (1976) for an excellent discussion of this subject (see also §§2.2.4.C and 2.8.1).

We must now ask *whether the rising filament can open the field of the overlying helmet*. This question can be addressed through a very crude MHD model of such an interaction (Pneuman, 1980a). Consider a magnetic flux tube oriented normal to the radius vector from the solar center as shown in Figure 7.15 defined by field lines labeled 1 and 2 at distances r_1 and r_2 from the solar

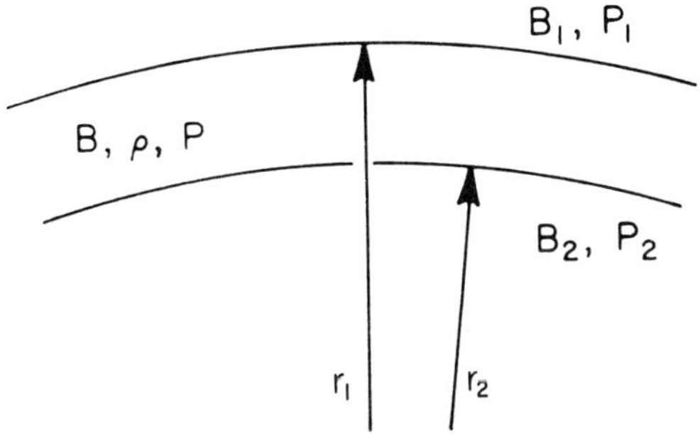

FIGURE 7.15 Schematic of field lines defining a magnetic flux tube. r_1 and r_2 denote the distances of the field lines from the solar center; the ambient magnetic field and gas pressure above and below the tube are denoted by B_1, p_1, B_2, and p_2 respectively. B, ρ, and p are the magnetic field strength, density, and pressure in the center of the tube.

center. We write the component of the MHD force equation normal to this flux tube in the form

$$\rho\frac{\mathrm{D}v}{\mathrm{D}t}=-\frac{\partial}{\partial r}\left(p+\frac{B^2}{2\mu}\right)-\frac{B^2}{\mu R}-\frac{GM_\odot}{r^2}\rho, \tag{7.21}$$

where ρ is the density, v the velocity of the tube normal to itself, p the gas pressure, B the field strength, R the radius of curvature, and r the radial distance from the solar center. If all quantities vary smoothly across the tube and mesh continuously with the external quantities at the boundary, then the derivatives such as, for example, $\partial p/\partial r$, can be approximated by

$$\frac{\partial p}{\partial r}\approx\frac{2(p-p_i)}{(r_1-r_2)},$$

where p is the pressure in the center of the tube and p_i $(i-1, 2)$ is the ambient pressure just outside the ith field line. Now, Equation (7.21) can be written separately for field lines 1 and 2 in the form

$$\rho\frac{\mathrm{D}v_i}{\mathrm{D}t}=\pm\frac{2(p-p_i)}{(r_1-r_2)}\pm\frac{(B^2-B_i^2)}{\mu(r_1-r_2)}-\frac{B^2}{\mu R}-\frac{GM_\odot}{r^2}\rho, \qquad (7.22)\ (7.23)$$

where $i=1$, 2 and the plus and minus signs refer to field lines 1 and 2, respectively.

Rather than use Equations (7.22) and (7.23), two more instructive and useful equations can be derived simply by taking the linear combinations [(7.22) +(7.33)]/2 and (7.22)-(7.23). Noting also that $v_i=\mathrm{D}r_i/\mathrm{d}t$, we obtain

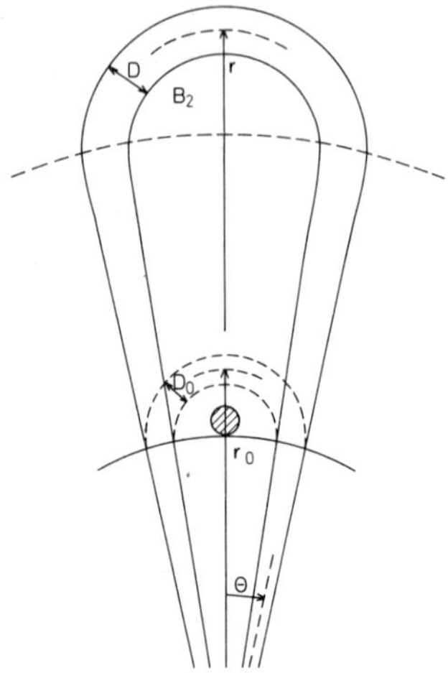

FIGURE 7.16 Idealized transient geometry in a vertical section normal to the arcade axis. The dashed curves represent the boundary of the flux tube in its equilibrium configuration with its top at a distance r_0 from the solar center and with a width D_0. The solid curves show the flux tube at some later time with displacement r and width D. B_2 is the driving field behind the transient and θ the half-angle between the legs. The initial location of the filament, seen edgewise, is shown as the shaded region.

$$\rho \frac{\mathrm{D}^2 r}{\mathrm{D}t^2} = \frac{1}{D}(p_2 - p_1) + \frac{1}{2\mu D}(B_2^2 - B_1^2) - \frac{B^2}{\mu R} - \frac{GM_\odot}{r^2}\rho, \tag{7.24}$$

$$\rho \frac{\mathrm{D}^2 D}{\mathrm{D}t^2} = \frac{2}{D}(2p - p_1 - p_2) + \frac{1}{2\mu D}(2B^2 - B_1^2 - B_2^2), \tag{7.25}$$

where $r = (r_1 + r_2)/2$ and $D = r_1 - r_2$. Clearly, r is the translational distance of the center of the tube or arcade and D is its width. We see that the translation depends on the ambient pressure gradient across the tube, both gas and magnetic, the restoring curvature force or tension, and gravity. The width is determined, on the other hand, by the interaction of the tube's internal gas and magnetic pressures with that of its surroundings.

Let us apply Equations (7.24) and (7.25) to a simple geometrical model. Consider a magnetic arcade whose cross-section is as shown in Figure 7.16. The cross-section is circular at the top with radial legs. The equilibrium configuration is shown dashed in the figure with an initial width D_0 with the top at a distance r_0 from the solar center. At some later time, the configuration is shown solid with width D at a translational distance r. Here, we are interested in a magnetically dominated process so we *neglect forces due to gas pressure*. Also, since the field is open above the helmet, we *neglect the ambient field B_1 in front*. Furthermore, the magnetic field B and the density ρ inside the tube can be written in terms of r and D according to the conservation of magnetic flux and mass (*assuming no motion along the field*), i.e.,

$$B = B_0\left(\frac{D_0}{D}\right)\left(\frac{r_0}{r}\right), \qquad \rho = \rho_0\left(\frac{D_0}{D}\right)\left(\frac{r_0}{r}\right)^2,$$

where B_0 and ρ_0 are reference values at $t=0$ and $r=r_0$, $D=D_0$. The radius of curvature, R, from geometry can be seen to be given by $R = r\tan\theta/(1+\tan\theta)$ where θ is half the angle between the legs.

B_2, of course, is to represent the field associated with the prominence which lies under the helmet and perpendicular to it. This may be the field which supports the prominence itself, the internal field of the prominence or, perhaps, emerging flux (Heyvaerts *et al.*, 1977) which disrupts both the prominence and the arcade. Another possibility is that magnetic reconnection beneath the filament could provide new flux which drives the configuration outward Pneuman, 1979, 1980a). If we assume that this field expands and conserves its flux in doing so, then B_2 will be given approximately by

$$B_2 = B_{2_0}\left(\frac{r_0}{r}\right)^2, \tag{7.26}$$

where B_{20} is its initial field strength. This formula is independent of the orientation of the field since it applies to both longitudinal and transverse components.

We now substitute the above expressions for B, ρ, R, and B_2 into Equations (7.24) and (7.25) and obtain two coupled, second order, non-linear, ordinary differential equations for S and D as a function of time,

$$\frac{d^2r}{dt^2} = \frac{B_{20}^2}{2\mu\rho_0 D_0}\left(\frac{r_0}{r}\right)^2 - \frac{D_0 B_0^2(1+\tan\theta)}{\mu\rho_0 D r\tan\theta} - \frac{GM_\odot}{r^2}, \tag{7.27}$$

$$\frac{d^2D}{dt^2} = -\frac{B_{20}^2}{\mu\rho_0 D_0}\left(\frac{r_0}{r}\right)^2 + \frac{2D_0 B_0^2}{\mu\rho_0 D^2}. \tag{7.28}$$

The equilibrium conditions can be easily found by setting the left-hand side of Equations (7.27) and (7.28) equal to zero. This gives

$$B_{20} = \sqrt{2}B_0 \tag{7.29}$$

and

$$D_0 = \frac{r_0}{\dfrac{1+\tan\theta}{\tan\theta} + \dfrac{2\mu\rho_0 GM_\odot}{r_0 B_{20}^2}}. \tag{7.30}$$

Note that an equilibrium solution exists for any value of r_0.

A close examination of Equations (7.27) and (7.28) reveals a very important property of the physical process. Note that an asymptotic solution of the system is given by

$$D \propto r, \qquad v \propto \left(1 - \frac{K}{r}\right)^{1/2},$$

where $v(=\mathrm{d}r/\mathrm{d}t)$ is the velocity and K is a constant. Hence, the velocity approaches a constant value at large distances and the width of the arcade increases linearly with radial distance from the Sun. The most important point to note, however, is that if $D \propto r$ then *all* the terms on the right-hand side of Equation (7.27) fall off with r at the same rate, namely, as $1/r^2$. This means that the *ratio of outward to inward forces is independent of radial distance from the Sun*, indicating that, if the driving force associated with the lifting prominence initially exceeds the restoring forces of gravity and tension, then it will always do so and the whole configuration will coast easily out to infinity.

This property can be demonstrated in another way — through a standard stability analysis of Equation (7.27) and (7.28). Letting $r = r_0 + r' \mathrm{e}^{i\omega t}$ and $D = D_0 + D' \mathrm{e}^{i\omega t}$, ω being the angular frequency and S' and D' small perturbations on the equilibrium values S_0 and D_0, we can find solutions for ω from the differential equations. These are given by

$$\omega^2(\omega^2 - \omega_1^2) = 0,$$

where

$$\omega_1^2 = \frac{B_{20}^2}{2\mu\rho_0 r_0^2}\left\{4\left[\frac{1+\tan\theta}{\tan\theta} + \frac{2\mu\rho_0 GM}{B_{20}^2 r_0}\right]^2 + \frac{(1+\tan\theta)}{\tan\theta}\right\}.$$

Note that ω_1^2 is positive. Thus, there are two solutions for ω. One is magnetogravitational mode and the other, most importantly, is one of neutral or marginal stability ($\omega = 0$). This shows that, once the system is set into motion, it will continue indefinitely. The situation is analogous to a ball rolling on a frictionless plane.

The formal, exact solution of Equations (7.27) and (7.28) is straightforward and is described in detail elsewhere (Pneuman, 1980a). A typical velocity profile resulting from such a solution is shown in Figure 7.17. We see that this result confirms the asymptotic behavior described earlier. Hence, we conclude that a filament eruption *can* open the overlying magnetic field structure — a necessary prerequisite for the reconnection phase which follows.

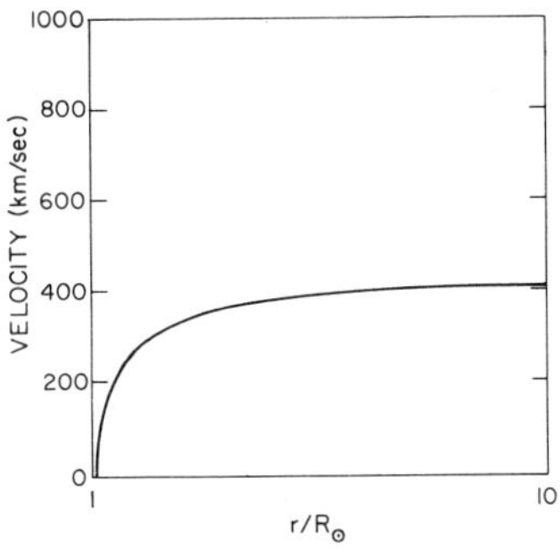

FIGURE 7.17 Typical velocity versus radial distance profile resulting from the solution of Eqs. (27) and (28) for a magnetic arcade propelled outward by an erupting field beneath. For this case, $B_0 = 5$ gauss (5×10^{-4} T), $\rho_0 = 1.7 \times 10^{-15}$ gm cm^{-3} (1.7×10^{-12} kg m^{-3}), $r_0 = 1.2\ R_\odot$, and $\theta = 20°$. This gives, from Eqs. (29) and (30), $B_{20} = 7$ gauss and $D_0 = 0.23\ R_\odot$. Note that the velocity increases sharply close to the Sun but quickly approaches a constant asymptotic value.

7.4.2 Relaxation and reconnection

For the same boundary conditions, an open radial magnetic field contains more energy than the corresponding closed potential field by about a factor of two. This can be seen, for example, by comparing the magnetic energies of the field given by Equations (7.10) and (7.11) for $r_1 = r_0$ (open field) and $r_1 = \infty$ (closed field). In this case, the difference is exactly a factor of two. Thus, the eruptive process described in Section 7.4.1 produces an increase in the magnetic energy of the corona by about this amount. Since filament eruptions occur quite often, we must conclude that, unless some complementary process which *decreases* the coronal magnetic energy density occurred, the energy of the corona would increase monotonically with time. In other words, since magnetic loops are never observed returning to the Sun, we must look for a process which dissipates magnetic energy. One such process is that of reconnection which we believe is manifested in the post-flare and post-transient loops formed in the lower corona. That the expulsion of a closed loop system sets up forces which tend, quite naturally, to precipitate reconnection can be demonstrated by the following crude analytic model.

Suppose the helmet or arcade has been driven out to a considerable distance and that, close to the Sun, the legs of the system are now essentially radial. Let L be the horizontal distance from the neutral sheet between the legs to the leg's center. We denote the width of the leg as D and field inside as B_2. The ambient field outside will be denoted as B_a. Then, neglecting the gas pressure, the balance of forces in the horizontal direction yields approximately

$$\rho \frac{d^2 L}{dt^2} = \frac{1}{2\mu D}(B_2^2 - B_a^2). \tag{7.31}$$

where ρ is the density. At a given height, let us assume that the gas moves only in the lateral direction. Hence ρ is given by the formula

$$\rho = \rho_0 \left(\frac{D_0}{D} \right)$$

If the field outside is approximately radial its strength varies as $B_a = B_{a0}(R_\odot / r^2)$ where B_{a0} is a reference value. Here, we must keep in mind the distinction between S and r. S is the translational distance to the top of the arcade and varies with time, whereas r is just the radial coordinate and hence is constant. Using the expression for B_2 from Equation (7.26) and the above equations for ρ and B_a we obtain from Equation (7.31)

$$\frac{d^2 L}{dt^2} = \frac{1}{2\mu\rho_0 D_0} \left[B_{20}^2 \left(\frac{S_0}{S} \right)^4 - B_{a0}^2 \left(\frac{R_\odot}{r} \right)^4 \right]. \tag{7.32}$$

Now, as the arcade rises, the first term in Equation (7.32) will decrease rapidly as compared to the second term which remains constant. Eventually, only the second term is important and

$$\frac{d^2 L}{dt^2} \approx - \frac{B_{a0}^2 R_\odot^4}{2\mu\rho_0 D_0 r^4}$$

or

$$v_{\text{leg}} \equiv \frac{dL}{dt} = - \frac{B_{a0}^2 R_\odot^4 \, t}{2\mu\rho_0 D_0 r^4} \tag{7.33}$$

assuming that $v_{\text{leg}} = 0$ at $t = 0$.

It is important to note from Equation (7.33) that, at any given height, the velocity of collapse towards the neutral sheet increases with time. Also for a specified time the velocity is a *decreasing* function of radial distance from the Sun. Hence we expect the reconnection to begin at the coronal base and proceed upward. This, of course, is what is observed. For a post-flare loop system, we take $B_{a0} = 10^{-3}$ tesla (10 gauss), $t = 10$ min (about when they are first observed), $\rho_0 = 1.7 \times 10^{-11}$ kg m^{-3} ($n_e \approx 10^{16}$ m^{-3}), $D_0 = 0.1\ R_\odot$, and $r = 1.1\ R_\odot$. This gives $v_{\text{leg}} = 137$ km s^{-1}. Thus, the collapse is quite rapid; however, we must remember that this is a strict upper limit since all forces opposing the collapse (such as a pressure gradient) have been neglected. Also, this velocity should not be confused with the speed at which the loop system rises, i.e., the velocity of upward motion of the neutral point. This latter velocity depends, not only upon how fast the field lines reach the neutral sheet but, also, on how fast reconnection can take place.

The velocity at which the neutral point rises might be expected to be proportional to the Alfvén speed. Roy (1972) used this assumption to reproduce theoretically the observed rate of rise of the large magnetic loop system of 18 November 1968. He assumed the growth-rate was determined by the Petschek mechanism (Petschek, 1964) as

$$v_{\text{growth}} - v_a/\ln R_m,$$

where v_a is the Alfvén velocity and R_m the magnetic Reynolds number. He estimated v_a by taking a density of 2.5×10^{18} particles m^{-3} and used a field strength calculated for the tops of the loops by a potential field extrapolation from the observed photospheric magnetic fields. This procedure yielded field strengths ranging from 78 gauss for the lowest loops down to 4 gauss for the highest. This yields values for v_a between 108 km s^{-1} and 5.5 km s^{-1}. The magnetic Reynolds number was taken to be 10^{10}. Hence, in his model, the reconnection proceeded at 1/10 the Alfvén speed. With this crude model, Roy was able to reproduce the observed curve of growth quite well.

7.5 MASS SUPPLY

If we are to view the post-flare loop phenomenon as a progressive reconnection process, then the problem of the origin of the large amounts of mass seen in these systems can be narrowed down to relatively few specific mechanisms. For example, *condensation from the corona can probably be ruled out*, both because of the difficulty in transporting material across field lines of the reconnected loops and also because the coronal volume needed to produce this amount of mass is quite large. However, past estimates of the mass in loop prominence systems have been of a relatively crude nature and reliable, at most, to an order of magnitude. Now, with the superior soft X-ray observations taken aboard Skylab, it would perhaps be useful to study a particularly well-observed event in detail in order to obtain a more precise estimate of the mass content. To this end, we will use data obtained for the class 3b flare-loop system of 29 July 1973.

The electron density in this system, as it varied with time, has been estimated by Moore *et al.* (1980) in two different ways. Firstly, using computational results of Tucker and Koren (1971) for thermal X-ray emission, Moore derived the emission measure M_e of the prominence system from the 1–8 Å and 8–20 Å X-ray fluxes measured by SOLRAD 9. Assuming constant density across the loop, n_e is related to M_e by the formula

$$n_e = (M_e/V)^{1/2},$$

where V is the volume. Moore took the volume V to be given by

$$V = l^2 d,$$

where l is the separation distance of the flare ribbons and d is the ribbon thickness. The time-variation of n_e estimated in this way is shown as the squares in Figure 7.17.

Using the time-variation of the effective temperature T and assuming the loops cool solely by radiation (we will show in the next section that this is a reasonable assumption), Moore deduced n_e in another way from the radiative cooling time (see Figure 7.6). The radiative cooling time, τ_r is given by

$$\tau_r = \frac{3k_b T}{\Lambda_r n_e}, \tag{7.34}$$

where Λ_r is the radiative cooling coefficient. For temperatures in the range 4×10^6 K to 1.5×10^7 K, Λ_r is relatively constant and has the value 4×10^{-36} J m^3 sec^{-1}, so that Equation (7.34) yields

$$n_e \simeq 1.0 \times 10^3 \frac{T}{\tau_r} \mathrm{m}^{-3} \tag{7.35}$$

The cooling time, τ_r, has been estimated in various ways for the 29 July flare and the reader is referred to Moore *et al.* (1980) for a detailed discussion. The temporal variation of n_e obtained in this way is shown as the open and filled circles in Figure 7.18. We see that n_e obtained by these two completely different methods agrees to within a factor of 2. Therefore, we will consider these results as good enough for our purposes here.

To estimate the mass of the loop system, we will make use of two observed characteristics. Firstly, the observations suggest that, for a given loop, the gas density is relatively independent of height — being about the same at the top as in the legs. Secondly, as can be seen by comparing Figures 8.3 and 8.5 of Moore *et al.* (1980), the flare ribbon separation distance and loop height are about equal at all times. This means that, at any time, the illuminated loop can be represented as an arcade of length L(Lbeing the length of the ribbons) and with a semi-circular cross-section of radius h, where h is the loop height or one-half the ribbon separation. Since, in our reconnection picture, the loops fill all space sequentially up to some maximum height $h_{\max}$, the total mass of the system is just

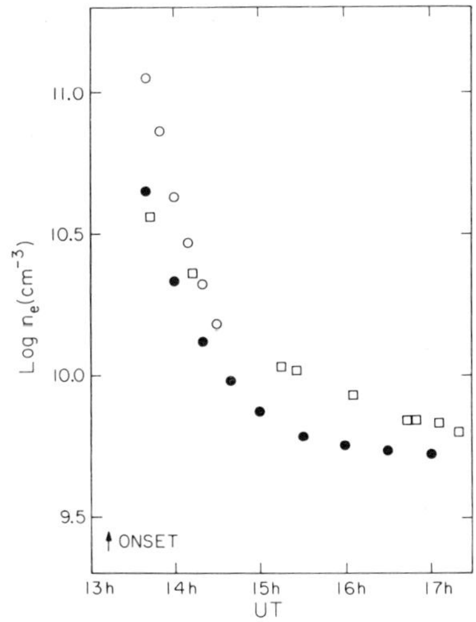

FIGURE 7.18 Maximum electron density in the flare loops of the 29 July 1973 flare from Moore *et al.*, 1980. ● — deduced from the radiative cooling time (see Figure 7.6) estimated from ribbon separation. ○ — from the radiative cooling time based upon extrapolated curves of loop-growth. □ — deduced by Moore from the emission measure and an assumed geometrical shape of the loops.

$$M_t = \pi m_p L \int_0^{h_{max}} n_e h dh, \tag{7.36}$$

where m_p is the proton mass.

Using the available data, the integral in Equation (7.36) can be evaluated quite easily. From Hα photographs we estimate the ribbon length L as 1.5×10^8 m. n_e as a function of time is taken from Figure 7.18 by drawing a mean curve through the points. $n_e(h)$ can then be evaluated using the observation of $h(t)$ as shown in Figure 7.6. This procedure yields

$$M_t = 7.5 \times 10^{13}\ kg$$

This is an *extremely large mass* —considerably greater than the 10^{12}–10^{13} kg loop prominence masses deduced previously from Hα observations alone (cf. Kleczek, 1964; Jefferies and Orrall, 1964). One reason for this might be that Hα loops are never observed up to the great heights seen in X-rays. Hence, the material in these upper portions is not accounted for in the Hα observations. This may be due to the fact that the upper loops never cool down to Hα temperatures but only to coronal temperatures. One should also keep in mind that Equation (7.36) does not take into account possible filamentary structure along the length of the arcade which would reduce the mass estimate. However, working in the other direction is the fact that the heights shown in Figure 7.6 have recently been revised upward (Švestka, 1979).

The total mass of the corona is approximately

$$M_{\text{cor}} \approx 4\pi R_\odot^2\, m_p \int_0^\infty n_e \mathrm{d}h$$

Assuming an isothermal corona, hydrostatic equilibrium yields

$$n_e = n_0 \mathrm{e}^{-(GM_\odot m_p/2R_\odot^2 k_B T)h},$$

where n_0 is the reference density at the base of the corona. Hence

$$M_{\text{cor}} = \frac{8\pi k_B T n_0 R_\odot^4}{GM_\odot}.$$

Taking $T = 1.5 \times 10^6$ K and $n_0 = 3 \times 10^{14}$ m^{-3} (Allen, 1963) we find

$$M_{\text{cor}} = 2.8 \times 10^{14} \text{ kg}.$$

This is only about three times the mass of the prominence system, meaning that if the prominence material were to be supplied by condensation of the corona, it would have to drain almost a whole hemisphere of the solar atmosphere. Since the filling time of the corona by the solar wind is a little over a day, somewhat longer than the prominence lifetime, one should observe a significant coronal depletion coincident with these events. Such is not the case; therefore, on the basis of the similarity between M_t and M_{cor}, *a coronal origin for the prominence mass can probably be safely ruled out.*

Kopp and Pneuman (1976) have shown that the forces generated by the dynamical motion of the field lines during the reconnection process can *enhance the injection of mass into the newly formed closed flux tubes by over a factor of ten* over that produced by the ambient solar wind. Although this process may occur in post-flare loop systems, it appears to fall short of accounting for the required mass for this particular event. For example, assuming the mass flow is enhanced by a factor of ten, we obtain an injection rate of 6.7×10^{16} kg m^{-2} sec^{-1}. Taking the base area of the system to be the ribbon length L times the maximum ribbon separation distance, we find that this process can inject approximately 7.4×10^7 kg s^{-1}. The total lifetime of this event is about 12 hours, so that the total mass accumulated during this time through this mechanism is 3.2×10^{12} kg. Although this is a significant amount of material, it appears to fall short of the $>10^{13}$ kg needed.

Having ruled out condensation from the corona and enhanced expansion as possible sources for the prominence mass, we are left with but one alternative — the *material must be supplied from the chromosphere by increasing the local pressure there.* Moreover, since the primary energy supply in our picture is at the tops of the loops, the energy needed to lift the mass probably does not originate locally in the chromosphere but must propagate downward from above. This process of chromospheric evaporation has been suggested by many authors (Hudson and Ohki, 1972; Sturrock, 1973; Hirayama, 1974; Lin and Hudson, 1976; Antiochos and Sturrock, 1978; Colgate, 1978; Withbroe, 1978) and is described in Section 4.5. The chromospheric material can be heated by two mechanisms — either by energetic particles propagating downward from above and colliding with chromospheric gas (Sturrock, 1973; Lin and Hudson, 1976; Kopp and Martin, 1980) or by thermal conduction (Hirayama, 1974; Antiochos and Sturrock, 1978; Colgate, 1978; Withbroe, 1978). It must be remembered that, by far, most of the energy resides in the thermal plasma so that it must be questioned whether non-thermal processes are sufficiently efficient.

The energy required to lift the amount of mass observed in the 29 July 1973 post-flare loop system can be estimated as

$$E = \frac{\pi G M_{\odot} \, m_p L}{R_{\odot}^2} \int_0^{h_{max}} n_e(h) h^2 \mathrm{d}h.$$

Here again we are using the same considerations as were employed in estimating the mass content, namely, that the density is constant along a given loop and that the ribbon separation and loop height at a given time are the same. Using the values for $n_e(t)$ from Figure 7.16 and $h(t)$ from Figure 7.5, we find that

$$E = 1.2 \times 10^{24} \text{ J } (= 1.2 \times 10^{31} \text{ erg}).$$

This is comparable with but probably somewhat less than the total energy content of the thermal plasma so that *conduction-driven evaporation as a mechanism for mass supply seems to be compatible with the observations* (§5.5.3). For example Withbroe (1978), for a similar flare, that of 7 September 1973, has estimated the conductive input rate to the lower transition region as 10^{20} J s^{-1}. At this rate, the required energy to lift the material can be delivered in about 3 hours, which is shorter than the lifetime of the flare.

Something about the rate at which the mass is delivered to the loop system can be inferred from the observed pressure distribution along a given loop. Since the temperature is highest at the top of the loop and the density is roughly uniform along its length, the plasma pressure is greatest at the top and the pressure gradient is *downward*. Assuming, as a first approximation, steady state conditions (this is probably not the case), we can write the equation of motion along the loops as

$$\rho v \frac{\mathrm{D}v}{\mathrm{D}r} = -\frac{\mathrm{D}p}{\mathrm{D}r} - \frac{GM_\odot}{r^2}\rho, \tag{7.37}$$

where v is the upflow velocity, p the gas pressure, ρ the density, and r the radial distance from the solar center. Since the density is uniform, Equation (7.37) can be immediately integrated from the bottom of the loop to the top yielding

$$v_b^2 = 2\left[\frac{p_t - p_b}{\rho} + \frac{GM_\odot}{r_b}\left(1 - \frac{r_b}{r_t}\right)\right] \tag{7.38}$$

$$= 2\left[\frac{2k_B}{m_p}(T_t - T_b) + \frac{GM_\odot}{r_b}\left(1 - \frac{r_b}{r_t}\right)\right],$$

where the subscripts b and t refer to the loop bottom and top respectively. Here, we have set $v_t = 0$.

The temperature difference between the loop tops and footpoints for the 29 July flare has been measured by Petrasso *et al.* (1979). Although T_t and T_b vary during the course of the flare, we take as average values, $T_t = 5 \times 10^6$ K and $T_b = 3.5 \times 10^6$ K. Also, taking an average value for h from Figure 7.3 to be 7×10^4 km, we find $r_t = 7.7 \times 10^5$ km. Substitution of these values into Equation (7.38) yields

$$v_b = 290 \text{ km/sec}.$$

This is quite a large velocity which is supersonic even with respect to the sound speed at the loop top. Since the flow must come to rest at the top, a standing shock must exist somewhere along the loops. We conclude, then, that *chromospheric evaporation models require sizable upflow speeds* and wonder whether doppler shifts might be observable.

7.6 HEATING AND COOLING OF THE LOOPS

Let us begin by estimating the cooling time for a single loop based upon losses due to both radiation and thermal conduction. Such calculations have been made for post-flare loop systems by several authors (Culhane *et al.*, 1970; Moore and Datlowe, 1975; Kopp and Pneuman, 1976; Antiochos and Sturrock, 1976; Krieger, 1977; Withbroe, 1978; Moore *et al.*, 1980). The time rate of change of energy due to radiative losses is (Cox and Tucker, 1969)

$$\frac{De}{Dt} = 6 \times 10^{-35}\, n_e^2,$$

where e is the energy density of the gas. Taking $e = 3n_e k_B T$, we have

$$\frac{De}{Dt} = \frac{D}{Dt}(3n_e k_B T) \approx \frac{3n_e k_B T}{t_r} = 6 \times 10^{-35}\, n_e^2.$$

Hence $t_r = 5 \times 10^{34}\, k_B T / n_e$. Using density and temperature estimates for the 29 July flare from Moore *et al.* (1980), we find that t_r varies from about 3 min at the onset to almost one-half hour in the later phases. This compares with the calculations of Withbroe (1978) for the 7 September event who finds t_r ranging from 18 to 38 min.

The cooling time by thermal conduction is given by

$$t_c \approx 5.5 \times 10^{-8} \frac{h^2 n_e}{T^{5/2}},$$

where h is the loop height. Again, using values for h, n_e, and T from Moore *et al.* (1980), we find that t_c increases from about 3 min to 20 min during the course of the flare. Hence, the radiative and conductive relaxation times are about the same, being of the order of one-half hour or less. Antiochos and Sturrock (1976) have suggested that the conductive cooling time could be increased significantly if the flux tube cross-section were constricted at the base as, for example, is expected where it enters the chromospheric network. This is probably the case; however, radiative cooling would then dominate giving roughly the same cooling times.

The most important fact which emerges from the above calculation is that the cooling time for a single loop is less than one-half hour whereas the duration of the loop prominence event is about 10–20 hours; i.e. *the cooling time for the individual loops is much shorter than the life-time of the prominence system.* The logical conclusion from this is that energy is being continuously fed into the loops long after the flash phase of the flare (Pallavicini *et al.*, 1975; Kahler *et al.*, 1976; Gerassimenko *et al.*, 1978; Withbroe, 1978; MacCombie and Rust, 1979; Petrasso *et al.*, 1979; Moore *et al.*, 1980; Pneuman, 1980b). Since the energy density in the loops is highest at their tops, it seems likely that the energy input occurs at these locations.

High-energy particles which arrive from below and thermalize at the loop tops have been suggested by Jefferies and Orrall (1964, 1965) and Newkirk (1973). However, for this process to be effective, inordinately strong magnetic fields seem to be needed to contain the required number of particles. Shock heating as the open flux tubes close during the reconnection process has been investigated by Kopp and Pneuman (1976). However, the temperatures which can be achieved by this process are in the range 3–4 × 10^6 K, somewhat lower than those observed in soft X-rays. Shock heating due to mass motions has also been proposed by Craig and McClymont (1976).

In line with the reconnection picture of the flare loop phenomenon, however, the most effective heating mechanism by far is via *ohmic dissipation of magnetic field energy* in the vicinity of the reconnection site (Sweet, 1958a, b, 1969; Parker, 1963, 1973; Petschek, 1964; Sturrock, 1968; Moore and Datlowe, 1975; Pneuman, 1980b) — see §3.3.4). Since the reconnection site is just above the most recently formed loops, this process deposits the energy at the loop tops where it is observed to reside. As we shall see, Joulean heating is a very effective mechanism in this context and the conversion of magnetic energy to thermal energy need not be even extremely efficient to account for the thermal plasma energy observed.

The overall effectiveness of ohmic heating as applied to the post-flare loop process can be quantitatively estimated using a simple analytic model of reconnecting magnetic field geometry. For a more detailed account of this approach, the reader is referred to Pneuman (1980b). Such a model, having all the desired physical characteristics, is given by Equations (7.10) and (7.11). We choose the origin of our coordinate system to be located at a distance r_0 beneath the solar surface which is chosen to be the scale of the active region, i.e., $r_0 = 2 \times 10^7$ m. Thus, B_0 denotes the radial field strength at the surface. θ here is the angle measured from the horizontal. Hence, the direction vertically outward from the active region is at $\theta = \pi/2$. The neutral point position, denoted as $r_1(t)$, separates two geometrically distinct field topologies. (This notation is similar to that of Figure 7.10, except there r_0 was taken to be a solar radius.) Below r_1, the field consists of closed field lines which are potential. Both field components are continuous ar $r = r_1$ (see Figure 7.9). At time $t = 0$, we set $r_1(t) = r_0$. Thus, initially, the field is completely open, radial, and non-potential. As r_1 increases, closed loops form below r_1 and the field becomes increasingly more potential. This is what we physically expect in the post-flare loop phenomenon.

At any time, t, the total magnetic energy of this system is

$$E_t = \frac{1}{2\mu} \iiint\limits_{\text{Vol.}} (B_r^2 + B_\theta^2) \mathrm{d}v = \frac{r_0^3 B_0^2}{3\mu} \frac{2x^6 + 8x^3 - 1}{(2x^3 + 1)^2}, \qquad (7.39)$$

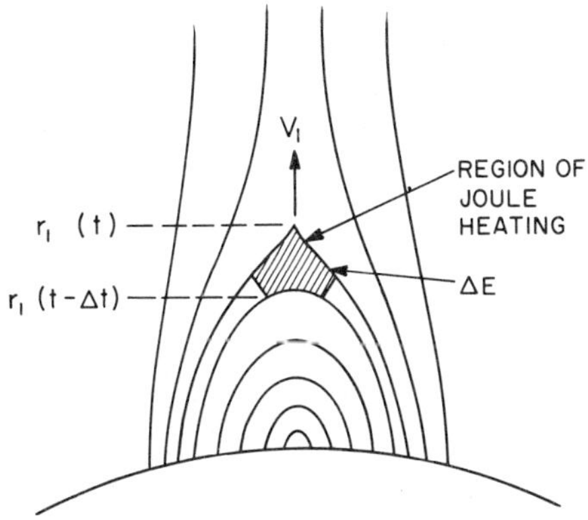

FIGURE 7.19 Schematic of the energy deposition region (shown shaded) in the loop created as the neutral point rises from position $r_1(t-\Delta t)$ to position $r_1(t)$.

where E_t is the total energy in the magnetic field per unit angle perpendicular to the plane of the loops and $x=r_1/r_0$. Note that, when the field is totally open ($x=1$), $E_t=r_0^3B_0^2/(3\mu)$, but when the field has completely relaxed ($x=\infty$), $E_t=r_0^3B_0^2/(6\mu)$ — exactly one-half the energy of the original radial field. The time rate of change of E_t is

$$\frac{\mathrm{D}E_t}{\mathrm{D}t}=\frac{12B_0^2r_0^3}{\mu}\frac{(x^3-1)x^2}{(2x^3+1)^3}\frac{\mathrm{D}x}{\mathrm{D}t}=-\frac{12B_0^2r_0^2}{\mu}\frac{(x^3-1)x^2}{(2x^3+1)^3}\frac{\mathrm{D}r_1}{\mathrm{D}t}. \tag{7.40}$$

Referring now to Figure 7.19, we assume that, as the neutral point rises from position $r_1(t-\Delta t)$ to $r_1(t)$, the fractional decrease in E_t, denoted as $-\Delta E_t$, is deposited as thermal energy into the gas contained in the newly formed closed flux tube lying between these two positions. This amount, ΔE_t, of course, must be regarded as an upper limit since we have neglected other possible sinks for the magnetic energy, such as a) mass motions, b) turbulence and waves, c) energy deposited in the upper disconnected inverted loops lost to interplanetary space, d) kinetic energy in downward motion of the reconnected loop, e) non-thermal particles, and f) energy which might be deposited at large distances from the neutral point. Now, we can write

$$-\Delta E_t=-\frac{\mathrm{D}E_t}{\mathrm{D}t}\Delta t=-\frac{\left(\dfrac{\mathrm{D}E_t}{\mathrm{D}t}\right)}{\left(\dfrac{\mathrm{D}r_1}{\mathrm{D}t}\right)}\Delta r_1=\frac{12B_0^2r_0^3}{\mu}\frac{x^3(x^3-1)}{(2x^3+1)^3}\left(\frac{\Delta r_1}{r_1}\right).$$

We will specify the volume of the energy deposition region (per unit longitude) as $\beta r_1^2 \Delta r_1$ where βr_1 is the lateral extent of the region (we expect this length to scale as r_1) shown as the shaded portion in Figure 7.18. With these considerations, the energy density, e, achieved in a given loop is

$$e=\left(\frac{12B_0^2}{\mu\beta}\right)\frac{x^3-1}{(2x^3+1)^3}. \tag{7.41}$$

Using Equation (7.41), we can now model the time-behavior of the energy density in a rising loop system through its entire lifetime.

Suppose we have an observational curve, $e(t)$, for the energy density $(=3n_e k_B T)$ at the top of a rising loop system. Choosing two times t_a and t_b relatively late in the event when reconnection is most likely to be the sole energy supply, we denote the observed energy densities as e_a and e_b, respectively, and the corresponding dimensionless positions of the neutral point as x_a, x_b. From Equation (7.41) we have

$$\frac{(x_b^3-1)(2x_b^3+1)^3}{(x_a^3-1)(2x_a^3+1)^3}=\frac{e_b}{e_a}. \tag{7.42}$$

However, x_a and x_b are also related through the velocity v_1 at which the neutral point rises into the corona. Assuming, for simplicity, that $v_1 = const.$, we have $r_1 = r_0 + v_1 t$ or

$$t=\frac{r_0}{v_1}(x-1),$$

and so

$$\frac{(x_b-1)}{(x_a-1)}=\frac{t_b}{t_a}. \tag{7.43}$$

Equations (7.42) and (7.43) can now be solved for x_a and x_b. Then, substituting either x_a and e_a or x_b and e_b into Equation (7.41), we can find the magnitude of the constant $12B_0^2/(\mu\beta)$. We can also find r_0/v_1 by applying the same procedure to Equation (7.43).

The results of this process when applied to the 29 July 1973 flare are shown in Figure 7.20. The dashed curve is constructed from temperature and density estimates provided by Švestka *et al.* (1978) using the relation $e=3n_e k_B T$. The solid curve represents the model calculation. We see that the theoretical curve fits the observations adequately in the later stages of the event but falls short during the first 4 hours. Three possible reasons for this discrepancy during the

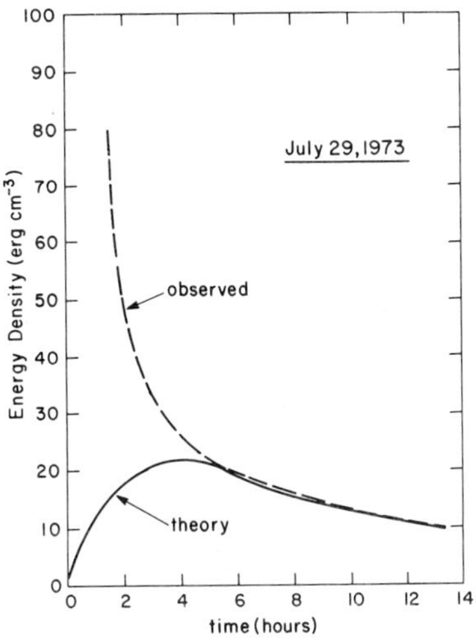

FIGURE 7.20 Predicted (solid) and observed (dashed) energy density versus time curves for the X-ray event of 29 July 1973. The observed curve is from the data of Švestka, *et al.*, 1978.

early phases come to mind:

1) An additional source of energy is present in the early stages.

2) For some reason, the efficiency of conversion of magnetic to thermal energy is higher at earlier times when the neutral point is close to the surface.

3) As the observations suggest, the neutral point velocity, v_1, is probably higher during the earlier part of the flare and progressively decreases. If the assumption $v_1 = const.$ were relaxed and a higher value used at the early phases, this could raise the theoretical curve for small times.

Taking $r_0 = 2 \times 10^7$ m, the value of v_1 needed to produce the fit in Figure 7.20 is $v_1 = 0.3$ km/sec. This velocity compares with that observed at about 23:20 UT, roughly 10 hours after flare onset (see Figure 7.7). Hence, we consider this a favorable comparison. The magnitude of the constant $12B_0^2/(\mu\beta)$ was found to be 2.6×10^{-5} tesla^2, which for $\beta \approx 0.1$ (estimated crudely from X-ray photographs) gives $B_0 = 1.6 \times 10^{-3}$ tesla. Since the observations suggest the actual fields are greater than 5×10^{-3} tesla at the loop tops (Rust and Bar, 1971; Moore *et al.*, 1980) and therefore even greater than 10^{-2} tesla at the coronal base, we see that the ohmic dissipation mechanism need not be very efficient to account for the necessary heating of flare-loop systems. For example, if the actual value of B_0 were 100 gauss, then only about 3% of the magnetic energy dissipated would have to be converted into thermal energy to account for the observed energy density at the loop tops.

It must be stressed that each point on the energy density curves of Figure 7.20 refers to a *different* magnetic flux tube. Hence, the slow decay seen here does not reflect the cooling of a single loop but, rather, the rate at which energy is being added to the newly formed loops. Presumably, according to our estimates of the radiative and conductive cooling times, the individual loops cool rapidly. In fact, if they didn't, one would not observe discrete loops at all but, rather, an amorphous cloud of brightness rising into the corona.

7.7 SUMMARY AND CONCLUSIONS

The morphological description of the flare-loop phenomenon has been greatly clarified during the Skylab period in which combined Hα, EUV, and X-ray observations were obtained for several large two-ribbon flares. We summarize a typical event as follows. The first indication of an impending flare is the activation and eventual eruption of a filament lying along a magnetic neutral line in the active region. We do not know whether a filament eruption is a necessary condition for two-ribbon flares; however, such an eruption was known to precede the four events best observed by Skylab — those of 29 July, 13 August, 21 August, and 7 September in 1973. No known contradictory examples have been reported. Very soon after the prominence disappears, a system of loops is seen rising into the corona. The system does not resemble a continuously rising flux tube but, rather, new stationary loops appearing at progressively higher levels. The hottest loops are observed in thermal X-rays and appear at the top of the system. Cool Hα loops are seen beneath the X-ray loops. When the Hα loops appear, material is seen streaming down both legs at speeds of about 150 km s^{-1}. Both the cool and hot loops are rooted in the chromospheric flare ribbons which expand slowly outward as the loop system grows. Consistent with the height variation of loop temperatures, the Hα loops are observed to emanate from the inside edges of the flare ribbons whereas the X-ray loops are rooted in the middle to outside edges. The velocity of growth of the system is greatest at the beginning of the event (being about 10–20 km s^{-1}) and steadily decreases until, when the loops reach heights of $>$50,000 km, about 10 hours later, the speed has decreased to less than 1 km s^{-1} — as if the system were approaching an equilibrium configuration.

The hottest part of a given loop is at the top whereas the density is roughly uniform. As seen in X-rays, the peak temperature at the top of the system decreases from over 10^7 K at the beginning of the event to a few times 10^6 K at the end. The plasma electron density decreases from almost 10^{17} m^{-3} to about 10^9 m^{-3}. These numbers yield high thermal energy densities which, when integrated over the total flare volume, yield total energies of the order 10^{25} J (10^{32} erg).

An integration of the observed density in the post-flare loop system of 29 July 1973 over the lifetime of the event shows that the total mass of the system is considerably greater than the traditionally reported 10^{12}–10^{13} kg estimated from Hα observations alone. In this case, using the X-ray observations, we obtain a mass of 7.5×10^{13} kg. The reason for this is probably that the higher loops are never seen in Hα since they only cool down to coronal temperatures. Since it is quite unlikely that this amount of material can be supplied from the surrounding corona, it must be supplied from the chromosphere below. This occurs through chromospheric evaporation driven either by energetic particles propagating downward from the loop tops or by thermal conduction. Since the energy required to lift this mass to the required height is about 10^{24} J (10^{31} erg), thermal conduction seems to be the most probable mechanism. In Hα loops, material is always seen streaming downward. Therefore, the mass must be injected into the system while the loops are still at X-ray temperatures. A crude estimate of the injection velocity shows that it must be quite high at the base of the corona — of the order of 200 km s^{-1}. We wonder whether such speeds should produce observable doppler shifts.

All of these observed characteristics of the flare-loop phenomenon seem to be consistent with the concept that the loops are formed at progressively higher levels in the corona via magnetic reconnection in which an open magnetic field configuration is relaxing to a closed one. The opening of the field is probably related to the eruption of the prominence which precedes the flare. Since the open field has a total magnetic energy which is roughly twice that of the potential field corresponding to the same boundary conditions, this relaxation releases magnetic energy by Joulean dissipation, heating the flare plasma to the high temperatures observed.

In conclusion, we think that the two-ribbon flare process can be understood in terms of the reconnection picture discussed in this chapter, *provided* we postulate an open bipolar magnetic field configuration as a starting point. To complete the picture, however, we must explain how this open unrelaxed field is produced. For possible answers to this question the reader is referred to Chapter 6.

Acknowledgement

Thanks is extended to Hermann Schmidt for reading the manuscript and offering many helpful suggestions. The author also wishes to thank J. Beckers, A. Bruzek, S. Martin, W. Mattig, R. Moore, and Z. Švestka, for their help in obtaining some of the figures shown here.

REFERENCES

Allen, C. W., *Astrophysical Quantities*, Athlone Press, London (1963).

Antiochos, S. K. and Sturrock, P. A., "Influence of magnetic field structure on the conduction cooling of flare loops," *Solar Phys.* **49**, 359 (1976).

Antiochos, S. K. and Sturrock, P. A., "Evaporation cooling of flare plasma," *Astrophys. J.* **220**, 1137 (1978).

Barnes, C. W. and Sturrock, P. A., "Force-free magnetic field structures and their role in solar activity," *Astrophys. J.* **174**, 659 (1972).

Beckers, J., private communication to R. Moore (1977).

Bruzek, A., "On the association between loop prominences and flares," *Astrophys. J.* **140**, 2 (1964a).

Bruzek, A., "Case histories of flares: The large flares of 11, 12, 18, and 20 July 1961," *AAS-NASA Symposium on the Physics of Solar Flares*, NASA SP-50 (1964b).

Bruzek, A., *Plasma Instabilities in Astrophysics*, Pacific Grove, Gordon and Breach, New York (1969).

Carmichael, H., "A process for flares," *AAS-NASA Symposium on the Physics of Solar Flares*, NASA SP-50 (1964).

Cheng, C. and Widing, K. G., "Spatial distribution of XUV emission in solar flares," *Astrophys. J.* **201**, 735 (1975).

Colgate, S. A., "A phenomenological model of solar flares," *Astrophys. J.* **221**, 1068 (1978).

Correll, M. and Roberts, W. O., "Atmospheric magnetic fields above the active solar region of April 13, 1950," *Astrophys. J.* **127**, 726 (1958).

Cox, D. P. and Tucker, W. H., "Ionization equilibrium and radiative cooling of a low density plasma," *Astrophys. J.* **157**, 1157 (1969).

Craig, I. J. D. and McClymont, A. N., "Mass motions in a heated flare filament," *Solar Phys.* **50**, 133 (1976).

Culhane, J. L., Vesecky, J. F. and Phillips, K. J. H., "The cooling of flare produced plasmas in the solar corona," *Solar Phys.* **15**, 394 (1970).

Datlowe, D. W., Hudson, H. S. and Peterson, L. E., "Observations of solar X-ray bursts in the energy range 5–15 keV," *Solar Phys.* **35**, 193 (1974).

De, B. R., "On the mechanism of formation of loop prominences," *Solar Phys.* **31**, 437 (1973).

Dulk, G. A., Smerd, S. F., MacQueen, R. M., Gosling, J. T., Magun, A., Stewart, R. T., Sheridan, K. V., Robinson, R. D. and Jacques, S., "White light and radio studies of the coronal transient of 14–15 September 1973," *Solar Phys.* **49**, 369 (1976).

Foukal, P., "The development and flaring of an active region exhibiting unusual magnetic structure. II Active regions," *Solar Phys.* **13**, 330 (1970).

Gerassimenko, M., Solodyna, C. V. and Nolte, J. T., "Observational evidence of continual heating in X-ray emitting coronal loops," *Solar Phys.* **57**, 103 (1978).

Gergely, T. E., Kundu, M. R., Munro, R. H. and Poland, A. I., "Radio and white light observations of the 21 August 1973 coronal transient", *Astrophys. J.* **230**, 575 (1979).

Heyvaerts, J., Priest, E. R. and Dust, D. M., "An emerging flux model for the solar flare phenomenon," *Astrophys. J.* **216**, 123 (1977).

Hirayama, T., "Theoretical model of flares and prominences," *Solar Phys.* **34**, 323 (1974).

Hirayama, T. and Endler, F., Paper presented at the AAS meeting, Boulder, Colorado (1975).

Hudson, H. S., *Symp. on High Energy Phenomena on the Sun*, NASA GSFC preprint, SP-342 (1973).

Hudson, H. S. and Ohki, K., "Soft X-ray and microwave observations of hot regions in solar flares," *Solar Phys.* **23**, 155 (1972).

Hyder, C., "A phenomenological model for disparitions brusques followed by flare-like chromospheric brightenings: I. The model, its consequences, and observations in quiet solar regions," *Solar Phys.* **2**, 49 (1967a).

Hyder, C., "A phenomenological model for disparitions brusques followed by flare-like chromospheric brightenings: II. Observations in active regions," *Solar Phys.* **2**, 267 (1967b).

Jefferies, J. T. and Orrall, F. Q., *The Solar Spectrum*, Utrecht (1963a).

Jefferies, J. T. and Orrall, F. Q., "On the interpretation of prominence spectra: VI. Temperature determination and a model for quiescent prominences," *Astrophys. J.* **137**, 1232 (1963b).

Jefferies, J. T. and Orrall, F. Q., "Solar flares and the loop prominence mechanism," *ASS-NASA Symposium on the Physics of Solar Flares*, NASA SP-50 (1964).
Jefferies, J. T. and Orrall, F. Q., "Loop prominences and coronal condensations: II. The source of mass and energy and a model of the loop prominence mechanism," *Astrophys. J.* **141**, 519 (1965).
Kahler, S. W., "Determination of the energy or pressure of a solar X-ray structure using X-ray filtergrams from a single filter", *Solar Phys.* **48**, 255 (1976).
Kahler, S. W., Krieger, A. S. and Vaiana, G. S., "Morphological evolution of X-ray flare structures from the rise through the decay phase," *Astrophys. J.* **119**, L57 (1975).
Kahler, S. W., Petrasso, R. S. and Kane, S. R., "The quantitative properties of three soft X-ray flare kernels observed with the AS&E X-ray telescope on Skylab," *Solar Phys.* **50**, 179 (1976).
Kippenhahn, R. and Schlüter, A., "Eine Theorie der Solaren Filamente," *Z. Astrophys.* **43**, 36 (1957).
Kleczek, J., *Bull. Astron. Inst. Czech.* **8**, 120 (1957).
Kleczek, J., *Bull. Astron. Inst. Czech.* **9**, 115 (1958).
Kleczek, J., *Bull. Astron. Inst. Czech.* **14**, 167 (1963).
Kleczek, J., "Mass balance in flare loops," *ASS-NASA Symposium on the Physics of Solar Flares*, NASA SP-50 (1964).
Kopp, R. A. and Martin, S. F., paper in preparation (1980).
Kopp, R. A. and Pneuman, G. W., "Magnetic reconnection in the corona and the loop prominence phenomenon," *Solar Phys.* **50**, 85 (1976).
Krieger, A. S., "The decay of coronal loops brightened by flares and transients," *Solar Phys.* **56**, 107 (1977).
Lin, R. P. and Hudson, H. S., "Non-thermal processes in large solar flares," *Solar Phys.* **50**, 153 (1976).
Lüst, R. and Zirin, H., "Condensation of prominences from the corona," *Z. Astrophys.* **49**, 8 (1960).
MacCombie, W. J. and Rust, D. M., "Physical parameters in long-decay coronal enhancements," *Solar Phys.* **61**, 69 (1979).
Martin, S. F., "On the interpretation of post-flare loops with the flare of 29 July 1973," (paper in preparation) (1980).
Martin, S. F. and Ramsey, H. E., *Solar Activity, Observations and Predictions*, M.I.T. Press, Cambridge, Mass. (1972).
Moore, R. L. and Datlowe, D. W., "Heating and cooling of the thermal X-ray plasma in solar flares," *Solar Phys.* **43**, 189 (1975).
Moore, R. L., McKenzie, D. L., Švestka, Z., Widing, K. G., Antiochos, S. K., Dere, K. P., Dodson-Prince, H. W., Hiei, E., Krall, K. R., Krieger, A. S., Mason, H. E., Petrasso, R. D., Pneuman, G. W., Silk, J. K., Vorpahl, J. A. and Withbroe, G. L., "The thermal X-ray plasma," Chapter 7 of *Solar Flares*, P. Sturrock (ed) Col. Ass. Uni. Press (1980).
Neupert, W. M., Thomas, R. J. and Chapman, R. D., "Spatial distribution of soft X-ray and EUV emission associated with a chromospheric flare of importance 1B on August 2, 1972," *Solar Phys.* **34**, 349 (1974).
Newkirk, G., Jr., "Coronal magnetic fields and energetic particles," *High Energy Phenomena on the Sun*, NASA SP-342 p. 453 (1973).
Nolte, J. T., Gerassimenko, M., Krieger, A. S., Petrasso, R. D. and Švestka, Z., "Study of the post-flare loops on 29 July 1973, II. Dynamics of the X-ray loops," *Solar Phys.* **62**, 123 (1979).
Olson, C. A. and Lykoudis, P. S., "Solar loop prominences: A theoretical model," *Astrophys. J.* **150**, 303 (1967).
Pallavicini, R., Vaiana, G. S., Kahler, S. W. and Krieger, A. S., "Spatial structure and temporal development of a solar X-ray flare observed from Skylab on June 15, 1973," *Solar Phys.* **45**, 411 (1975).
Parker, E. N., "Dynamics of the interplanetary gas and magnetic fields," *Astrophys. J.* **128**, 664 (1958).
Parker, E. N., "The solar flare phenomenon and the theory of reconnection and annihilation of magnetic fields," *Astrophys. J. Suppl.* **8**, 177 (1963).
Parker, E. N., "The reconnection rate of magnetic fields," *Astrophys. J.* **180**, 247 (1973).

Petrasso, R. D., Nolte, J. T., Gerassimenko, M., Krieger, A. S., Krogstad, R., Seguin, F. H. and Švestka, Z., "Study of the post-flare loops on 29 July 1973, II. Physical parameters in the X-ray loops," *Solar Phys.* **62**, 133 (1979).
Petschek, H. E., "Magnetic field annihilation," *AAS-NASA Symposium on the Physics of Solar Flares*, NASA SP-50 (1964).
Pneuman, G. W., "Reconnection driven coronal transients," *Proc. IAU Symposium No. 91, Solar and Interplanetary Dynamics* (1979).
Pneuman, G. W., "Eruptive prominences and coronal transients," *Solar Phys.* **65**, 369 (1980a).
Pneuman, G. W., "Energetics of two-ribbon solar flares," *Solar Phys.* (submitted) (1980b).
Roy, J., "The magnetic configuration of the November 18, 1968 loop prominence system," *Solar Phys.* **26**, 418 (1972).
Roy, J. and Tang, F., "Solar X-ray bursts with flares and filament disruption," *Solar Phys.* **42**, 425 (1975).
Rust, D. M., "An active role for magnetic fields in solar flares," *Solar Phys.* **47**, 21 (1976).
Rust, D. M. and Bar, V., "Magnetic fields, loop prominences and the great flares of August, 1972," *Solar Phys.* **33**, 445 (1973).
Rust, D. M. and Roy, J., "Coronal magnetic fields above active regions," *IAU Symp.* **43**, 569 (1971).
Rust, D. M. and Webb, D. F., "Soft X-ray observation of large-scale coronal active region brightenings," *Solar Phys.* **54**, 403 (1977).
Schmidt, H. U., "Theoretical aspects of the flare phenomenon," *COSPAR Symp. on Solar Flares and Space Research*, North-Holland Publishing Co., p. 331 (1969).
Sturrock, P. A., "A model of solar flares," *IAU Symp.* **35**, 471 (1968).
Sturrock, P. A., "A classification of magnetic field configurations associated with solar flares," *Solar Phys.* **23**, 438 (1972).
Sturrock, P. A., "Mass motion in solar flares," *High Energy Phenomena on the Sun*, NASA SP-342, p. 3 (1973).
Švestka, Z., Dodson-Prince, H. W., Martin, S. F., Mohler, O. C., Moore, R. L., Nolte, J. T. and Petrasso, R., *Solar Phys.*, in press (1980).
Švestka, Z., private communication (1979).
Sweet, P. A., "The neutral point theory of solar flares," *Proc. IAU Symp. on Electromagnetic Phenomena in Cosmical Physics*, No. 6, Stockholm (1958a).
Sweet, P. A., "The production of high energy particles in solar flares," *Nuovo Cimento Suppl.* **8**, 188 (1958b).
Sweet, P. A., "Mechanisms of solar flares," *Ann. Rev. Astron. Astrophys.* **7**, 149 (1969).
Thomas, R. J., "Solar activity observed in X-rays and EUV from OSO-7," *IAU Symp. No.* **68**, 25 (1975).
Tucker, W. H. and Koren, M., "Radiation from a high-temperature, low-density plasma: The X-ray spectrum of the solar corona," *Astrophys. J.* **168**, 283 (1971).
Vorpahl, J. A., Gibson, E. G., Landecker, P. B., McKenzie, D. L. and Underwood, J. H., "Observations of the structure and evolution of solar flares with a soft X-ray telescope," *Solar Phys.* **45**, 199 (1975).
Vorpahl, J. A., Tandberg-Hanssen, E. and Smith, J. R., Jr., "Coronal plasma parameters in a long duration X-ray event observed by Skylab," *Astrophys. J.* **212**, 550 (1977).
Widing, K. G., "Fe XXIV emission in solar flares observed with the NRL/ATM XUV slitless spectrograph," *IAU Symp.* **68**, 153 (1975).
Widing, K. G. and Cheng, C., "On the Fe XXIV emission in the solar flare of 1973 June 15," *Astrophys. J.* **194**, L111 (1974).
Widing, K. G. and Dere, K. P., "Multiple loop activations and continuous energy release in the solar flare of June 15, 1973," *Solar Phys.* **55**, 431 (1977).
Withbroe, G. L., "The thermal phase of a large solar flare," *Astrophys. J.* **225**, 641 (1978).
Zirin, H. and Tanaka, K., "The flares of August, 1972," *Solar Phys.* **32**, 173 (1973).

CHAPTER 8

Particle acceleration in solar flares

J. HEYVAERTS

Observatory of Meudon and University of Paris VII

8.1 OBSERVATIONAL EVIDENCE FOR ACCELERATED PARTICLES

8.1.1 Radiation

Because we cannot make *in situ* measurements, the existence of fast particles at the flare site on the Sun is only inferred from secondary phenomena produced by them. Broadly speaking, these manifestations are electromagnetic, continuum and line radiation over the whole spectrum from the radio to the gamma-ray range, and accelerated particles detected far away in the interplanetary space. The basic difficulty that faces the solar observer is then to link correctly the measured quantities with the spatial, spectral and temporal characteristics of the particles. This step involves theoretical considerations and usually some hypotheses as well. Improving these is then also of great interest.

8.1.2 Detection in space

The direct detection of fast particles of solar origin can be made in space. One can usually measure energy spectra over some range, pitch-angle distributions of the particles and, in the case of nuclei, the mass distribution. In some experiments mass spectroscopy is also used, so that the charge state can be determined. Electrons are detected using electrostatic analysers, which take advantage of charged particle deflection in an electric field, or using scintillation detectors. (Observations of proton events are described in $2.9.1.)

8.1.3 Particle propagation

A major problem in relating the observational data to the spectrum at the Sun, or even in determining accurately the instant of acceleration, is that the propagation of particles from the Sun to the satellite is not a mere free flight in a given magnetic structure. Intense scattering occurs along the path in the solar wind, due principally to the interaction of fast protons and nuclei with MHD turbulence (of wave-length comparable to the particle gyroradii) as well as (in some instances) acceleration by interplanetary disturbances. In the case of electron streams, the effect of the electrostatic turbulence created by the stream itself is very important. The spectrum observed at the satellite depends also on time, being harder at the beginning when only fast particles have reached the detector; it depends also on whether the magnetic connection from spacecraft to flare is direct or not, and to a large extent on storage, propagation and release in the corona itself. For all these reasons, the inference of a spectrum "at the source" is difficult and requires some skill. Van Hollebecke *et al.* (1975) have shown that the interplanetary effects can be minimized if the energy spectra are observed at the time of maximum particle intensity, under good magnetic connection conditions; such an analysis performed on 32 protons event in the 20–80 MeV range revealed that spectra could be expressed as a power law in kinetic energy $N(E) \propto E^{-s}$, with a rather small dispersion in s, between 2.5 and 3.7. Spectra observed under less carefully selected conditions exhibit a much larger scatter in the result, and this may be taken as an indication that the results are indeed representative of conditions at the acceleration site.

8.1.4 Electron beams

The propagation of electron streams has received also considerable attention. However, because of the complexity of the problem, in which plasma turbulence (perhaps strong) and inhomogeneity effects come into play together, the theory has not yet reached a state in which it could be used to infer source conditions from observations. It is quite probable that the subject will make significant progress in the future. (See §2.9.3.)

8.1.5 Number of particles

The number of particles involved is difficult to estimate, because of the spatial extent of the electron flux and the filling factor of this region, both of which are largely unknown. Nevertheless, numbers in the range of 10^{33}–10^{34} particles seem reasonable (Lin and Hudson, 1976) for so called pure electron events (Lin, 1974). More details concerning the detection of flare particles in space can be found in §2.9 and §8.2.

8.1.6 Electromagnetic signatures

Electromagnetic signatures of accelerated particles include a number of phenomena and wavelengths, such as radio bursts (§2.7), hard X-ray and microwave bursts (§2.5) and even γ-rays (§2.9.4).

8.1.7 Purposes of this chapter

In §§8.2–8.10 we review in some detail some of these various pieces of observation. The rest of the chapter develops more theoretical arguments. These will be mainly restricted to those aspects most closely related to the acceleration process itself. First-phase acceleration is covered in §§8.11–8.20 and the second-phase process in §§8.21–8.26. A review of diagnostics such as those concerning radio emission would fill at least another chapter of this book and have recently been treated in the book by Melrose (1980).

8.2 ENERGETIC PARTICLES AT 1 AU

8.2.1 Spectra

Proton energy spectra, when analysed as described in §8.1.3, usually show a power-law shape from 20 MeV to 80 MeV with a spectral index s between 2.5 and 3.7. Steeper spectra are observed in flares which are not well-connected to the observer, and this may be the result of an energy-dependence of particle escape from the corona. Below 10 MeV, the spectrum is flatter than at higher energies. Because a flattening appears also in analysis of the γ-ray emission, one is presently inclined to think that this feature is characteristic of particle acceleration at the Sun rather than an interplanetary propagation effect. The spectrum at energies higher than 100 MeV is not yet known with sufficient accuracy; it may steepen in this range. Protons from solar flares are then seemingly subrelativistic, with Lorentz γ factors of the order of 1.1 corresponding to velocities between $c/10$ and $c/3$.

Electron spectra have been studied by Lin (1971), who found them to decrease with a spectral index between 3 and 4.5, in the energy range 22–45 keV (see §2.9.3).

8.2.2 Electron–proton relationships

A further analysis (Lin, 1974) revealed the existence of so-called *pure electron events* (which are not accompanied by energetic protons) and of mixed events. A demonstration of the effect is given in Figure 8.1, which shows a correlation

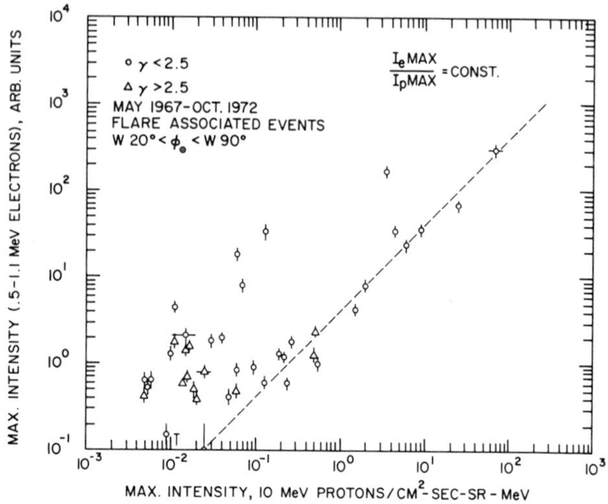

FIGURE 8.1 Correlation plot between 0.5 to 1.1 MeV electrons and 10 MeV protons for some 40 flare-associated events detected by the IMP 4/5 GSFC cosmic ray experiment. All these events are associated with flares well-connected to the observer. The flux has been measured at its maximum for both protons and electrons. (From Ramaty *et al.*, 1980.)

plot of electron (0.5, 1.1 MeV) versus proton (10 MeV) intensity. It is seen that for low proton intensity the correlation with electron intensity disappears. This indicates that any event is the superposition of a pure electron component (for which almost no protons are accelerated) and of a component for which electrons and protons are accelerated in a given fixed proportion, namely

$$\frac{Ie(0.4\text{ MeV})}{Ip\ (10\text{MeV})} = 100$$

or, using a spectral index $s=3$ for electrons,

$$\frac{Ie\ (10\text{ MeV})}{Ip\ (10\text{ MeV})} = 10^{-2}.$$

It is moreover found that pure electron events exhibit an abrupt spectral steepening at 100 keV. This is not so for mixed events, which have also a smaller spectral index and a change of spectral character around 100 keV.

A composit spectrum of energetic electrons observed in the 7 Sept, 1973 flare has been prepared (Lin *et al.*, 1978). It indicates that two different acceleration mechanisms are at work in a fully-developed solar flare: an ever-present *first phase*, which accelerates essentially electrons to energies of the order of 100 keV (but not much above 1 MeV), and a *second phase*, which accelerates both type of particles in such a way that at a given energy (10 MeV, say) the ratio of protons to electrons is much larger than 1.

8.3 HARD X-RAY EMISSION IN FLARES

8.3.1 Information available

Since their first detection, X-rays from flares have been regularly observed in space (see §2.5.1). As will soon appear, the hard X-ray emission is an almost ideal signature of electrons in the 100 keV range, being very prompt and related in a very direct way to the accelerated particles, though the characteristics of the ambient medium are unfortunately also important in determining the radiation. This radiation is due to *bremsstrahlung* of electrons on the ambient plasma ions. *Gyrosynchrotron* radiation is also expected, and will be studied below. Hard X-rays suffer negligible radiation-transfer effects. (The length of a unit optical thickness for Thomson scattering under coronal conditions is not less than 10^{10} km.) They are particularly well suited for a study of impulsive phenomena, where high resolution, both spatial and temporal, is needed. Spatially-resolving detectors have been launched on the Solar Maximum Mission. It has unfortunately not been possible to achieve a resolution better than 8″. Time-resolved spectral observations, on the other hand, have been carried out for some time, in particular in the observations by the ESRO TD1A satellite (Van Beek *et al.*, 1973) and by OSO5 satellite (Frost and Dennis, 1971). A complete analysis of the results has been carried out by Hoyng *et al.* (1976). The time-resolution in these observations was 1.2s and the spectral range above 25 keV was covered with 12 logarithmically spaced channels.

8.3.2 Impulsive and extended bursts

For some events (called *impulsive*) the duration is not more than 10 to 100 sec (see §2.5.1); they exhibit either a single peak or a few peaks, the duration of which is some 10s (Figure 8.2). For some other events (*extended bursts*), such as the three large events of 2, 4, 7 August 1972, the hard X-ray emission lasts 10^3 s and the flux is an order of magnitude larger than for impulsive events. Figure 8.3 displays the time-profile for the 4 August 1972 event in three different spectral ranges: it is seen that the spectrum becomes significantly harder (relatively more energetic photons) as the flare develops. The converse is true for impulsive bursts. These large differences should indicate again the existence of two different acceleration mechanisms. Actually, these two

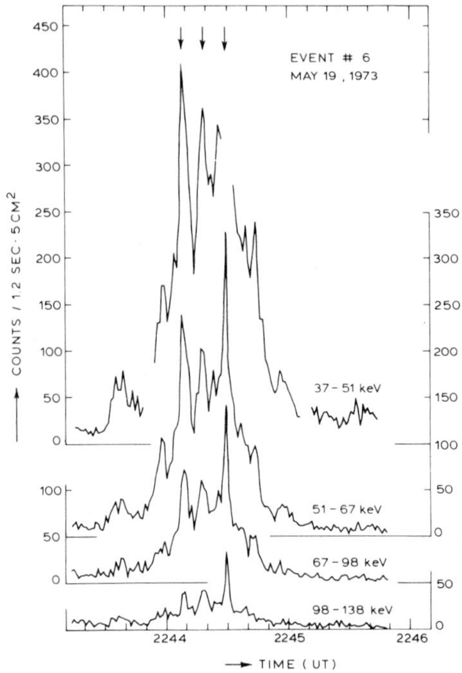

FIGURE 8.2 Event of 19 May, 1973. The X-ray emission shows a precursor and a few pronounced spikes. The arrows identify the spikes. (From Hoyng *et al.*, 1976.)

mechanisms should operate one after the other, as shown by Frost and Dennis (1971), who observed an event in which two physically distinct components were clearly apparent (Figure 8.4); an impulsive "first phase" and a more gradual "second phase". We mean by this the existence of a first prompt mechanism producing 100 keV electrons followed later by a second mechanism giving rise to protons, nuclei and much higher energy electrons. It is, in our opinion, still a matter of discussion as to whether the second phase electrons are first accelerated in the first phase, or whether the first phase produces some ion preheating, or (more generally) whether the first phase is causally related to the second phase. Up to now, we have evidence only for the separate existence of the two phases and their temporal order.

8.3.3 Source size

Time resolution can be used to infer some information about the source of the X-ray radiation. This has been already described by Švestka (Chapter 2). The source size l cannot be larger than the rise-time τ of the burst multiplied by the velocity of the agent triggering either the emission or the acceleration of the

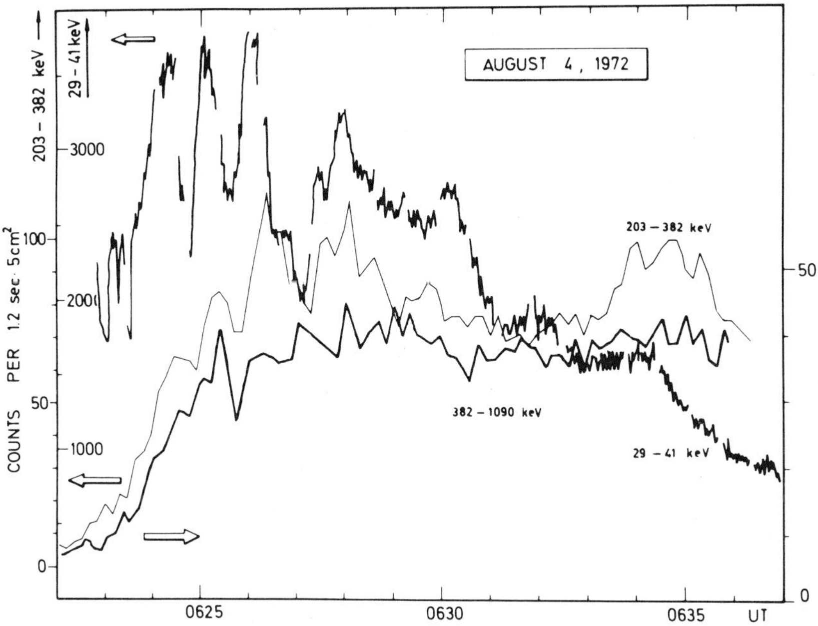

FIGURE 8.3 Correlation of low-energy (29–41 KeV) with high-energy photon counts. For the range 203–382 KeV, two channels (and, for the range 382–1090 KeV, four channels) have been integrated for a time 9.6 s. (From Benz, 1977.)

radiating particles. This velocity is less than c, so the source size is less than $c\tau$. For $\tau \sim 2$ to 4 sec, one obtains:

$$l \leqslant 600{,}000\text{–}1{,}200{,}000 \text{ km, (light)} \tag{8.1}$$

which is of course quite large! If one takes the more definite view that the trigger is carried by an MHD signal, then the velocity of the trigger is the Alfvén velocity, whose determination needs a knowledge of the density in the X-ray source. This may be bounded from below by the fact that, once the injection of fast particles in the X-ray emitting source has ceased, the radiation should not last more than the observed duration of the burst τ_x. From this Hoyng *et al.* (1976) obtain

$$n(\text{source}) > 10^{16} \text{ m}^{-3} \; (10^{10} \text{ cm}^{-3}). \tag{8.2}$$

For a reasonable field of 100 Gauss, one obtains then an upper limit on the source size l of

$$l < 10{,}000 \text{ Km (MHD)}. \tag{8.3}$$

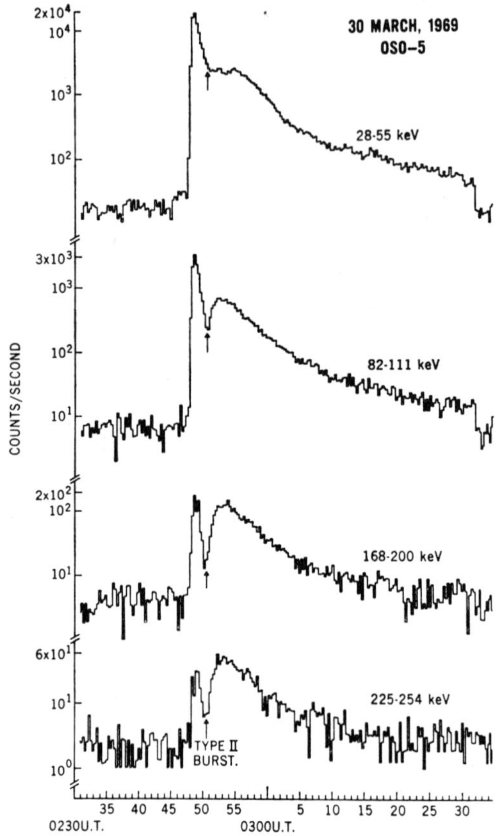

FIGURE 8.4 Intensity-time profiles which cover the energy intervals indicated. Vertical arrows indicate the time of appearance of a type II burst in the metric band. (From Frost and Dennis, 1971.)

From the high densities involved (8.2) and from the small scales (8.3) one sees that the *source should be compact and dense.* (This idea may be in error because the more rigorous limit (8.1) is much larger.) Lin *et al.* (1978) report the detection of a seemingly impulsive burst from an event having occurred behind the limb, whose altitude was then very much larger than (8.3). If (8.2) and (8.3) are valid for this event, a small dense spot should be sitting high in the corona. If (8.3) is wrong, a point which could be tested from the planned spatially-resolved observations, then the nature of the trigger will have to be explained. If (8.2) turns out to be wrong, the energy budget problem will become even more difficult than we now think it is.

Information concerning the size and location of the second-phase acceleration source can only be obtained by spatially-resolved observations. The associations of these emissions with various high-corona radio emission makes it quite likely that the X-ray emission also occurs there. This has been convincingly established recently by Hudson (1978), who observed in OS07 (on 14 December 1971) a typical long duration event (42 min) associated with type II and IV radio emission with no observable impulsive phase and no observable soft X-ray burst. This is because the parent flare, which occurred more than 20° behind the limb was occulted. This is compelling evidence that the non-observed phenomena occurred below the occultation altitude of 70,000 Km. The observable part of the emission, however, could not be associated with specific radio manifestations. For this event, a lower limit of 10^{37} electrons responsible for the second-phase radiation could be derived on purely observational grounds, since the less efficient X-ray radiation was found to be produced by 10^{38} particles.

Recently, an extremely interesting analysis of an observation of an impulsive X-ray burst caused by a flare behind the solar limb as viewed by two distinct spacecraft (Kane *et al.*, 1979) has shown that the "coronal" part of the X-ray source was 600 times less intense than the lower altitude part (below 25,000 km, say).

8.3.4 Time variations

The time-resolution of the TD1A observations was only 1.2 sec (Hoyng *et al.*, 1976). A finer observation with 0.1 s time-resolution for a 100 s duration burst on 4 July 1974 was reported by Hurley and Duprat (1977); it did not show any significant change shorter than 1 s, which seems to indicate that the characteristic times presently quoted are representative of the true solar variations. Searches for regular pulsations have not revealed significant periodicities for impulsive or short events. An interesting correlation between the spectral indices of radio microwave and X-ray emission with the radio and X-ray flux has been reported by Benz (1977), as shown in Figure 8.5. Actually, the X-ray intensity and spectral index anticorrelate and even better with a small time-lag, of the order of 10 sec. Anticorrelation and phase shift lead to inclined ellipses in the (I_x, γ) plane, as reported by Hoyng *et al.* (1976), whose results are shown in Figure 8.6. A time-shift between the temporal features is also observed by Benz (1977), the higher energy features lagging the lower energy ones by several seconds. Correlation between microwave intensity and X-ray intensity is best for X-ray energies between 100 and 200 keV, whereas radio and X-ray spectral indices correlate (as expected) for X-ray bremsstrahlung and microwave synchrotron emission by a power-law electron distribution.

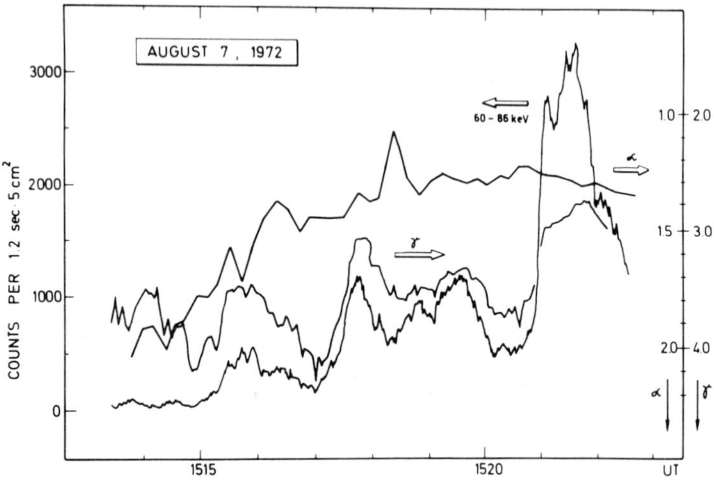

FIGURE 8.5 Correlation between the hard X-ray and the high-frequency radio spectral indices (γ and α resp., scale downwards). Note the anticorrelation between flux densities and their spectral indices. (From Benz, 1977.)

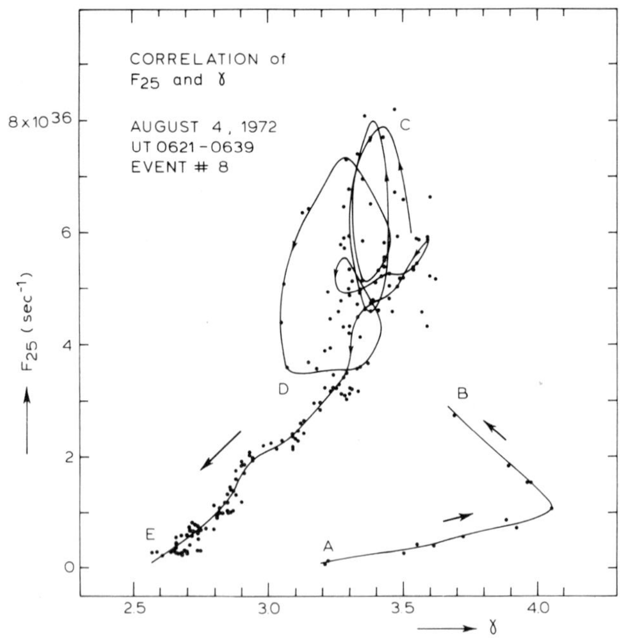

FIGURE 8.6 Correlation diagram for the August 4, 1972 flare, showing variations of the flux $F_{25}(t)$ versus the spectral index $\gamma(t)$. The solid line indicates the smoothed path of the event development. (From Hoyng *et al.*, 1976.)

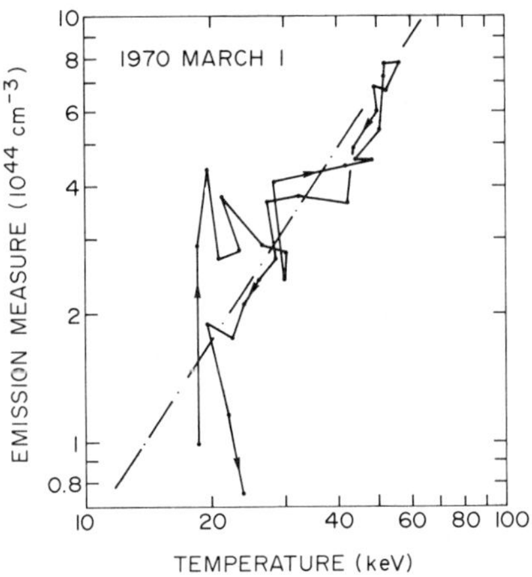

FIGURE 8.7 Trajectory of the 1st March 1970 flare in the (log T–log $n_e^2 V$) plane. (From Mätzler *et al.*, 1978.) Note that 1 keV = 1.16×10^7 K.

8.3.5 Evidence of a very hot plasma

Near its maximum the burst emission appears to be fairly compatible with optically-thin radiation by a high temperature plasma (of volume V and density N, say). Actually, the corresponding emission formula, including a proper approximation of the Gaunt factor

$$I(h\nu)_{\text{photons}}\ \text{cm}^{-2}\ \text{s}^{-1}\ \text{keV}^{-1} = 1.3\ 10^3\ n_e^2\ V \frac{e^{-h\nu/T}}{(h\nu)^{1.4} T_{\text{keV}}^{0.1}} \tag{8.4}$$

defines for each pair of energy channels (i, j) a temperature T_{ij}, which appears to be essentially independent of (i, j). The event of 1st March 1970 so analysed, with corrections for photospheric albedo, would be correctly represented late in the event by (8.4), with a time-dependent temperature $T(t)$ and *emission measure* $(n_e^2 V)(t)$. The very late phase could not agree with that picture however. Interesting enough, the trajectory in the $[T, n_e^2 V]$ plane revealed a remarkable, reversible behaviour, shown in Figure 8.7. The evolution during the reversible, brightest part of the event is compatible with the idea that the source suffers an *adiabatic compression*. In such a process, one has:

$$\begin{cases} T V^{2/3} = \text{constant}, \\ n_e V = \text{constant}, \end{cases}$$

so that

$$n_e^2 V / T^{3/2} = \text{constant}.$$

This relation is shown in Figure 8.7 by the oblique straight line. Similar, though not so clearly-demonstrative, results have been found for other bursts as well. One may then think that, at least during the brightest part of the event, the source consists essentially of a compressed (and then depressed) very hot plasma. This is not so during the very beginning and end of the burst: simple adiabatic compression cannot explain the energization of the X-ray emitting plasma, and deviations from thermal behaviour occur late in the burst, the emission being relatively too high for a thermal distribution at high energies.

Some conclusions can be drawn concerning the *source size* by considering the microwave radio-emission that takes place at the same time. The time profile of X-rays in the channels 28–55 keV and 55–82 keV and of the 10.5 GHz radio emission are quite similar. This makes it reasonable to hypothesize that both types of radiation are emitted from the same region with volume V and projected area $A = V^{2/3}$. For temperatures above 1 keV and typical solar magnetic fields, the emission and absorption processes are dominated by gyrosynchrotron emission and absorption. Usually, the source is optically-thick at low frequencies, giving rise (for thermal populations) to a flux that scales with frequency f, emitting area A and temperature T as (Af^2T). The surface A can then be deduced from the observations: Mätzler *et al.* (1978) find, using microwave data and the temperature deduced from their X-ray analysis, an area

$$A = 5 \times 10^{14}\ \mathrm{m}^2, \tag{8.5}$$

or a source size when the temperature reaches 57 keV of

$$l \simeq 20{,}000\ \mathrm{km}\ \text{(thermal interpretation)},$$

quite consistent with (8.3). An estimate of the electron density in the source can then be found from the observed X-ray emission measure: Mätzler *et al.* found

$$V = 10^{22}\ \mathrm{m}^3$$

$$n_e^2 V = 8 \times 10^{50}\ \mathrm{m}^{-3},$$

$$n_e = 3 \times 10^{14}\ \mathrm{m}^{-3}\ (= 3 \times 10^8\ \mathrm{cm}^{-3}),$$

$$n_e V = 3 \times 10^{36},$$

Thermal energy $= 8 \times 10^{22}\ \mathrm{J} = 8 \times 10^{29}$ ergs (if $T_i = T_e$),

$$B = 80\ \text{Gauss}$$

(from limit frequency of optically thick radio emission),

$$\frac{B^2}{2\mu} = 26 \text{ J m}^{-3} \ (= 260 \text{ ergs cm}^{-3}),$$

$$\text{Thermal en. density} = 8 \text{ J m}^{-3} \ (= 80 \text{ ergs cm}^{-3}).$$

It is to be noted that the thermal energy density does not differ very much from the magnetic energy density.

8.4 IMPLICATIONS OF X-RAY OBSERVATIONS: HOW MANY ELECTRONS?

8.4.1 Bremsstrahlung

Hard X-ray continuum is emitted by bremsstrahlung. Two interesting limiting cases are of special interest: the *thin-target* case and the *thick-target* case. In the thin case, it is assumed that the radiating electrons lose a negligible part of their energy in the plasma before leaving it; the result is that a power-law spectrum

$$\varphi(\nu) = A(h\nu)^{\gamma} \text{ photons m}^{-2} \text{ s}^{-1}$$

is produced at Earth by an electron spectrum with index $(\gamma - \frac{1}{2})$. In the thick-target case, the target density is high enough for the energy-loss time to be much smaller than the injection time-scale and so the same photon spectrum needs an electron injection flux with spectral index $(\gamma + 1)$. In both cases the number of electrons can also be deduced (Hoyng *et al.*, 1976).

8.4.2 Number and energy involved

In a convincing paper Lin and Hudson (1976) argued that virtually all of the flare manifestations could be explained if the *bulk of the flare energy is first deposited in a population of high-energy electrons* (accelerated high in the corona by some anomalous current dissipation) and if they later *impinge on the photosphere and chromosphere*. If this energy deposition can be balanced by an efficient enough loss, one would observe optical, UV and EUV radiation. If not, the bombarded regions would evaporate and form the soft X-ray flare plasma. Their study shows that the whole picture is tenable on the basis of energy and mass budget. There are, however, difficulties with it: the number of electrons that impinges on the solar atmosphere is so large that a coherent flow to the photosphere is impossible, as we shall see below. The thick-target parameters derived vary over a considerable range but are actually very large,

TABLE 8.1

Event UT	Date	Nature	"Thick target" Number of electrons	Energy of electrons >25 keV
0103	21 March 1972	simple	2.3×10^{36}	1.3×10^{22} J
1406	18 May 1972	simple	5.0×10^{37}	2.3×10^{23} J
1617	18 May 1972	simple	4.8×10^{37}	2.3×10^{23} J
1839	2 August 1972	simple	4.0×10^{37}	2.2×10^{23} J
0252	7 August 1972	simple	3.6×10^{37}	2.0×10^{23} J
2244	19 May 1973	simple	1.4×10^{38}	7.2×10^{23} J
0620	4 August 1972	extended	3.5×10^{39}	2.0×10^{25} J
1513	7 August 1973	extended	2.8×10^{39}	1.6×10^{25} J

as seen in Table 8.1, which shows the total number of more-than-25 keV electrons produced according to this picture (note that $1\ \text{J} = 10^7$ erg). One must realize that the number of electrons quoted in the list above is incredibly large: it represents approximately the *total* number of electrons present in a region of size l (Eq. 8.3) with a density of $10^{16}\ \text{m}^{-3}$. The thermal interpretation of the event studied by Mätzler *et al.* (1978) does not lead to small numbers either if we consider it (in view of its short duration) as a simple event, though it gave rise to a type II burst. Nevertheless, we shall discuss below some reasons to believe that thermal models need less electrons. We stress again, however, that the *number of particles estimated to be released in space* (10^{32}–10^{34}) *is much smaller*. This discrepancy is usually attributed to a small probability of escape, which is, in our opinion, not entirely convincing. For the 1st March 1978 event, the fact that the type III burst was visible up to decameter wavelengths indicates a good connection to the high corona. Comparison of thermal and target interpretations for the same event would be invaluable. Other clues to the number problem can be provided by a reliable theory of radio emission, in particular microwave impulsive bursts, so closely associated with hard X-rays, and by a comparison of hard X- and EUV-impulsive bursts.

8.5 MICROWAVE IMPULSIVE BURSTS

8.5.1 Mechanism involved?

Microwave radiation is described in §2.5.2. It refers to radiation in the several tens GHz band (10.5 GHz, 19 GHz, 35 GHz), i.e. in the centimeter wavelength range. The density corresponding to a frequency of this order of magnitude is of the order of some $10^{18}\ \text{m}^{-3}$, which means that the radiation is refracted only in very dense regions of the corona. It is, however, worth recalling that densities of this order of magnitude are observed in the flare plasma, so that coherent processes are not excluded *a priori* as a source of this radiation. An origin low in the chromosphere or photosphere is, however, certainly excluded. Two different emissions occur in this frequency range: the *impulsive microwave emission*, which is more specifically the subject of this chapter, and a longer duration emission, which sometimes follows it and is called a *microwave type IV burst*. These distinct types are apparently the counter-part of the first- and second-phase X-ray emissions of Figure 8.4. Due to their similarity, both for impulsive and long-lasting events, microwave and hard X-ray radiation should be explained as being produced by one and the same population. One likely mechanism is *gyrosynchrotron radiation*, which must occur as soon as energetic particles are present in a magnetic field. Benz (1977) obtained the correlation shown in Figure 8.5, which is especially good for 80 keV X-energies. The radio and X spectral indices (γ_R and γ_X) should be related by:

$$\gamma_R = \tfrac{1}{2}\gamma_X - 0.75, \qquad (8.6)$$

which seems approximately true (Figure 8.3) after 06.27 UT. A model in which X-rays and microwaves are produced at the same location in the corona seems *a priori* better. The synchrotron hypothesis should be tested from spectral features and polarization data. The dispersive effects of the medium on the one-particle emission and the effects of finite or large optical thickness must be taken into account.

8.5.2 Synchrotron radiation

For a detailed discussion of *synchrotron radiation*, the reader is referred to basic review papers and text-books (Melrose, 1968; Bekefi, 1966). A striking effect of the ambient plasma is the *Razin effect*, which suppresses the radio emission at low frequencies for synchrotron radiation (Ramaty and Lingenfelter, 1967). Negative absorption is virtually impossible (Wild *et al.*, 1963; Heyvaerts, 1968), except due to medium-effects at low frequencies (Zhelezniakoff, 1967). For mildly relativistic energies, the emissivity has been studied by Ramaty (1969) and Trulsen and Fejer (1970); the absorption coefficient is strongly frequency-dependent (Mätzler, 1978b).

8.5.3 Existing models and some observations

Mätzler (1978a, 1978b) examines the consequences of a thermal, hot plasma on microwave radiation and compares with observational data. The source would become optically thin in each polarization at well-defined frequencies,

f_{0+} and f_{0-}, say. The temperature of the source is not uniform, nor is the emitting plasma. The source is characterized by a distribution of turnover frequencies f_{0+} or f_{0-}. As the emission drops drastically when the source is optically thin, the power emitted at each frequency is essentially a black body spectrum multiplied by a frequency- and polarization-dependent "effective area" $A(f)$, which represents the actual surface of that part of the source which is optically thick at that frequency. The dependence of $A(f)$ on f is determined by geometrical details which are not at hand. It is nevertheless interesting to see that a rough approximation to $A(f)$ can still give a good account of burst variation for the 1 March 1970 event (Mätzler, 1978a), for which the high-frequency slope of the spectrum indicated that at any time one could approximate $A(f)$ by:

$$A_{\pm}(f) = A_0(f_{0\pm}/f)^3,$$

where $f_{0\pm}$ is the turnover frequency of the burst at time t, which can be determined theoretically from a knowledge of the magnetic field, size and temperature of the source as a function of time. These quantities may be derived from lower frequency observations, (Eq. 8.7), and from the X-ray burst analysis (Eq. 8.4). The trend of the high-frequency variations can be predicted. This has been done for the 1st March 1970 event, and is shown for the 10.5 GHz frequency in Figure 8.8 (Mätzler, 1978a). On the other hand, the good fit of this estimate, which seems to hold everywhere except late in the burst (where the thermal picture for X-ray emission also fails), does not rule out some non-thermal models, and *a clear-cut decision between these two models (non-thermal or thermal) will eventually be possible only when enough geometrical details of the source are known.*

It is worth recalling the model calculated by Böhme *et al.* (1977) which pictured a non-thermal core surrounded by a maxwellian halo. Gyrosynchrotron reabsorption and ray tracing was included, and parameters of the model were chosen to fit hard and soft X-ray emission. It was found that the radio microwave radiation so calculated was consistent with observations of the events of August 4th 1972 and 18th May 1972. Geometry is very important here and can only be determined by high-resolution instruments. Alisandrakis and Kundu (1978) have used the Westerbork synthesis radio telescope, which is able to give 6″ resolution at 6 cm wavelength in one-dimension, to analyse the spatial structure of a sufficiently long burst (the time-resolution was 30 s only). Because of the fast time-variations, aperture synthesis techniques are not easily usable. Figure 8.9 shows the result of their observations for one rare case of an impulsive event shorter than 2 min. A very interesting feature of this event is that it consisted of two spatially-distinct bright regions, one of them lying very close to a constant source of cm wave

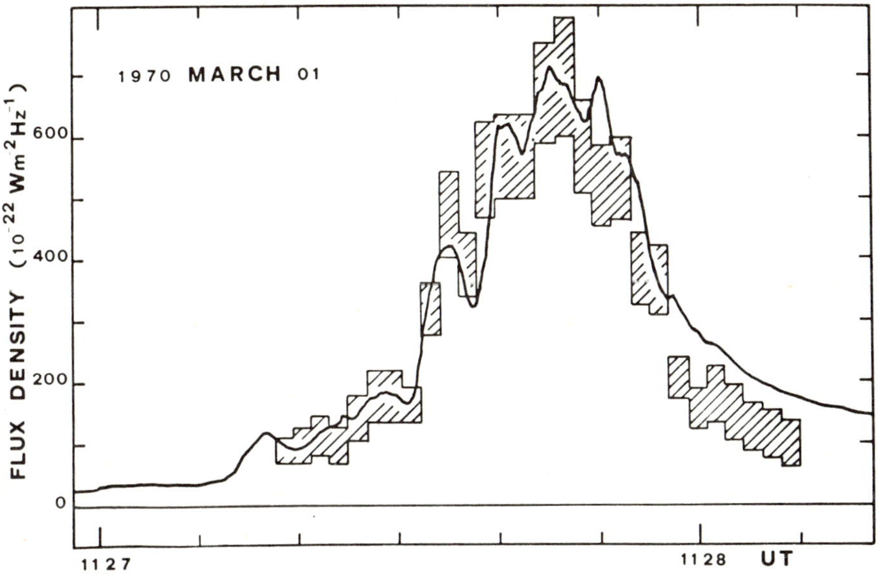

FIGURE 8.8 The solid line shows the microwave time-profile of 10.5 GHz, with a time-constant of 0.05 s (from A. Magun). The hatched region indicates the microwave flux computed from the temperature variation $T(t)$ deduced from X-ray observations, for constant magnetic field strength. The vertical extent of the computed values indicates the uncertainties. (From Mätzler 1978a.)

radiation in the active region (S source). Another important feature is that both sources are polarized in the same sense, though spatial indications show that they occur near a reversal of the line-of-sight component of the magnetic field. A word of caution must be introduced here, however: the absence of simultaneous X-ray data makes it difficult to decide how far the reported burst is really of the impulsive type, and we must be cautious that impulsive and long-duration events may have quite different physical characteristics, as far as thermal nature, altitude, and mean energy of the electrons is concerned.

Spatially-resolved observations also give information concerning brightness temperatures, and then also the plasma temperature if the thermal model is correct. Measurements of this sort have been reported by Kundu *et al.* (1974), who found $T_{\text{brightness}} \geqslant 10^9$ K at 3.7 cm and $\geqslant 1.65 \times 10^8$ K at 11 cm. These brightness temperatures are not in contradiction with a thermal model. More puzzling, and perhaps more able to reveal details of the elementary acceleration mechanism, is the discovery by Slottje that a microwave outburst at 2.65 GHz on 11th April 1978 showed many spikes with half-power duration smaller than 20 ms. All spikes were 100% polarized, with flux densities as high as 6×10^{-19} W m^{-2} Hz^{-1}. Estimating the source size to be at most $c\tau$, where $\tau \simeq 10^{-4}$ sec, implies, according to Slottje (1978), a lower limit on the brightness

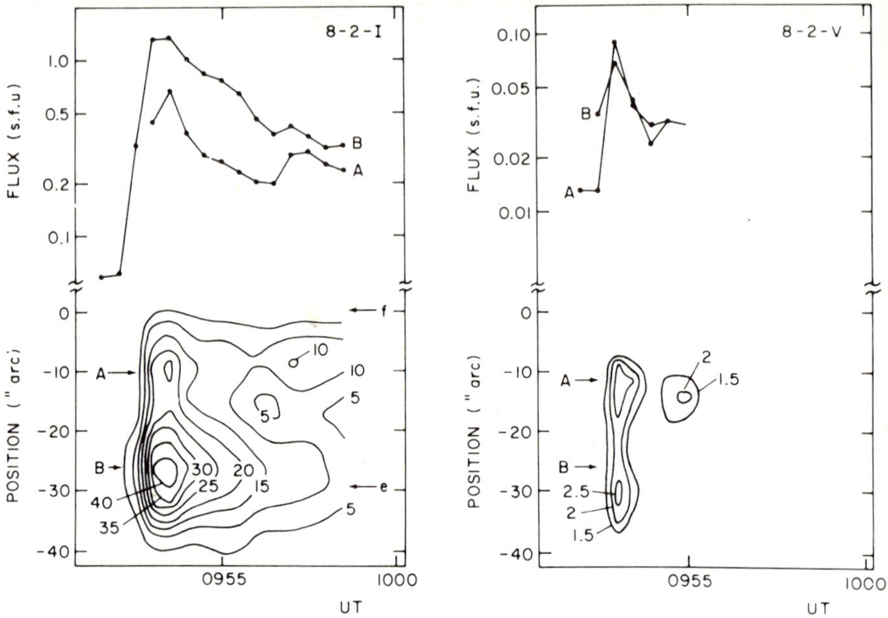

FIGURE 8.9 Total intensity I and circular polarization V of a microwave burst. The upper part shows the total quantity versus time, while the lower part gives a one-dimensional map as a function of position and time. Contour labels are in units of 10^6 K. Negative polarization values (dashed contour lines) correspond to left-hand polarization. The position of the constant sources near each burst is shown by arrows on the I contour map. All constant sources have been subtracted from this figure. (From Alisandrakis and Kundu, 1978.)

temperature of 10^{15} K, definitely non-thermal! This certainly will be a major piece of information, especially if these features are a regular component of microwave bursts.

Radio radiation beyond the microwave range is a major diagnostic tool, but it is a very elaborate subject and so cannot be considered in this review, since it would lead us too far from our main topic. We insist, however, that a theoretical and observational analysis of this radiation is of the utmost importance for diagnosing fast particles and their acceleration (see §2.7).

8.6 GAMMA-RAY EMISSION

Solar gamma-ray radiation was of course expected, due to the presence of a number of accelerated nuclei pervading the solar plasma giving rise to *γ-ray line emission* as well as *bremsstrahlung* from high energy electrons. However, these radiations were not detected until 1972. The Figure 2.17 of

Chapter 2 shows the results obtained by Hudson *et al.* (1980) for the large solar flare of July 11, 1978. Co^{60} lines are calibration lines, but the 0.51 MeV, 2.2 MeV, 4.4 MeV and 6.1 MeV lines are of solar origin. They are due respectively to:

0.511 MeV $\quad e^+ + e^- \rightarrow \gamma + \gamma \quad \text{or} \quad \gamma + \gamma + \gamma,$

2.22 MeV $\quad p + n \rightarrow D^* \rightarrow D + \gamma_{2.22},$

4.44 MeV $\quad p + {}^{12}\mathrm{C} \rightarrow p + {}^{12}\mathrm{C}^* \rightarrow p + {}^{12}\mathrm{C}_+ + \gamma_{4.44},$

6.13 MeV $\quad p + {}^{16}\mathrm{O} \rightarrow p + {}^{16}\mathrm{O}^* \rightarrow p + {}^{16}\mathrm{O} + \gamma_{6.13}.$

Flare γ-rays have also been observed by other observers (Chupps *et al.*, 1975), and also on board the Soviet Prognoz satellite (Chambon *et al.*, 1978) for a flare on 22 Nov. 1977, including a high time resolution device which will allow the determination of onset time with some precision; similar observations are reported by Hudson *et al.* (1980) for the event of 11th July 1978. These lines and the γ-ray continuum are very useful diagnostic tools of the high energy nuclei produced in the flare region. The information which it is possible to gain from data of Chupp *et al.* (1975) has been already reviewed by Dr Švestka in §2.9.4 and will not be repeated in detail. It suffices to recall that Ramaty and Crannel (1976) have established that the 4.4 and 6.1 MeV lines are emitted soon. The other two lines are delayed, but more prominent. Crannel *et al.* (1978) discuss the formation of the 0.511 MeV line and that of a possible positronium continuum. Bai and Ramaty (1976) have concluded (on the basis of timing data of the 2.2 MeV line, γ-ray continuum, hard X-ray radiation and 37 GHz microwave radiation) that the protons are accelerated in a second phase, some minutes after the impulsive phase together with the highest energy electrons. The spectrum of the electrons has been estimated from the X- and γ-ray bremsstrahlung continuum giving:

$$n_i N(E) = 3.3 \times 10^{50}\ \mathrm{E}^{-2.4}\ \text{electrons m}^{-3}\ \text{MeV}$$

for $E < 100$ keV,

$$n_i N(E) = 2.6 \times 10^{49}\ \mathrm{E}^{-3.5}\ \text{electrons m}^{-3}\ \text{MeV}^{-1}$$

for $E > 100$ keV, where n_i is the plasma ion density. Information on the proton spectra has been discussed by Ramaty *et al.* (1977), and it has been established by Ramaty *et al.* (1980) that the γ-ray continuum itself imply that the fast nuclei are high-Z enriched. Again, the reader is referred for details on these important data to §2.9.4.

8.7 ELEMENTAL AND ISOTOPIC ABUNDANCES OF SOLAR COSMIC RAYS

8.7.1 He^3/He^4

As is apparent from the γ-ray spectrum, the solar flare cosmic-ray composition may be quite different from the normal solar composition. This was clear as early as 1962, when Schaeffer and Zarringer (1962) measured the ^3He abundance of energetic solar particles and found the $^3He/^4He$ *ratio to be of the order of* 0.2. Subsequent measurements have revealed the existence of a class of solar particle events in which the $^3He/^4He$ ratio is substantially larger than in the ambient solar atmosphere (Hsieh and Simpson, 1970; Anglin *et al.*, 1977; Dietrich, 1973; Garrard *et al.*, 1973). Normally this ratio is between 10^{-4} and 10^{-3} in the solar wind. There is no direct observation of He^3 in the photosphere, but the 2.2 MeV γ-ray line sets upper limits on the photospheric He^3 abundance consistent with that measured in the solar wind (Wang and Ramaty, 1974). A detailed summary of isotopic composition of solar cosmic rays has been made at the Skylab Workshop on Solar Flares (Ramaty *et al.*, 1980). The characteristics of ^3He-rich event can be summarized as follows. The $^3He/^4He$ ratio (from 1 to 10 MeV per nucleon) varies from 10^{-2} up to more than 1, but the enhancement in ^3He is not accompanied by a similar enhancement in deuterium or tritium. Large He^3 enhancements are found in small proton events, sometimes even not reported as flares; this is shown in Figure 8.10, while Figure 8.11 shows that the $^4He/^1H$ ratio is similarly enhanced for these same events. *This is certainly a stringent condition for building successful acceleration models.* Two different types of explanation can be envisaged for these anomalies: *nuclear reaction effects or plasma mechanisms which might differentiate among different species* and charge states for injection in the acceleration process. It is very important to reach a definitive answer, because these may be a clear indication as to the nature of the second-phase acceleration process.

8.7.2 Nuclei composition

Similarly, the recognition of systematic effects in the elemental and isotopic composition of solar cosmic rays would be invaluable. This search is presently underway, though it is somewhat obscured by difficulties associated with rather small counting rates and with a large variability from event to event. *Conspicuous is the existence of Fe-rich flares.* The largest Fe-enhancements are observed at low energies (smaller than 10 MeV/nucleon), where Fe/He = 1 have been seen. At higher energies, however, (15–30 MeV/nucleon) Bertoch and Reames (1977) report Fe/O $\simeq$ 0.4, while Dietrich and Simpson (1978)

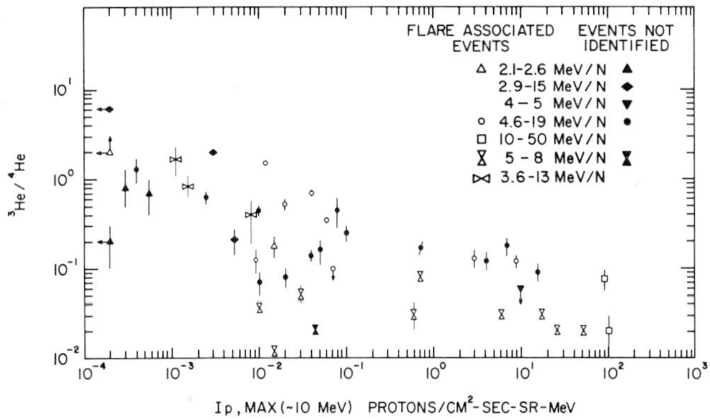

FIGURE 8.10 The ratio $^3He/^4He$ measured at various energies as a function of Ip_{max} (10 MeV), the proton intensity at 10 MeV, and at the maximum intensity of the event. (From Ramaty *et al.*, 1978.)

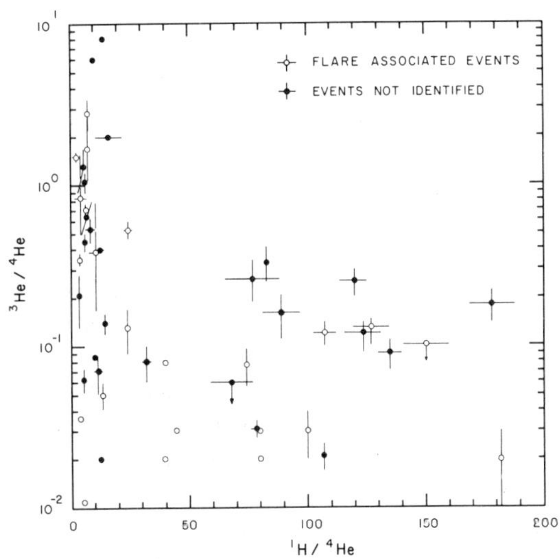

FIGURE 8.11 The ratio He^3/He^4 as a function of H/He^4 for all available energies. (From Ramaty *et al.*, 1980).

report Fe/O $\simeq 1$ for the 24th Sept. 1977 event. In this latter paper, an attempt at a systematic analysis of Z-dependence of cosmis ray abundances has been made, and the ratio

$$Q(E, Z) = \frac{\text{abundance of acc. nucleus of charge } Z \text{ at energy } E}{\text{abundance of nucleus } Z \text{ in the solar system sample}}$$

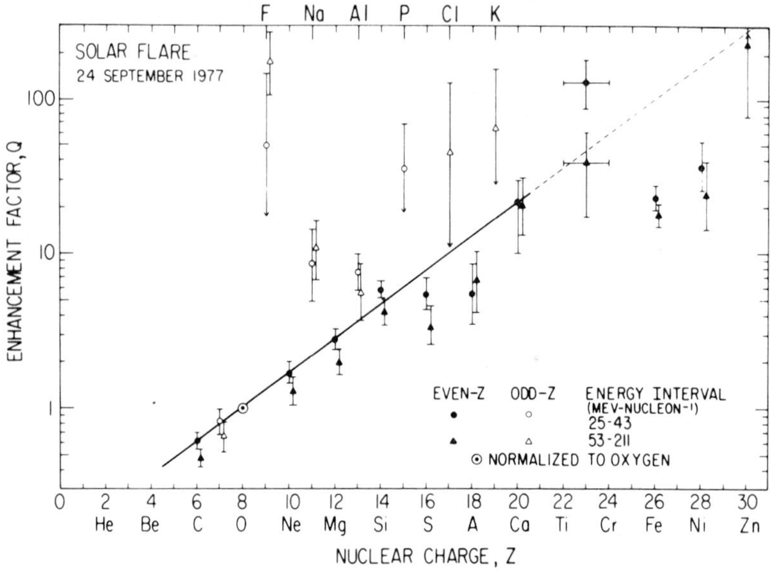

FIGURE 8.12 The enhancement factor Q versus nuclear number Z. All ratios are corrected for differences in measured energy intervals by assuming a $E^{-2.5}$ spectrum for all species. Error bars represent 1σ uncertainties. Errors associated with the solar system abundances are not represented. Arrows extending below a point indicate upper limit measurement. (From Dietrich and Simpson, 1978.)

has been determined. $Q(\mathrm{Fe}) \simeq 10^2, 10^3$ had previously been observed (Anglin *et al.*, 1977), but the determination of this ratio for some other elements suffers from a poor determination of the denominator, in particular for elements Ne($Z=10$), Mg($Z=12$), Si($Z=14$). Dietrich and Simpson chose to use for these elements a reference abundance which is not that of the regular solar system sample, but rather that observed in a class of flare which does not display strong abundance peculiarities. It has been argued since then that some of the regular solar system abundances, such as Ne, were more reliable than previously thought (Meyer, 1978 and private comm.), so that the normalization made by Dietrich and Simpson could have been unfortunate. It remains that a systematic enhancement of Q for large Z appears in the results shown in Figure 8.12, though the points Ne, Mg, Si, still controversial, should be regarded with some care.

It will be interesting to have in the near future the results of more complete studies involving a more systematic normalization on the one hand, as well as a careful comparative analysis of the Z-dependence of Q for Fe-rich events as compared to non-Fe-rich events. In particular, it will be important to test with some precision whether solar cosmic rays exhibit a correlation between abundance of an element and the first ionization potential of the correspond-

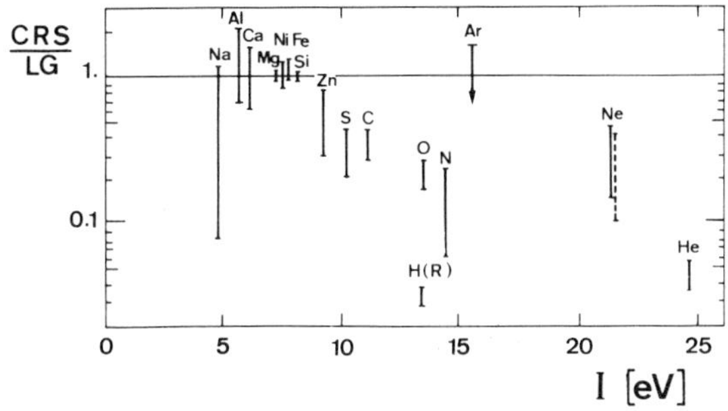

FIGURE 8.13 CRS/LG versus first ionization potential, *I*. CRS = abundance in cosmic ray sources; LG = local group abundances. (From Meyer *et al.* 1979.)

ing atom, implying that the source of the cosmic ray is ultimately a cool gas. Such a correlation is known to exist for galactic cosmic rays (Weber *et al.*, 1972; Shapiro *et al.*, 1975; Cassé, 1976; Cassé *et al.*, 1975) as compared to the local group abundances. This is shown in Figure 8.13 (Meyer *et al.* 1979), where the most easily ionizable elements are preferentially accelerated or injected in the acceleration process. Actually, a reasonable fit to the galactic cosmic-ray abundances is obtained with an exponential temperature distribution $\exp(-T/T_0)$ (Cassé and Goret, 1978). An additional mass effect, however, is apparent from the fact that electrons and protons (presumably present in equal proportions in the gas from which galactic cosmic rays are extracted) appear in an unequal proportion after the acceleration ($e^-/p^+ = 10^{-2}$ in galactic cosmic rays). This might be understood in terms of a preheating mechanism sorting out charged particles in a weakly-ionized gas, and injecting them in an acceleration mechanism which would work in some mass-dependent way (for example, be dependent on the Larmor radius of particles).

Recent work by McGuire *et al.* (1978) has pursued a similar analysis for solar cosmic-ray elemental abundances. They also found here a selection effect according to the first ionization potential operating in all solar events: all elements with an ionization potential smaller than approximately 10 eV are enhanced by a factor 3 with respect to those with higher ionization potential. This is shown in Figure 8.14. The same effect seems to be present in all large flares (Meyer, J. P., private communication). On top of this selection mechanism there is a Z-dependent enhancement, variable from event to event, which is especially apparent in Fe-rich events (Dietrich and Simpson, 1978; Meyer, 1978). Preliminary analysis (McGuire *et al.*, 1979) does not reveal a drastic difference between even and odd nuclei, as seemed to appear in Figure 8.12 (this was perhaps due to the possible bias mentioned before). Figure 8.15 reproduces the results of a comparison of abundances recorded for the Fe-rich

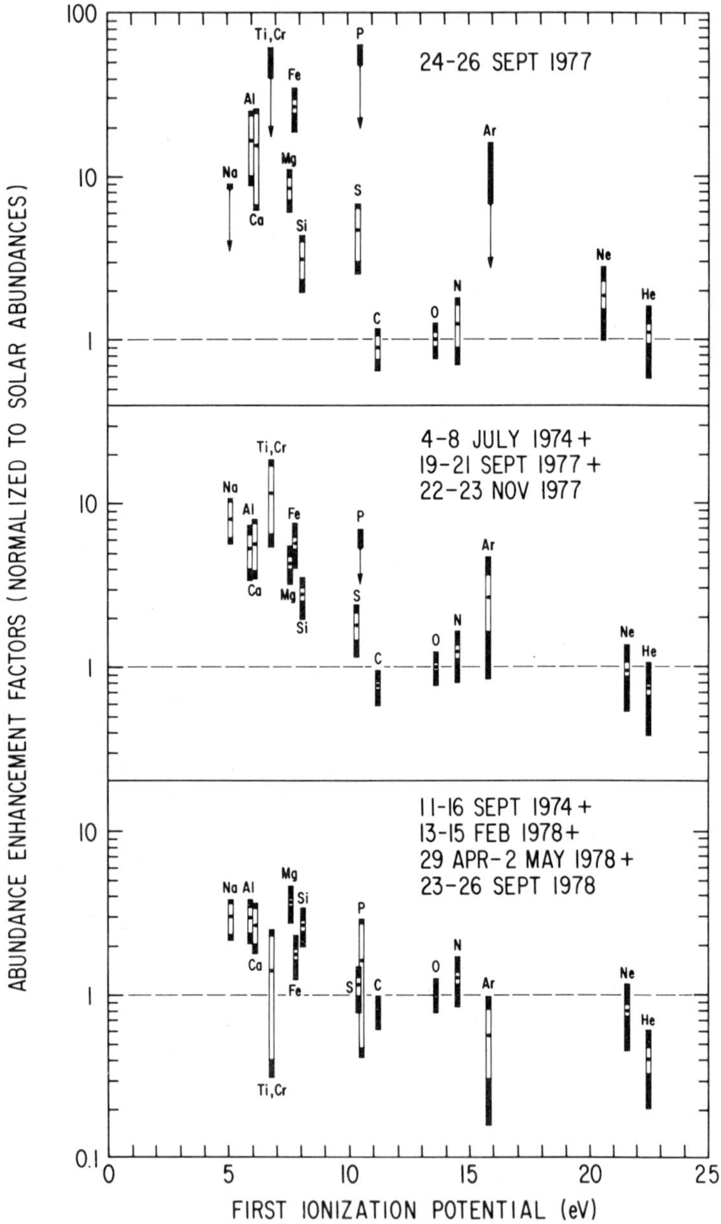

FIGURE 8.14 Abundance enhancements relative to solar abundances as a function of first ionization potential. (From McGuire *et al.*, 1979.)

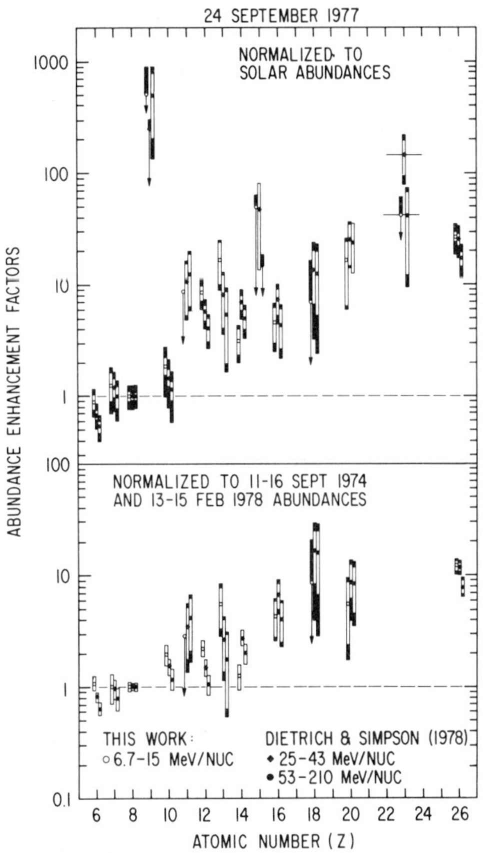

FIGURE 8.15 Abundance enhancements from the Dietrich and Simpson (1978) data (25–43, 53–210 MeV/nucleon). Normalization is indicated in each part of the figure. The Goddard data (6.7–15 MeV/nucleon) is shown for comparison. (From McGuire *et al.*, 1979.)

flare (24th Sept. 1977) studied by Dietrich and Simpson (1978) and McGuire *et al.* (1979) to those recorded in a typical non-Fe-rich event. The possible correlation with first ionization potential, if real, would imply that *second-phase cosmic rays should be accelerated or pre-accelerated in a rather cool plasma* (10^4 *K*), *much cooler than the solar corona itself and even more than the very hot gas* (10^8 *K*) *in which first-phase electrons are probably accelerated. A word of caution, however! In the sample of elements used, most of those which have a low ionization potential I* (*less than* 10 *eV, say*) *have also Z* larger than 10, except Argon, while on the other hand, C, N, O, Ne, which have Z smaller than 10, have ionization potentials larger than 10 eV. Any correlation between Q and Z would then be very much reproduced as a Q–I correlation (except for the

Argon point for which one has only upper limits). It would then be premature to conclude that one correlation is more significant from the physical standpoint than the other. Obviously, developments in this field will be of the utmost interest.

8.8 FAST PARTICLE IMPACT ON THE CHROMOSPHERE AND PHOTOSPHERE FROM EUV AND Hα BURSTS

Whereas X-ray radiation in the 10 keV range obviously reveals the existence of particles of higher energy, we can also consider more indirect (but still important) evidence for their existence. This is provided by the emissions occurring in fair temporal simultaneity with impulsive hard X-ray and microwave bursts, namely *EUV and Hα impulsive emissions and white-light emission*. The former emissions are thought to be due to electrons, and their intensity bears some relation to the number of accelerated electrons; it does not say much, however, about the form under which the energy is transmitted to the lower atmosphere.

Other indirect phenomena are of some interest in this respect such as flare white-light emission and Hα line profiles; these provide evidence for *heating deep in the atmosphere at the photospheric level*. The problem with these phenomena is that they involve interpretation to a rather large extent. Nevertheless, timing considerations and large energy deposition deep in the atmosphere give important constraints for selecting models.

8.8.1 EUV bursts and number of accelerated electrons

Our observational knowledge of EUV *impulsive emission* is described in some detail by Donnelly and Kane (1978). Impulsive solar EUV and hard (greater than 10 keV) X-ray bursts are well-correlated in occurrence, time-dependence and intensity. The impulsive EUV flux rises and decays with a time-structure most like that of the lowest energy hard X-rays (10–20 keV), but slower than X-rays harder than 30 keV. Timing and estimates of the EUV flux were obtained by measurement of an ionospheric effect (Donnelly, 1973). The timing accuracy achieved is between 1 and 6 seconds. Excellent simultaneity has been observed between X- and EUV-bursts to this accuracy. Very important also is the confrontation of hard X-ray fluxes with EUV fluxes (and fluxes in other wave bands) (Emslie *et al.*, 1978; Brown *et al.*, 1978). These comparisons strongly suggest that approximately *one to two orders of magnitude fewer electrons thermalize in the lower atmosphere to produce the EUV and Hα impulsive emission than are required to produce the hard X-ray burst*, according to the non-thermal thick-target bremsstrahlung process. These conclusions require further verification, including high resolution (in space and time) measurement. However, the number of accelerated electrons needed to explain this EUV emission is more consistent with the number of electrons observed in space. We then have some reason again to believe that the *thick-target bremsstrahlung picture overestimates the electron number*; however, were the thick target model true, one could explain the discrepancy by establishing that a smaller amount of electrons thermalize to produce EUV radiation than impinge on the thick-target X-ray source (small precipitation efficiency in brief) or that most of the energy dumped in the atmosphere by fast particles escapes in forms other than EUV and Hα radiation (small EUV efficiency). Overall estimates at the Skylab Workshop of the energy released over all wavelengths (Canfield *et al.*, 1980) for the flare of September 5th, 1973 showed that this overall radiated energy falls short of the total "thick-target" electron number by a similar amount. Ways out of the discrepancy involve another interpretation of Hα emission (Zirin, 1978) or a theory which does not explain X-rays by thick-target emission.

8.8.2 Deep energy penetration

We finally just mention here the observation (§2.3.5) that substantial photospheric heating occurs in flares. A complete interpretation will certainly require an estimation of energy transmission by radiation and it will perhaps be necessary to consider photospheric heating produced by early proton bombardment.

8.9 DIRECTED FLOW OF FAST PARTICLES TO THE CHROMOSPHERE OR HOT FIREBALL: OBSERVATIONS

8.9.1 Distribution function

The reader should now be convinced that the capabilities required from the acceleration mechanism differ very much according to whether *fast electrons "freely" leak by mere streaming to the chromosphere or form in the corona a compact hot plasma, a "fireball"*. However, the possibility of maintaining an extremely hot plasma against all sorts of losses, and the conditions under which its bulk "heating" could occur deserve detailed examination. Attempts to decide from observation whether X-ray radiating electrons have a maxwellian rather than a power-law distribution have failed because an inhomogeneous temperature in a thermal source produces effects similar to power-law distributions. Davis and Rogerson (1977) have shown in detail that such a determination is virtually impossible.

8.9.2 Particle flux directivity

Of more relevance is the search for a particle directivity signature. It may appear in X-ray emission directivity: though they are not ultra-relativistic, 100 keV electrons are fast enough for aberration effects to beam sensitively the emission forwards. A statistical center-limb effect on burst-occurrence was sought, but not found, possibly because the trajectory of the supposed beam is far from reproducible from one event to the other (Datlowe *et al.*, 1977).

8.9.3 Polarization

A decisive signature of beaming would be the observation of a high degree of polarization in hard X-ray emissions, as is expected from bremsstrahlung produced by directed high-energy beams (Kortchak, 1967). Perfect beaming of mono-energetic electrons would produce 100% polarization, but magnetized streams with a unique pitch-angle give a smaller polarization (Elwert and Haug, 1970). Many factors may reduce the expected polarization. These include effects of the dispersion in electron energies, averaging on the configuration of the magnetic field in the flare region (Kortchak, 1974), isotropization of the electrons by Coulomb scattering in a thick-target source (Brown, 1972). The effect of photospheric albedo must also be included (Henoux, 1975). In total, the *maximum expected polarization is at most* 30%.

X-ray polarization may be detected by using the fact that a perfect crystal reflecting at the Bragg angle of 45° transmits only that polarization which has its electric vector perpendicular to the plane of incidence, or by taking advantage of the dependence of the cross-section for non-coherent Thomson scattering in the plane at 90° angle from the direction of incidence on the angle made in this plane between the direction of polarization of incident photons and the direction of scattered photons, namely $d\sigma/d\Omega = r_0^2(1-\cos^2\varphi)$. The first method (Novick *et al.*, 1972) has extreme spectral selectivity but a poor signal/noise ratio. Thomson-type polarimeters, used in Russian experiments (Tindo *et al.*, 1972), are more efficient over the wide range 5–90 keV. Two essential sources of systematic error in these measurements must be kept in mind: a possible anisotropy of the background can bias measurements (Novick and Wolff, 1971); also, drifts in the sensitivity of counters could give rise to time-dependent effects. For this reason perhaps, high levels of polarization have been measured in the early Intercosmos experiments. In more recent experiments the system of detectors was periodically rotated around the optical axis of the experiment, thus eliminating this cause of errors. Experiments using this technique on board later Intercosmos satellites seemed to indicate as a rule lower polarization percentages. For example, Intercosmos 11 detected no polarization larger than 5% in July 1974. We present below a summary of results as they were known in 1977, adapted from a review by Tindo and Somov (1977).

Date	Flare	Maximum polarization	Spacecraft	Mode of operation
20.10.69	10–49			Thomson with
23.10.69	05–13	20% < P < 40%	Intercosmos 1	stationary
30.10.69	09–36			detectors
24.10.70	04–51	16%		
26.10.70	02.40	5%	Intercosmos 4	Same
5.11.70	03–16	21%		
5.07.74	21–56	5%		Thomson with
6.07.74	18–48	1.9%	Intercosmos 11	rotating mode of
6.07.74	19–06	2%		operation

Were the measurements never to exceed the figures quoted for the 1974 experiment, it could not be concluded that beams of electrons are implied. Actually, Henoux (1975) has shown that a few percent level of polarization can be produced by the Compton albedo of the photosphere. Also, in a situation of strong heat conduction, which shares many aspects with a beam situation (more energetic electrons going preferentially in one direction rather than in the other), a few percent polarization from emitted X-rays is expected (Emslie, 1979). *A positive large polarization detection with a sophisticated (rotated) detector, however, would be an almost unambiguous proof of beaming.*

8.10 DIRECTED FLOW OF FAST PARTICLES TO THE CHROMOSPHERE OR HOT FIREBALL: THEORY

8.10.1 Effects of beams: self-field and return current

From a theoretical point of view, the directed flow of a significant number of electrons through the coronal plasma between the site of primary release and the chromosphere is not without difficulties. This will be soon realized if one notes that a flux of 10^{36} electrons per second flowing through a surface of 10^{14} m^2 represents a current of 10^{13} A, which creates at its border an absurdly high self magnetic field of 10^8 Gauss! There is, of course, the possibility that this high current is shielded by *return-current* in the plasma, though even this would be impossible if the electron motion is not ballistic but driven by an electric field. We can then conclude that, *in the source where the electrons receive their energy, they cannot all* (10^{38} *or* 10^{36}) *be accelerated together in the same direction* by a DC electric field. Either the electric field must be itself random, or the effective resistivity of the plasma must be so high that the organized part of the motion of the electrons in the DC electric field is much smaller than the random part, or the electrons must be accelerated (one small bunch after the other) by DC electric fields restricted to spatially-small regions. Once, however, the electrons have left this acceleration region, shielding of the current they represent by a return-current driven in the background plasma is possible, but this is not without consequences, which should be observable.

The relevant physical picture is essentially that of a black-box source of high-energy electrons leaking a certain number flux of them at energy higher than, say, 20 keV towards a high-density target through a low-density plasma, the temperature and density of which is known at time $t=0$. Electron and energy flow are assumed to be channeled by the magnetic field. In a stationary state the total number flux vanishes, to keep the charge of the black box constant, so that

$$n_b\mathbf{v}_b + n_0\mathbf{v}_0 \simeq 0 (\text{i.e.} \ll n_b\mathbf{b}_b). \tag{8.7}$$

Subscript b refers to beam and 0 to ambient plasma. The energy flux does not vanish since the black box is a source of energy. Beam streaming appears as a mode of energy transport which does not differ qualitatively from thermal conduction. The transitory state of growth of the beam gives rise to an electric field induced by the growing self magnetic field of the beam. This field first sets up the return-current. After a very short time, everything is stationary and the return-current is maintained against dissipation by an electrostatic field $\mathbf{E}_0$ which results from a slight charge imbalance between target and black box. $\mathbf{E}_0$ pushes the return-current and brakes the beam. Return-current losses are "paid" by the beam (dot product (8.7) by $\mathbf{E}_0$). So, if for some reason the plasma resistivity becomes large, the toll price may be so high that the beam never reaches the target.

8.10.2 Beam losses

Losses and relevant microinstability thresholds depend on plasma temperature, which should be self-consistently evaluated from an energy equation. This has not yet been entirely done, but the stability of the return-current against ion-acoustic wave-growth has been analyzed (Brown and Melrose, 1977; Hoyng *et al.*, 1978). Let the cross-section of the beam be $A_{14}\ 10^{14}\ \text{m}^2$, the number flux above 25 keV be $F_{25}\ \text{s}^{-1}$, the plasma density and electron temperature be $n_{16}\ 10^{16}\ \text{m}^{-3}$ and $T_7\ 10^7$ K. Then the ratio of return-current drift-velocity v_0 to ion-sound velocity c_s is:

$$\frac{v_0}{c_s} = (3.5)\frac{F_{25}}{10^{36}} n_{16}^{-1} A_{14}^{-1} T_7^{-1/2}.$$

Fluxes as high as 10^{36} s^{-1} will almost certainly trigger microturbulence. In stable situations the beam loses energy by Coulomb collisions and by sustaining the return-current against Ohmic losses. For particles of energy E_b the collisional losses per unit beam length amount to

$$P_{coll} = \sigma_{coll} F n_0 E \text{ watts m}^{-1},$$

where σ_{coll} is the Coulomb collision cross-section

$$\sigma_{coll} = 2\pi r_0^2 \left(\frac{mc^2}{E}\right)^2 \ln \Lambda,$$

where $\ln \Lambda$ is the Coulomb logarithm and r_0 the classical electron radius given by

$$r_0 = e^2/(4\pi\varepsilon_0 mc^2).$$

Similarly, return-current losses are given in terms of the plasma conductivity σ by

$$P_{ret} = \frac{e^2 F^2}{\sigma} \text{ watts m}^{-1}.$$

It is found (Hoyng and Melrose, 1977) that

$$\frac{P_{coll}}{P_{ret}} = \left(\frac{k_B T_e}{E_b}\right)^{3/2} \frac{n_0}{n_b} = 2^{5/2} \frac{v_{T_e}^3}{V_0 v_b^2}.$$

Putting together the stability condition and the condition which determines the dominant losses, one concludes that there are essentially three beam regimes, as shown in Figure 8.16. Hoyng *et al.* (1978) discuss which observations are needed to determine where on this diagram actual solar X-ray sources should find themselves. Region I corresponds to *a beam able to reach the chromosphere* and dump its energy there (Brown, 1973; Shmeleva and Syrovatsky, 1973). In region II the *beam suffers heavy return-current losses and heats the corona* (Knight and Sturrock, 1977). In region III the *further beam-evolution depends on the instability saturation.* For practical purposes, however, this is a case where the beam simply cannot penetrate the plasma, the black box being "closed" by a turbulence wall to beam electrons. Thermal

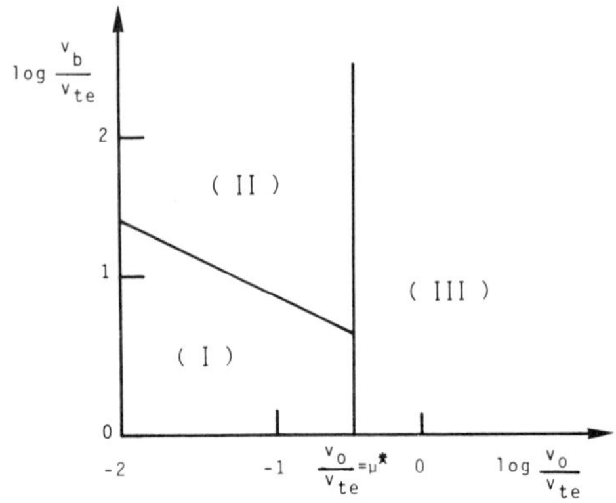

FIGURE 8.16 Different regimes of beam-plasma interaction. Region I, beam collisions dominate. Region II, reverse-current dissipation dominates. Region III, turbulent return-current. (Adapted from Hoyng *et al.*, 1978.)

conduction is similar in many respects to the situation described here and also gives rise to microinstabilities. Return-current instability, driven by beam injection, has been studied in the laboratory in some detail, in complete agreement with the above conclusions. The interested reader is referred to the paper by Stenzel (1978), where it is shown that, under conditions which are not completely different from coronal ones (long mean-free-path, $\omega_p/\omega_b > 1$), a beam driving an unstable return-current does not travel more than 2 meters; the frequency spectrum of the saturated instability peaks at 0.2 ω_{pi}, with an $\omega^{-5/2}$ high-frequency decrease and a low-frequency decrease (presumably due to the destruction of resonance by electron-scattering on shorter waves (Horton *et al.*, 1976)).

Beam-return-current interaction is thought to be a simpler problem under conditions applicable to region II of Figure 8.16. This, we think, depends to a large extent on the stability of the Langmuir wave modes to the two-stream instability, and the latter depends on geometrical aspects such as the relative size of the black box with respect to the length of the coronal loop in which the beam is injected. Actually, free streaming in one dimension gives as a solution of the single-particle Louville equation

$$f(x, v, t) = f(x - vt, v, 0).$$

If the injection time of fast particles is much less than the time of interest, this relation is a good representation of the particle distribution. It develops positive slopes in a typical time a/v_0, where a is the source size and v_0 the typical fast-particle velocities, at distances of the order of a. If injection is long-lasting, the one-particle Louville equation cannot be treated as an initial-value problem. Let us schematize the acceleration by a source term:

$$\frac{\partial f}{\partial t}+v\frac{\partial f}{\partial x}=s(x,\, v,\, t).$$

Assuming $s=\varphi(v)\exp(-t^2/\tau^2)\delta(x)$ for $t>0$, and 0 for $t<0$, the solution for short times $t\ll\tau$ and $x\ll v\tau$ is:

$$f=\varphi(v)v^{-1}\ \exp-\frac{(t-x/v)^2}{\tau^2}\simeq\varphi(v)v^{-1}.$$

For large sources or sustained injection, electrostatic instabilities may not develop spontaneously as a result of a particle's ballistic motions, though collective effects may restore a distribution with positive slope (Papadopoulos *et al.*, 1977). For $v=c/3$ and $x=$ a loop length, the time τ must be longer than 0.1 sec for positive slopes to be absent. Similarly, the fact that type III bursts are not observed below 600 MHz (8000 km, say) could be understood if the injection time-scale is longer than 0.08 sec, compatible with radio emission time-scales as well as with the observations of Hurley and Duprat (1977). In these conditions it makes sense to examine the beam motion and energy losses under conditions of region II, including only return-current losses as a brake and ohmic losses as a plasma heating term (Knight and Sturrock, 1977). It suffices here to say that the problem may be solved analytically, and that an efficient heating results, as shown in Figures 8.17a and 8.17b from Shapiro and Knight (1978): 8.17a represents a stable return-current, and 8.17b is unstable to ion-cyclotron waves. A specific assumption has been made concerning the saturation level of these waves.

These various effects are then, as we have seen, *very efficient processes for converting high-graded, directional, 25 keV, electron energy into heat*, and at the same time blocking the energy in the corona, provided the flux is high enough. It remains then to examine which energy-loss mechanism could replace free-electron streaming. It is worth mentioning here that impulsive plasma heating (expected if return currents become microunstable) could be diagnosed thanks to a lack of equilibrium between electron temperature and ion ionization states, which would drastically affect the emergent X-ray spectrum (Shapiro and Moore, 1977) in the form of a burst-like enhancement of soft X-ray lines. Enhancement factors have been calculated by Shapiro and Knight (1978).

8.10.3 Fireball–plasma interface

We have now to examine *how long a hot plasma can remain bottled-up in the solar corona.* The most likely loss processes would be heat conduction and

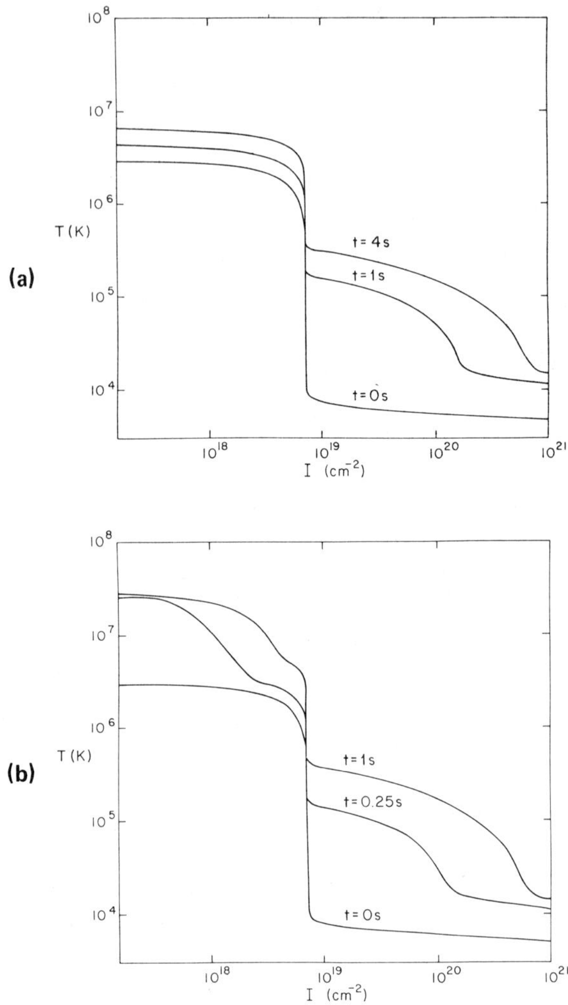

FIGURE 8.17 (a) Temperature T as a function of integrated number density I from the injection point for an energetic electron number flux 1.414 10^{17} cm^{-2} s^{-1} with an energy of 25 keV. The temperature is displayed for times before the beam is injected ($t = 0$), and after the beam is injected. (b) Same as (a) for a flux 5.656 10^{17} cm^{-2} s^{-1}. $I = \int_0^s n(s')ds'$ replaces the height variable s. (From Shapiro and Knight, 1978.)

magnetohydrodynamical motions. Actually, characteristic radiation loss-times are fairly long, usually of the order of 10^3 s (Heyvaerts, 1974) and possibly much longer here. The conduction time-scale, based on the classical Chapman thermal conductivity,

$$x_{//}(T)=3.16\ x_0 T^{5/2}(\ln\Lambda)^{-1}\ \mathrm{J\ m^{-1}\ s^{-1}\ (K)^{-1}},$$

$$x_0=0.585\times10^{-10}\ \mathrm{J\ m^{-1}\ s^{-1}\ (K)^{-1}},$$

gives rise to a time-scale

$$\tau_{\mathrm{cond}}=\frac{(nk_B T_e)}{(x_{//}\nabla T)S},$$

which depends somewhat on the geometry of the hot volume. If the heat is conducted along the magnetic field only this may have some quantitative effect (Antiochos and Sturrock, 1976). The conduction perpendicular to field lines is dramatically smaller than along. Let us assume that the flare plasma is contained in a cylindrical loop, 10,000 km long, then

$$\tau_{\mathrm{cond}}\simeq10^{-2}\left(\frac{n(\mathrm{m}^{-3})}{10^{16}}\right)\left(\frac{10^8}{T}\right)^{5/2}\ \mathrm{sec.}$$

This calculation of conduction losses, however, is valid only under the following conditions. *First*, the temperature-gradient scale-length must be much longer than the collisional mean-free-path, l:

$$l=5.7\times10^8\ \frac{T^2}{n\ln\Lambda}\ \mathrm{m\ (MKS)};$$

for the conditions $T=10^8$ K and $n=10^{16}\ \mathrm{m}^{-3}$, l is precisely of the order of 10,000 km. Hot electrons are then collision-free, both in their environment and in the cooler plasma around. *Second*, the heat flux must not be dominated by collective collision processes, which may very well happen, as we see below; and *third* it can anyway not be larger than uncompensated free-flight of all the plasma electrons from the hot to the cool region (Spicer, 1977b)

$$q\leqslant q_{\mathrm{sat}}=\tfrac{3}{2}n_e m_e v_e^3.$$

For a 10^8 K plasma, the velocity of thermal transport by free-flying electrons is 40,000 km/s. However, when this situation occurs, the nearby cold plasma is permeated by a directed flux of electrons, and, in order to ensure a zero net number flux, a return-current must flow in the plasma. We are then back again to the previous situation. It has already been shown that this situation may lead to instability, in which case interaction between the electrons and the turbulent waves dominates the transport properties. Spicer (1977b) used results of microinstability analysis for the skewed electron distribution characteristic of thermal conduction in collisional plasmas, as derived by the Chapman–Enskog technique,

$$f(\mathbf{x})=\frac{n_e}{(2\pi v_e^2)^{3/2}}\exp-\left(\frac{(v_\parallel-u)^2+v_\perp^2}{2v_3^2}\right)$$

$$\left\{1+\frac{Q_\parallel(v_\parallel-u)}{3n_e m_e v_e^3}\left(\frac{(v_\parallel-u)^2+v_\perp^2}{v_e^3}-\frac{3}{v_e}\right)\right\}.$$

Under collisionless conditions, a beam-plus-return-current distribution is more appropriate (Feldman, 1974), but the results for these two situations are quite similar. The heat flow described above could drive *ion-cyclotron waves* unstable, as well as *ion-acoustic waves* if $T_e \gg T_i$ in the downstream plasma. It seems to be believed, though this depends very much on details of the saturation mechanism of the instabilities (see below), that the ion-cyclotron waves saturate at a low enough level for their macroscopic effects to be negligible. This assumption may be unjustified. For this reason, mainly ion-acoustic instability has been considered. The growth-rate (Mannheimer, 1977) is given, for a wave vector parallel to a return-current with drift speed u, by

$$\gamma(k)=\sqrt{\frac{Mm_e}{8M_i}}\left(|k|(u-v_{Ti}-c_s)+|k|\frac{Q}{n_e m_e v_{Te}^2}-|k|c_s\sqrt{\frac{M_i}{m_e}}\left(\frac{T_e}{T_i}\right)\exp\left(\frac{(T_e/2T_i)+\frac{3}{2}}{1+k^2/k_D^2}\right)\right)$$

where c_s is the ion-sound speed. The instability occurs roughly when

$$Q>n_e k_B T_e c_S = Q_{AN}.$$

The actual reduction of thermal conduction depends on the turbulence level reached at saturation. If the instability reaches saturation due to ion trapping (Mannheimer, 1977), we get

$$Q=-3n_e k_B v_{T_e}\lambda_D\left(\frac{e\varphi}{k_B T_e}\right)^{-2}(\nabla T_e).$$

Taking $e\varphi/(k_B T_e)\simeq 10^{-1}$ at saturation, one obtains an anomalous thermal conduction of:

$$\kappa^*_{\text{Ion Ac}}\simeq 3\times 10^2\ k_B n_e\frac{k_B T_e}{m\omega p}\sim 10^{-15}\ n_e^{1/2} T_e\ \mathrm{J\ m^{-1}\ s^{-1}}\ \text{(MKS)}.$$

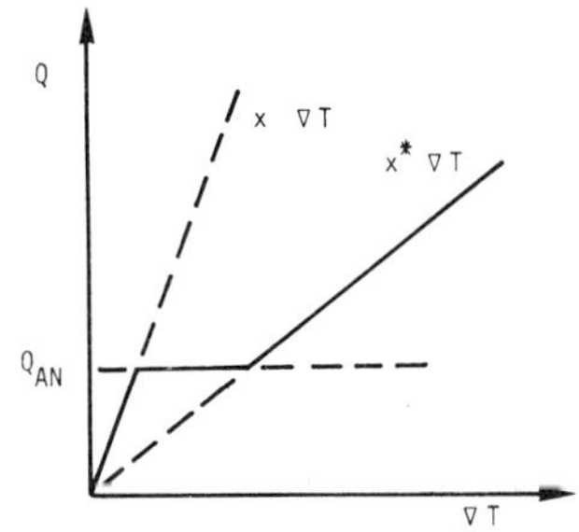

FIGURE 8.18 A simple form of the heat flux as a function of ∇T. (Adapted from Smith and Lilliequist, 1978.)

For $T=10^8$ K and $n_e \sim 10^{16}$ m^{-3}, this is an enormous reduction as compared to the classical value, and effectively limits the heat flux.

It may be, however, that the instability is not driven to its point of maximum saturation, and then the heat flux will be larger than this limit. Smith and Lilliequist (1978) consider that, in that case Q is blocked at the marginal instability value Q_{AN} (see Figure 8.18). Once the turbulence has developed at the boundaries of the hot plasma, the latter is insulated by a relatively thin *conduction-front*, which starts to propagate. The saturated heat flux Q_{AN} allows in one second the transfer of the heat necessary to raise a length c_S of cold plasma to the hot temperature. Hence the propagation velocity is approximately c_S. The thickness of the conduction front is of the order of one electron mean-free-path for energy exchange. Computer simulation by Lampe *et al.* (1972) gives a momentum scattering-frequence ω_{pi}^{-1}. The heat exchange-time is $(m_i/m_e)^{1/2}$ larger, and so the front thickness is the corresponding distance travelled by random walk, namely

$$l \simeq \frac{v_{T_e}}{\omega_{p_i}}\left(\sqrt{\frac{m_i}{m_e}}\right)^{1/2} = \left(\frac{m_i}{m_e}\right)^{3/4} \lambda_{D_e}.$$

This turbulent boundary is leaky, because in ion-sound turbulence (as in an ordinary plasma) the collision cross-section scales as $(v_{T_e}/v)^3$. The corresponding mean-free-path equals l for a rather modest value of the velocity v_{esc} given by

$$\frac{v_{T_e}}{\omega_{p_i}}\left(\frac{v_{\text{esc}}}{v_{T_e}}\right)^3 = l \quad \text{or} \quad v_{\text{esc}} = v_{T_e}\left(\frac{m_i}{m_e}\right)^{1/6}$$

Nevertheless, if the leaks are so large that the return-current associated with them is unstable, exactly the same phenomena as described before will occur. *There is then a self-regulation mechanism, by which the leaks self-consistently adapt precisely to allow the escape of a fast electron flux giving rise to the maximum stable return-current in the plasma downstream.* As a result, we see that a hot plasma confined in the corona by turbulence walls is not at all inconsistent with immediate chromospheric bombardment, and allows fast electrons to be found in a volume somewhat larger than the one where the bulk of the hot electrons are retained (Brown *et al.*, 1978), a point which may be relevant to the modelling of microwave impulsive emission. Recently Vlahos and Papadopoulos have given more details for a model based on this idea (Figure 5.3). It is shown there that the major energy losses of the hot region are expansion losses and energetic particle losses. They also show that, even if it develops a positive slope, the beam of escaping particles can reach the chromosphere thanks to the same stabilization mechanism which operates for type III bursts. They estimate that about 10^{-2} of the hot spot leaks.

In summary, *collective turbulent effects are able to adjust self-consistently the fast-particle flow to the chromosphere to the maximum stable value of the corresponding plasma return-current. The excess number of electrons energized, if any, is trapped between two ion-acoustic turbulent walls,* which expand downwards at the ion-sound speed and reach the chromosphere after a time $\tau = L/c_S$ which is typically some 3 to 10 seconds.

A certain number of oversimplifications, however, have been made in the above description. One of the most important is the neglect of hydrodynamical motions. This aspect has been considered by Smith and Lilliequist (1978), who treated the problem numerically with a one-fluid, two-temperature (T_e and T_i) model, taking into account heat conduction (as in Figure 8.18), radiation losses and ion-electron Coulomb-collision heat-exchange. The ratio of power input to radiated power and to power used in cold plasma heating was found, and it was concluded that the model is 22 times more efficient than the thick-target model and of course free of its inconsistencies. (A discussion of purely-hydrodynamical effects can be found in Chapter 5.)

8.11 FIRST-PHASE ACCELERATION PROCESS: AN OVERVIEW

Having discussed how the particles behave in the coronal environment once they have received energy, we have now to specify the mechanism by which acceleration really occurs. Here we shall concentrate on mechanisms able to meet the requirements of first-phase acceleration, but shall defer to later sections a discussion of the origin of the accelerating fields, DC or turbulent. Because flare energy derives from the conversion of magnetic energy, the agent which feeds this energy to particles must ultimately be an induced electric field developed in some resistive MHD process, i.e. some sort of DC field. The energy gained by electrons, however, may be shared with turbulent modes, both interacting to provide the observed spectra. As an intermediate step, stochastic acceleration may then be of importance.

8.11.1 DC electric field acceleration

For velocities larger than v_{T_e}, the Coulomb-collision cross-section decreases as v^{-3}. If the electric field component along the magnetic field is high enough to accelerate electrons to v_{T_e} in less than a collision-time, electrons will cease to be collisionally-coupled to ions. The electric field required is the so-called *Dreicer* (or *runaway*) *field* E_D given by

$$eE_D = \frac{e^2}{4\pi\varepsilon_0 \lambda_D^2} \frac{\ln \Lambda}{\sqrt{2}} \quad \text{or} \quad E_D = 2.15 \times 10^{-13} \ln \Lambda \frac{n(\mathrm{m}^{-3})}{T^\circ} \mathrm{V\ m}^{-1}.$$

In such a high field, however, the electrons do not run away in a straightforward manner, because instabilities provide the electron-ion coupling by collective effects. For fields smaller than E_D, some electrons of the tail of the maxwellian distribution already run away. Their relative number is given by (Nroman and R. Smith, 1978)

$$\frac{N_{\text{runaway}}}{N_{\text{total}}} \simeq \frac{1}{2} \exp - \frac{1}{2}\left(\sqrt{\frac{E_D}{E}} - \frac{E}{E_D}\right)^2 \simeq \frac{1}{2} \exp - \frac{1}{2} \frac{E_D}{E}.$$

It is a sensitive function of E/E_D, and, if this number is small enough, some electron acceleration may occur without producing prohibitive levels of electrostatic turbulence, though the distribution will be unstable to the two-stream instability. The beam may still form if conditions for stabilization as quoted in paragraphs 8.1 and 8.10 are met.

Norman and R. Smith (1978) have considered the interaction of the primary beam of accelerated particles with the neighbouring plasma, and show that, if magnetic energy is stored in this vicinity, the acceleration process can develop explosively. Actually, the scenario for the interaction could be as described in the preceding paragraph, namely a regular *beam-plasma interaction* coupled with an *ion-acoustic instability* of the *return-current* (as in the experiment by Stenzel (1978)); or, if the return-current is stable, the beam-plasma interaction may develop in a non-linear way through the so-called *oscillating-two-stream instability* (OTS), an interaction which, while reshaping the Langmuir wave spectrum, at the same time produces ion-density fluctuations not unlike ion-acoustic waves. In both cases, the ion fluctuations, be they acoustic or stationary, give rise to an *anomalous resistivity* (Papadopoulos and Coffey, 1974). This increase in the local dissipation will give rise locally to magnetic

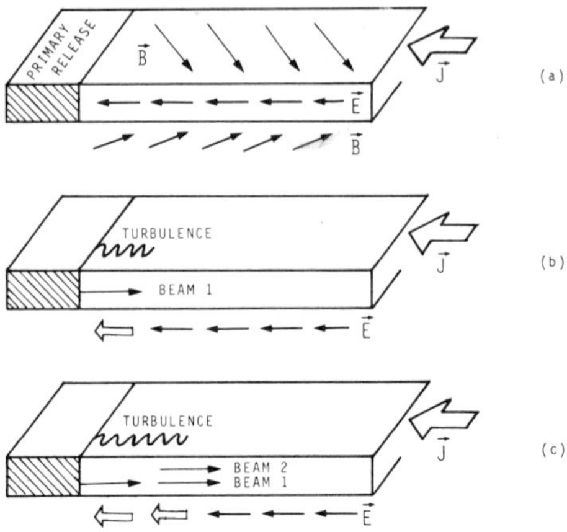

FIGURE 8.19 A schematic view of *dissipation-spreading* in a current-carrying structure.

energy conversion which was not previously possible, part of which will go into accelerated particles, which will join the primary ones to propagate the anomalous dissipation further.

Papadopoulos (1977) proposes an interesting explicit mechanism for spreading this explosive dissipation, which is illustrated in Figure 8.19. Suppose a primary beam from the trigger region enters the current-carrying neighbourhood and creates there (losing some of its own energy) fluctuations providing a high resistivity. Because the inductive time-scales are long, the current density determined by the local value of rot **B** will remain constant, while the electric field will adapt and take locally a high value (Figure 8.19b). As a result, bulk heating of the local plasma will occur and hot electrons will enter the neighbouring still quiet region (Figure 8.19c) and so propagate the dissipative region; or else the high electric field will give rise (as explained above) to a partial runaway beam. The process is able to continue from one place to the next, until all the free magnetic energy (which is responsible for the existence of the DC current and hence of the large DC electric field) is exhausted. This works much like the conduction front process described above, with the additional feature that the medium in which the "front" progresses contains free energy. This idea of *dissipation spreading* appears to various degree of precision in a number of other papers (Heyvaerts *et al.*, 1977; Norman and Smith, 1978). This scenario, however, is not unique, and Norman and Smith (1978) indicate an interesting variant in which the anomalous resistivity is spread by a beam of waves emitted in the trigger region.

The case where E would be almost equal to E_D, and in which all the electrons would runaway, is perhaps a more academic one, because many instabilities arise at lower thresholds, as will be reviewed below. The next paragraphs summarize conditions under which DC acceleration may operate in the flare plasma (§8.12), as well as results concerning current-driven instabilities of interest (§§8.13, 8.14).

8.11.2 Stochastic acceleration

Sections 8.15, 8.16, and 8.17 will deal with Langmuir wave-particle interactions and models based on this idea. These models have been conceived to remedy the relative inefficiency of low-frequency turbulence (which results from current-driven instabilities) at accelerating particles, while it was earlier largely thought that DC-field acceleration (which essentially leads to bulk heating) could not be made consistent with the existence of "suprathermal" particles. Thanks to the arguments developed above, the bulk heating picture, whatever its details, is now considered more viable, while selective acceleration by high-frequency turbulence remains an alternative to be studied seriously. Former arguments favouring selective acceleration have been summarized by Kahler (1975).

For *bulk energization there is no need for anything else than anomalous resistance provided it is not prematurely quenched.* The only things a theory based on this idea has to prove are essentially the *possibility of triggering*, the *absence of effective heat removal*, the *self-sustaining of the turbulent conditions* and, in some instances, the *spreading of the phenomenon* (if the primary trigger-region is too small, as it usually is). One has then just to extract out of the processes quoted in §8.12 and §8.13 those which meet our requirements. The reader will notice, however, that most of the current-driven instabilities operate under constant current density and self-quench (by electron or ion heating): once the instability threshold is reached, for example $v_D > c_S$, the electron heating just adjusts in such a way that marginal stability conditions are restored. The *hydrodynamic state is then one of marginal stability*, as shown in detail by Heyvaerts and Kuperus (1978) with

$$|\mathbf{J}| \leqslant |J^*|.$$

Any mechanism based on a slow approach of the critical conditions cannot meet the requirements of impulsiveness. One needs either some exterior force strongly driving the system to instability or a demonstration that the process under consideration develops and extends in space explosively. For this reason the *spreading problem* is a crucial one (Heyvaerts *et al.*, 1977; Norman and Smith, 1978) and the scenario of sustained dissipation (Papadopoulos, 1977) by progressive propagation of a "dissipation front" is extremely promising.

Selective acceleration, however, also leads to severe complications. The point is that electrons cannot be readily accelerated from thermal energies to high energies by low-frequency turbulence, especially ion-sound, because this interaction is elastic or almost so. Heyvaerts *et al.* (1977) estimate the characteristic acceleration time by these waves to be of the order of

$$t_{\rm acc}(v) \sim 10\ \omega_{pi}^{-1} \left(\frac{M_i}{m_e}\right)\left(\frac{v}{v_{T_e}}\right)^5,$$

which may be as long as 10^3 seconds. MHD turbulence is also a slow accelerator and requires pre-injection (§8.22 ff.). An exception must be made for low-frequency waves propagating almost normal to the magnetic field, because their phase velocity along the field may be high, and they can effectively resonate with electrons (Lampe and Papadopoulos, 1977). Apart from this ill-studied possibility, selective acceleration requires high-frequency turbulence. Langmuir waves have been considered in a number of papers. Apart from the difficulty of generating them, the problem with this acceleration mechanism is as follows: because of the resonant interaction condition $\omega = \mathbf{k} \cdot \mathbf{v}$, the spectrum should contain high phase velocity waves in order to be able to resonate with fast particles; but, in order to tap the pool of thermal particles, waves with epithermal phase velocities are also required; in brief, the *spectrum must be very broad*, and specific mechanisms able to give it the right shape must be present!

8.12 RUNAWAYS THROUGH TURBULENCE

8.12.1 Behaviour of an electron in low-frequency turbulence

The possibility of runaway acceleration through turbulence is dependent on the nature and spectral shape of the turbulence, because the coupling of particles to waves depends on the resonance condition ($\omega = \mathbf{k} \cdot \mathbf{v}$) and hence of the map of turbulence in **k**-space. In the presence of a magnetic field the resonance condition becomes

$$\omega = k_{//} v_{//} + s\Omega^*,$$

which differs very much from the unmagnetized case, which we shall study in this paragraph. For low-frequency turbulence, in an unmagnetized plasma, the phase velocities of waves are much smaller than the electron velocity and the resonance condition implies that the velocities of the electron and of the wave with which it interacts are inclined at very nearly 90°. This has several consequences.

First, the momentum $h\,\mathbf{k}$ exchanged between the electron and the wave changes the direction of the electron momentum $\mathbf{p}$ but not its modulus: this is an elastic collision. Acceleration effects are of first, or even sometimes second, order in the small parameter $v_{\text{phase}}/v_{\text{electron}}$, and consequently *such interactions are not suited for fast acceleration*, although for ion-acoustic turbulence the characteristic time depends very much on the ratio (T_e/T_i) (Lacombe and Mangeney, 1969).

Second, the conditions for interaction depend on the angular distribution of unstable waves; for example, if waves are beamed in a cone, only those electrons which have their velocity in the complementary cone are coupled to the turbulence; the others do not resonate. All electrons are coupled in the limit when the cone of unstable waves opens to a half-space, or when there is some agent (for example, a weak magnetic field) which turns the electron velocity in phase-space, so that they all resonate at some time (cyclotron mixing, Galeev and Sagdeev, 1969).

Third, the collision frequency decreases with increasing velocity. This may be understood because the collision frequency is the characteristic time required to turn, by collisions with turbulent waves, the momentum of the particle by 90°. In each interaction, the momentum is changed by a small amount Δp, but the corresponding angular deflection is only $\Delta\psi = \Delta p/p$. Detailed calculations show that, in the case of ion-acoustic turbulence saturated with a Kadomtsev spectrum (Kadomtsev, 1965), the collision-frequency scales as v^{-3}, as for Coulomb collisions (Kaplan and Tsytovitch, 1973). *Runaway through such turbulence is then possible*, even if electrons are all coupled to the waves. One can introduce a concept of a *turbulent Dreicer field*, E_D^*, defined as the field needed to accelerate the particles to a thermal velocity in one turbulent collision period τ_*^{-1}:

$$eE_D^* = \tau^* m v_{T_e}.$$

8.12.2 Changes in the ratio E/E_D

Interestingly enough, for a large inductance system as ours, the ratio of the actual field to the Dreicer field does not change when the plasma makes a transition from a non-turbulent to a turbulent state. This is because **j** does not change. Let E and E^* be the field before and after the transition, then:

$$|\mathbf{j}| = \sigma E = \sigma^* E^*.$$

But σ scales as ν^{-1}, where ν is the collision-frequency; then

$$\frac{E}{\nu} = \frac{E^*}{\nu^*},$$

and, because E_D scales as ν as well:

$$\frac{E}{E_D} = \frac{E^*}{E_D^*}$$

As far as the number of runaway particles is concerned, it then makes no difference whether the plasma is low-frequency turbulent or not, but of course the energy acquired by the partial runaways is much larger in the turbulent system.

A key remark here is that *bulk runaway does not occur*, but perhaps a substantial number of *partial runaways* is produced, because of either the v^{-3}-dependence or the "uncoupled" angular region. Numerical simulations of the Buneman instability by Boris *et al.* (1970) have been made with a constant electric field, and have shown runaway acceleration of the bulk of the electrons to roughly one-half of the free-streaming value. This is because, under a constant electric field, the Buneman instability does not self-quench. Actually, after the electrons have reached the threshold velocity v_0 a burst of turbulence appears, which heats the electrons. At that stage, they are almost all at v_0, and receive thermal energy in such a way that v_0 is less than the new threshold v_1 and the instability shuts off; but the new Dreicer field is smaller now and the applied electric field accelerates the electrons again up to v_1, a new turbulence burst occurs and so on ...

We stress that this picture is only correct when the electric field is given. When the *current* is fixed by boundary conditions, the electric field adapts to the actual resistivity. Assuming that j and n remain constant during the turbulence burst, we obtain (subscripts 1 and 2 referring to before and after heating)

$$\begin{cases} j_1 = j_2 \Rightarrow \sigma_1 E_1 = \sigma_2 E_2 \Rightarrow T_1^{3/2} E_1 = T_2^{3/2} E_2, \\ n_1 = n_2 \Rightarrow E_{D_1} T_1 = E_{D_2} T_2, \end{cases}$$

$$E_2/E_{D_2} = (T_1/T_2)^{1/2} (E_1/E_{D_1}).$$

The ratio of E to the Dreicer field has decreased, so the instability will be definitively quenched. We see that the *persistence of the acceleration is as much a matter of boundary conditions as of internal microphysics.* This reasoning supposes, however, that all electrons couple to the turbulence and heat up. For ion-acoustic instability, Orens and Dawson (1974) obtained the same results as Boris *et al.* (1970). Biskamp and Chodura (1971) explored a situation in which T_e/T_i is large at the beginning and in which the current density is maintained constant. The instability saturates by ion-trapping, heating them substantially; the electrons which are not in the resonance cone accelerate freely (runaway), while the bulk of the electrons takes a slower drift ($v_D < v_{T_e}$). The instability in this case does not quench, because of turbulence anisotropy. The electron temperature and runaway kinetic energy are found to increase with time at approximately the same rate at the expense of the exterior system which keeps the current constant.

8.12.3 Limits to DC electric field acceleration

A limitation on runaways is provided by the self magnetic field of the beam, which in the presence of a DC electric field cannot be shielded by a return current! This limitation forbids bulk runaway of some large fraction of the plasma. Requiring the self-field to remain smaller than the bias field for stability, D. Smith (1979) obtains the condition

$$n_b v_b R_b \leqslant n v_D l_B,$$

where n_b, v_b, R_b are the runaway beam density, velocity and sectional size; n is the ambient plasma density, v_D the current drift velocity of electrons, and l_B is the magnetic field scale-length. For v_b to greatly exceed v_D, one must have either a weak runaway beam ($n_b \ll n$) or a small strong runaway beam ($R_b \ll l_B$).

8.13 CURRENT-DRIVEN MICRO-INSTABILITIES: WEAK MAGNETIC FIELD

8.13.1 Current-Driven instabilities

As noted earlier, the *Buneman instability* is most unlikely to be the first current-limitation encountered in practice. The development of other instabilities, however, depends to a large extent on plasma parameters, direction of the current, and inhomogeneity of the plasma. Let us consider here the case when the magnetic field vanishes, and model the electron current flow by a velocity v_D. Standard linearized Vlasov equation manipulations (Stix 1962) performed in the ion rest-frame yield the dispersion relation for electrostatic waves:

$$0 = \varepsilon_l(\mathbf{k}, \omega) = 1 + \frac{k_D^2}{k^2}\left(\left(\frac{\omega - \mathbf{k}\cdot\mathbf{v}_D}{|k| v_{T_e}}\right) Z\left(\frac{\omega - \mathbf{k}\cdot\mathbf{v}_D}{|k| v_{T_e}}\right)\right)$$

$$+ \frac{T_e}{T_i}\left(1 + \frac{\omega}{|k| v_{T_i}} Z\left(\frac{\omega}{|k| v_{T_i}}\right)\right),$$

where Z is the usual plasma dispersion function

$$Z(z) = \frac{1}{\sqrt{\pi}} \int_{-\infty}^{+\infty} \frac{e^{-x^2}}{x - z} dx, \quad Im z > 0, \text{ (analytical continuation elsewhere).}$$

In particular, for real z

$$Z(z) = -2e^{-z^2}\int_0^z e^{x^2}dx + i\sqrt{\pi}e^{-z^2}.$$

Almost-real solutions in the low-frequency range $\omega < \omega_{pi}$ exist only in the case $T_e \gg T_i$, with phase velocities in the interval $(v_{T_i}, v_{T_e} + v_D)$. The dispersion relation simplifies in this limit where one can use the large-argument approximation for Z (ions) and small-argument approximation for Z (electrons)

$$Z(z) = -z^{-1}(1 + z^{-2}/2 + \ldots) + i\sqrt{\pi}\sigma e^{-z^2}, \quad |z| \gg 1,$$

$$\sigma = 0,\ 1,\ 2 \text{ for } Imz >,\ =,\ <0, \text{ resp.},$$

$$Z(z) = -2z(1 - \tfrac{2}{3}z^2 + \ldots) + i\sqrt{\pi}e^{-z^2}, \quad |z| \ll 1;$$

and it can be found that a wave $\mathbf{k}$ becomes unstable for

$$v_D > c_S \frac{\left(1 + \left(\dfrac{T_e}{T_i}\right)^{3/2}(M_i/m_e)^{1/2}\exp - (\tfrac{3}{2} + T_e/[2T_i(1 + k^2/k_D^2)]\right)}{(1 + k^2/k_D^2)^{1/2}}, \tag{8.8}$$

where c_S is the ion-sound velocity given by

$$c_S^2 = k_B T_e/m_i.$$

For T_e/T_i very large, the second term dominates and large-k waves become unstable first; but for moderate T_e/T_i, small k's become unstable first. In the former case, we have the *ion plasma wave instability* and in the second case the *ion-acoustic instability*, which is in most situations the relevant one. The instability criterion is then:

$$v_D > c_S\left(1 + \left(\frac{T_e}{T_i}\right)^{3/2}\left(\frac{m_i}{m_e}\right)^{1/2}\exp - \left(\frac{3}{2} + \frac{T_e}{2T_i}\right)\right).$$

The Figure 8.20 summarizes these various situations. The growth rate is

$$\gamma(\mathbf{k}) = -\sqrt{\frac{\pi}{8}}\frac{k_D^2}{k^2}\frac{\omega^3}{\omega_{pi}^2}\left(\frac{\omega - \mathbf{k}\cdot\mathbf{v}_D}{|k|v_{T_e}} + \frac{T_i}{T_e}\exp - \frac{\omega^2}{2k^2 v_{T_e}^2}\right)$$

and the dispersion relation in the same limit is approximately

$$\omega^2 = k^2 c_S^2/(1 + k^2/k_D^2).$$

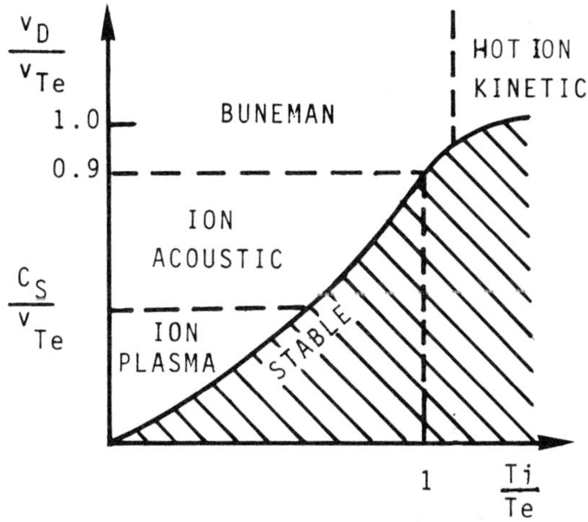

FIGURE 8.20 Instability diagram for electron-ion drift. (Adapted from Papadopoulos, 1977.)

8.13.2 Heating and resistance in the presence of turbulence

The effects produced by fluctuations in the plasma, and in particular turbulence, may be expressed in terms of the *fluctuation spectrum* of some quantity, usually the electric field, but sometimes the magnetic field or the densities. It is convenient to introduce *correlation functions* and *spectral power densities* as follows (indices α and β refer to particle species):

$$\langle E(0, 0)E(\mathbf{r}, t)\rangle = \int \frac{d^3k}{8\pi^3} \int d\omega \quad \mathscr{E}(\mathbf{k}, \omega) e^{i(k \cdot \mathbf{r} - \omega t)}, \tag{8.9a}$$

$$\langle n_\alpha(0, 0) n_\beta(\mathbf{r}, t)\rangle = \int \frac{d^3k}{8\pi^3} \int d\omega N_{\alpha\beta}(\mathbf{k}, \omega) e^{i(\mathbf{k} \cdot \mathbf{r} - \omega t)}. \tag{8.9b}$$

In terms of these quantities the friction force excerted on electrons and the heating rates (including kinetic energies due to sloshing motion in waves) are given by (Tange and Ichimaru, 1974)

$$\mathbf{F}_{ei} = -\omega_p^2 m_e \int \frac{d^3k}{8\pi^3} \int d\omega \frac{\mathbf{k}}{k^2} \frac{Im N_{ei}(\mathbf{k}, \omega)}{n^2}, \tag{8.10}$$

$$\frac{\mathrm{d}}{\mathrm{d}t}(nk_BT_e) = -\omega_p^2 nm_e \int \frac{\mathrm{d}^3k}{8\pi^3} \int \mathrm{d}\omega \frac{\omega - \mathbf{k}\cdot\mathbf{v}_D}{k^2} \frac{ImN_{ei}(\mathbf{k}, \omega)}{n^2},$$

$$\frac{\mathrm{d}}{\mathrm{d}t}(nk_BT_i) = -\omega_p^2 nm_e \int \frac{\mathrm{d}^3k}{8\pi^3} \int \mathrm{d}\omega \frac{\omega}{k^2} \frac{ImN_{ei}(\mathbf{k}, \omega)}{n^2}.$$

N_{ei} and $\mathscr{E}$ are related by the Poisson equation $\operatorname{div} D = e(n_i - n_e)$, which introduces the (in general non-linear) polarizability $\tilde{\chi}_\sigma$ of species σ. In general,

$$-4\pi\omega_p^2 \frac{m_e}{nk^2} N_{ei}(\mathbf{k}, \omega) = \mathscr{E}(\mathbf{k}, \omega)\tilde{\chi}_e(\mathbf{k}, \omega)\tilde{\chi}_i^*(\mathbf{k}, \omega).$$

In the case of weak electrostatic turbulence, the dominant fluctuations are proper modes, which obey the dispersion relation $1 + \tilde{\chi}_e + \tilde{\chi}_i = 0$. Then

$$ImN_{ei}(\mathbf{k}, \omega) = +\frac{nk^2}{m_e\omega_p^2}\varepsilon_0\, \mathscr{E}(\mathbf{k}, \omega) Im\tilde{\chi}_e(k, \omega). \tag{8.11}$$

The estimate of heating and drag requires the calculation of the turbulence spectrum and non-linear polarizability $\tilde{\chi}$, which must self-consistently contain the turbulence effects on particle orbits. These are the difficulties.

8.13.3 Instability saturation

One has to identify the process which is responsible for the saturation of turbulence growth. The *quasilinear theory* of the ion-acoustic instability has been considered by Rudakov and Korablev (1966) in the presence of a constant electric field, assuming quasi-elastic collisions between electrons and ion sound waves. In this over-simplified theory, the current saturates at $|\mathbf{j}| = n_e e C_S$ and, slowly, the electron distribution function develops a runaway tail. This is not so simple, because one must introduce in the picture non-linear effects which modify the wave spectrum, weak magnetic fields (which may on a long time-scale influence the trajectory of particles, preventing runaway even though not changing the dispersion relation ($\Omega_e \ll \omega_{pi}$)) and finally one must consider ion dynamics and heating.

Kadomstev (1965) introduced the interaction of ion-acoustic waves beats with ions (also called *non-linear Landau damping* or induced scattering on ions). For proper modes, excitations $\mathscr{E}(\mathbf{k}, \omega)$ in (8.9) depend on $\mathbf{k}$ with $\omega = \omega(\mathbf{k})$ and the kinetic equation for non-linear Landau damping has the following structure:

$$\frac{\partial}{\partial t}\log \mathscr{E}(\mathbf{k})=\gamma(\mathbf{k})-\int d^3k'\ \mathscr{E}(\mathbf{k}')p(\mathbf{k},\mathbf{k}')\int d\mathbf{v}\delta(\omega_k-\omega_{k'}-(\mathbf{k}-\mathbf{k}')\cdot\mathbf{v})\left((\mathbf{k}-\mathbf{k}')\cdot\frac{\partial f_i}{\partial \mathbf{v}}\right);$$

the delta-function in the second term expresses the resonant interaction of the ion of velocity $\mathbf{v}$ with the beat $(\omega_k-\omega_{k'}, \mathbf{k}-\mathbf{k}')$. It can be shown that, in a time-averaged sense, this equation predicts the existence of a turbulence spectrum confined around the electron drift velocity $\mathbf{v}_D$ in an angle $\cos^{-1}(c_S/v_D)$. This spectrum is the so-called *Kadomtsev spectrum* (Kadomtsev, 1965; Galeev and Sagdeev, 1969; Papadopoulos, 1977), which scales with k as $k^{-5}\ \ln(k_D/k)$, and is divergent for $k\to 0$, but this divergence is limited by collisions. A very important non-linear effect (*mode coupling*) which may convert two waves into a third one, when the beat between the two original waves turns out to be a proper mode of the plasma does not, *a priori*, contribute, because the combination relations

$$\mathbf{k}_1+\mathbf{k}_2=\mathbf{k}_3 \quad \text{and} \quad \omega_{k_1}+\omega_{k_2}=\omega_{k_3}$$

cannot be fulfilled by ion-acoustic waves. It can, however, be almost fulfilled at low frequencies $(\omega\ll\omega_{pi})$ (Tsytovitch, 1971) giving a spectrum similar to Kadomtsev's.

Choi and Horton (1975) have included in the calculation the modification of a particle's orbit by turbulence. This is a renormalization procedure which allows one to obtain an expression for the effective one-particle propagator from an integral equation. This gives rise to effects known as resonance broadening, equivalent to collisions on waves. They obtain a spectrum for the spectral power of the electric potential $I(\mathbf{k})$ given by

$$I_k=\frac{15}{64}\frac{T_e}{T_i}\frac{v_D}{v_{T_e}}\frac{1}{\sqrt{2\pi}}\left\{\delta\ \ln\left(\frac{1}{k\lambda_D}\right)-2\sqrt{\frac{\pi\nu}{\omega p_e}}\left(\sqrt{\frac{1}{k\lambda_D}}-1\right)\right\}\frac{1}{k^3},$$

where the frequency ν and the number δ can be determined from appropriate equations in Choi and Horton (1975). In this theory, allowance is made for a deformation of the ion distribution function, in the form of extended ion tails, and of turbulent elastic electron-wave collisions. The latter destroy Landau resonance for small k's and keep the electron distribution isotropic. Their theory seems to be supported by the results of Stenzel (1978), whose experimental results shown in Figure 8.21 exhibit a characteristic decrease at small periods due to destruction of long wavelength resonance by electron scattering and a fall-off at short wavelengths (large ω) which is the result of ion nonlinear Landau damping (Stenzel, 1978).

Some attempts have been made to describe particle-ion–acoustic-wave interactions in the case where electron runaways are present, as suggested by

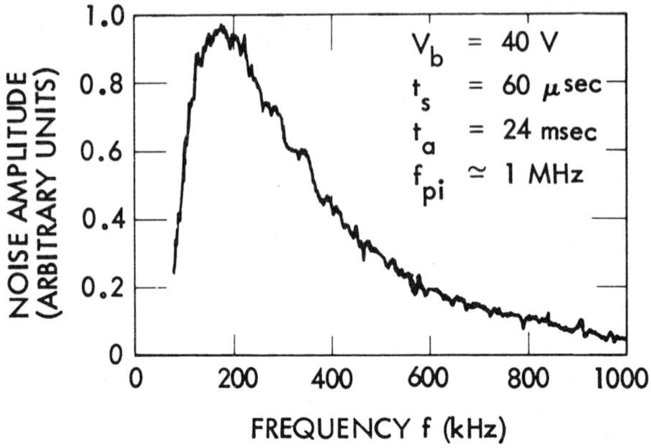

FIGURE 8.21 Frequency spectrum of the saturated current-driven ion-acoustic instability. $\delta n_e/n_e$ versus frequency. The absolute peak density fluctuation level at $\omega/\omega_{pi}=0.2$ is $\delta n_e/n_e>20\%$. The high-frequency fall-off is $\delta n_e/n_e \sim f^{-2.5}$. Krypton plasma at 2.3 10^{-4} Torr. (From Stenzel, 1978.)

Biskamp and Chodura (1971) on the basis of self-similar solutions, but the modelling (Vekshtein *et al.*, 1970) fails because of the neglect of electron drift motion (Balescu, private comm.). Simulations have revealed that the turbulence development is accompanied by intense ion heating (Dum *et al.*, 1974; Horton *et al.*, 1976). Choi and Horton (1975) have shown that the effect of turbulence is actually to produce a non-local diffusion of ions in velocity space, and is responsible for the rather important ion tails which appear. This high-energy tail drastically affects the ion Landau damping rate, and is ultimately responsible for the extinction of ion sound turbulence, as demonstrated by Dum *et al.* (1974), which disappears under constant current after some hundred ω_{pi}^{-1}.

One might wonder how much a zero-field theory is relevant to the physics of a highly magnetized plasma. As far as field-aligned waves and motions are both considered, the zero-field treatment is correct, but oblique wave effects are also very important. Nevertheless, when the turbulent collision-frequency becomes high enough, that is at high turbulence levels, the particle's cyclotronic motions may be so much altered that the field effect ceases to dominate.

8.14 CURRENT-DRIVEN MICRO-INSTABILITIES: MAGNETIZED MEDIUM

8.14.1 Current-driven instabilities

A list of current-driven instabilities has been compiled by Norman and R. Smith (1978), including only homogeneous plasma situations, disregarding a number of gradient-current-driven instabilities (which have up to now been ill-studied in flare physics, presumably because of our ignorance of what the actual gradients are). Of special interest are electrostatic, low-β modes, driven unstable by field-aligned currents, which would fit the usual force-free, low-β assumption made about coronal magnetic configurations. It is still worth bearing in mind that this assumption may very well fail in current sheets ($\beta \simeq 1$, inhomogeneous) and in filamentary matter ... A glance at Table 8.2, reproduced from Norman and Smith (1978), shows that a most interesting instability is the *low-frequency electrostatic ion-cyclotron one*.

8.14.2 Electrostatic ion-cyclotron instability

A detailed discussion of electrostatic low-frequency modes in a current-carrying plasma has been made by Kennel and Kindel (1971), who start from the kinetic theory dispersion relation

$$0 = k^2 + \sum_{\substack{\text{species}\\ \alpha}} \sum_{s=-\infty}^{+\infty} e^{-\mu_\alpha} I_S(\mu_a)\left(1 + \frac{\omega - k_\| v_{D_\alpha}}{|k_\|| v_{T_\alpha}} Z\left(\frac{\omega - k_\| v_{D_\alpha} + S\Omega_\alpha}{|k_\|| v_{T_\alpha}}\right)\right), \tag{8.12}$$

where $\|$ and $\perp$ refer to directions along and perpendicular to $\mathbf{B}$; Ω_α (or equivalently ω_{B_α}) is the α-species cyclotron frequency; ω is complex ($\gamma > 0$, growth); I_S is the modified Bessel function of order s; Z is the function introduced in §8.13 and

$$\mu_\alpha = \frac{k_\perp^2 \, v_{T_\alpha}^2}{2\Omega_\alpha^2}; \tag{8.13}$$

k_{D_α} is the Debye wave vector for species α, ω_{p_α} the corresponding plasma frequency

$$\omega_{p_\alpha} = k_{D_\alpha} v_{T_\alpha}$$

and v_{D_α} the drift velocity of this species.

For oblique waves, this relation simplifies in the limit of long wavelengths as compared to the electron gyroradius and intermediate parallel phase velo-

TABLE 8.2
Current-driven instabilities. (From Norman and R. Smith, 1978.)

Instability	Dispersion Relation	V_{crit}	
A) Field-aligned ($\mathbf{j} \| \mathbf{B}$)			
Buneman	$1-\frac{\omega_i^2}{\omega^2}-\frac{\omega_e^2}{(\omega-KV_D)^2}=0$	V_e	$\left(\frac{m}{M}\right)^{1/3}\omega_e$
Electrostatic Ion-Cyclotron	$1+\frac{\omega_i^2}{K^2V_e^2}\sum_{n=-\infty}^{\infty} e^{-\lambda_i}I_n(\lambda_i)\left[1+\frac{\omega}{\sqrt{2}K_\| V_i}Z\left(\frac{\omega-n\Omega}{\sqrt{2}K_\| V_i}\right)\right]$ $+\frac{\omega_e^2}{K^2V_e^2}\left[1-\frac{\omega-K_\| V_D}{2K_\| V_e}\left(\frac{\omega-K_\| V_D}{\sqrt{2}K_\| V_e}\right)\right]=0$	V_i	Ω_i
Electrostatic Ion-Acoustic	$1-\frac{\omega_i^2}{\omega^2}+\frac{\omega_e^2}{K^2v_e^2}-i\pi\frac{\omega_e}{K^2}\frac{K}{\|K\|}\frac{\partial f_0}{\partial v}\Big\|_{\omega/K}=0$	C_s	$\sqrt{\frac{\pi}{8}\frac{m}{M}}\,\omega$
Kink-like ($\beta\sim 1$, $KR_{Li}\ll 1$)	$K^2V_A^2\pm\frac{KV_D\Omega_t}{\omega^2}=0 \qquad V_D<V_A$	$\frac{KV_A^2}{\Omega_i}$	$\frac{V_D}{V_A}\Omega_t$
Whistler	$K^2C^2+\frac{\omega_t^2}{\Omega_t}(\omega-KV_D)-i\omega\frac{\omega_i^2}{\sqrt{2}KV_i}\pi\exp-((\omega-KV_i)^2/2V_e^2)=0$	V_i	Ω_i

B) Cross-field (**j** ⊥ **B**)			
Buneman	$1-\frac{\omega_i^2}{\omega^2}-\frac{\omega_e^2}{(\omega-KV_D)^2}=0$	V_e	$\left(\frac{m}{M}\right)^{1/3}\omega_e$
Modified 2-stream	$1-\frac{K^2\omega_e^2}{K^2\Omega_e^2}-\frac{\omega_i^2}{(\omega-KV_D)^2}-\frac{K_\parallel^2\omega_e^2}{K^2\omega^2}=0$	V_i	$\frac{\Omega_i}{(1+\omega_e^2/\Omega_e^2)}$
Electron Cyclotron Drift	(i) Fluid limit		
	$1-\frac{\omega_e^2}{\omega^2-\omega_e^2}-\frac{\omega_i^2}{(\omega-KV_D)^2}=0$	V_i	Ω_e
	(ii) $K^2\lambda_e^2+1=e^{-\lambda}I_0(\lambda)+\frac{T_e}{2T_i}Z\left(\frac{\omega-KV_D}{KV_i}\right)+2\omega^2\sum_{n=1}^{\infty}\frac{e^{-\lambda_e}I_n(\lambda_e)}{\omega^2-n^2\Omega_e^2}$	V_i	$n\Omega_e$
Electromagnetic Ion-Cyclotron	$\omega\sim K_\parallel V_D$	V_A	Ω_e
Electromagnetic Ion-Acoustic	$\omega\sim KC_S$	$\geqslant 8\cdot V_i$	$\sqrt{\frac{\pi}{8}\frac{m}{M}}\,\omega$

cities: we obtain after using some identities for I_n's (Kennel and Kindel, 1971) in the limit $\omega \simeq \Omega_i$

$$\omega = \Omega_i \left(1 + \frac{e^{-\mu_i} I_1(\mu_i)}{1 + \frac{T_i}{T_e} - e^{-\mu_i} I_1(\mu_i) - \frac{1 - e^{-\mu_i} I_0(\mu_i)}{\mu_i}} \right) = \Omega_i \left(1 + \Delta\left(\mu_i, \frac{T_i}{T_e} \right) \right). \tag{8.14}$$

In the limit $\mu_i \ll 1$ ($\perp$ wavelength longer than the ion gyroradius), we get

$$\omega^2 = \Omega_i^2 + k_\perp^2 \; c_S^2.$$

The drift velocity for which a mode becomes unstable can be calculated as a function of k and minimized to get the threshold drift velocity

$$v^* = v_{T_i} \frac{1 + \Delta(\mu^*)}{\Delta(\mu^*)} \left(\ln\left(2 \sqrt{\frac{M_i T_e}{m_e T_i}} e^{-\mu^*} I_1(\mu^*) \right) \right)^{1/2}, \tag{8.15}$$

where μ^* is the variable which minimizes $1 + \Delta^{-1}$ with respect to μ at fixed T_e/T_i (see Eq. 8.14). Numerically, $v^* = 13 \; v_{T_i}$ for $T_e = T_i$. The instability exists in the limit $T_e = T_i$ because for these waves the ion Landau damping is suppressed by the ion cyclotron motion, while the quasi-neutrality of the plasma is ensured by electron motions parallel to the line of force. The threshold velocity v^* in (8.15) becomes larger than the ion-acoustic threshold (8.8) when T_e/T_i becomes larger than 10. Both (8.8) and (8.15) are decreasing functions of (T_e/T_i).

8.15 PRODUCTION OF A LANGMUIR WAVE SPECTRUM SUITABLE FOR ACCELERATION

8.15.1 Generation of Langmuir waves

The simplest way of producing Langmuir waves is to excite them by means of some electron beam. This process is of course of no interest from a global point of view, since the beam has to receive its energy somewhere, and the best it can do is to share some of it with the medium which it penetrates. This, however, does not create more energy by itself; it may at best redistribute it. The only interesting question relevant to the energy release problem is then to know how effective a beam can be in triggering *additional* energy release. Stated another way: *what can a beam do for spreading fast dissipation*? We have already given partial and positive answer to that question: if strong enough, the beam may give rise to an unstable return-current, generating anomalous resistance and then *additional* dissipation if magnetic energy is stored there (Norman and Smith, 1978). We show in §8.16 that the dynamics of strong beam-generated turbulence could also lead by itself to anomalous resistance, even if the return-current is stable. It still remains of interest to study how the Langmuir turbulence spectrum forms, and how it evolves and interacts with particles. This has been the subject of many researches and, in our specific field, of some controversy too. Much effort has been devoted to understand *whether it is possible to produce particle acceleration without reaching anywhere the Dreicer field limit*, which is so high. Current-driven instabilities are somewhat easier to trigger, though this also is debatable. A major problem was then to understand *whether high-frequency turbulence could grow from low-frequency turbulence*. It seems now that this is virtually impossible, but it is worth recalling the arguments.

8.15.2 Inefficiency of Conversion up from low-frequency turbulence

Tsytovitch *et al.* (1975) suggested that electrons deflected in the field of resonant ion-acoustic waves should emit Langmuir waves as a result of what is essentially a bremsstrahlung process. Their calculation has been repeated for a more realistic case of maxwellian distributions of ions and electrons by Kuipers (1979), who finds a Langmuir wave growth rate:

$$\gamma^l_{\mathrm{Brem}}(\mathbf{k}) = -24\frac{\pi^2 e^2}{m_e^2}\int \mathrm{d}^3 q\, \frac{U_S(\mathbf{q})}{q^2}\,\frac{l}{\omega_l(\mathbf{k})}(\mathbf{q}\cdot\vec{\mathbf{k}})\int \mathrm{d}^3\mathbf{v}\left(q\cdot\frac{\partial F_e}{\partial \vec{\mathbf{v}}}\right)\delta(\omega_S(\mathbf{q})-\mathbf{q}\cdot\mathbf{v}), \tag{8.16}$$

where $U_S(\mathbf{q})$ is the spectral electric energy density of sound waves. For drifting Maxwellian distributions, this growth-rate becomes explicitly:

$$\gamma^l_{\mathrm{Brem}}(\mathbf{k}) = +\sqrt{\frac{\pi}{2}}\,\omega_{p_e}\int \mathrm{d}^3\mathbf{q}\,\frac{W_S(\mathbf{q})}{nk_BT_e}\,\frac{\omega_S^2(q)}{\omega_{p_i}^2}\,\frac{\omega_S(\mathbf{q})-\mathbf{q}\cdot\mathbf{v}_D}{|q|v_{T_e}}\,\frac{(\mathbf{k}\cdot\mathbf{q})}{q^2}, \tag{8.17}$$

where now $W_S(\mathbf{q})$ is the spectral (total) energy density of sound waves, and $\mathbf{v}_D$ is the mean electron drift velocity. It is seen that, for $v_D = 0$ and an isotropic wave spectrum, $\gamma^l_{\mathrm{Brem}}(\mathbf{k})$ vanishes. Some anisotropy of waves or particles is needed for growth.

Several objections against this mechanism have been raised, the least being that (taking account of linear Landau damping) the net growth-rate ($\gamma^l_{\mathrm{Brem}} + \gamma_{\mathrm{Landau}}$) peaks at a phase velocity which is several v_{T_e} but not comparable to that of 50 keV electrons (Hoyng, 1977b). Ulterior reshaping of the wave-spectrum by processes well-known from type III burst theory, such as non-linear Landau damping (Heyvaerts and de Genouillac, 1974 and earlier references therein) and strong turbulence processes, such as the modulational instability (Papadopoulos *et al.*, 1974), are described in §8.16. More important are the following criticisms:

a) the ion-acoustic wave-spectrum is beamed in the direction of the electron drift, $\mathbf{v}_D$, with the consequence that in Equation (8.17) $\mathbf{q}\cdot\mathbf{v}_D>0$, so that Langmuir waves grow for $(\mathbf{k}\cdot\mathbf{q})<0$; the net result (D. Smith, 1979) is seemingly to increase the momentum of the electron + waves system; though ions are a momentum reservoir, this property is still surprising;

b) the growth-rate (8.16) is proportional to the growth-rate of ion-acoustic waves (D. Smith, 1979) and should vanish identically at marginal stability;

c) in any case one must consider at the same time the reverse process, which is here Langmuir wave absorption by ion sound; this damping process is just the high frequency version of the anomalous resistance, and has been calculated by Dawson and Oberman (1963) and Dawson (1968). Let us call γ_{DO} the corresponding damping rate, k_S a typical sound wave vector and W_S the ion sound energy density. Vlahos and Papadopoulos (1979b) take, as an order of magnitude,

$$\gamma_{\mathrm{DO}}=\left(\frac{k_D}{k_S}\right)^4\left(\frac{W_S}{nk_BT_e}\right)\omega_{p_e}.$$

They show that bremsstrahlung emission and Dawson–Oberman damping combine to give a net Langmuir-wave-damping in general. This can be seen by writing down the rate equations for ion-sound and Langmuir waves, approximately (8.17), as

$$\gamma^l_{\mathrm{Brem}}(\mathbf{k})=\omega_{p_e}\left(\frac{W_S}{nk_BT_e}\right)\frac{k\ v_D}{k_S v_{T_e}}.$$

Let γ_i and γ_l be the linear Landau growth- and damping-rates of ion-sound and plasma waves respectively. The rate equations are

$$\frac{\partial W_S}{\partial t}=2\gamma_i W_S-2(\gamma^l_{\mathrm{Brem}}-\gamma_{\mathrm{DO}})W_l,$$

$$\frac{\partial W_l}{\partial t}=2(\gamma^l_{\mathrm{Brem}}-\gamma_{\mathrm{DO}})W_l-2\gamma_l W_l.$$

Langmuir waves grow provided $\gamma^l_{\mathrm{Brem}}>\gamma_{\mathrm{DO}}$ or

$$\frac{v_D}{v_{T_e}} > \left(\frac{k_D}{k_S}\right)^3 \left(\frac{k_D}{k_l}\right).$$

Since the right-hand side is systematically larger than one, *plasma waves cannot grow unless the drift velocity exceeds* v_{T_e}! All models based on the generation of plasma waves from ion-acoustic turbulence by this process were then in error on this point, which reduces somewhat their interest. Nevertheless, part of the analysis will remain useful if some other source of Langmuir turbulence is found in future. Some of these models consider the evolution of turbulence under strong-turbulence effects towards shorter wavelengths. We describe now the physics of these processes.

8.16 STRONG-TURBULENCE EFFECTS

8.16.1 Basic physics

They are based on two complementary mechanisms. *First*, when the level of turbulence is high enough, the fluctuating electric field gives rise to an electric pressure force, just as radiation exerts a radiative pressure. This force is called the ponderomotive force. The particles tend then to be pushed away from regions with a high level of turbulence. *Second*, Langmuir waves of frequency ω cannot propagate in those regions where the local plasma frequency exceeds ω. These plasma waves (plasmons) are then trapped in low-density regions. These two effects may conspire, if the ponderomotive force is large enough, to build up regions of high density surrounding regions of lower density where the turbulence level is high. The onset of this separation mechanism is the so-called *modulation instability*. Attempts to describe its further fate is a matter of present-day plasma physics research. Easy accounts of it have been given by Galeev *et al.* (1975) and D. Smith (1977b).

Let us describe the plasma by fluid equations for electrons and ions coupled by a self-consistent space-charge field. These are, with standard notations, and assuming one-dimensional motions for simplicity,

$$\frac{\partial n_i}{\partial t} + \frac{\partial}{\partial x} n_i v_i = 0, \quad n_i\left(\frac{\partial v_i}{\partial t} + v_i \frac{\partial v_i}{\partial x}\right) = \frac{n_i}{M_i} qE - \frac{k_B T_i}{M_i}\frac{\partial n_i}{\partial x},$$

$$\frac{\partial n_e}{\partial t} + \frac{\partial}{\partial x} n_e v_e = 0, \quad n_e\left(\frac{\partial v_e}{\partial t} + v_e \frac{\partial v_e}{\partial x}\right) = -\frac{n_e}{m_e} qE - \frac{k_B T_e}{m_e}\frac{\partial n_e}{\partial x}, \tag{8.18}$$

$$\varepsilon_0 \frac{\partial E}{\partial x} = q(n_i - n_e).$$

It is convenient to use dimensionless variables defined by:

$$t' = \omega_p^{-1} t, \quad n' = n_0^{-1} n, \quad v' = v_{T_e}^{-1} v, \quad x' = k_D x, \quad E' = \varepsilon_0 E (q n_0)^{-1},$$

where n_0 is some mean plasma density. The other notations are standard, and from now on we shall use these new variables while omitting the superscript "prime". In the spirit of the physical ideas presented above, we shall tread on two different footings the motions occurring on a time-scale ω_p^{-1}, (the Langmuir, fast-wave motions) from slower ones involving ions, which describe such effects as their hydrodynamical response to turbulent pressure. We write:

$$n_e = \tilde{n}_e + \langle n_e \rangle + 1, \qquad V_e = \tilde{v}_e + \langle v_e \rangle,$$

$$n_i = \langle n_i \rangle + 1, \qquad v_i = \langle v_i \rangle, \qquad E = \langle E \rangle + \tilde{E}.$$

(The $\sim$ refers to fast perturbations and the bracketting $\langle\ \rangle$ refers to slow ones.)

Inserting this in Eq. 8.18, linearizing with respect to fast variables, and assuming the plasma to remain quasi-neutral for slow perturbations ($\langle n_e \rangle = \langle n_i \rangle$), one gets the equations for fast electron motion:

$$\frac{\partial^2 \tilde{n}_e}{\partial t^2} - \frac{\partial^2 \tilde{n}_e}{\partial x^2} = \frac{\partial}{\partial x}(1 + \langle n_i \rangle)\tilde{E},$$

$$\frac{\partial \tilde{E}}{\partial x} = -\tilde{n}_e,$$

so that

$$\frac{\partial^2 \tilde{E}}{\partial t^2} - \frac{\partial^2 \tilde{E}}{\partial x^2} = -(1 + \langle n_i \rangle)\tilde{E}. \tag{8.19}$$

These are just the familiar equations for electron Langmuir waves, except for the $\langle n_i \rangle$ term, which represents the effect of the slowly-varying plasma density. This term represents the plasmon trapping.

The low-frequency part of the equations may be obtained in the same way: one must only average the basic equations (8.18) with respect to the fast time-scale, keeping in mind that the average of two rapidly varying quantities is not zero! Actually, the average $\langle \tilde{v}_e \partial \tilde{v}_e / \partial x \rangle$ in the electron motion equation just describes the effect of the ponderomotive force on them. This force is transmitted to the ions by space-charge coupling. Neglecting electron inertia and all non-linearities except this one, one is led (for slow ion motion) to the system of equations:

$$\frac{\partial\langle n_i\rangle}{\partial t}+\frac{\partial\langle v_i\rangle}{\partial x}=0,$$
$$\frac{\partial\langle v_i\rangle}{\partial t}=\frac{m_e}{M_i}\left(-\left\langle\tilde{v}_e\frac{\partial}{\partial x}\tilde{v}_e\right\rangle-\frac{\partial\langle n_i\rangle}{\partial x}\right). \tag{8.20}$$

From the fast equations (8.19) and (8.18) one may express as well:

$$\left\langle\tilde{v}_e\frac{\partial}{\partial x}\tilde{v}_e\right\rangle=-\frac{1}{2}\frac{\partial}{\partial x}|\tilde{E}|^2=-\frac{1}{2}\frac{\partial}{\partial x}\quad\mathscr{E}^2,$$

which clearly shows that this term represents the ponderomotive force.

The amplitude $|\tilde{E}|=\mathscr{E}$ is a slowly-varying function, by contrast with the complex $\tilde{E}$, which is rapidly varying. An equation for $|\tilde{E}|$ is obtained by demanding that the propagation equation for $\tilde{E}$, namely (8.19), holds true on the long time-scale. We put $\tilde{E}=|\tilde{E}|e^{-it}=\quad\mathscr{E}e^{-it}$ and obtain to first order in the slowness parameter

$$2i\frac{\partial\mathscr{E}}{\partial t}+\frac{\partial^2\mathscr{E}}{\partial x^2}=(\langle n_i\rangle-1)\mathscr{E}.$$

This equation and the pair of equations (8.20) for $\langle n_i\rangle$ and $(v_i\rangle$ constitute a closed system, which describes completely the ionic response of the plasma to the Langmuir turbulence. Coming back to dimensional quantities, we obtain the system (Zakharov, 1972) called *Zakharov's equations*:

$$\frac{\partial^2\langle n_i\rangle}{\partial t^2}-c_S^2\frac{\partial^2\langle n_i\rangle}{\partial x^2}=\frac{\varepsilon_0}{2M_i}\frac{\partial^2\mathscr{E}^2}{\partial x^2}, \tag{8.21a}$$

$$i\frac{\partial\mathscr{E}}{\partial t}+\frac{\omega_p\lambda_D^2\partial^2\mathscr{E}}{2\quad\partial x^2}=\frac{\omega_p}{2n_0}\langle n_i\rangle\mathscr{E}. \tag{8.21b}$$

Two regimes are to be considered, according to which terms dominate the first member of (8.21). A first regime occurs when the time-derivative term is negligible. In this case, the ions keep an equilibrium between their pressure gradient and the turbulence pressure. This is the so-called subsonic regime for which

$$\mathscr{E}^2=-\frac{2M_ic_S^2}{\varepsilon_0}\langle n_i\rangle. \tag{8.22}$$

Putting this in (8.21b), we get for $\mathscr{E}$ the "non-linear Schrödinger equation"

$$i\frac{\partial \mathscr{E}}{\partial t}+\frac{\omega_p\lambda_D^2}{2}\frac{\partial^2 \mathscr{E}}{\partial x^2}+\frac{\omega_p}{2n_0}\frac{\varepsilon_0}{2M_i c_s^2}\mathscr{E}^3=0. \tag{8.23}$$

This equation has been shown to have solutions in which $\mathscr{E}$ becomes a more and more peaked function of x when time progresses, provided that $\mathscr{E}$ is large enough at the beginning. In one dimension, the equation admit *soliton solutions*, travelling undistorted (Rudakov, 1973), but in more than one dimension no stationary solutions have been obtained in numerical solutions of (8.23). However, for too large a value of $\mathscr{E}$, the subsonic approximation (8.22) ceases to be valid. This occurs when:

$$\varepsilon_0 \quad \mathscr{E}^2 \gg \frac{m_e}{M_i} n_0 k_B T_e,$$

in which case the first term of (8.21) dominates. This is the supersonic regime; the ions move inertially in response to the electrostatic pressure (ponderomotive force). The system (8.21) becomes in this case:

$$\frac{\partial^2 \langle n_i \rangle}{\partial t^2}=\frac{\varepsilon_0}{2M_i}\frac{\partial^2 \mathscr{E}}{\partial x^2},$$

$$i\frac{\partial \mathscr{E}}{\partial t}+\frac{\omega_p\lambda_D^2}{2}\frac{\partial^2 \mathscr{E}}{\partial x^2}=\frac{\omega_p}{2n_0}\langle n_i \rangle \mathscr{E}.$$

It has been shown to have *self-similar solutions* (functions of the variable x/t alone), in which the characteristic dimension of the "cavity" in which $\mathscr{E}$ is large and n_i negative ever decreases in time according to the law (Galeev *et al.*, 1975):

$$l/(-t)^{2/s}=c^t \quad \text{for} \quad t<0, \tag{8.24}$$

where s is the dimensionality. The cavity collapses to zero thickness in a finite time! This of course does not occur physically because of a variety of effects (kinetic effects of wave-particle interaction, non-linear effects other than ponderomotive force, strong dispersion of plasma waves), all of which are omitted in (8.21). Whichever mechanism limits plasmon collapse, it is clear that a reduction of the size of the regions where turbulence exists will occur and that the *wavelengths of the turbulent waves will accordingly decrease*, the effect of which is to produce a transfer of waves towards low phase velocity regions of the spectrum where the coupling to particles is particularly effective. The effects which limit plasmon collapse and the role of this process is one of the main concerns of present plasma physics. Not included in Zakharov's equations are non-linear Landau damping of plasma waves, kinetic type interactions, radiation effects, radiation of sound by fast moving solitons, reflexion of particles. (See Smith (1977a) for a review of the theoretical ideas under study.)

8.16.2 Tentative models of developed strong turbulence

Some authors have tried to picture strong turbulence as a *superposition of solitons*, which evolve more or less independently or weakly interact. The soliton evolution may be towards collapse if the dimension is effectively larger than one, the usual case ($\omega_p \gg \omega_{Be}$), or may be free propagation in one dimension. Solitons are created, or exist, with a certain distribution of their strength. This point of view is developed by Galeev *et al.* (1975) and by Rudakov (1973). The wave-vector space contains a region where individual soliton dynamics dominates the evolution (the inertial region) and a dissipation region, in which interaction with particles etc. is important. The individual collapsing soliton dynamics is summarized in the equation which gives the rate of change of the inverse of the size, namely

$$\left.\frac{\mathrm{d}k}{\mathrm{d}t}\right|_{\text{soliton dynamics}} = \dot{K}(k).$$

Accordingly, wave energy in k-space evolves according to an equation which expresses that following the soliton collapse (collapsing solitons conserve their energy):

$$\frac{\partial W(k)}{\partial t} + \frac{\partial}{\partial k}(\dot{K}(k)W(k)) = 0. \tag{8.25}$$

The spectral slope of the stationary spectrum in the inertial range is then:

$$W(k)_{\text{Stat. inertial}} \sim (\dot{K}(k))^{-1}.$$

Galeev *et al.* (1975) show that $\dot{K}(k) \sim k^{1+s/2}$ where s is the dimensionality of electron motion. For 3D motion

$$W(k)_{\text{3D Stat. inertial}} \sim k^{-5/2}.$$

It is interesting to repeat their argument: the dynamics of one soliton collapse is given by (8.24) and possesses as invariants the total energy (electric + kinetic) and the total electrostatic energy

$$I_1 = \int d^3r |E^2(r)|.$$

A soliton of size l and field $E(l)$ evolves conserving

$$I_1 \sim E^2(l) l^s.$$

We now use a Kolmogoroff type of argument to state that, in a stationary state, the number flux of solitons in "size-space" must be constant. Let $N(l)$ be the soliton number density in this space. Then

$$N(l)\frac{dl}{dt} = C^{tc} = F$$

and from (8.24) we know that

$$\frac{dl}{dt} \sim (-t)^{\frac{2}{s}-1} \sim l^{1-\frac{s}{2}}$$

and so

$$N(l) \sim F l^{(\frac{s}{2}-1)}$$

It suffices now to change to $k = l^{-1}$ and to assume that all solitons are created with about the same energy (which they retain during their evolution) to see that the energy in the spectral range $(k, k+dk)$ is:

$$W(k)dk \sim N(l)dl \sim F l^{\frac{s}{2}-1} dl \sim \frac{F}{k^{\frac{s}{2}-1}} \frac{dk}{k^2} \sim k^{-(1+\frac{s}{2})} dk.$$

When interaction with particles plays a role, in the dissipative region, the energy is damped at a rate $\gamma(k)$ and the conservation equation (8.25) is modified to:

$$\frac{\partial W(k)}{\partial t} + \frac{\partial}{\partial k}(\dot{K}(k) W(k)) = -\gamma(k) W(k).$$

Galeev *et al.* (1975) use as a damping rate the linear Landau expression and couple it to a *quasilinear* equation for the distribution function. With this crude approximation they show that the epithermal distribution function and

$\gamma(k)$ have power-law dependences on k and v at stationarity, namely

$$W(k)_{\text{Stat. Dissipative}} \sim k^{-(3+\frac{s}{2})}$$

$$f(v)_{\text{Stat.}} \sim v^{-\frac{3s}{2}}, \qquad |v| \gtrsim v_{T_e}.$$

The formation of high-energy tails on the electron distribution has been observed in computer simulation (Galeev *et al.*, 1975; Papadopoulos and Coffey, 1974). It must, however, be clear that the concept of weakly-interacting, individualized, collapsing solitons is debatable, and that, even if interaction with particles turns out to dominate in the dissipation region, the use of Landau damping rates and quasilinear equations may be unjustified.

For one-dimensional motions, solitons do not collapse, but propagate and give rise to a field which is a solution of Zakhavov's equations (Rudakov, 1973), namely

$$E(x, t) = E_m \frac{\sin(kx - \omega t)}{ch(k_0(x - v_g t))};$$

$$\omega = \omega_p\left(1 + \tfrac{3}{2}k^2\lambda_D^2 - \frac{\varepsilon_0 E_m^2}{8nk_B T}\right); \qquad v_g = 3k\lambda_D^2\omega_p;$$

k_0, the inverse size of the envelope is related to E_m by

$$k_0^2\lambda_D^2 = \varepsilon_0 E_m^2/(12nk_B T).$$

Kingsep *et al.* (1973) consider these solitons as weakly-interacting and randomly distributed in one-dimensional space, and calculate the fluctuation spectrum. On the other hand, Kakimov and Tsytovitch (1975) argue that the solitons do not form weakly-interacting structures, but that the process of plasmon condensation is an element which yields a back-reaction on the turbulence outside the region of the "condensate". They have worked out a statistical, one-dimensional theory of waves in this outside medium, which they consider remains the principle place where turbulence is to be found, including the effect of the condensation process on spectral transfer outside condensates. They also find a k^{-2} spectrum like Kingsep *et al.* (1973). Spectra scaling as k^{-2} have been observed in numerical experiments (Thomson *et al.*, 1973).

8.16.3 Anomalous resistance from strong Langmuir turbulence?

Another aspect of strong turbulence deserves mentioning. It has been suggested (Papadopoulos and Coffey, 1974) that the *oscillating two-stream* *instability* could produce enhanced resistivity without meeting the conditions for current-driven instabilities. It suffices that a strong enough beam pervades the plasma. To see this, it is useful to convert the electron-ion drag force (8.10) into an effective collision frequency ν^* by defining

$$\mathbf{F}_{ei} = -\nu^* m_e \mathbf{v}_D,$$

where $\mathbf{v}_D$ is the drift velocity of electrons with respect to ions. Taking the dot product of this relation with $\mathbf{v}_D$ and using (8.10) and (8.11), we obtain:

$$\nu^* = \frac{\varepsilon_0}{n_0 m_e v_D^2} \int \frac{d^3k}{8\pi^3} \int d\omega \, (\mathbf{k}\cdot\mathbf{v}_D) \quad \mathscr{E}(\mathbf{k}, \omega) \quad Im(\tilde{\chi}_e(\mathbf{k}, \omega)).$$

Papadopoulos and Coffey (1974) estimated the parameter-dependence of this frequency under conditions when an electron beam of density n_b, velocity v_b and velocity spread Δv_b is passing through a current-carrying plasma. Neglecting the effect of a magnetic field

$$Im(\tilde{\chi}_e(\mathbf{k}, \omega)) = \frac{k_D^2}{k^2} \frac{\omega - \mathbf{k}\cdot\mathbf{v}_D}{\sqrt{2} k v_{T_e}} \exp - \left(\frac{\omega - \mathbf{k}\cdot\mathbf{v}_D}{\sqrt{2} k v_{T_e}} \right)^2.$$

They consider only low-frequency fluctuations, $\omega \ll \mathbf{k}\cdot\mathbf{v}_D$, the contribution of which dominates. With $\mathbf{v}_D \ll v_{T_e}$ and zero frequency field fluctuations $\mathscr{E}(\mathbf{k}, \omega) = S(\mathbf{k})\delta(\omega)$, this leads in a straightforward way to the expression

$$\nu^* = 4\pi \frac{\omega_p}{n_0 m_e v_{T_e}^2} \varepsilon_0 \int d^3\mathbf{k} \frac{k_D}{k} S(\mathbf{k}).$$

It is convenient to define a mean wave-vector of low-frequency fluctuations such that

$$\int d^3\mathbf{k} \frac{S(\mathbf{k})}{k} = \frac{1}{\langle k \rangle} \int d^3\mathbf{k} S(\mathbf{k}),$$

and to introduce the electrostatic energy density of low-frequency fluctuations, W_{lf}, which is $\varepsilon_0 \int d^3\mathbf{k} S(\mathbf{k})$. Then

$$\nu^* = 4\pi\omega_p \frac{k_D}{\langle k \rangle} \frac{W_{lf}}{n_0 k_B T_e}.$$

The difficult point is to relate W_{lf} to the parameters of the beam.

Papadopoulos and Coffey (1974) estimate it to be a fraction $\langle k\rangle^2/k_D^2$ of high-frequency fluctuations W_0, which are themselves a fraction $\alpha(\approx 1)$ of $m_e n_b v_b \Delta v_b$, a result which they take from simulation results. Then

$$\nu^* \simeq 4\pi\alpha\omega_p \frac{\langle k\rangle}{k_D}\frac{n_b}{n_0}\left(\frac{v_b}{v_{T_e}}\right)^2\left(\frac{\Delta v_b}{v_b}\right),$$

which is to be compared with the quiet plasma collision frequency

$$\nu = \frac{3}{\sqrt{2}}\omega_p \frac{\ln\Lambda}{\Lambda}, \qquad \Lambda = 12\pi n_0 \lambda_D^2,$$

$$\frac{\nu^*}{\nu} = \frac{\sqrt{2}\,48\pi^2}{3\ \ln\Lambda} n_0\lambda_D^3 \frac{\langle k\rangle}{k_D}\frac{n_b}{n_0}\frac{v_b^2}{v_{T_e}^2}\frac{\Delta v_b}{v_b}.$$

It is seen that for anomalous resistivity to be produced it suffices that the beam energy density exceeds a rather small threshold

$$\frac{n_b v_b^2}{n_0 v_{T_e}^2} > \frac{3\ln\Lambda}{48\pi^2\sqrt{2}}\frac{v_b}{\Delta v_b}\frac{k_D}{\langle k\rangle}\frac{1}{(n_0\lambda_D^3)}.$$

These are of course bold estimates, but they are sufficient to show that the idea is worth pursuing.

8.17 PROPOSED MICROTURBULENT ACCELERATION MODELS

8.17.1 Stationary spectra in the presence of Langmuir waves

Models which relate particle acceleration to some microturbulence process are almost all based on the same idea. A primary process is assumed to give rise to some plasma turbulence, either low or high frequency. Non-linear processes are then invoked, which reshape this turbulence spectrum into one which is adequate for particle acceleration, i.e. is broad enough (see §8.11) with phase velocities ranging from several v_{T_e} to some fraction of c. In many models the conversion of low-frequency turbulence into high-frequency turbulence by the bremsstrahlung process has been considered. We have seen earlier (§8.15) that such a mechanism is not possible, at least in the simple situations pictured up to now, and models based on this idea must now be considered as containing the ad-hoc assumption that Langmuir turbulence can indeed be generated (Pikelner and Tsytovitch, 1975; Benz, 1977; Hoyng, 1977b; Kuijpers, 1978).

Usually, the coupling to particles is described by the quasilinear approximation, or it is assumed that tails form, as described in §8.16. The various models differ sometimes in the geometry or physical situation, but these show up very little in the final equations, most of which balance a stochastic acceleration (represented by a quasilinear diffusion operator in velocity or energy space) with some loss process, like escape or Coulomb collisions.

The basic equation of such models can then be written

$$\frac{\partial f}{\partial t}=\frac{\partial}{\partial v}D(v)\frac{\partial f}{\partial v}-\mathscr{L}(f),$$

where $\mathscr{L}(f)$ is a loss operator. The equation may be converted to one for the energy spectrum $N(E)$, and takes the form:

$$\frac{\partial N}{\partial t}=\frac{\partial}{\partial E}D'(E)\frac{\partial}{\partial E}\frac{N(E)}{\sqrt{E}}-\mathscr{L}'(N). \tag{8.26}$$

For example, in the work of Pikelner and Tstovitch (1975) the site of acceleration is assumed to be a large (10,000 km size, 1 m thick) ion-acoustic and Langmuir turbulent current sheet. The Langmuir turbulence is assumed to have reached a power-law shape, with a $k^{-\nu}$ energy spectrum, as a result of strong turbulence processes, and losses are due to escape. The above equation may then be written

$$\frac{\partial N}{\partial t}=\frac{\partial}{\partial E}DE^{1+(\nu/2)}\frac{\partial}{\partial E}\frac{N(E)}{\sqrt{E}}-\frac{N(E)}{\tau_{\text{esc}}},$$

and a stationary solution is:

$$N(E)=\left(\frac{E}{E^*}\right)^{(2-\nu)/4}K_{\nu/(3-\nu)}\left(\frac{4}{3-\nu}\left(\frac{E}{E^*}\right)^{(3-\nu)/4}\right),$$

where $K_{\nu/(3-\nu)}$ is a McDonald function and E^* is some characteristic energy, which can be determined from the ratio of acceleration efficiency D to escape efficiency τ_{esc}^{-1}. This function approximates reasonably a power-law spectrum.

Recently, Pustilnik (1978) has emphasized that a consistent treatment of the model should include anomalous bulk heating by ion-sound turbulence, an effect which he finds to be more important that stochastic acceleration. In the work of Benz (1977), less details are given concerning the origin of Langmuir turbulence, though most of the ideas are basically the same as above. Only the case of a white-noise spectrum ($\nu=0$) is considered, but Coulomb collision losses are included as well as escape; the resulting Equation (8.26) becomes explicitly

$$\frac{\partial N}{\partial t}=\frac{\partial}{\partial E}(D\cdot E)\frac{\partial}{\partial E}\frac{N(E)}{\sqrt{E}}-\frac{1}{2}\frac{\sqrt{2E^{\frac{1}{2}}}}{m^{\frac{1}{2}}L}N(E)+b\frac{\partial}{\partial E}\frac{N(E)}{\sqrt{E}}$$

and a stationary solution is

$$N(E)=\left(\frac{E}{E_*}\right)^{-d+(1/2)}K_d\left(\frac{E}{E_*}\right),$$

where E_* is now another critical energy and d is a small number related to Coulomb losses, $d=b/(2D)$.

It is interesting in Benz's work to see that this picture is consistent with the observational data concerning the correlative variations of X-ray intensity and spectral index (§8.3). Actually, for energies large compared to E_*, the stationary solution of Benz has a local power-law index of

$$\delta\sim d+\frac{E}{E_*}\sim\frac{E}{E_*}.$$

Hence, for a given E, the local power-law index depends on E_* mainly, which scales with L and the turbulent energy density W_0 (Benz, 1977) as

$$\delta\sim\frac{E}{E_*}\sim(W_0L)^{-1/2};$$

whereas the X-ray emission in the thick-target picture is proportional to the electron flux divided by the photon spectral index $\gamma^{3/2}$. Benz finds:

$$I_x\sim(W_0L)^{7/4},$$

which anticorrelates correctly with the spectral index δ. It is fair to say, however, that this diffusion-loss theory is not the only one able to reproduce this anticorrelation; it also exists in the quasithermal model (Eq. (8.4) and Mätzler *et al.*, 1978). It is also shown by Benz that an observed small phase-shift (see §8.3) between these two quantities may be understood as due to a delayed variation of the spectral index, namely the small part d, which depends on ionization losses and on the slowly-varying temperature of the background plasma. The model of Hoyng (1977b) does not give many details on the particle spectrum reached, but rests rather heavily on the Langmuir bremsstrahlung process; this specific and rather detailed analysis of shock-wave acceleration is now in our opinion obsolete (see §8.15).

It still remains conspicuous that, in all the above-mentioned models, the generation of plasma microturbulence is a non-trivial problem, which will have to be solved in more specific descriptions of the flaring region. If these turn out to be far too difficult to build, it will be necessary to propose models which do not rely so much on microturbulence, involving DC electric fields. An alternative may be produced by double-layers, though their existence requires conditions almost as difficult to meet. We first give a description of these plasma states, which are particle accelerators to be considered on the same footing as turbulent states, and then proceed to a discussion of how and where such states could arise.

8.18 DOUBLE-LAYERS

8.18.1 Early concepts

In microturbulent acceleration processes, it is assumed that the instability of interest develops into a random state of fluctuation. It is not *a priori* evident that this should be so. A school of thought has grown up mainly under the influence of Alfvén, who examined the possibility that a plasma subject to a large enough current density develops an organized potential structure able to accelerate particles merely by the effect of a self-consistent DC electric field. The mechanism by which these coherent structures could form is by no means clear, even now. In their first communication on this question, Alfvén and Carlqvist (1967) argued that if (in a current flowing in a plasma) a density dip arises, while the inductance of the total circuit is large enough for the current to be maintained, a DC electric field must appear to adjust the velocity of the electron flow in such a way that the current density remains constant:

$$en(x)v(x)=j.$$

The corresponding electrostatic force density is found from the equation of motion and the potential drop between the edge and the centre of the density dip is then found by integration as

$$\Delta U=-\frac{mj^2}{2e^3}\left(\frac{1}{n_{\min}^2}-\frac{1}{n_0^2}\right).$$

Let L be the size of the dip. Poisson's equation implies a relation between ΔU, L, and the charge density near minimum. Keeping in mind that this charge imbalance cannot exceed $n_{\min}$ (complete charge separation), we come to the conclusion that (for given L, n_o, j) $n_{\min}$ cannot be less than a certain value. If this minimum is reached, one expects a time-dependent regime to begin, with a displacement current $\partial E/\partial t$. Under the influence of this field, the edges of the dip will acquire a net charge, and the potential drop ΔU will become asymmetrical. These arguments are only meant to justify intuitively the idea that large DC potential drops could form and remain in a current-carrying plasma. It has proved possible, however, to demonstrate rigorously the existence of self-consistent solutions of the Vlasov equations able to sustain large potential drops, and then to accelerate electrons and possibly ions to large energies.

8.18.2 BGK modes or "Buneman double-layers"

The most straightforward example consists of the so-called *BGK solutions* (Bernstein *et al.*, 1957). These describe stationary one-dimensional solutions to the Vlasov equation in a non-magnetized plasma. Let f_+ and f_- be the distribution functions of ions (of mass M and charge e) and electrons (of mass m and charge $-e$). The problem consists of finding a solution of the coupled system

$$v\frac{\partial f_+}{\partial x}+\frac{e}{M}E\frac{\partial f_+}{\partial v}=0, \tag{8.27a}$$

$$v\frac{\partial f_-}{\partial x}-\frac{e}{m}E\frac{\partial f_-}{\partial v}=0, \tag{8.27b}$$

$$\varepsilon_0\frac{\partial E}{\partial x}=n_0 e\left(\int f_+\mathrm{d}v-\int f_-\mathrm{d}v\right), \tag{8.27c}$$

where n_0 is some average density (f_+ and f_- are then not normalized). The first two equations are satisfied if we take for f_+ and f_- some functions of the first integrals of the motion, in particular, of the energy. Let us introduce a function ψ, quite similar to the electrostatic potential, such that

$$E=-\frac{\mathrm{d}}{\mathrm{d}x}\left(\frac{m}{2e}\psi(x)\right).$$

Also, the quantities ε_+ and ε_-, which are proportional to the energies of ions and electrons are

$$\varepsilon_-=v^2-\psi(x), \qquad \varepsilon_+=\frac{M}{m}v^2+\psi(x). \tag{8.28}$$

A solution of (8.27a) and (8.27b) is

$$f_+(x, v) = f_+(\varepsilon_+); \qquad f_-(x, v) = f_-(\varepsilon_-).$$

The equation (8.27c) then provides the equation which relates the functions $f_+(\varepsilon_+)$ and $f_-(\varepsilon_-)$ to the still unknown potential $\psi(x)$:

$$\frac{d^2f}{dx^2} = -2\omega_p^2(\int (f_+ - f_-)dv).$$

The integral on the right can be expressed as a function of ψ by changing to the variables $\varepsilon_\pm$. Some care must be exercised in defining the bounds between which this integral runs: this is because v^2 must be positive, and so ε_- can run only between $-\psi(x)$ and $+\infty$, while ε_+ runs between $+\psi(x)$ and $+\infty$.

$$0 = \frac{d^2\psi}{dx^2} - \omega_p^2\left(\sqrt{\frac{m}{M}}\int_{\psi}^{\infty} \frac{f_+(\varepsilon_+)d\varepsilon_+}{(\varepsilon_+ - \psi)^{1/2}} - \int_{-\psi}^{+\infty} \frac{f_-(\varepsilon_-)d\varepsilon_-}{(\varepsilon_- + \psi)^{1/2}}\right).$$

The calculation of ψ can be brought to a closed form by noting that the second term is actually an exact derivative with respect to ψ; we get

$$0 = \frac{d^2\psi}{dx^2} + 2\omega_p^2 \frac{d}{d\psi}\left(\sqrt{\frac{m}{M}}\int_{\psi}^{\infty} f_+(\varepsilon_+)\sqrt{\varepsilon_+ - \psi}\,d\varepsilon_+ - \int_{-\psi}^{\infty} f(\varepsilon_-)\sqrt{\varepsilon_- + \psi}\,d\varepsilon_-\right),$$

which integrates, after multiplication by $d\psi/dx$, to

$$\left(\frac{d\psi}{dx}\right)^2 + 2\omega_p^2\left(\sqrt{\frac{m}{M}}\int_{\psi}^{\infty} f_+(\varepsilon_+)\sqrt{\varepsilon_+ - \psi}\,d\varepsilon_+ - \int_{-\psi}^{\infty} f_-(\varepsilon_-)\sqrt{\varepsilon_- + \psi}\,d\varepsilon_-\right) = C^{tc}. \tag{8.29}$$

The second term is a known function of ψ once f_+ and f_- are known, and the problem of defining the self-consistent potential profile is then reduced to some quadrature of the form:

$$\left(\frac{d\psi}{dx}\right)^2 = V(\psi) \quad \text{or} \quad dx = \pm d\psi(V(\psi))^{-1/2}. \tag{8.30}$$

The reader will easily convince himself that the shape of $\psi(x)$ depends essentially on the number and multiplicity of the zeroes of $V(\psi)$. Two single zeroes give a periodic (non-sinusoidal) ψ, one single and one double zero gives a single hump or dip, and two double zeroes give a monotonic shape, changing from one value ψ_1 at $-\infty$ to another ψ_2 at $+\infty$. This is called a *double-layer*, because the space charge, proportional to $d^2\psi/dx^2$, essentially constitutes two sheets of opposing polarity. One could say that a double-layer is a "self-consistent capacitor".

Any Bernstein–Green–Kruskal (*BGK*) solution to the Vlasov equation involves different populations of particles: passing (or free) particles, for which v never vanishes ($\varepsilon_+ > \psi_{\max}$ or $\varepsilon_- > -\psi_{\min}$, respectively) and trapped (or reflected) particles, for which v vanishes at some point. In the case of a double-layer with a potential increasing from 0 at $-\infty$ to ψ_1 at $+\infty$, ions with energy less than ψ_1 will be confined to some left-hand side region of the axis (see Eq. (8.28)), while electrons with total energy between $-\psi_1$ and 0 will be found only in some right-hand side region of the axis (see Eq. (8.28)). The phase-space trajectories for electrons and ions will then look as shown in Figures 8.22a and 8.22b. It is seen that the boundary conditions on one side of the double-layer (say $x = -\infty$) do not suffice to determine entirely $f_+(\varepsilon_+)$ and $f_-(\varepsilon_-)$; for example, here one would miss the distribution of trapped electrons. It is possible to show that this indetermination can be exploited to build a self-consistent solution for almost any prescribed potential $\psi(x)$. It suffices to show that one can obtain some trapped particle distribution from an integral equation deduced from (8.29) (Bernstein *et al.*, 1957).

8.18.3 Conditions for the existence of accelerating double-layers

The double-layer becomes an accelerator if the potential drop greatly exceeds the thermal energy of electrons and ions in both "left" and "right" media. This implies that no particle is able to pass the layer against the field, so that the Figures 8.22a and 8.22b should contain essentially no "left free-ions" or "right free-electrons". They should appear more like Figures 8.23a and 8.23b. It is seen that the system necessarily carries a net current in that case (net electron flow to the right, and net ion flow to the left). A double-layer is said to be *strong* when

$$e\varphi_1 \gg k_B T.$$

General properties of double-layers have been reviewed by Block (1978).

A first condition for double-layers to exist under natural conditions is, of course, that the system be just able to provide the reflected particle distributions. This is not always the case, though back-scattering or magnetic mirroring may provide the mechanism. It is nevertheless worth mentioning

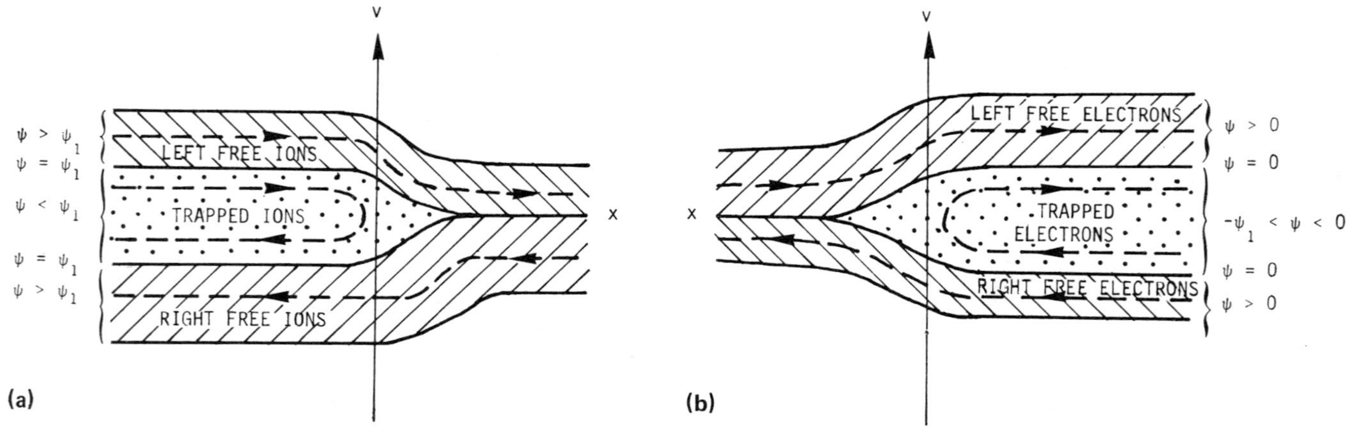

FIGURE 8.22 (a) Phase-space diagram of ions in a double-layer. (b) Phase-space diagram of electrons in a double-layer.

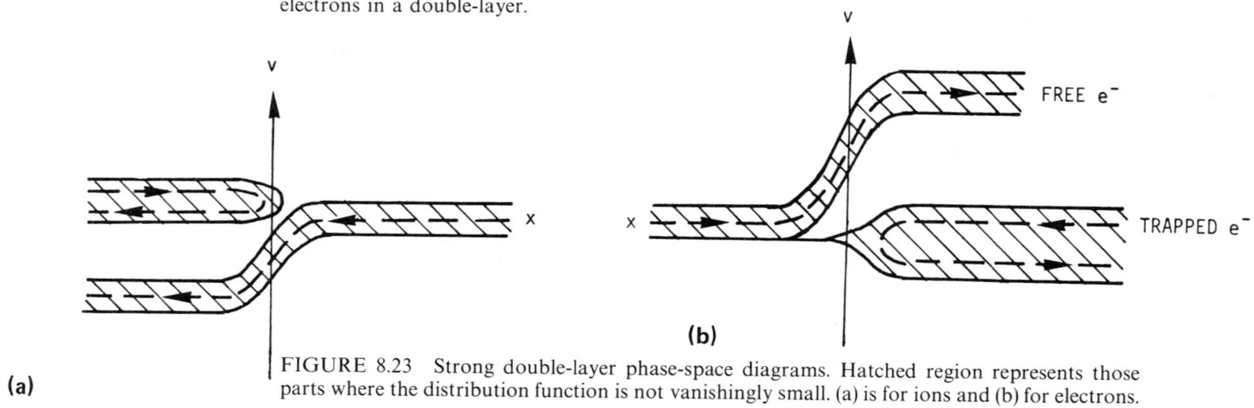

FIGURE 8.23 Strong double-layer phase-space diagrams. Hatched region represents those parts where the distribution function is not vanishingly small. (a) is for ions and (b) for electrons.

that, in the experiments of Quon and Wong (1976), reflected ions had to be supplied by the biased end-cap in their device in order for the layer to form: while, in the case of numerical simulations, the formation of a double-layer seems sensitive to the way the boundary conditions are handled.

A second condition is implied by the fact that the electric field vanishes at infinity, so that the double-layer, though it sustains a charge separation by a balance between inertia and electrostatic forces should still be globally neutral. This implies that in strong layers there is a relation between the ion and electron flux through it. This is easily seen by neglecting pressure forces so that the passing particles can be treated by fluid equations, which for species s are

$$\frac{\mathrm{d}}{\mathrm{d}x} n_s v_s = 0,$$

$$n_s m_s v_s \frac{\mathrm{d}v_s}{\mathrm{d}x} = n_s q_s E,$$

$$\varepsilon_0 \frac{\mathrm{d}E}{\mathrm{d}x} = \sum_s n_s q_s,$$

from which it can be deduced that

$$\frac{\mathrm{d}}{\mathrm{d}x}\left(\sum_s n_s m_s \frac{v_s^2}{2} - \frac{1}{2}\varepsilon_0 E^2\right) = 0.$$

Using this equation far from the layer, where $E=0$, one gets

$$(n_e m_e u_e^2)_{+\infty} = (n_i M_i u_i^2)_{-\infty}.$$

But, because the particles have been accelerated in the same potential,

$$(m_e u_e^2)_{+\infty} = (M_i u_i^2)_{-\infty},$$

so that

$$(n_e u_e)_{+\infty} = \sqrt{\frac{M_i}{m_e}}(n_i u_i)_{-\infty}.$$

This is the *Langmuir condition* (Langmuir, 1929; Block, 1978), from which one can calculate the current density

$$j = -e n_e u_e (1 + \sqrt{m_e/M_i}).$$

A *third condition* for the existence of the double-layer involves the temperature of the particles. Consider, for example, the left-hand side negatively-charged part of the layer. The most dense populations there are the trapped ions and the free electrons, which have not yet been accelerated. In this region, the number of these particles varies as:

$$n_{it} \sim \exp - \frac{e\varphi}{k_B T_{it}},$$

$$n_{e_{\rm free}} \sim u_e^{-1} \sim \left(u_e^2(-\infty) + \frac{2e\varphi}{m_e} \right)^{-1/2}.$$

To maintain a negative charge, n_{it} must decrease faster than $n_{e_{\rm free}}$, though they are roughly equal for $\varphi = 0$. An expansion for small φ gives

$$1 - \frac{e\varphi}{k_B T_{it}} < 1 - \frac{e\varphi}{m_e u_e^2(-\infty)},$$

or

$$m_e u_e^2(-\infty) > k_B T_{it}.$$

This shows that the electron drift speed $u_e(-\infty)$ must exceed the thermal threshold, or, because the electron flux dominates, one can state that, the electric current driven through the layer must exceed

$$j^* = n_e e \left(\frac{k_B T_{ti}}{m_e} \right)^{1/2}.$$

This is called the *Bohm condition.* In this sense, the *condition for the existence of the double-layer is similar to that for the occurrence of the Buneman instability.* It has therefore been conjectured that this electrostatic structure could develop as a result of the Buneman instability. This threshold is *very high* under solar conditions, and for this reason it is *very unlikely that such a double-layer could ever form,* if only because some current-driven microinstabilities have lower thresholds. It is then interesting to know whether these could develop similar non-linear structures.

8.18.4 Oblique (or "ion-cyclotron") double-layers

We quote now some important results obtained in this direction by Swift (1975), who considered the possibility of the existence of electrostatic double layers forming a plane whose normal makes some angle θ, near 90°, with the magnetic field. This is a natural question to ask because, in a state of perfect conductivity, magnetic field lines should be equipotential. The theory is in principle identical to that outlined above, except for the ion motion which has a drift component. Swift has obtained approximate first integrals of the motions, under the assumption that they are adiabatic. The electrons are simply assumed to move along field lines. In that case $(v_{\|}^2 - 2e\varphi/m_e)$ is a constant of the motion, and the calculation of electron density proceeds just as in the simple one-dimensional theory. For the ions, the motion is expanded around the guiding centre. Swift assumes that all quantities in the double-layer (he uses the term *electrostatic shock*) are functions of the distance η to the layer

$$\eta = z - x \operatorname{tg} \theta,$$

where $\mathbf{e}_x$, $\mathbf{e}_y$, $\mathbf{e}_z$ form an orthonormal frame, with $\mathbf{e}_z$ along $\mathbf{B}$ and $\mathbf{e}_x$ in the plane of $\mathbf{B}$ and the shock normal $\mathbf{n}$. The electric field is then entirely in the $(\mathbf{e}_x, \mathbf{e}_z)$ plane. The shock thickness is assumed to be much larger than an ion Larmor radius. Expansion to first order in the corresponding small parameter gives the ion motion (Swift, 1975). The motion of the guiding centre consists of two drifts, while the gyration motion proceeds at constant Larmor radius, but with a slightly time-dependent gyration frequency.

$$\tilde{\Omega} = \left(\Omega^2 - \frac{e}{M_i}\frac{\partial E_x}{\partial x}\right)^{1/2}.$$

An obvious first-integral of the motion is then the Larmor radius:

$$\rho = v_\perp \Bigg/ \left(\Omega^2 - \frac{e}{M_i}\frac{\partial E_x}{\partial x}\right)^{1/2}.$$

The total energy of the ion is also conserved. It includes the gyration motion, but can be translated (using the constancy of ρ) into a law of conservation following the guiding centre, namely

$$\tfrac{1}{2}M_i v_{\|}^2 + \tfrac{1}{2}M_i \frac{E_x^2}{B^2} + e\varphi = K.$$

In the adiabatic approximation it is then possible to write the solution of the Vlasov equation for ions as:

$$f(v_{\|}, v_\perp, x, z) = f_+\left(\left[v_{\|}^2 + 2e\frac{\varphi}{M_i} + \left(\frac{\partial\varphi}{\partial x}\right)^2 B^{-2}\right]^{1/2}, v_\perp\left(1 - \frac{e}{M_i z}\frac{\partial E_x}{\partial x}\right)^{1/2}\right).$$

The ion charge density can then be obtained, not only as a function of the potential φ, but also of its first and second derivative in x. These dependences can be treated as a perturbation, which means that the drift velocities are considered small with respect to $v_{\|}$ and $v_{\perp}$. It is then possible to express the Poisson equation explicitly as a differential equation relating the potential $\varphi(\eta)$ to the functions f_+ and f_-, which describe the electron and ion distributions. One obtains the equation (Swift, 1975)

$$(1+tg^2\theta\ (1+c^2/v_A^2(\varphi))\frac{d^2\varphi}{d\eta^2}+\frac{1}{2}tg^2\theta\frac{d\varphi}{d\eta}\frac{d}{d\eta}(1+c^2/v_A^2(\varphi))$$

$$=-\frac{e}{\varepsilon_0}(n_+(\varphi)-n_-(\varphi)), \qquad (8.31)$$

where

$$n_+(\varphi)=n_0\int dv_{\|}\int v_{\perp}dv_{\perp}f_+\left(\sqrt{v_{\|}^2+2e\frac{\varphi}{M_i}},\ v_{\perp}\right),$$

$$n_-(\varphi)=n_0\int dv_{\|}\int v_{\perp}dv_{\perp}f_-\left(\sqrt{v_{\|}^2-2e\frac{\varphi}{m_e}},\ v_{\perp}\right), \qquad (8.32)$$

$$v_A^2(\varphi)=B^2/(\mu M_i n_+(\varphi)).$$

The Equation (8.31) can be reduced to the form (8.30), and then again double-layer solutions can be constructed. These solutions have the interesting property that the layer thickness is several ion gyroradii, unlike the "Buneman-type" double-layers of the one-dimensional theory, which are only some Debye-lengths broad; the potential jump can take arbitrarily high values as before. Swift discusses the nature of the unstable waves which could possibly give rise to this sort of double-layer, and finds that the nonlinear equation (8.31) has much in common with the dispersion equation of electrostatic ion cyclotron waves. To our knowledge, however, it has never been proved (even numerically) that such a structure could be the result of the corresponding current-driven instability.

8.18.5 Double-layer formation

A most difficult problem is the formation of double-layers. Detailed numerical studies have been made for the one-dimensional case. Some support the idea that such structures spontaneously develop as a result of the Buneman instability. The coherent development of the instability of two counter-streaming electron populations has been observed by Berk and Roberts (1967) in a water-bag numerical experiment, which approximates the distribution function by constant values in variable domains of phase-space. On the other hand, the results of Buneman (1959) do not clearly show any such trend. Recent results with light ions ($M/m=16$) made by Goertz and Joyce (1975) definitely indicate that a double-layer forms for an electron drift velocity larger than the thermal velocity. Smith and Priest (1972) on the other hand have criticized the hypothesis that the instability would develop coherently. They noted that laboratory experiments using completely-ionized plasmas rather show intense electron heating and anomalous resistivity, and expressed the opinion that in any case the Buneman instability should turn into an ion-acoustic one once the electrons are heated and no particles could be accelerated. We know that this is not so (see §8.11). Moreover, double-layers have now been observed in some experiments (Quon and Wong, 1976).

It still remains quite a puzzle to understand *how an instability which would encompass a rather large fraction of the current-carrying circuit could evolve into such a localized structure.* We recall that the one-dimensional theory predicts the layer to be only some ten Debye lengths broad. The answer of Swift (1975) to that question is that, if the electrostatic shock is of the kind he described, the size along the field lines need not be infinitesimal, being of the order of several tens ion gyroradii divided by $\cos\theta$. Another idea for explaining this localization has been investigated by R. Smith and Goertz (1978), who studied the possibility of double-layer formation in the Io–Jupiter current circuit. These authors include in their analysis the inhomogeneity of the medium where the current instability is to take place. This inhomogeneity causes the current-driven instability to occur in only a portion of the current circuit, say for $x>0$ in terms of some curvilinear coordinate x along a current line. Let W be the level of waves and W_{sat} the level reached at saturation, v_* the instability threshold of the drift velocity of electrons at $x=0$ where $n=n_0$, and $v(x)$ and $n(x)$ the drift velocity and density elsewhere. One then has $n(x)\,v(x)=n_0 v_*$ and the growth of the instability can be pictured using an amplitude-dependent growth rate

$$\gamma=\gamma_0\left(1-\left(\frac{v^*}{v}\right)^{\alpha}\right)\left(1-\frac{W}{W_{sat}}\right)H(x),$$

where H is the Heaviside function. Assuming the density to change slowly as $(1+\varepsilon x)^{-1}$, and the group velocity of the unstable waves to scale as the drift velocity of electrons ($v_g(x)=bv(x)$), it is possible to write an equation for the level of waves which qualitatively includes inhomogeneity and saturation effects

$$\frac{\partial W}{\partial t}+bv(x)\frac{\partial W}{\partial x}=\gamma_0\left(1-\frac{1}{(1+\varepsilon_x)^\alpha}\right)\left(1-\frac{W}{W_{\text{sat}}}\right)H(x)W.$$

This equation is shown by R. Smith and Goertz (1978) to evolve $W(x)$ more and more like some Heaviside function, developing very steep gradients on the unstable side near the origin. As a consequence, a ponderomotive force (8.6) will be exerted on ions

$$\mathbf{F}_{\text{pond}}=-\frac{1}{m}\boldsymbol{\nabla}(\varepsilon_0\langle E^2\rangle).$$

Electrons respond readily, and ions follow by an ambipolar field effect. At equilibrium the electrostatic force in the region of turbulence gradient balances the ponderomotive force. In other words, a DC charge separation is set up. This structure is not yet a double-layer (the electric force is not at all balanced by inertia in this region) and it remains to be studied how it could evolve into such an electrostatic shock.

8.18.6 Relevance to Flares?

In conclusion, it may be said that double-layers, as they appear in the present state of the theory, *are not very likely to be relevant to flares, because of their high thresholds, because of stability problems and because their formation is rather obscure.* Future developments along the lines opened by Swift (1975) and R. Smith and Goertz (1978), however, could prove more interesting. In this respect, it is worth remembering that the presence of such structures in the high auroral ionosphere is very probable, though the controversy still exists on the question of whether anomalous resistance alone can explain the observations. More can be read on this in Block (1978).

8.19 APPROACHING MICROTURBULENT CONDITIONS: A DIFFICULTY

8.19.1 Numerical values of instability thresholds are high

As has now been realized, most of the models discussed previously require the existence of a *very strong current density* either for *ion sound turbulence* triggering (in neutral sheets or in any other structure) or for reaching the large current densities needed for the existence of *double-layers.* All require electron drift velocities between v_{T_e} and v_{T_i}. The fact that the actual value may vary by a factor 40 is not extremely important, as we shall soon see. Let us adopt for the sake of definiteness the value v_{T_e}, the critical current being now

$$J_* = n_e q v_{T_e};$$

with $n_e = 10^{17}\,\mathrm{m}^{-3}$ $(=10^{11}\,\mathrm{cm}^{-3})$ and $T = 10^6$ K, we find $J_* \simeq 6.4 \times 10^4\,\mathrm{A\,m}^{-2}$ and is weakly dependent on temperature as $T^{-1/2}$ for given pressure. This current density is huge as compared to those which have been measured to leave the photosphere ($10^{-1}\,\mathrm{A\,m}^{-2}$ according to Moreton and Severny (1968) at most 1 to 10 $\mathrm{A\,m}^{-2}$ (Title and Andelin, (1971)). It is fair to say, however, that the more spatial resolution, the higher these figures could be. From Ampere's law J_* corresponds to a field-reversal of approximately 1000 Gauss on a distance scale of 1 meter only! This has an extremely important consequence: *if current-driven instabilities are to be the primary source of particle acceleration, the currents must be extremely concentrated and cannot pervade a large fraction of the magnetic structure of interest.*

The recognition of this leads to different kinds of primary triggering theories:

1) If the currents are initially spread through the structure, one has to produce a specific theory for their concentration.

2) Concentrated currents may exist from the start: this is so, for example, in all models involving shocks or current layers, either in a state of stationary reconnection or not (Chapter 3).

3) Current concentrations may occur naturally during the development of some global MHD or resistive instability. The cause of triggering in this sort of model is the MHD dynamic process itself. Such models include the various kinds of tearing mode theories, as well as the global equilibrium theories reviewed by Birn and Schindler in Chapter 6.

Little has been done in models of the third kind, and in particular for global equilibrium theories, to specify how particles could be accelerated. It may involve a reconnection process driven by the MHD motion; both situations 2) and 3) then amount to a discussion of current layers in various geometries.

8.19.2 Mechanisms for current concentration

Few mechanisms for concentrating currents if they are not already concentrated initially have been discussed in the literature. Most likely is perhaps a slow concentration mechanism of a thermal nature, related to the fact that the hotter a plasma is, the more conducting it is. Currents tend then to concentrate in small more-conducting threads, thus enhancing the dissipation.

This instability is the so-called *superheating instability* (Kadomtsev, 1966). Its application in the context of solar physics (i.e. including specific heating and loss processes relevant for that case) has been examined in detail by Heyvaerts (1974), who has thoroughly analysed the modes of an MHD current-carrying plasma, including heat conduction, Joule and wave heating and optically-thin radiation losses of a collisionally-excited plasma (Raju, 1968) in the local short wavelength approximation (§3.2.4.B). The analysis does not include stability studies of coronal arches as a whole, a subject which is considered in Chapters 4 and 6, but it gives a fair account of the MHD and thermal microstructure of the plasma. It is found that the usual MHD modes (Alfvén, fast and slow) are modified by thermal and finite-resistivity effects and that new modes essentially describe thermal effects (Field, 1965) and magnetic diffusivity.

The mode of interest to us here is the so-called *Joule mode*, which describes the relaxation of a temperature inhomogeneity. Its frequency is given by (Heyvaerts, 1974).

$$\omega = -i(\gamma-1)\left(\rho^2 \frac{Q(T)}{p}\frac{d\log Q}{d\log T} + \frac{\mathbf{k}\cdot\bar{\bar{\boldsymbol{\kappa}}}\cdot\mathbf{k}}{p}T - \frac{J^2}{\sigma p}\frac{d\log\sigma}{d\log T}(\sin^2\theta - \cos^2\theta)\right),$$

where Q is the temperature-dependent part of radiative losses ($\rho^2 Q(T)$), $\bar{\bar{\boldsymbol{\kappa}}}$ is the tensor of thermal conduction, J the current density pervading the medium and p the pressure. Growth would correspond to $\mathrm{Im}(\omega)$ positive. This mode is dominated by heat conduction if $(\mathbf{k}\cdot\mathbf{B})$ is not very small, but its physical character changes in perpendicular propagation when field-aligned thermal conduction becomes ineffective. It represents in that case the combined effects of temperature relaxation by radiation and Joule heating (i.e. field diffusion). The frequency becomes then

$$\omega = -i(\gamma-1)\left(\rho^2 \frac{Q(T)}{p}\frac{d\log Q}{d\log T} - \frac{J^2}{\sigma p}\frac{d\log\sigma}{d\log T}\right).$$

The second term is destabilizing, but for growth to occur the supplement of Joule heating must overwhelm the supplement of radiative cooling represented by the first term. This occurs only for large enough current densities, or in situations when the first term is by itself destabilizing. Assuming, as is reasonable, that the current-carrying plasma has already reached a state of thermal stability, say $T = 10^5$ K, it can be seen, for chromospheric conditions and $J_0 \sim 10^{-1}$ Am^{-2} (already quite high), that $J^2/(\sigma p)$ is much less than the first term (by a factor 10^2). One then needs very high current densities, though admittedly smaller than the large micro-instability thresholds, for this process to occur; any even modest increase in effective resistivity, however, may change the situation. The growth-rate, when the Joule effect acts alone may be of the order of $10^{-4}\ s^{-1}$ for normal conductivity. We insist, however, that it would be a circular argument to invoke already anomalous resistivity to justify the effectiveness of this current-concentration mechanism, which is itself supposed to realize the conditions for microinstability to occur! This current-filamentation will ultimately be saturated by perpendicular conduction if it ever turns unstable, or by temperature-gradient-driven microinstabilities. The physics described here is the solar version of the superheating instability of Kadomtsev (1966), and its possible role in solar flares have been incorporated in the model of Spicer (1978), who, however, reaches more optimistic conclusions than ours. A non-linear study has been undertaken by Sokolov and Kosovichev (1978). We feel it difficult to believe that the peaking of the current density profile could proceed up to the very high current-concentrations needed for microinstabilities, though perhaps it could lead up to a situation where tearing-mode instability becomes fast enough, as advocated by Spicer (1978); but this remains to be shown in detail.

The suggestion that microturbulence may be triggered as a consequence of fast MHD or thermal charges, particularly in current sheets, is discussed in §3.5.

8.20 TEARING MODES AND FAST PARTICLES

8.20.1 Energy budget considerations

Little is known concerning the way by which magnetic energy is transferred into particle energy in a *tearing-mode instability* §4.5, and even more on how preferential acceleration of a select number of particles could arise. Normally, the magnetic energy is converted during the development of the instability mainly into heat and kinetic energy, but at the same time large DC-electric fields appear in the structure, which may operate either along the magnetic field or in neutral regions (where the particle's motion is not adiabatic). One can conceive of two different possibilities for energizing particles in such a process: DC acceleration or thermalization of liberated free energy. Thermalization can even itself be quite an effective process, as discussed already earlier (§§8.3, 8.9, 8.10), though this depends to some important extent on the boundary conditions and not only on the "local" physics. The general argument is as follows. Assume that, as a result of the tearing-mode instability, the magnetic field is rearranged in the volume V, and that in this process a fraction ε of the total magnetic energy is dissipated and shared between a certain number N of particles. We assume for the moment, consistent with the heating picture, that this energy is equally-shared between the electrons, by contrast with what occurs in a properly-called acceleration mechanism, for which the sharing is unequal. In that case, each electron receives a mean energy

$$\langle E\rangle = V\varepsilon\frac{B^2/2\mu}{N}.$$

All the discussion is on how to estimate ε and N. An efficiency of several percent seems typical, say 10%. The number of particles involvedin the dissipation depends on the geometry and on loss processes, in particular thermal conduction. The energy gain is optimized if we admit that no heat is lost, which was shown in §8.10 to be possible evrn for extreme heating rates, thanks to the anomalous conduction effect. In that case

$$\langle E\rangle = \varepsilon V\frac{B^2/2\mu}{V\langle n\rangle} = \varepsilon\frac{B^2}{2\mu}\frac{1}{\langle n\rangle},$$

where $\langle n\rangle$ is a mean value of the density in the region of interest, which may be strongly inhomogeneous, however. It is seen, that, if the region is more or less homogeneous,

$$\langle E\rangle = \varepsilon\frac{k_B T}{\beta},$$

where T is the average temperature and β is the ratio of kinetic to magnetic pressure. In the normal coronal plasma $T = 10^6$ K (i.e. 100 eV approximately). In order to rise to 100 keV average energy, the dissipation should then proceed in a low-β plasma, namely

$$\beta < 10^{-5}\varepsilon \simeq 10^{-6}.$$

This implies a rather high average magnetic field, of the order of 500 Gauss,

$$B \simeq (10^6 2\mu n k_B T)^{1/2} \simeq 0.5\ T_6^{1/2} n_{16}^{1/2}\ \text{Tesla}.$$

This reasoning is quite general, in the sense that it holds for any heating process, but it is valid in an average sense only. It does not exclude the possibility that some plasma with $\beta = 1$ participates in the reconnection process, but such regions must be a minor fraction of the region concerned by magnetic rearrangement. For example, Heyvaerts and Priest (1976) estimate that, if the reconnecting current sheet of a Petchek flow suddenly switches to a state of anomalous resistivity, σ^*, the field rearrangement will involve a region around the current sheet comparable to the anomalous diffusion thickness, say Kl where l was the thickness prior to the change and $K = \sigma/\sigma^*$. In that case, the number of electrons involved in the field dissipation is, for a perturbulent current sheet of thickness l and surface L^2,

$$\mathcal{N} = n_0 L^2 l + n_\infty L^2 l(K-1),$$

while the energy available for dissipation is

$$\mathcal{E} = \varepsilon K L^2 l(B^2/2\mu).$$

Then

$$\langle E \rangle = \mathcal{E}/\mathcal{N} = \varepsilon \frac{B^2}{2\mu} \frac{K}{n_0 + (K-1)n_\infty}.$$

According to the value of K, $\langle E \rangle$ may be between $\varepsilon k_B T_0$ for $K=1$ (where T_0 is the temperature of the perturbulent current sheet) and $\varepsilon k_B T_\infty/\beta_\infty$ for large K, as in the low $-\beta$ case. If one estimates σ^* from an effective collision-frequency of twice the Buneman frequency, $\omega_p(m_e/M_i)^{1/3}$, it is found that:

$$K\frac{n_\infty}{n_0} = \frac{\sigma}{\sigma^*}\frac{n_\infty}{n_0} = \frac{2\sqrt{2}}{3}\left(\frac{m_e}{M_i}\right)^{1/3} \frac{\Lambda}{ln\,\Lambda}\frac{n_\infty}{n_0}$$

$$= \frac{2\sqrt{2}}{3}\left(\frac{m_e}{M_i}\right)^{1/3} \frac{\lambda_D^3 n_\infty}{\ln \Lambda(n_0)} = \frac{\Lambda_\infty}{\ln \Lambda(n_0)}\left(\frac{T_0}{T_\infty}\right)^{3/2}.$$

This number is then necessarily very large, so the estimates above still hold, with all quantities referring to the medium faraway from the neutral sheet in which $\beta = 1$

$$\langle E \rangle \sim \varepsilon \left(\frac{k_B T_\infty}{\beta_\infty}\right).$$

(It was incorrectly stated in Heyvaerts *et al.* (1977) that the internal energy of the current sheet was enough to explain keV energization: a broadening of the dissipation is necessary, as claimed in an earlier paper (Heyvaerts and Priest, 1976).)

We then see that *bulk heating may perfectly explain the acceleration of keV electrons, provided it occurs in an essentially low-β plasma with a strong enough field.* Tearing modes, or more generally any type of localized reconnection process, may achieve this. For the tearing-mode theory of bulk heating to be satisfactory, however, it is necessary that the structure be driven to instability without noticeable effects occurring before, and also that the dynamic evolution (which liberates energy) operates fast enough.

8.20.2 Tearing modes and Dreicer field

It still remains to understand why and how the observed particle spectra are formed in this process. The evidence for power-law spectra existing at the source is not compelling, as discussed in §8.9, though some observations in space are so strictly power-law-shaped that one sometimes find it difficult to explain them by a superposition of thermal spectra (Lin, 1979, private comm.). Perhaps such power-law distributions could be the result of the rearrangement of some beam distribution by a collective plasma effect, as discussed earlier (Smith, 1977a, 1977b). Now the question arises *whether such beams may be formed as a result of the tearing instability.*

Tearing perturbations have the property to create induced electric fields parallel to the magnetic field, or, if a reverse sheet pinch is considered, to create intense electric fields in the region where there is no magnetic field. If one finds oneself in the collisionless regime of particle motion, such situations lead directly to particle acceleration. On the other hand, if the collisional regime is the relevant one, one may discuss the possibility of DC acceleration by comparing the induced field to the Dreicer field E_D. For fast modes, which exponentiate up to very high amplitudes, it is possible to exceed the Dreicer field. In both cases the problem with these mechanisms is that they arise from microinstabilities which will surely develop and modify in a way for which it is not easy to estimate both the growth-rate of the tearing mode and the particle's motion. Moreover, the total volume where large electric fields give rise to DC acceleration instead of $\mathbf{E} \times \mathbf{B}$ drift is very small.

The volume may range from an insignificant size if the reconnection has evolved to a stationary Petchek flow at high temperature, up to large sizes if *magnetic braiding* has occurred (Rochester and Stix, 1976), but in this latter case the estimate of the volume is almost impossible to make. The weak point of many such theories resides then in the fact that the number of particles which can be preferentially accelerated is either small or extremely uncertain. The possibility that the induced electric field in a stationary reconnection exceeds the Dreicer field has been discussed in §8.19. The idea that this threshold be exceeded during a time-dependent process has also been quoted repeatedly, in particular by Syrovatsky (1970) and also by Coppi (1974), Priest and Heyvaerts (1974) and Spicer (1977b). It is, nevertheless, fair to say that no definite answer has been reached on this question, because a sound comparison of the actual field to the Dreicer field requires a consideration of the *energy balance* of the plasma, in order to calculate the temperature which enters in the expression of E_D.

Up to now, no detailed calculation of the tearing evolution (including a simultaneous consideration of temperature and resistivity variations) is available. Partial results by the authors cited above, however, seem to indicate that it should not be very difficult to reach E_D. The electric field in the central part of a tearing magnetic neutral sheet can be estimated from Faraday's law to be, in order of magnitude,

$$E_{\mathrm{T.M.}} = \gamma \lambda B,$$

where B is the component of the perturbating magnetic field normal to the current sheet, λ the wavelength and γ the growth-rate. We may write

$$E_{\mathrm{TM}} = \frac{B}{B_\infty} \frac{\lambda}{a} \gamma a B_\infty,$$

where now B_∞ is the field faraway from the current sheet of thickness a.

We can roughly express E_{TM} as a function of the island width Wreached in the process. Near the central plane the field has magnitude B and scale λ, while at the border of the island, we can estimate it to be $B_\infty W/a$ with a scale W; the vanishing of div $\mathbf{B}$ gives then $B \sim B_\infty W^2 a^{-1} \lambda^{-1}$ and in order of magnitude

$$E_{\mathrm{TM}} \simeq \frac{W^2}{a^2} \gamma a B_\infty.$$

This expression assumes that the linear growth-rate still holds at amplitude W. The ratio E_{TM}/E_D can then easily be obtained, and expressed in a number of ways. If the growth rate γ is given by $(\tau_R^\alpha \tau_A^{1-\alpha})^{-1}$ and if the expression for the Dreicer field is arranged in terms of the Reynolds number S, it is possible to obtain the following formula:

$$\frac{E_{\mathrm{TM}}}{E_D} = \frac{W^2}{a^2} S^{1-\alpha} \frac{c}{a\omega_{p_e}} \sqrt{\frac{B_\infty^2}{\mu_0 n_0 k_B T_\infty}}.$$

For a reverse sheet-pinch, the last term is of order $\sqrt{2}$; let us denote it $\sqrt{2/\beta_{\infty,0}}$, so that numerically

$$\frac{E_{TM}}{E_D} = \sqrt{\frac{2}{\beta_{\infty,0}}} \frac{W^2}{a^2} 5.31 \times 10^{-5} (B_{10^2G})^{1-\alpha} a_{Km^{-2}} T_6^{3/2(1-\alpha)} (n_{16}(m^{-3}))^{-1+(\alpha/2)} \quad (8.33)$$

which gives for the cases of interest:

Plane slow tearing modes,

$$\alpha=\tfrac{3}{5},\qquad \frac{E_{TM}}{E_D}=\sqrt{\frac{2}{\beta_{\infty,0}}\frac{W^2}{a^2}}\,0.42\,\frac{B_2^{2/5}T_6^{3/5-7/10}n_{16}}{a_{Km}^{3/5}}.$$

Plane fast tearing modes,

$$\alpha=\tfrac{1}{2},\qquad \frac{E_{TM}}{E_D}=\sqrt{\frac{2}{\beta_{\infty,0}}\frac{W^2}{a^2}}\,3.5\frac{B_2^{1/2}T_6^{3/4-3/4}n_{16}}{a_{Km}^{1/2}}.$$

Cylindrical $m=1$ tearing mode,

$$\alpha=\tfrac{1}{3},\qquad \frac{E_{TM}}{E_D}=\sqrt{\frac{2}{\beta_{\infty,0}}\frac{W^2}{a^2}}\,126\,\frac{B_2^{2/3}T_6n_{16}^{-5/6}}{a_{Km}^{1/3}}.$$

It is seen that slow tearing modes (the growth-rate of which also drastically decreases at high amplitudes) certainly will not lead to any DC field acceleration, but that an $m=1$ tearing instability (see §3.5), which keeps its large growth-rate up to $W\sim a$ almost certainly could produce fields in excess of the Dreicer value. In that case, the plasma electrons will immediately heat up as a result of the microinstabilities and the energy balance will come drastically into play. The electron heating will have several effects: one is to allow ion-acoustic waves to exist (but the damping of these waves depends on the ion distribution too!); another will be to reduce the growth-rate of the tearing mode by affecting the parameter S and hence decreasing the electric field developed by the instability, while, on the other hand, the Dreicer field (proportional to T^{-1}) will decrease also. It is difficult to assess without a complete study how these two effects will compete. The problem is complicated by the fact that resistivity gradients will be produced, and these will be unstable to the rippling mode, a short wavelength perturbation in the current layer (§3.2.4.C) which tends to wash them out, while, because the tearing mode is slow on the MHD time-scale a readjustment of the structure to the heat impulse will take place too. If we assume that the change proceeds at constant nT, then we can see from (8.33) that E_{TM}/E_D will scale as $T^{(5/2-2\alpha)}$, with an always positive power ($\alpha<1$), while if, for some reason, one proceeds at constant n, it scales as $T^{1-3\alpha/2}$, which is positive only if $\alpha<\tfrac{2}{3}$. The runaway situation would, in cases of positive exponent, develop explosively and the reasoning above will soon cease to be valid, because it is based on the idea that resistivity remains classical. It is quite probable, however, that runaway production will continue in the turbulent stage, because, as shown by Boris *et al.* (1970), a plasma maintained under constant electric field evolves through a succession of phases of DC field acceleration and heating in such a way that the net acceleration proceeds at a constant fraction of the free acceleration. It seems that, unlike static situations, the conditions in a tearing perturbation are more like E being kept constant than J being kept constant, though strictly speaking neither holds.

8.20.3 Characteristics of accelerated particles

Energies that can be reached by DC-field acceleration, as well as particle spectra, have been calculated by some authors. Biskamp and Schindler (1971) have evaluated the order of magnitude of the energies of particles accelerated in the merging of neighbouring current threads, a situation which may be relevant to the nonlinear stage of the collisionless, tearing instability. They find electrons to be accelerated typically to $(a/\rho_{gi})\sqrt{m_e//M_i}$ times the thermal energy, but small groups of particles, whose initial velocity is aligned with the X neutral lines can reach larger energies, $(a/\rho_{gi})^2 k_B T_e$. However, the collisionless situation, which uses the collisionless growth-time is rarely our case.

Bukanov and Sasorov (1975) have made a one-particle analysis of the spectra formed near the neutral point of a tearing layer, which they model by two semi-infinite current layers separated by a gap of width $a(t)$ which increases with time at a prescribed rate. They distinguish a region of drift motion, and a region of free acceleration near the neutral point. They solve the one-particle Liouville equation in that region and obtain the spectra of particles near the border of the free acceleration region; the spectra are exponentials of the energy to some power that depends on the regime of acceleration. They conclude that the acceleration proceeds in two steps. First, acceleration of particles near the neutral line, and then in the drift region, where the energy of particles increases in proportion to their initial energy. Spectral indices vary between 3 and 5, though the spectra are not strictly power-law (Syrovatsky, 1975).

In general, the number of particles directly accelerated is difficult to estimate. The region where particles do not simply drift in the presence of the electric and magnetic field, but are really accelerated, is very small. A discussion of processes which limit the size of this region is made by Drake and Lee (1977), but it is restricted to the linear situation and has little to do with cases when the field becomes greater than the Dreicer field. In such a case, rather than the collisional thickness

$$\Delta = a(ka)^{-3/5} S^{-2/5},$$

the collisionless value for this layer size would be a more appropriate estimation:

$$\Delta_K = a(ka)^{-1}\left(\frac{c^2}{a^2\omega_p^2}\right).$$

This thickness is defined precisely in such a way that in one growth-time a thermal particle at Δ_K from the central plane is displaced by one wavelength in the direction transverse to the current layer. Any consistent tearing theory which neglects line-tying effects must assume that this length is smaller than the size of the tearing region along the electric current flow direction. It takes then more than γ^{-1} for the particles to escape from the acceleration region, which means that not much more than a relative number (Δ_K/a) of particles is accelerated, which is insignificant. However, the calculation of Δ_K on a linear theory does not take into account the fact that particles are substantially accelerated. A non-linear theory would be needed. In the non-linear regime the tearing mode develops singular lines. In the strictly two-dimensional problem considered by Biskamp and Schindler, when islands have been formed particles suffer $\mathbf{E}\times\mathbf{B}$ drift everywhere, except in the vicinity of neutral point, in a region of radius $\sqrt{a\rho_{ge}}$, which is again much smaller than a. A two-dimensional structure, however, is a singular case in the sense that there is no component of the magnetic field parallel to the electric field of the perturbation. In brief, it needs much more work than presently available to have an idea of the number of particles subject to DC-acceleration at high energies in a tearing instability.

Various experimental reports point to effective acceleration process in tearing modes (or time-dependent reconnection), but *care must be taken when comparing with solar flares, since the regime may be different.* For example, Altyntsev *et al.* (1979) have measured the formation of isotropic distribution functions in collisionless tearing modes ($a\omega_p/c=10$); acceleration here was apparently related to plasma turbulence and anomalous resistivity. However, similar conditions in the solar atmosphere imply sheet thickness of the order of 0.5 m ($c/\omega_{pe}\simeq 5$ cm). Experiments by Baum and Bratenahl (1976, 1977) and Syrovatsky *et al.* (1973) are rapidly driven. For example, in the experiment of Baum and Bratenahl the time-scale of the experiment is 10 μs, not much larger than the Alfvén time-scale $\tau_A=1$ μs, while the ion gyroradius (Baum and Bratenahl, 1976) is not much smaller than the current-concentration size (Baum and Bratenahl, 1977). In both experiments anomalous resistivity appears, which shows that very high degrees of compression have been reached. It is not certain that such high inflow velocities and degrees of compression are reached in solar flares, where they need some justification.

8.21 SECOND-PHASE ACCELERATION: AN OVERVIEW

We have shown in §§8.3, 8.5, 8.6 that in large flares a second-phase, delayed, acceleration process is likely to be present. This is also apparent from the spectral characters of the two parts of the event, as well as from the time-profile of hard X-ray bursts (§8.3). The close association between type II bursts and proton events suggest a *collisionless shock-wave acceleration* in the corona. Up to now, this idea seems to have been ill-studied, and has not yet met a high level of success, though modern galactic cosmic-ray acceleration theories use shock-wave acceleration a great deal. The study of similar processes in the Sun's atmosphere, where various limitations (like size or injection mechanisms) are more severe than in the interstellar space, is in progress. The rather long time-scale for this acceleration suggests we consider *acceleration by MHD perturbations* in the corona. Such theories use the concept of particle diffusion on magnetic irregularities, and should demonstrate two things: first, that the correct number of particles can be accelerated in an adequate time and, second, that one can explain the origin of the MHD fluctuations. The first aspect has met with a reasonable degree of success. The second aspect poses no problem *a priori*, because there is a host of MHD fluctuations in the quiet corona, even more near a flare site. Nevertheless, it has been shown that short-wavelength perturbations (necessary to scatter accelerated particles) can be self-consistently generated by longer wavelength perturbations. To our knowledge, however, it has never been explained in detail why and how the fluctuation level able to produce the acceleration (and/or the preheating process) is to be found only in large flares. Actually, it has become apparent recently that even weak flares are capable of accelerating nuclei, with some anomalies in the abundances (§8.8). We shall report below how this peculiarity can be understood in terms of a specific preheating mechanism. If, as is very likely, this theory is correct, then it implies that second-phase acceleration is also going on in weak events. The correlation of abundances with the first ionization potential, however, has not been explained at all at present. As far as electrons are concerned, they can be accelerated by the same processes as protons and nuclei, but the requirements for injection in the acceleration process are more difficult to fulfil. This may be the reason why they are under-abundant at the same energy (§8.6). On the other hand, because of their rather small gyroradii, electrons do not have in general the same interaction as protons with shock waves. Some theories have considered microscopic electron-acceleration processes in shock-fronts or shock wakes.

Below we review these various ideas. The presentation will be biased somewhat, with most emphasis put on the *Fermi mechanism*. This only reflects the fact that in the present literature this process has been much more discussed, maybe because of its relative simplicity, but it does not mean that shock phenomena play a negligible role. It is probable that work during the next few years will complete the examination of possibilities for shock processes. Only then will it be possible to make a statement on the relative efficiency of these two mechanisms.

8.22 SECOND-PHASE FERMI ACCELERATION IS COMPATIBLE WITH OBSERVATIONAL DATA

8.22.1 Theoretical predictions of a Fermi-acceleration model

By Fermi acceleration we mean any process which yields a mean acceleration rate proportional to the momentum, that is, proportional to the energy in the relativistic limit (Korchak, 1978). Originally, Fermi (1949) explained the acceleration by a mechanism of collision with “moving magnetic mirrors”; he conceived one situation in which the collisions were always head-on (*first order*) and another in which the relative velocity of particles and mirrors is random (*second-order Fermi mechanism*). In the first-order mechanism, the mean rate of energy gain is proportional to the velocity of mirrors, while in the second case it is proportional to the square of this velocity. Obviously, Fermi’s mechanism must be a particular case of the interaction of particles with MHD turbulence. We return below to the question of the connection of Fermi’s point of view with modern wave-particle interaction concepts. Also, the second-order process is likely to be the most relevant in practice, though we shall show when discussing shock acceleration that situations which essentially correspond to the first order process do exist. For the moment, however, we are interested in the compatibility of Fermi’s mechanism with observational data, and a sketchy picture of the acceleration (for example, random collisions with randomly-moving infinitely-heavy scatterers) will suffice. An excellent discussion has been given by Ramaty (1979), which we follow rather closely here.

Such a mechanism can be safely expressed by equations. Let $4\pi p^2 f(p)\mathrm{d}p$ be the number of particles whose momentun *is between* p and $p+\mathrm{d}p$. The distribution function is assumed to be, and then remains, isotropic. Its equation of evolution is (Tverskoi, 1967)

$$\frac{\partial f}{\partial t}=\frac{1}{p^2}\frac{\partial}{\partial p}p^2\mathrm{d}(p)\frac{\partial f}{\partial p}, \tag{8.34}$$

where the diffusion coefficient $\mathrm{d}(p)$ is given by

$$\mathrm{d}(p)=\frac{1}{3}\frac{u^2}{lc}\frac{p^2c}{v(p)}, \tag{8.35}$$

u being the blob's velocity, l the mean-free-path for collisions with these blobs, and $v(p)$ the velocity of a particle of momentum p. Changing to kinetic energy variables, Equation (8.34) can be written equivalently as

$$\frac{\partial}{\partial t} N(E) = -\frac{\partial}{\partial E} AN + \frac{\partial^2}{\partial E^2} DN, \tag{8.36}$$

$$A(E) = \frac{1}{p^2}\frac{\mathrm{d}}{\mathrm{d}p} p^2 v(p) \mathrm{d}(p) = \frac{4u^2}{3\,lc} pc = \frac{4}{3}\alpha pc,$$

$$D(E) = v^2(p) \mathrm{d}(p) = \frac{1}{3}\frac{u^2}{lc}\frac{v}{c} p^2 c^2 = \frac{1}{3}\alpha\beta p^2 c^2.$$

It can readily be checked from (8.36) that the time-derivative of $\langle E \rangle$ is proportional to $\langle p \rangle$. The model in this form is still too crude, because it does not include particle energy losses, nor particle losses due to escape out of the acceleration region. These two phenomena are necessary for a stationary state. Moreover, energy losses have the effect of quenching the mean-energy growth if the initial energy is too small. We can then speak of a loss-threshold or loss-injection energy. With both these effects included in the model, Equation (8.36) transforms to:

$$\frac{\partial}{\partial t} N(E) = -\frac{\partial}{\partial E}(A(E) - L(E))N(E) + \frac{\partial^2}{\partial E^2} D(E)N(E) - \frac{N(E)}{T(E)}, \tag{8.37}$$

where $T(E)$ is the characteristic escape- (or disappearance-) time at energy E, while $L(E)$ is the rate of energy loss. This function depends on the nature of the particle and on the conditions in the ambient medium. Most important are the Coulomb energy losses, which vary as $E^{-1/2}$ for non-relativistic energies larger than $k_B T$. Detailed descriptions of these functions can be found in various references (Ginzburg and Syrovatsky, 1964; Northcliff and Schilling, 1970; Berger and Seltzer, 1964; Trubnikov, 1965). As a rule the rate of energy gain $A(E)$ becomes larger than the rate of energy loss at some critical energy E_l. This is not the only injection threshold. We anticipate here that any resonant interaction with MHD turbulence requires the particle momentum to be larger than $p_A = M_i v_A$.

It is possible to solve analytically (8.37) in the particular cases when gains overwhelm the losses. $L(E)$ will then be neglected and we consider $T(E) = \infty$

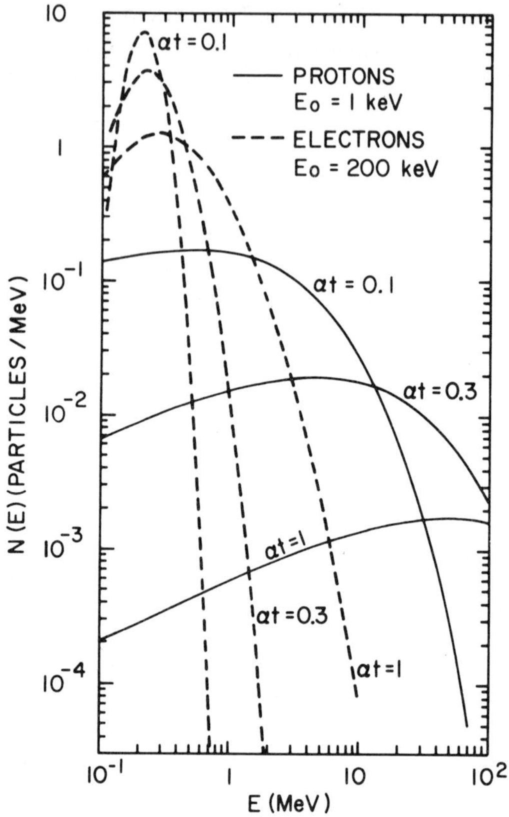

FIGURE 8.24 Proton and electron spectra from Fermi acceleration with impulsive injection. E_0 is the injection energy. (From Ramaty, 1979.)

(i.e. no escape). The solution is then (Lee, 1978) expressible in terms of modified Bessel functions for an injection of p_0 at time $t=0$:

$$N(E, t)=\frac{3}{\alpha t p_0 c} I_2\left(\frac{6\sqrt{p_0 p}}{\alpha t m c}\right) \exp-\left\{3\frac{p_0+p}{m c \alpha t}\right\} \tag{8.38}$$

for $E \ll mc^2$, a case applicable to protons, and

$$N(E, t)=\frac{1}{E}\sqrt{\frac{3}{4\pi\alpha t}} \exp-\left\{3\frac{(\ln(E/E_0)-\alpha t)^2}{4\alpha t}\right\}$$

for $E \gg mc^2$, a case applicable to electrons. The result is shown in graphical form in Figure 8.24, where it can be seen that the acceleration proceeds much faster for protons than for electrons. This is a consequence of the larger acceleration rate of protons (Morrison *et al.*, 1954):

$$A_i(E) = \frac{4}{3}\alpha mc^2 \sqrt{\frac{2M_i}{m_e}} \sqrt{\frac{E}{m_e c^2}},$$

$$A_e(E) = \frac{4}{3}\alpha mc^2 \frac{E}{m_e c^2}.$$

It is also possible to obtain stationary solutions with a finite escape time T, which are given by:

$$N(E) = \frac{1}{p_0 c\alpha T}\frac{6}{} I_2\left(2\left(\frac{3p_0}{mc\alpha T}\right)^{1/2}\right) K_2\left(2\left(\frac{3p}{mc\alpha T}\right)^{1/2}\right); \quad E \ll mc^2, \qquad (8.39)$$

$$N(E) = \frac{1}{E_0}\frac{3}{\alpha T}\left(g + \frac{12}{\alpha T}\right)^{-1/2} \exp\left(\frac{1}{2}\ln\left(\frac{E}{E_0}\right)\left(1 - \left(9 + \frac{12}{\alpha T}\right)^{1/2}\right)\right); \quad E \gg mc^2.$$

Both solutions are shown in Figure 8.25.

8.22.2 Requirements from observations

Satisfactory fits to observed proton spectral indices (from 2.5 to 3.7 between 20 and 80 MeV) are obtained when αT in Equation (8.39) or αt in (8.38) lies between 2×10^{-2} and 5×10^{-2}. Moreover, observations indicate that T is of the order of 100 sec, so that $2 \times 10^{-4} \leqslant \alpha \leqslant 5 \times 10^{-4}$ approximately. On the other hand, collision losses must not impede acceleration. From the condition $A(E) > L(E)$, one can therefore derive a constraint on α/n, which Ramaty (1979) finds to be

$$\alpha/n(m^{-3}) \geqslant 3 \times 10^{-18}.$$

As α is bracketed in a certain interval, we find, for consistency

$$n \leqslant 10^{15}\ \text{m}^{-3}.$$

It is found that electron spectra calculated with the same values of αT are too steep. This seems to indicate that, if a Fermi process is also responsible for their acceleration, α or T must differ from those appropriate to protons. From this discussion, one can conclude that the *Fermi mechanism is consistent with*

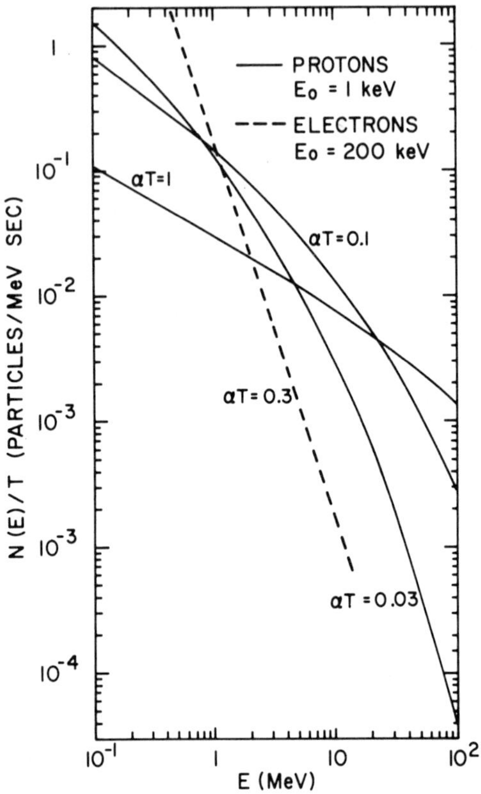

FIGURE 8.25 Proton and electron spectra from Fermi acceleration with continuous steady-state injection at E_0. (From Ramaty, 1979.)

observational data if and only if it takes place in the high, tenuous corona.

Gamma-ray lines and spectra provide useful indications concerning the proton spectrum at the Sun, and these can also be compared for consistency with theoretical or interplanetary spectra, which we have just seen to be possibly consistent. A sensitive test is the ratio of the 4.44 MeV line, which is due to carbon excitation by proton impact, a process which depends on the proton energy, and the 2.2 MeV line, which is due to neutron capture. The intensity of this line may be deduced using a conversion efficiency factor (Wang and Ramaty, 1974) from the neutron density, and the latter in turn can be deduced by a calculation from the proton spectrum. Of course the neutron yield depends in a different way on the proton spectrum from the 4.4. MeV γ-ray yield. The measured ratio of these lines can be compared to a calculated ratio based on a guessed proton spectrum like that of Equation (8.39) (Ramaty *et al.*, 1975). Calculations with a *thin-target* model were found to give observed line ratio for values of αT inconsistent with those deduced above from a fit to the interplanetary spectrum. On the other hand, a *thick-target* calculation yields consistent results. We can conclude from this that proton acceleration by the Fermi mechanism with $\alpha T \simeq 3 \times 10^{-2}$, say, in a tenuous coronal plasma, and thick-target bombardment of the solar atmosphere to yield γ-ray emission would be a scenario consistent with observations. The abundance composition of bombarding particles should be high-Z enriched (§8.6) and the electron escape-time somewhat longer than that applicable to protons. Nevertheless, the proton/electron ratio would still be large, as observed. A pre-acceleration is needed, which may be provided by the first phase, or by some independent mechanism. Nuclei acceleration during the first phase is an almost untouched problem for which some clues will be discussed below. Colgate (1978), however, expresses the view that their acceleration occurs early in the flare.

8.23 FERMI MECHANISM AND RESONANT MHD PROCESSES

8.23.1 The various particle–MHD-turbulence interaction processes

In its original formulation, the Fermi mechanism is conceived as the acceleration resulting from random or systematic reflection on moving magnetic mirrors. These mirrors may of course be viewed as a set of MHD waves propagating on a mean field. For reflection to occur, however, one must conceive of perturbations which are not infinitesimal, so that Fermi's mechanism cannot strictly speaking be viewed as a weak-turbulence mechanism: the orbits of the particles have to be altered with respect to the mean field motion in an important way. Collisions are strong, though almost-elastic and visualized as discrete events, so that in momentum-space a Boltzman equation would be most appropriate to describe the interaction process (Parker and Tidman, 1958), but as the momentum's modulus changes little, a Fokker Planck description is appropriate for this variable (Tverskoi, 1967). But, of course, this Boltzman-like interaction with strong localized mirrors is in no way the only mode of interaction that a particle could have with MHD turbulence; for example the collisions could be weak and the particle suffer only small velocity changes. Another important mechanism is the *betatron effect*, which is an increase in perpendicular energy of a particle sitting in a time-increasing magnetic field (Swann, 1933). For slow field-variations, the first adiabatic invariant is conserved, and the perpendicular energy varies in proportion to the magnetic field intensity. In the presence of a random time-variation of the field, particles may diffuse in energy space, provided an efficient scattering process exists (Davis, 1956; Berger *et al.*, 1958).

8.23.2 Role of scattering

The necessity of a scattering mechanism is obvious for the betatron process because otherwise the energy gain would be reversible. An efficient one is also needed for Fermi acceleration, because each reflection adds only to the parallel velocity, not to the perpendicular energy. After acquiring too much parallel velocity, the particle could not even be reflected! For it to keep its chances of being reflected, a scattering mechanism is needed (Tverskoi, 1967), which is an essential element of the acceleration process. It can be provided either by scattering on the MHD waves themselves, in particular the short-wavelength waves, which are most efficient in scattering (Jokipii, 1966; Kulsrud and Pearce, 1969; Wentzel, 1969), or on whistlers. Scattering waves may have an independent origin or may owe their existence to the fast particles themselves (Melrose, 1974). A general theory including all these effects should be a statistical one describing the behaviour of fast particles in the presence of a broad spectrum of (possibly intense) MHD turbulence, under the effect of an efficient scattering mechanism which may or may not be self-consistent. In the case where the turbulence is not strong, it might seem that a quasilinear theory of MHD-wave–particle interaction is adequate. Though such theories may lead to a certain degree of success (Melrose, 1968, 1969), it is clear that they cannot be completely satisfactory, because the motion on long scale-lengths is heavily influenced by scattering on the short wavelength part of the spectrum. A renormalized quasilinear theory, in the spirit of the resonance broadening theories (Dupree, 1968; Weinstock, 1969), would be more appropriate. Such a complete theory is not yet available, but a fairly complete study of particle motion in the presence of a given long-wavelength MHD turbulence and a given scattering process has been made by Kulsrud and Ferrari (1971). Various quasilinear results have been obtained for MHD waves (Tverskoi, 1968; Lacombe, 1977, 1979; Achterberg, 1979) including more-or-less explicitly the effect of intense scattering by establishing or assuming an isotropic particle distribution. Renormalized quasilinear theory has been considered in order to overcome the difficulty that the angular diffusion coefficient predicted by quasilinear theory approaches zero when $p_{\|}\rightarrow 0$ (Völk, 1973).

8.23.3 Weak-turbulence theories of acceleration

Quasilinear theories involve resonant interactions satisfying

$$\omega - k_{\|}v_{\|} = s\Omega^*,$$

which imply that the Döppler-shifted frequency of the wave in the gyration-centre frame is an integral multiple of the relativistic gyrofrequency Ω^*. One can then speak of a *Čerenkov resonance* for $s=0$, or of a *cyclotron resonance* for

$s = \pm 1$ (Tverskoi, 1968), and in general of resonant interaction processes. Fermi acceleration is a non-resonant interaction, which strictly speaking does not exist for weak turbulence, as emphasized above; it involves strong perturbations of the constant field motion and so contradicts the concept of weak turbulence. Nevertheless, similar processes already appear in weak turbulence theory; for instance, Achterberg (1979) has established that the Čerenkov resonance interaction of fast magnetosonic long waves in a low-β plasma containing relativistic particles leads to a $p_{\|}$ diffusion of this population due to the stochastic small changes suffered by this variable in interacting with fluctuations of magnetic field intensity. No reflection, however, is involved here. His result, a quasilinear equation appropriate to such waves, is

$$\frac{\partial}{\partial t} f(p_{\perp}, p_{\|}) = \frac{\partial}{\partial p_{\|}} \mathrm{D}_{\|\|} \frac{\partial}{\partial p_{\|}} f(p_{\|}, p_{\|}),$$

$$\mathrm{D}_{\|\|} = \frac{\pi p_{\perp}^4 c^4}{2E^2} \int \mathrm{d}^3 k \int_0^{\infty} \mathrm{d}\omega k_{\|}^2 I(\mathbf{k}, \omega) \delta(\omega - k_{\|} \mathrm{v}_{\|}),$$

$$\langle \tilde{B}_{\|}(\mathbf{k}, \omega) \tilde{B}_{\|}(\mathbf{k}', \omega') \rangle = B^2 I_{\|}(\mathbf{k}, \omega) \delta(\mathbf{k} + \mathbf{k}') \delta(\omega + \omega').$$

If pitch-angle scattering is (non-consistently) included in the equation, so that it maintains the distribution function isotropic, the gain in parallel energy is shared with the other degrees of freedom, and a diffusion equation is obtained (Achterberg, 1979), for which the diffusion coefficient in momentum-space scales as p^2, as in the Fermi mechanism described in §8.22, namely

$$\frac{\partial f}{\partial t} = \frac{1}{p^2} \frac{\partial}{\partial p} p^2 D_0(p) \frac{\partial f}{\partial p},$$

$$D_0(p) = \frac{\pi}{4} p^2 \int \mathrm{d}^3 k \int_0^{\infty} \mathrm{d}\omega \frac{\omega^2}{|k_{\|}||v|} \left(1 - \frac{\omega^2}{\mathrm{k}_{\|}^2 \mathrm{v}^2}\right) I_{\|}(\mathbf{k}, \omega), \tag{8.40}$$

A Čerenkov resonant interaction with fast magnetosonic modes appears as a weak-turbulence limit of the Fermi process (Kulsrud and Ferrari, 1971; Achterberg, 1979), though the particle motions are qualitatively different, as already stressed. This is reflected in the diffusion coefficients $D_0(p)$ in (8.40) and $\mathrm{d}(p)$ in Equation (8.35). Both coincide in the ultra-relativistic limit if by analogy with (8.35) we define a "weak Fermi process" mean-free-path by

$$D_0(p)=\frac{1}{3}\frac{p^2 v}{c}\frac{v_A^2}{l_0 c}, \qquad (8.41)$$

with

$$l_0^{-1}=\frac{3\pi}{4}\int d^3k\int_0^\infty d\omega\frac{k^2}{k_\parallel}I_\parallel(\mathbf{k},\omega). \qquad (8.42)$$

However, the Fermi theory does not make any assumption on the nature and velocity of propagation of "magnetic mirrors" nor on the dependence of $I_\parallel(\mathbf{k}, \omega)$ (which is not necessarily proportional to $\delta(\omega-kv_A)$), nor does it suppose any resonance to occur, a fact which is already hidden in (8.40). In the weak-turbulence, resonant, "Fermi" process, the interaction is only possible if $p\geqslant p_A=m_i v_A$. No such condition holds for Fermi acceleration in its classical version.

The interaction of particles with MHD modes different from fast magnetosonic ones also leads to a process of acceleration, which Tverskoi (1968) names *cyclotron resonance* and many other authors name *acceleration by Alfvén waves.* A momentum-space diffusion equation can be obtained as well using the methods described by Tsytovitch (1970) and summarized in Melrose (1968). Recently, Lacombe (1979) has reviewed the results concerning quasilinear interaction of particles and Alfvén waves. The resonance condition for such interactions becomes

$$\omega\equiv|k_\parallel|v_A=k_\parallel v_\parallel+s\Omega^*;$$

and, for $v_\parallel\gg v_A$, it can be written in the form

$$\frac{\omega}{v_A}\equiv|k_\parallel|=\left|\frac{s\Omega^*}{v_\parallel-v_A}\right|.$$

The resonant particles must have a large enough velocity, namely

$$|v_\parallel-v_A|=v_A\left|\frac{s\Omega^*}{\omega}\right|>v_A\frac{\Omega^*}{\omega_{\max}}>v_A\frac{\Omega^*}{\Omega_i},$$

where $\omega_{\max}$ is the highest turbulent frequency of the spectrum, which cannot be larger than Ω_i. This relation shows that $v_\parallel\gg v_A$ for not-too-relativistic electrons, and $|v_\parallel-v_A|\gtrsim v_A$ for protons. Usually, we shall be interested in particles with $v_\parallel\gg v_A$, so that the resonance condition implies the following injection threshold, valid both for electrons and ions:

$$|p_{\|}| > M_i v_A = p_A.$$

In the same approximation, the quasilinear equation (Melrose, 1968; Lacombe, 1979) for linearly polarized Alfvén waves, which is the most general case (Melrose, 1968), can be expanded in powers of v_A/v. The magnetic fluctuations are represented by a power spectrum S_A, which depends on frequency and direction $x = \cos(\mathbf{k}, \mathbf{B}_0)$. We choose here a normalization which makes it dimensionless:

$$2\pi \int_{-1}^{+1} \mathrm{d}x \int_{\omega_{\min}}^{\omega_{\max}} \frac{\mathrm{d}\omega}{|\Omega_i|} S_A(x, \omega) = \frac{\langle \delta B^2 \rangle_{\mathrm{Alf}}}{B_0^2}, \tag{8.43}$$

where $\langle \delta B^2 \rangle_{\mathrm{Alf}}/2\mu$ is the energy density of the fluctuations. Following Lacombe (1979) and Tverskoi (1968), we give the resulting equation for an isotropic distribution function, in the limits $k_\perp \; p_\perp \; /m\Omega \ll 1$ (magnetized particles) and $k_\perp \; p_\perp \; /m\Omega \gg 1$ (unmagnetized particles). We obtain a diffusion equation of type (8.40). In the following, m and Ω refer to the particle of interest, an electron or an ion, while M_i and Ω_i refer to a proton. The diffusion coefficient D takes the following form for magnetized particles:

$$D_M(p) = \frac{\pi^2}{4} |\Omega| \frac{M_i}{m} \left(\frac{v_A^3}{c^3}\right) m^2 c^2 \frac{E(p)}{pc} \frac{\langle \delta B^2 \rangle_{\mathrm{Alf}}}{B_0^2} \int_{Ev_A/pc^2}^{1} \mathrm{d}\mu \frac{1-\mu^2}{\mu} \left\{ \int_{-1}^{-\sqrt{1-\mu^2}} \mathrm{d}x S_A\left(x, \frac{\Omega_i p_A}{\mu p}\right) \right.$$

$$\left. + \int_{+\sqrt{1-\mu^2}}^{1} \mathrm{d}x S_A\left(x, \frac{\Omega_i p_A}{\mu p}\right) \right\}; \tag{8.44}$$

and for unmagnetized particles:

$$D_{UM}(p) = \pi |\Omega| \frac{M_i}{m} \frac{v_A^3}{c^3} m^2 c^2 \frac{E(p)}{pc} \frac{\langle \delta B^2 \rangle_{\mathrm{Alf}}}{B_0^2} \int_{Ev_A/pc^2}^{1} \mathrm{d}\mu \, \mu^2 \int_{-\sqrt{1-\mu^2}}^{+\sqrt{1-\mu^2}} \mathrm{d}x$$

$$\frac{|x|^3}{(1-x^2)\sqrt{1-\mu^2-x^2}} \int_{\max(\omega_{\min}, \, \Omega_i p_A/\mu p)}^{\omega_{\max}} \frac{\mathrm{d}\omega}{\omega} S_A(x, \omega). \tag{8.45}$$

Just as for Eq. (8.41), we can define corresponding mean-free-paths

$$l_M^{-1} = \frac{3\pi^2 |\Omega_i| p_A}{4 \quad v \quad p} \frac{M_i c^2}{E} \frac{\langle \delta B^2 \rangle_{\text{Alg}}}{B_0^2} \int_{v_A/v}^{1} d\mu \frac{1-\mu^2}{\mu} \left\{ \int_{-1}^{-\sqrt{1-\mu^2}} dx S_A\left(x, \frac{\Omega_i p_A}{\mu p}\right) + \int_{+\sqrt{1-\mu^2}}^{1} dx S_A\left(x, \frac{\Omega_i p_A}{\mu p}\right) \right\}, \tag{8.46}$$

$$l_{UM}^{-1} = 3\pi \frac{|\Omega_i|}{v} \frac{p_A}{p} \frac{M_i c^2}{E} \frac{\langle \delta B^2 \rangle_{\text{Alf}}}{B_0^2} \int_{v_A/v}^{1} d\mu \mu^2 \int_{-\sqrt{1-\mu^2}}^{+\sqrt{1-\mu^2}} dx \frac{|x|^3}{(1-x^2)\sqrt{1-\mu^2-x^2}} \int_{\max(\omega_{\min}, \Omega_i p_A/\mu p)}^{\omega_{\max}} \frac{d\omega}{\omega} S_A(x, \omega).$$

We stress that these formulae are valid for isotropic distribution functions. In the case of Alfvén fluctuations, this may be justified by the fact that, to lowest order, the quasilinear operator reduces to a scattering operator which is proportional to (v_A/c) and so is two orders of magnitude (in this small parameter) larger than the acceleration operator (see 8.41, 8.45), which is proportional to $(v_A/c)^3$. The full quasilinear operator contains terms proportional to $(v_A/c)^2$, which vanish for strict isotropy. But for an anisotropy of order (v_A/c) only, some of these terms which are neglected in (8.44, 8.45) become comparable to those retained. It becomes then very important to know exactly the small degree of anisotropy of particles. Skilling (1975) tried to calculate this by retaining certain terms in the full quasilinear operator, but Lacombe (1979) criticized this choice as arbitrary, some important terms of order $(v_A/c)^3$ having been left out unduly. Though unsatisfactory in this respect, Skilling's acceleration rates exhibit a dependence on the forward and backward Alfvén wave levels, which gives a vanishing result when either forward or backward waves are lacking. This is highly satisfactory from the physical standpoint, because Alfvén perturbations travel undistorted at their group velocity ($\pm v_A$ along the magnetic field). If only one direction of waves is excited, the perturbation is purely magnetostatic in a rest frame moving at velocity v_A along the field. In this rest frame there is no electric field and hence no energy exchange. The particles just tend to isotropize in this frame, but no acceleration occurs. Moreover, non-linear effects have a tendency to accumulate waves in one and only one direction (Lifschitz and Tsytovitch, 1970; Dobrowolny *et al.*, 1980). When these processes dominate (i.e. mainly at low frequencies (Lacombe, 1977)), the acceleration effect may be quenched by this directional effect.

8.23.4 Relative efficiency

In order to compare these various mechanisms, it is useful to compare the mean-free-paths l_F, l_0, l_M (or l_{UM}). For this, it is convenient to introduce a dimensionless fluctuation spectrum, as we did in (8.43) for Alfvén waves. According to Achterberg (1979), the correlator $I(\mathbf{k}, \omega)$ is given in terms of the plasmon number $N(\mathbf{k})$ for fast MHD waves by

$$I(\mathbf{k}, \omega) = \frac{1}{4}\frac{k_\perp^2}{k^2}\frac{2\mu_0\hbar\omega_k N(\mathbf{k})}{B_0^2\,8\pi^3}\delta(\omega - |k|v_A).$$

Using the relation between $N(\mathbf{k})$ and turbulent energy density W

$$W = \int \frac{d^3k}{8\pi^3}\hbar\omega_k N(\mathbf{k})$$

and normalizing W as in (8.43), we obtain from (8.42):

$$l_0^{-1} = 3\pi\frac{\Omega_i}{v_A}\frac{\langle\delta B^2\rangle_{\text{fast}}}{B_0^2}\int_{\omega_{\text{min fast}}}^{\omega_{\text{max fast}}}\frac{\omega d\omega}{\Omega_i^2}\int_{v_A/c}^{1} d\mu\,\frac{1-\mu^2}{|\mu|}\left(1 - \frac{v_A^2}{v^2\mu^2}\right)^2 S_{\text{fast}}(\mu, \omega).$$

Comparing with (8.46), it is seen that l_0^{-1} differs from l_M^{-1} by a factor

$$\frac{l_0^{-1}}{l_M^{-1}} = \frac{\langle\delta B^2\rangle_{\text{fast}}}{\langle\delta B^2\rangle_A}\frac{v^2}{v_A^2}\frac{E}{M_i c^2}.$$

The relative efficiency of cyclotron resonance acceleration by Alfvén waves and weak-turbulence "Fermi-type" resonant acceleration by fast magnetosonic waves depends then on the relative level of these two types of wave, which may be extremely large indeed (the emission efficiency strongly favours the Alfvén waves, and non-linear effects tend to reduce to zero any level of compressive waves (Lifschitz and Tsytovitch, 1970)). Fast-mode acceleration does not necessarily dominate, then. The comparison between Alfvén wave acceleration and non-resonant Fermi-acceleration is more difficult. Lacombe (1977) assumes that the mean-free-path is equal to the longest wavelength of the turbulent spectrum, and finds that the dominance then depends on the slope of the turbulent spectrum, the Fermi mechanism being more important for steeper spectra. Tverskoi (1968) reaches the same conclusion: he shows that the second-order Fermi mechanism dominates for wave energy spectra steeper than ω^{-2}, while for high-frequency-rich spectra the resonance process should be more effective. In the case when Fermi acceleration is more effective, resonant interaction with Alfvén waves may provide the scattering mechanism which is needed (see below).

8.23.5 A quasilinear theory with scattering

Kulsrud and Ferrari (1971) have developed an interesting relativistic quasi-linear theory of acceleration, valid for magnetized particles in the weak-turbulence limit, in the presence of an efficient scattering mechanism, which is assumed to be given. They discuss the acceleration rate so obtained, and show that previously-known results, which seem to be physically different, are actually different regimes of the same basic mechanism. Their procedure is to carry out two successive expansions of the Vlasov equation by taking the limit of small gyroradii and slow-wave frequencies, analogous to the CGL procedure (Chew *et al.*, 1956), and by performing the usual quasilinear expansion in wave amplitude. These calculations are, however, complicated by the introduction of the ad hoc scattering operator,

$$\mathcal{O}_{\text{scatt}} \equiv \nu_S \frac{\partial}{\partial \mu} \frac{1-\mu^2}{2} \frac{\partial}{\partial \mu},$$

where ν_S is the scattering frequency and μ the cosine of the pitch-angle. This complication makes it impossible to solve for the perturbed distribution function in terms of the field fluctuation. Here, the relation between them is implicitly given by a linear differential equation

$$(\mathcal{O}_{\text{scatt}} + i(\omega - kv\mu))\ f_{1\,k\omega}(v, \mu) = C\frac{1-\mu^2}{2} B_{1(k\omega)} \frac{\partial f_0}{\partial p}.$$

Assuming the solution to be known, it is possible to obtain a generalized form of the quasilinear equation (Kulsrud and Ferrari, 1971), which has again the form (8.40), the diffusion coefficient being now given by

$$D_{KF}(p) = p^2 \int \mathrm{d}^3k \int_{-\infty}^{+\infty} \mathrm{d}\omega \gamma(\omega, k_{\parallel}, v) I(\mathbf{k}, \omega),$$

where γ is a frequency defined by the system of equations

$$\gamma(\omega, k_{\parallel}, v) = -\frac{\omega^2}{2} \int_{-1}^{+1} \mathrm{d}\mu \frac{1-\mu^2}{2} z_{\omega, k_{\parallel}, v}(\mu),$$

$$\frac{d}{d\mu}v_S\frac{1-\mu^2}{2}\frac{d}{d\mu}z_{\omega,\,k_\parallel,\,v}(\mu)+i(\omega-k_\parallel v\mu)z_{\omega,\,k_\parallel,\,v}(\mu)=\frac{1-\mu^2}{2};$$

and $I(k, \omega)$ is the correlator of the amplitude of field fluctuations:

$$\langle B(\mathbf{k},\,\omega)B^*(k',\,\omega')\rangle=B_0^2\delta(\mathbf{k}-\mathbf{k}')\delta(\omega-\omega')I(\mathbf{k},\,\omega).$$

It is possible to obtain explicit expressions for γ in certain limiting cases, the most interesting being those when the parallel wavelength is much longer or much shorter than the collision mean-free-path $l=vv_S^{-1}$. The case of long wavelength represents a situation in which inhomogeneities in the direction of the field are relatively unimportant and the main effect is due to the betatron process. In that case,

$$\gamma=\frac{1}{3}\frac{\omega^2k_\parallel^2v^2/v_S}{(k_\parallel^4v^4/v_S^2)+9\omega^2}+\frac{1}{15}\frac{\omega^2 v}{\omega^2+9v_S^2};\qquad \frac{k_\parallel v}{v_S}\ll 1.$$

Physically, the particles are accelerated by the magnetic pumping effect: particles gain or lose energy by the betatron effect during a compression or a dilatation of the field, but escape out of the region (first term) or collide (second term) before the sense of compression reverses. The change in energy is then not reversible. A diffusion in momentum-space results. In the opposite case of wavelength much shorter than the mean-free-path, the motion is collisionless, and we have the quasilinear Cerenkoff resonance-interaction described before, or the "weak-field Fermi" process (Hall and Sturrock, 1967; Achterberg, 1979). The value of γ reproduces the result $D_0(p)$ (8.40), namely

$$\gamma=\frac{\pi}{8}\frac{\omega^2}{|k_\parallel|v}\left(1-\frac{\omega^2}{k_\parallel^2v^2}\right)^2;\qquad \frac{k_\parallel v}{v_S}\gg 1.$$

8.23.6 Requirements from observations

Observations seem to indicate that the acceleration time should be of the order of 100 sec, in a medium where $v_A/c\simeq 10^{-2}$ and $\delta B/B$ is at most 1. The accelerating waves should have periods around 1 sec, $\delta B/B\gg 10^{-2}$ and $n_1/n_e > 10^{-4}$ if we estimate the field to be 10 Gauss. Oscillations with a period of the order of one second may be evidenced by regularly-varying radio emissions, so-called pulsating structures (Abrami, 1970).

In brief, we conclude from this and the preceeding paragraph that *the weak-field Fermi mechanism could well be at work in the second phase* of particle acceleration in solar flares. The existence of the required level of MHD turbulence should not be a problem, though it remains to prove it in detail. On the other hand, one has to explain how particles have received their injection energy, in first phase perhaps. For electrons resonating with whistlers

$$p_{\text{inj e}^-} = \sqrt{m_e M_i}\, v_A \simeq 4 \times 10^{-1}\ m_e c; \qquad E_{\text{inj e}^-} \simeq 40\ \text{keV}.$$

For protons interacting with MHD waves

$$p_{\text{inj ions}} = M_i v_A \simeq 10^{-2}\ M_i c; \qquad E_{\text{inj ions}} \simeq 50\ \text{keV}.$$

We have good evidence that electrons of these energies are indeed accelerated during the impulsive phase. We discuss later some evidence concerning a preheating mechanism for nuclei (§8.25).

8.24 FIRST-ORDER FERMI MECHANISM NEAR SHOCKS

8.24.1 Basic physics of bouncing through a shock

It has been recently recognized that particles can be accelerated in the vicinity of shock fronts by the combined action of scattering centres, such as magnetic irregularities imbedded both in the upstream and down-stream plasma. These centres are converging on one another, and *particles bouncing between upstream and downstream suffer a systematic first-order Fermi acceleration.* To see this more precisely, let us consider a parallel shock, and a subrelativistic particle which scatters on the upstream side. In the rest frame S_- of upstream scattering centres moving at velocity u_- with respect to the shock, the particle suffers only elastic collisions. Its energy keeps constant. When the particle with momentum p in S_- crosses to the down-stream region, making an angle θ with the shock normal, it will appear in the rest frame S_+ of downstream scattering centres to have an energy

$$E_+ = E_- + (\mathbf{u}_+ - \mathbf{u}_-) \cdot \mathbf{p}. \tag{8.47}$$

Conversely, in making a transition from S_+ to S_-, a particle with momentum $\mathbf{p}$ and energy E_+ in S_+ will appear in S_- to have an energy

$$E_- = E_+ - (\mathbf{u}_- - \mathbf{u}_+) \cdot \mathbf{p}. \tag{8.48}$$

Now $|u_-|$ is larger than $|u_+|$ in a shock and $\mathbf{p}$ is oriented on each crossing in such a way that the second terms are positive; explicitly, in place of (8.47) and (8.48) we have, respectively,

$$E_+ = E_- + |\mathbf{u}_- - \mathbf{u}_+||p\cos\theta|,$$

$$E_- = E_+ + |\mathbf{u}_- - \mathbf{u}_+||p\cos\theta|.$$

The particle then gains energy at each cycle of crossing and back-crossing. Due to scattering, some particles may suffer many crossings and be strongly accelerated. If they suffer ν_1 crossings per second, their rate of energy increase is given by

$$\left\langle \frac{\mathrm{d}E}{\mathrm{d}t} \right\rangle = \nu_1 \sqrt{\tfrac{1}{2} m |u_- - u_+|^2 E} = \frac{\nu_1}{\sqrt{2}} \frac{|u_- - u_+|}{c} \sqrt{mc^2 E}.$$

The same reasoning can be made for ultra-relativistic particles (Bell, 1978a) and gives

$$\left\langle \frac{\mathrm{d}E}{\mathrm{d}t} \right\rangle = \frac{\nu_1}{2} \frac{|u_- - u_+|}{c} E.$$

The process then makes $\mathrm{d}E/\mathrm{d}t$ proportional to both the particle momentum and also the first power of the velocity with which scattering centres converge. It is then a first-order Fermi mechanism. For this mechanism to work it is necessary that the shock really looks like a discontinuity to the particles of interest, which implies in the case of oblique shocks that the shock thickness must be less than the particle gyroradius. In brief, the following ordering holds between the shock thickness δ, the Larmor radius ρ_g of particles to be accelerated, the size L of the part of the wake in which particles are diffused, and the overall geometrical size $\mathscr{L}$ of the structure in which the process is taking place (such as the distance to the Sun or the shock-curvature radius):

$$\delta \ll \rho_g \ll L \ll \mathscr{L}\ .$$

It is to be expected that the last inequality leads to some difficulties if one is to consider this mechanism as working in the flare region itself, where $\mathscr{L}$ can be rather small. Probably, it will be easier to satisfy for Type II-generating shocks which have already expanded largely in the corona, but the time-scale requirement that protons be accelerated within a time τ_{acc} (which we think $\simeq 100$ s but may be shorter) could turn out to set difficult limits on $\mathscr{L}$ ($\leqslant v_A \tau_{\mathrm{acc}}$).

As a second condition, it is necessary that the particles do indeed cross the shock many times before escaping (because of some geometrical effect) or disappearing (by impact on the photosphere, for example).

As a third condition, it is necessary that the scattering centres do really exist, both up- and down-stream. In earlier considerations of similar ideas, reflection was assumed to be produced by magnetic mirrors far away (Jokipii, 1966), or the nature of scattering centres was unspecified (Fisk, 1971). Being more specific, Bell (1978b) argues that the MHD waves responsible for scattering result from the shock process itself downstream, and are driven unstable upstream by reflected particles travelling faster than the Alfvén speed (Wentzel, 1974). The origin of these scattering waves is nevertheless postulated in some later works (Blandford and Ostriker, 1978), and no study of their generation near shocks nor examination of their possible observational manifestations has appeared up to now for the solar flare problem. Finally, it is also necessary that the particles be effectively coupled to the scattering centres. If these are MHD waves interacting resonantly with particles, the same injection threshold appears as discussed in §8.25 for interaction with MHD waves and whistlers. The relative efficiency of acceleration by a shock-crossing, first-order Fermi-process and by weak MHD turbulence depends on the shape of the turbulence spectrum, just as for second-order Fermi versus weak MHD turbulence processes (Lacombe, 1977).

8.24.2 Spectra

It is interesting to show now that the systematic acceleration process leads to power-law energy spectra. Bell (1978a) calculates this spectrum, averaging over the random elements, which are the angle of shock-crossing and the number of crossings. He first remarks that in the downstream region the density is constant in a stationary regime, because it is the only bounded solution of the equation balancing convection and diffusion; for positive x

$$u_+ f - D\frac{\partial f}{\partial x} = 0.$$

The flux of particles crossing from left to right is

$$F = \int_0^{\pi/2} \sin\theta \mathrm{d}\theta \int_0^{2\pi} \mathrm{d}\varphi v \cos\theta f(0, v), \tag{8.49}$$

where polar coordinates around the shock-normal (x-axis) have been used, and the (isotropic) distribution function is normalized to

$$\int_0^{\pi} \sin\theta \mathrm{d}\theta \int_0^{2\pi} \mathrm{d}\varphi f(0, v) = n(0, v).$$

One finds in (8.49), $F = (vn_0(v)/4)$. The flux of particles escaping far downstream at $+\infty$, where diffusion has ceased because gradients have vanished, is $u_+ n_0(v)$. Hence any particle crossing from up- to down-stream has a chance

$$\eta = (u_+ n_0)/(\tfrac{1}{4} v n_0) = 4u^+/v \tag{8.50}$$

of escaping. Treating each crossing as independent of the preceding one, the probability for a particle to perform at least n such crossings is $(1-\eta)^n$. For each pair of up–down and down–up crossings, the energy of a particle which is initially E_0 is increased to a value E_1, such that (assuming $E_1 - E_0 \ll E_0$)

$$E_1 = E_0 \left(1 + \frac{|u_- - u_+| p_0}{E_0} |\cos\theta_1|\right)\left(1 + \frac{|u_- - u_+| p_0}{E_0} |\cos\theta_2|\right),$$

where θ_1 and θ_2 are the angles made with the shock-normal at the moment of forward and backward crossings, respectively. After a great many such cycles, n say, the distribution of E, for a given injection E_0, will be strongly peaked around the mean (central limit theorem). One can then say that after n cycles the energy will be

$$E_n = E_0 \left\langle \left(1 + \frac{\Delta u p_0}{E_0} |\cos\theta_1|\right)\left(1 + \frac{\Delta u p_0}{E_0} |\cos\theta_2|\right)\left(1 + \frac{\Delta u p_1}{E_1} |\cos\theta_3|\right) \cdots \right\rangle.$$

For ultra-relativistic particles, this quantity is easily estimated for an isotropic distribution of particles directions. Bell (1978a) obtains

$$\ln\left(\frac{E_n}{E_0}\right) = n \frac{4}{3}\left(\frac{u_- - u_+}{c}\right). \tag{8.51}$$

The probability that, starting from E_0, the energy grows over E_n is $P(n)$. $P(n)$ is then the relative average number of particles having energies above $E(n)$. Then, using (8.50) and (8.51):

$$\ln(P(E > E_n)) = \ln(1-\eta)^n \cong -n\eta = -4n \frac{u_+}{c} = \frac{-3u_+}{u_- - u_+} \ln\left(\frac{E_n}{E_0}\right).$$

In brief, the proportion $\mathcal{N}(E)$ of particles having energies larger than E is

$$\mathcal{N}(E) \sim \left(\frac{E}{E_0}\right)^{\frac{3u_+}{u_- - u_+}}$$

and the energy spectrum is a power law

$$N(E)\mathrm{d}E=\mathrm{d}\mathcal{N}\sim\left(\frac{E}{E_0}\right)^{-\frac{u_-+2u_+}{u_--u_+}}\mathrm{d}E.$$

For a compression factor r in the shock defined by $u_-=ru_+$ (assuming roughly that the scattering centres do not move with respect to the fluid) one obtains

$$N(E)\sim\left(\frac{E}{E_0}\right)^{-\frac{r+2}{r+1}}. \qquad (8.52)$$

This result is identical with that obtained by Blandford and Ostriker (1978), who obtain (under essentially the same assumptions) a momentum distribution function down given by

$$f(p)\sim p^{-\frac{3r}{r+1}}. \qquad (8.53)$$

Keeping in mind that $N(E(p))\sim p^2f(p)$ and that in the ultra-relativistic limit $E=cp$, the result (8.52) above is recovered. Blandford and Ostriker arrive at this result by solving the proper kinetic equation for particles and take full account of the obliquity of the shock, which, as concluded by Bell (1978a), does not change the result provided the electric field can be transformed away by a Lorentz boost parallel to the plane of the shock. This is always possible except for nearly-perpendicular shocks (Hudson, 1965). We do not go here into a detailed description of the very elegant formulation and solution of the problem given by Blandford and Ostriker (1978). They start from a kinetic equation for the isotropic part of the distribution function of cosmic rays coupled by pitch-angle scattering to a background medium with a velocity field $\mathbf{u}$ (Hasselmann and Wibberenz, 1968; Skilling, 1975):

$$\frac{\partial f}{\partial t}+(\mathbf{u}\cdot\underset{\sim}{\nabla})f-\nabla\cdot\bar{\bar{\mathrm{D}}}\cdot\nabla f=\frac{1}{3}(\boldsymbol{\nabla}\cdot\mathbf{u})p\frac{\partial f}{\partial p}; \qquad (8.54)$$

and note that in the problem at hand $(\boldsymbol{\nabla}\mathbf{u})$ reduces to a delta function, namely $(\boldsymbol{\nabla}\cdot\mathbf{u})=(u_+-u_-)\delta(x)$. A solution can easily be found then by matching independent solutions obtained for $x>0$ and $x<0$ by jump conditions obtained by integrating (8.54) near $x=0$. The result (8.53) is obtained avoiding any probabilistic consideration, already included in (8.54), in agreement with Bell's results and proving its validity for oblique shocks.

It is then established that, provided the conditions for this mechanism to operate are met, *power-law distributions of accelerated particles can be formed as a result of shock-wave propagation.* It remains to know whether these conditions are actually met in solar-flare conditions, and how exactly the scattering centres are produced. Hopefully, this will be better known, both from the theory and observation, in the next few years. As stressed in this paragraph, the first-order Fermi acceleration by shock crossing also occurs above a certain injection energy too. We discuss in the next paragraph some evidence concerning the preheating mechanism of nuclei, before coming back briefly to acceleration processes occurring in the shock-front proper.

8.25 NUCLEI PREHEATING

8.25.1 First-phase heating?

It is difficult to have direct evidence of rather low-energy ions in the first phase (impulsive phase) of the flare. The fact that a thermal plasma at 10^9 K seems to be present (as judged from hard X-ray emission) is not a sufficient argument to conclude that ions are at the same temperature. For this reason, it is interesting to measure the ratio of electron to ion temperature at various phases of the flare. The characteristic energy-exchange-time for $T_e \gg T_i$ from Coulomb collisions (if no other exchange mechanism is at work) can be very long. However, the ions are heated to some extent in the ion-acoustic turbulence, as calculated by Manheimer (1977) at a rate comparable to that of the electrons. The details of ion heating and of its observational effects is still a matter for future investigation. Direct observational evidence for the existence of low-energy protons is difficult to obtain. It has, however, been pointed out by Orrall and Zirker (1976) that such fast nuclei impacting on the solar chromosphere should produce (by charge exchange) fast neutrals which will radiate Doppler-shifted photons, and this effect could be observed on the hydrogen Lyman-α line, before the flare heating blurs out completely this non-thermal effect. No completely satisfactory observation of this kind is yet available.

8.25.2 He^3 puzzle and nuclei preheating

By contrast, interesting clues to the heating mechanism which makes it possible for nuclei to be injected in the acceleration process of the second phase can be gained from a comprehensive analysis of He^3 abundance anomalies. In an early approach to this problem, Colgate *et al.* (1977) attempted to explain it as a result of thermonuclear reactions occurring in a dense and hot plasma. The difficulty with this point of view is to explain the absence of deuterium and tritium due, in their view, to the destruction of these elements by spallation and thermonuclear reactions taking place after the 100 MeV protons have caused the complete spallation of He^4. In this theory (Colgate, 1978), protons of high energy are claimed to be accelerated during the first phase of the flare. It was found necessary to have a thermal (200 keV) plasma at a density of 3×10^{21} m^{-3}. According to Colgate *et al.* (1977), this *flare plasma should be confined in thin filaments* (0.7 m thick). In view of the small scales involved, and of the fact that such high densities have never been reported, this view has been up to now considered as speculative.

An alternative explanation has been advanced by Fisk (1978), which makes use of more conservative parameters, but is rather definite as far as the interaction suffered by nuclei is concerned; nuclei should be preheated by some interaction taking place with electrostatic ion-cyclotron turbulence. Fisk remarks that, due to the existence of an injection threshold for second-phase acceleration mechanisms, the abundances of accelerated particles could differ very much from those of the ambient plasma if the mechanism which brings particles to the threshold is selective. If such a selective preheating could favour He^3 and certain heavies such as Fe, it would provide an explanation of the anomaly completely free of the deuterium–tritium difficulty. In this respect, it is noteworthy that completely-ionized He^3 is one of the very rare neutron-poor isotopes whose gyrofrequency is between the gyrofrequency of protons Ω_i and the gyrofrequency of completely-ionized He^4, the next most abundant element in the solar plasma after hydrogen; other neutron-poor isotopes are insignificantly abundant. In a pure hydrogen plasma, the ion-cyclotron waves all have frequencies larger than Ω_i, as was quoted already in §8.14. However, in the presence of a noticeable amount of helium, waves with frequencies smaller than Ω_i may propagate too, and may be excited as a result of current-driven instabilities in a frequency range in the vicinity of Ω_{He^3}, the gyrofrequency of completely-ionized He^3; this would lead to preferential heating of this isotope, and consequently to preferential injection in any second-phase mechanism with a high enough threshold. In cases of very strong preheating, almost all particles would reach the threshold, and the selection effect would become much less apparent than for weaker heating; this would be consistent with the fact that more modest events have more striking anomalies.

Taking into account hydrogen and helium ions, both completely-ionized, as major constituents of the plasma, the dispersion relation for electrostatic waves propagating in the presence of an electron current ($n_e e v_D$) is (§8.14 and Kennel and Kindel, 1971)

$$-(k_\perp^2 + k_\parallel^2)\lambda_D^2 = 4\frac{n_{H_e}}{n_e}\frac{T_e}{T_{H_e}}\sum_{n=-\infty}^{+\infty} I_n\left(\frac{k_\perp^2 v_{TH_e}^2}{2\Omega_{H_e}^2}\right) e^{-k_\perp^2 v_{TH}^2/2\Omega_{H_e}^2}$$

$$\times\left\{1+\frac{\omega}{k_\parallel v_{TH_e}} Z\left(\frac{\omega+n\Omega_{H_e}}{k_\parallel v_{TH_e}}\right)\right\} \ldots/\ldots$$

$$+\frac{n_H T_e}{n_e T_i}\sum_{n=-\infty}^{+\infty} I_n\left(\frac{k_\perp^2 v_{T_i}^2}{2\Omega_i^2}\right)e^{-k_\perp^2 v_{T_i}^2/2\Omega_i^2}\left\{1+\frac{\omega}{k_\| v_{T_i}} Z\left(\frac{\omega+n\Omega_i}{k_\| v_{T_i}}\right)\right\}$$

$$+\sum_{n=-\infty}^{+\infty} I_n\left(\frac{k_\perp^2 v_{T_e}^2}{2\Omega_e^2}\right)e^{-k_\perp^2 v_{T_e}^2/2\Omega_e^2}\left\{1+\frac{\omega-k_\| v_D}{k_\| v_{T_e}} Z\left(\frac{\omega-k_\| v_D+n\Omega_e}{k_\| v_{T_e}}\right)\right\},$$

where I_n is the modified Bessel function of order n, Z the complex plasma dispersion function, v_{TH_e}, v_{T_i}, v_{T_e} the thermal velocities of He^4, ions and electrons, respectively, and ω the complex eigenfrequency corresponding to wave vector $\mathbf{k}$, with components $k_\perp$ and $k_\|$ transverse to and along the constant magnetic field, respectively. It is sometimes convenient to introduce short notations for clarity, like (8.13):

$$\mu_e=\frac{k_\perp^2\ v_{T_e}^2}{2\Omega_e^2};\quad \mu_{He}=\frac{k_\perp^2\ v_{THe}^2}{2\Omega_{He}^2};\quad \mu_i=\frac{k_\perp^2\ v_{T_i}^2}{2\Omega_i^2};\quad z_e(n)=\frac{\omega-k_\| v_D+n\Omega_e}{k_\| v_{T_e}}.$$

The general equation may be simplified by the usual assumptions of low frequency ($\omega \ll \Omega_e$) waves longer than electron gyroradius ($k_\perp v_{T_e}/\Omega_e \ll 1$) and Debye length are $(k_\|^2+k_\perp^2)^{1/2}\lambda_D \ll 1$) and small electron drift speeds ($v_D \ll v_{T_e}$). It is, moreover, assumed (as in §8.14) that:

$$\left|\frac{\omega}{k_\|}-v_D\right| \ll v_{T_e};\quad \left|\frac{\omega+n\Omega_{He}}{k_\|}\right| \gg v_{THe};\quad \left|\frac{\omega+n\Omega_i}{k_\|}\right| \gg v_{T_i}.$$

The real part of the dispersion equation can be reduced then to a cubic for the real part of the frequency, ω, by retaining terms $-3 \leqslant n \leqslant +3$ for helium, $-1 \leqslant n \leqslant +1$ for ions (which contribute only to the second helium cyclotron resonance) and taking the small μ_e limit for electrons ($n=0$ term). With the same approximations, the growth-rate may be obtained from the imaginary part of the dispersion relation as a function of k and ω (known as a function of $\mathbf{k}$). Explicitly:

$$\gamma=\frac{\sqrt{\pi}}{v_{T_e}}\frac{T_{H_e}}{T_e}\frac{(v_D-v^*)}{D}$$

where

$$v^*(\omega,\ k_\perp,\ k_\|)=\frac{\omega}{k_\|}\left(1+\frac{n_i}{n_e}\sqrt{\frac{M_i}{m_e}}\frac{T_e^{3/2}}{T_i^{3/2}}\varphi_H(\omega,\ k_\perp,\ k_\|)+4\frac{n_{He}}{n_e}\sqrt{\frac{4M_i}{m_e}}\varphi_{He}(\omega,\ k_\perp,k_\|)\right)$$

$$D = 16\frac{n_{He}}{n_e}\left(I_1(\mu_{He})e^{-\mu_{He}}\frac{\omega\Omega_{He}^2}{(\omega^2-\Omega_{He}^2)^2} + I_2(\mu_{He})e^{-\mu_{He}}\frac{4\omega\Omega_{He}^2}{(\omega^2-4\Omega_{He}^2)^2}\right.$$

$$\left. + I_3(\mu_{He})e^{-\mu_{He}}\frac{9\omega\Omega_{He}^2}{(\omega^2-9\Omega_{He}^2)^2}\right) + 16\frac{n_H}{n_e}I_1(\mu_i)e^{-\mu_i}\frac{\omega\Omega_{He}^2}{(\omega^2-4\Omega_{He}^2)^2},$$

$$\varphi_H = I_1(\mu_i)\exp(-\mu_i)\exp -\left(\frac{\omega-\Omega_i}{k_{\|}v_{T_i}}\right)^2,$$

$$\varphi_{He} = I_1(\mu_{He})e^{-\mu_{He}}\exp-\left(\frac{\omega-\Omega_{He}}{k_{\|}v_{T_{He}}}\right)^2 + I_2(\mu_{He})e^{-\mu_{He}}\exp-\left(\frac{\omega-2\Omega_{He}}{k_{\|}v_{T_{He}}}\right)^2$$

$$+ I_3(\mu_{He})e^{-\mu_{He}}\exp-\left(\frac{\omega-3\Omega_{He}}{k_{\|}v_{T_{He}}}\right)^2.$$

A wave grows or damps according to whether v_D is larger or smaller than $v^*(\omega, k_{\|}, k_{\perp})$ for given plasma parameters. The game consists in determining under which conditions waves near the He^3 resonance grow first. For this, Fisk (1978) has calculated the smallest $v^*(\omega, k_{\perp}, k_{\|})$ such that some wave above $\Omega_i = 2\Omega_{He}$ becomes unstable, let us call it v_H^{**}, and the smallest $v^*(\omega, k_{\|}, k_{\|})$ such that some wave below Ω_i grows, let us call it v_{He}^{**}.

$$v_H^{**} = \inf v^*(\omega, k_{\perp}, k_{\|}), \quad \text{for } \omega > \Omega_i,$$

$$v_{He}^{**} = \inf v^*(\omega, k_{\perp}, k_{\|}), \quad \text{for } \omega < \Omega_i.$$

These critical drift speeds are still functions of the plasma parameters, in particular of the He^4 abundance n_{He}/n_H, which in a typical solar plasma should be near 0.1. This parameter however has never been directly measured in the corona itself, where the helium (completely ionized) emits no optical or UV line. The other important parameter is the ratio of electron to proton temperature (Fisk assumes equal helium and proton temperature). Of course, T_e/T_i must keep moderately large, because otherwise (namely if $T_e/T_i > 10$) the ion cyclotron waves do not grow unstable first, the ion-acoustic instability being excited prior to them (Kennel and Kindel, 1971). The requirements for efficient helium 3 heating are then that v_{He}^{**} be smaller than v_H^{**}, i.e. that waves below Ω_i grow first for reasonable values of n_{He}/n_H and T_e/T_i smaller than 10. Moreover, the frequency ω^* which is first excited below Ω_i can be numerically calculated as well. For efficient He^3 heating it must be resonant with He_2^3 ions,

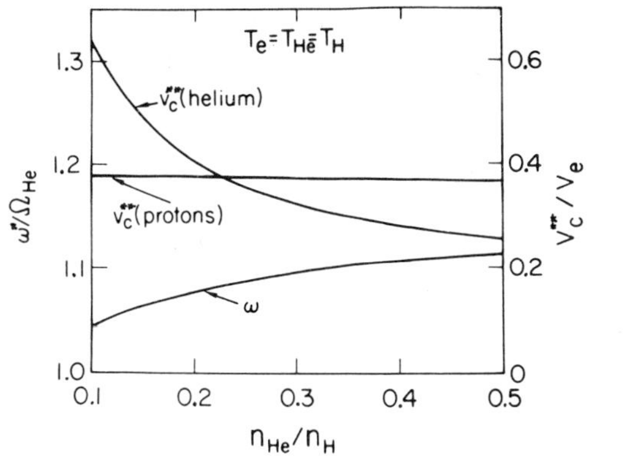

FIGURE 8.26 The minimum critical drift speed v_{He}**, the minimum critical drift speed v_H** and the frequency ω^* of the waves associated with v_{He}** plotted versus the relative concentration of $^4He^{+2}$ to protons. (From Fisk, 1978.)

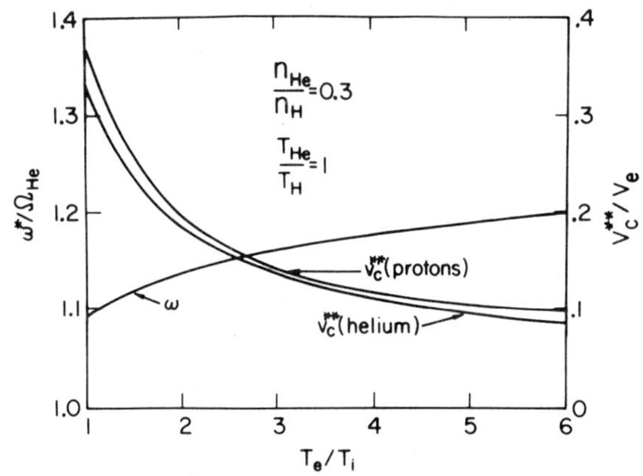

FIGURE 8.27 Same quantities as 8.26 plotted versus T_e/T_i. (From Fisk, 1978.)

namely:

$$|\omega^* - \Omega_{He_2^3}| < k_{\parallel} v_{THe_2^3}. \tag{8.55}$$

Note that $\Omega_{He_2^3} = \frac{4}{3}\Omega_{He_2^4} = \frac{2}{3}\Omega_i$. Figures 8.26 and 8.27 show the run of the critical velocities v** and of $\omega^*/\Omega_{He_2^4}$ as functions of the plasma parameters. It is seen that a noticeable, but not unreasonable, excess of helium is needed in the coronal plasma (namely n_{He}/n_H larger than 0.2 or 0.3) together with a large enough temperature ratio for the most unstable frequency to approach Ω_{He^3}, namely T_e/T_i larger than 5 (but still less than 10).

The calculation of the heating rate of helium 3, which interacts linearly as a minor ion with the turbulence, as well as the heating rate of protons and He^4, which may interact with the turbulence thanks to non-linear effects, requires the determination of the turbulence spectrum, that is, of the saturation mechanism. As already reported in §8.14 the non-linear evolution of this turbulence is still uncertain. Fisk estimated the heating rate semiqualitatively, assuming that the turbulence saturates by ion-resonance broadening (Dum and Dupree, 1970) and finds

$$\frac{1}{T_i}\frac{\mathrm{d}T_i}{\mathrm{d}t} \sim \frac{1}{2}\langle\gamma\rangle_i.$$

$$\frac{1}{T_{He^4}}\frac{\mathrm{d}T_{He^4}}{\mathrm{d}t} = \frac{1}{2}\langle\gamma\rangle_{He^4},$$

where the symbols $\langle\ \ \rangle_i$ and $\langle\ \ \rangle_{He^4}$ denote averages taken over frequencies larger and smaller than Ω_i respectively. He evaluates these growth-rates as if the temperature were still the initial one at $t=0$. The heating rate of He^3 is calculated taking into account the resonance condition (8.55), and is found to be given by:

$$\frac{1}{T_{He^3}}\frac{\mathrm{d}T_{He^3}}{\mathrm{d}t} = \frac{8\sqrt{\pi}}{3} I_1\left(\frac{\langle k^2\rangle v_{THe^3}^2}{2\Omega_{He^3}^2}\right) e^{-(\langle k_\perp^2\rangle v_{THe^3}^2)/(2\Omega_{He^3}^2)}\left(\frac{(\langle\omega\rangle-\Omega_{He^3})^2}{\langle k_\parallel\rangle v_{THe^3}}\right)\exp - \frac{(\langle\omega\rangle-\Omega_{He^2})^2}{\langle k_\parallel\rangle 2v_{THe^3}^2}.$$

The results of these calculations is shown in Figure 8.28. It has been assumed that the heating proceeds only while He^4 is heated, because the instability in the frequency range below Ω_i ceases when v_{He}^{**} exceeds v_D as a result of helium heating. He^3 reaches higher thermal speeds than either He^4 or protons. As a result, the ratio of He^3 to He^4 ions whose velocity exceeds the second-phase threshold may be raised to values of order unity, though normally the isotopic ratio for the full populations is near 10^{-4}. It seems then that the mechanism could be able to meet the requirements for effective pre-acceleration. It is even more interesting to remark that it can also explain high levels of preheating for heavier ions, which could interact by a second-harmonic resonance, with a heating rate for an ion s:

$$\frac{1}{T_s}\frac{\mathrm{d}T_s}{\mathrm{d}t} = \frac{8\sqrt{\pi}}{3} I_2(\langle\mu_s\rangle) e^{-\langle\mu_s\rangle}\left(\frac{Z_s}{Z_{He^4}}\right)^2\left(\frac{T_{He^4}}{T_s}\right)^2 \frac{(\langle\omega\rangle-\Omega_{He^4})^2}{\langle k_\parallel\rangle v_{T_s}} e^{-(\{\langle\omega\rangle-2\Omega_s\}/\{\langle k_\parallel\rangle v_{T_s}\})^2}.$$

This heating rate is larger for heavier nuclei because of its Z_s^2 dependence, and, because of the resonance condition, it will favour charge states able to resonate with $\langle\omega\rangle/2$, namely those ions for which

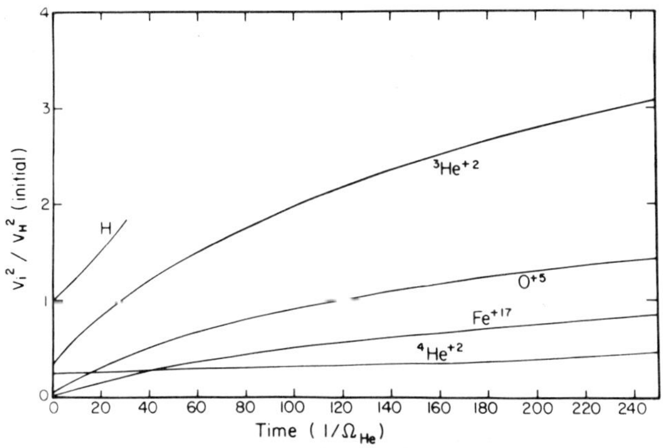

FIGURE 8.28 The increase in the square of various ion speeds with time that result from the ion-cyclotron heating mechanism. Ion thermal speeds are compared to v_{Ti} (initial). The various ions are assumed to be at equal temperatures initially. (From Fisk, 1978.)

$$2\Omega_s = \langle\omega\rangle \simeq \Omega_{He_2^3}; \quad \text{or} \quad \frac{Z_s}{A_s} \simeq \tfrac{1}{3}.$$

For the most abundant elements after helium, the heating should act most effectively on $^{12}C^{+4}$, $^{16}O^{+5}$, $^{56}Fe^{+17}$. Provided these charge states are reasonably abundant in the ambient plasma, a preferential injection of these ions is to be expected, especially for iron, because the ratio of final-to-initial thermal speed is larger for this nucleus than for $^{16}O^{+5}$, and the percentage of $^{56}Fe^{17}$ may be more important than $^{16}O^{+5}$ in the preflare coronal plasma.

It is interesting to know that recent measurements of charge states of low-energy solar cosmic rays by Gloeckler seem to confirm the over-abundance of precisely these ions (Fisk, 1979, private communication). It is clear that, when completely confirmed, *this piece of observational evidence will have to be considered as one of the most important clues to first-phase ion acceleration.* It implies not only that an *electrostatic ion-cyclotron instability is excited* (which requires inequalities mentioned above concerning n_{He}/n_H and T_e/T_i) but also that the *heating proceeds in a low β-plasma* carrying a large enough current density. We summary these conditions

$$\frac{n_{He}}{n_H} > 0.2,\ 0.3; \qquad 5 \leqslant \frac{T_e}{T_i} \leqslant 10; \qquad \beta \leqslant 10^{-3}; \qquad v_D \geqslant 0.1\ v_{T_e}.$$

The β-condition implies $n \leqslant 10^{15}\ \text{m}^{-3}$ for $T = 1.5 \times 10^6$ K and $B = 75$ G. The condition on the drift velocity still implies a very thin current sheet if the instability is to grow in a current-carrying structure. It is possible, however, that the instability is the result either of a return-current (with no net current) or of some thermal-conduction instability. It is not yet known whether other instabilities working under different conditions concerning the temperature ratio, the plasma's β or the He/H ratio could meet the same degree of success.

8.26 ACCELERATION IN AND NEAR SHOCK FRONTS

8.26.1 Conditions near collisionless shocks

Shocks propagating in the corona are collisionless. The structure of *collisionless shock waves* has been reviewed by various authors, in particular Sagdeev (1966) and Tidman and Krall (1971). Particular attention has been paid to *perpendicular shocks* for which the direction of the normal to the front is perpendicular to the magnetic field. Their laminar structure has been described by Sagdeev, who shows that (in the absence of any dissipation) solitons of width c/ω_p can propagate for Alfvén mach number less than 2. At this critical mach number, ions are reflected by the potential barrier of the solitary wave, and the laminar motion as such ceases to exist. In the presence of dissipation, the solitary wave turns into a succession of oscillations of the magnetic field described by the equation (Sagdeev, 1966)

$$-\frac{c^2}{\omega_p^2}\frac{\mathrm{d}^2H}{\mathrm{d}x^2}=H_0-H+H\frac{H^2-H_0^2}{2\mu(n_0M_iu^2)}+\frac{c^2}{\omega_p^2}\frac{v\mathrm{d}H}{u\,\mathrm{d}x},$$

where v is the effective collision-frequency, u the velocity of the shock, and H_0 is the magnetic field in the non-shocked region. The solution settles down to a higher value of H, namely

$$H^*=H_0(\sqrt{M^2+\tfrac{1}{4}}-\tfrac{1}{2}),$$

and peaks at a maximum $H_{\max}\simeq H_0(2M-1)$. For high enough mach numbers, ion reflection occurs (as for the dissipationless soliton) by the electrostatic potential of the first solitary wave of the train. In any case, this potential reflects at least those particles which have in the shock rest-frame a velocity small enough in the direction of propagation of the wave. Sagdeev (1966) first noted that these ions will be turned again to the shock by the magnetic field H_0 downstream and reflected again. These particles are, so to speak, tied to the shock rest-frame and suffer acceleration parallel to the shock front (y-direction) by the static electric field which exists in that frame.

8.26.2 DC-electric-field acceleration by collisionless shocks

Reflected particles must have in the shock frame a velocity v_x smaller than $2v_A\sqrt{M-1}/M$. Acceleration cannot proceed indefinitely, because, when v_y has grown large enough, the Lorentz force will pull the accelerated ion through the potential barrier. The maximum energy of such an ion is then

$$\frac{M_i}{m_e}\left(M_i\frac{u^2}{2}\right)$$

(Sagdeev, 1966), where u is the shock velocity. For parameters appropriate to the solar corona, this limit energy is of the order of 0.3–30 MeV, as discussed recently by Gubchenko and Zaitsev (1979). These authors have made a detailed analysis of reflected ion motion, assuming mirror-like reflection from the shock.

8.26.3 Wave-particle interaction?

It must first be stressed that collisionless shock waves sustain very high current densities and are very likely to give rise to microinstabilities, although this requires some specific study in each case. For example, Sagdeev (1966) shows that in a perpendicular shock the drift velocity of electrons would exceed the thermal velocity if the Alfvén mach number (calculated with parameters of the unperturbed plasma) exceeds

$$M^* = 1 + \tfrac{3}{8}\beta^{1/2}.$$

The waves that are able to grow in the shock at a high enough rate for them to reach high amplitudes in less than the time it takes for the front to pass by are essentially those which result from current-driven instabilities, such as Buneman waves and ion-acoustic waves. Care must be taken to include properly in the stability analysis the inhomogeneous nature of the shock (e.g. Priest and Sanderson (1972) for ion-acoustic instability and Sanderson and Priest (1972) for Bernstein instability). As far as ion-acoustic and Buneman turbulence are concerned, it has been shown by Krall and Smith (1975) that their interaction with electrons leads only to a broadening of the distribution function, not to the acceleration of tails up to high energies. The idea developed at some length by various authors (Tsytovitch *et al.*, 1975; Hoyng, 1977) that high-frequency Langmuir turbulence could be produced near shock fronts as a result of non-linear interactions with ion-acoustic waves has already been mentioned and criticized in §8.15.

8.26.4 Role of lower hybrid waves

The task of examining the role of the third prevalent instability in shock waves, namely the *modified two-stream instability* and *lower-hybrid drift instability* has begun recently with a paper by Lampe and Papadopoulos (1977), who argue that, due to their long parallel wavelengths along the field, such waves would be able to accelerate particles up to MeV energies under coronal conditions. Let us briefly recall that the modified two-stream instability is due to a drift of electrons with respect to ions perpendicular to a magnetic field. Strictly speaking, the current is associated with field gradients and also with gradients in the other plasma parameters (density and temperature). Under certain conditions, the perturbations may be analysed in the quasi-electrostatic approximation and also in the local approximation, neglecting all gradients. It is found in that case that the dispersion relation is very similar in form to the hydrodynamic Buneman dispersion relation for drifting electrons, as can be judged from Table 8.2. The instability grows at some small angle to the direction of the electron drift, but the growth-rate vanishes in directions strictly perpendicular to the magnetic field.

This zero-gradient approximation has been discussed in detail by McBride *et al.* (1972), who established from kinetic theory a dispersion relation, which reduces, in the fluid limit, in the plane containing the magnetic field and drift velocity, and in the electron rest frame, to:

$$1-\frac{k^2}{k_\perp^2}\frac{\Omega_e\Omega_i}{(\omega-\mathbf{k}\cdot\mathbf{u})^2}-\frac{k_\parallel^2}{k_\perp^2}\frac{M_i}{m_e}\frac{\Omega_e\Omega_i}{\omega^2}=0,$$

where $\mathbf{u}$ is the drift velocity of electrons with respect to ions. As compared to the Buneman dispersion relation, the ion plasma frequency is approximately replaced by $(\Omega_e\Omega_i)^{1/2}$, the lower hybrid frequency, Ω_{LH}. while the electron plasma frequency is replaced by $k_\parallel/k_\perp(M_i/m_e)^{1/2}\Omega_{LH}$. These two quantities are not in the ratio $(m_e/M_i)^{1/2}$ as for the field-free case. Everything is as if the electron-ion mass ratio had been multiplied by $k_\perp^2/k_\parallel^2$. It is often said that the electrons acquire an effective mass $(k_\parallel/k_\perp{}^2\, m_e=\bar{m}_e$. The effect is due to the fact that electrons perceive only a small fraction of the electric force, being tied to the magnetic field lines. The growth-rate becomes in this case of the order of Ω_{LH}, just of the same order as the frequency of unstable oscillations, and it peaks for $k_\parallel/k\sim(m_e/M_i)^{1/2}$.

Kinetic effects of interaction with electrons limit this growth-rate severely at larger angles. For large T_e/T_i ratio, the damping becomes less effective, just as in the case of ion-acoustic waves. Electromagnetic effects yield an equation of the same form (McBride *et al.*, 1972), but with a different k-dependence of the characteristic frequencies. The effect is stabilizing for $|u|\geqslant(v_A^2+c_S^2)^{1/2}$, while low phase velocity waves are damped by ions. The instability in this simple local form occurs only for

$$c_S < |\mathbf{u}| < (v_A^2 + c_S^2)^{1/2}.$$

The situation changes somewhat if gradients need to be included. The modified two-stream instability, or lower hybrid two-stream instability, which has a wavelength longer than electron gyroradius, competes with a drift-wave instability, also driven by the electron current. These two regimes have been described by Krall and Liewer (1971). The instability in the last case of shorter wavelength is flute-like ($k_{\|}=0$). It is called the lower-hybrid drift instability. Krall and Liewer (1971) show that, in the limit of weak but non-negligible gradients, the electrostatic waves with frequency much larger than Ω_i have the dispersion relation

$$k^2\lambda_D^2 = -\frac{T_e}{T_i}\left(1+\frac{\omega}{kv_{T_i}}Z\left(\frac{\omega}{kv_{T_i}}\right)\right) - 1 + \frac{1}{\pi^{3/2}}\frac{1}{v_{T_e}^3}\sum_{l=-\infty}^{+\infty} 2\pi\int_0^{\infty} v_\perp \mathrm{d}v_\perp \int_{-\infty}^{+\infty} \mathrm{d}v_\|$$

$$\left\{\omega - \frac{k_y v_{T_e}^2}{2\Omega_e}\left(\varepsilon' + \alpha\frac{v_\perp^2}{v_{T_e}^2}\right)\frac{1}{\omega - k_y v_d - k_\| v_\| - l\Omega_e} J_l^2\left(\frac{k_\perp v_\perp}{\Omega_e}\right) e^{-v^2/v T_e^2}\right\},$$

where ε' and α are defined as gradient-scales in the inhomogeneity direction, **B** being oriented in the z-direction, and

$$\alpha = -\frac{1}{T_e}\frac{\mathrm{d}T_e}{\mathrm{d}x}; \qquad \varepsilon = -\frac{1}{B}\frac{\mathrm{d}B}{\mathrm{d}x}; \qquad \beta = \frac{nk_B T_e}{B^2/2\mu};$$

$$v_d = \varepsilon\frac{v_\perp^2}{2\Omega_e} - \frac{E}{B}.$$

For propagation purely perpendicular to the magnetic field, and neglecting the gradients, the above equation describes short wavelength waves of frequency larger than Ω_e (electron-cyclotron waves, Gary and Sanderson, 1970) and ion-acoustic waves. Retaining the gradients (or a small $k_\|$) opens a branch of low-frequency long-wavelength waves, with $\Omega_i \ll \omega \ll \Omega_e$ and $k_\perp\ \rho_{ge} \ll 1$. The dispersion relation simplifies in that limit. Drift modes appear alone for $k_\|=0$; in the hydromagnetic limit of phase velocity larger than v_{T_i} the dispersion relation becomes

$$(\omega - k_y v_E)\left(\frac{\omega^2}{\omega_{pi}^2}\left(1+\frac{\omega_{pe}^2}{\Omega_e^2}\right) - 1\right) = -k_y\frac{v_\Delta}{k^2\lambda_D^2}\frac{\omega^2}{\omega_{pi}^2},$$

$$v_E = \frac{E}{B}; \qquad v_\Delta = -\frac{1}{2}\frac{v_{T_e}^2}{\Omega_e}\left(\frac{1}{n_e}\frac{\mathrm{d}n_e}{\mathrm{d}x} - \frac{1}{B}\frac{\mathrm{d}B}{\mathrm{d}x} - \frac{k^2\rho_{ge}^2}{2}\frac{1}{T_e}\frac{\mathrm{d}T_e}{\mathrm{d}x}\right).$$

For negligible gradients, we obtain a stable drift wave $\omega = k_y v_E$ and a stable lower-hybrid wave $\omega^2 = \omega_{pi}^2(1+(\omega_{pe}^2/\Omega_e^2))^{-1} \simeq \Omega_e\Omega_i$. The gradients couple these two stable modes, leading to instability. In the weak drift limit, $v_E v_\Delta \ll c_S^2$, Krall and Liewer obtain

$$\omega = \omega_R + i\gamma; \qquad v_E v_\Delta \ll c_S^2,$$

$$\omega_R = \omega_{pi}\left(1 + \frac{\omega_{pe}^2}{\Omega_e^2}\right)^{-1/2} \simeq \Omega_{LH}; \qquad \gamma = \omega_R \frac{k_y}{k}\left(\frac{v_E v_\Delta}{c_S^2}\right)^{1/2};$$

and in the opposite limit

$$\omega_R = \omega_{pi}\left(1 + \frac{\omega_{pe}^2}{\Omega_e^2}\right)^{-1/2}\left(1 + \frac{1}{22}\left(\frac{k_y}{k}\left(\frac{v_\Delta v_E}{c_S^2}\right)^{1/2} - 6\right)\right),$$

$$\gamma \simeq \omega_{pi}\left(1 + \frac{\omega_{pe}^2}{\Omega_e^2}\right)^{-1/2}\sqrt{2}; \qquad v_E v_\Delta \gg c_S^2.$$

Instability requires $v_\Delta v_E > 0$ in both cases.

When the gradients are weak this instability competes with the modified two-stream instability described above. Gladd (1976) has included both effects in a single numerical calculation of growth-rates and shown that the presence of gradients has the result of shifting the largest growth-rate towards $k_\parallel = 0$, while the wavelength of the most unstable wave becomes comparable to ρ_{ge}. It is found that both instabilities in fact merge into a single process, in which the lower-hybrid drift instability makes the most important contribution.

The role of these instabilities in collisionless shock waves, in particular in oblique shock waves (McBride *et al.*, 1972), may be major, because conditions are just adequate for them to grow and because, as hydrodynamic instabilities, they are expected to be less easily saturated than kinetic instabilities. Their role as particle accelerators may be quite effective because their parallel phase velocity is large, so that they are able to resonate quite effectively with the tail of the electron distribution function, or, stated another way, they give rise to field-aligned electric fields having a rather long coherence time, of the order of Ω_{LH}. The effect of this turbulence depends on the saturation mechanism, which is still at present poorly known. McBride *et al.* (1972) show that the quasilinear interaction leads to almost equal heating of ions in the direction perpendicular to **B** and electrons in the direction parallel to **B**. These thermal energies grow at the expense of the streaming velocity. Their numerical simulation, however, shows evidence of strong trapping effects, though the initial development is compatible with a quasilinear prediction, in particular the equality of perpendicular ion and parallel electron heating rates.

According to Lampe and Papadopoulos (1977), the level of saturation trapping given by McBride *et al.* (1972) is excessive. Keeping in mind that the free energy source for the instability is the electron kinetic energy, they argue that saturation must occur at lower levels by quasilinear interaction, reducing the drift velocity u to the marginally stable value c_S. Their turbulent shock-front model consists of a thin layer, where lower hybrid waves grow at a very large rate and saturate, followed by a thicker region at marginal stability, where the effective coupling between ions and electrons is provided by the lower-hybrid turbulence. The spectrum consists then of stable lower-hybrid waves

$$\omega=\Omega_{LH}\left(1+\frac{k_\parallel^2}{k^2}\frac{M_i}{m_e}\right)^{1/2},$$

excited in the whole range where electron Landau damping is not excessive, namely

$$\frac{l}{\mathscr{L}}<k_\parallel\ll\frac{\Omega_{LH}}{v_{T_e}},$$

where $\mathscr{L}$ is the size of the shock front measured along the field. The level of turbulence can be guessed from the marginal stability analysis. McBride *et al.* (1972) find that the quasilinear rate of streaming-energy loss is

$$\frac{\mathrm{d}}{\mathrm{d}t}\left(\tfrac{1}{2}nmu^2\right)=-3\frac{\omega_{pe}^2}{\Omega_e^2}\frac{\mathrm{d}}{\mathrm{d}t}W_{\text{waves}},$$

which can be translated into a friction force

$$\mathscr{F}=\frac{3}{u}\frac{\omega_{pe}^2}{\Omega_e^2}\frac{\mathrm{d}}{\mathrm{d}t}W_{\text{waves}}=\frac{3}{u}\frac{\omega_{pe}^2}{\Omega_e^2}\gamma W_{\text{waves}}\simeq\frac{3}{u}\frac{\omega_{pe}^2}{\Omega_e^2}\Omega_{LH}W_{\text{waves}},$$

which itself implies a collision-frequency

$$\nu_{\text{coll}}^*=3\frac{\omega_{pe}^2}{\Omega_e}\sqrt{\frac{m_e}{M_i}}\frac{W_{\text{waves}}}{nm_eu^2}.$$

Ohm's law written in the shock front, where the electric field is $v_A \Delta B$, and the current density fixed by marginal stability $n_e e c_S$ yields an independent estimation of the collision-frequency ν^*_{coll}, and then of the wave level:

$$\frac{n_e e^2}{m_e \nu^*_{\text{coll}}} v_A \Delta B = n_e e c_S.$$

The two preceeding equations give then (correcting what we think is an algebraic error in the Lampe and Papadopoulos (1977) derivation of their equation (10) from (6) and (9))

$$\frac{W_{\text{waves}}}{n m_e u^2} = \frac{1}{3} \frac{\Omega_e^2}{\omega_p^2} \frac{v_A}{c_S} \frac{\Delta B}{B} \sqrt{\frac{M_i}{m_e}} = \frac{1}{3} \frac{\Delta B}{B} \left(\frac{M_i}{m_e}\right)^{3/2} \frac{v_A^3}{c^2 c_S}.$$

Assuming this energy to be equally shared among the excited waves, the quasilinear equation for parallel momentum can easily be written:

$$\frac{\partial}{\partial t} f(p_\parallel) = \frac{\partial}{\partial p_\parallel} D_{\parallel\parallel} \frac{\partial f}{\partial p_\parallel} = \frac{\partial}{\partial p_\parallel} 2\pi \frac{\omega_{pe}^2 m_e u^2}{n v_\parallel^3} \frac{W_{\text{waves}}}{\Omega_{LH}/v_{T_e} - \mathscr{L}^{-1}} \frac{\partial f}{\partial p_\parallel}.$$

For relativistic electrons, $v_\parallel = c$, this equation is readily solved, and shows that the particle energy reached after a time t is on average

$$\langle E \rangle = 2c(D_{\parallel\parallel} t)^{1/2}. \tag{8.56}$$

The interaction time is restricted by wave escape along the field, which is of the order of $\mathscr{L}/v_A$. Putting this value of t, together with the expressions for $D_{\parallel\parallel}$ and W_{waves} obtained above, in the expression (8.56) for $\langle E \rangle$

$$\langle E \rangle = m_e v_{T_e}^2 \left(8\pi \mathscr{L} \frac{\omega_p}{c} \frac{\omega_p^2}{\Omega_e^2} \frac{v_A^3}{c^3} \frac{M_i}{m_e} \right)^{1/2}.$$

For the figures adopted by Lampe and Papadopoulos, $\omega_{pe} = 1.5 \times 10^8\ \text{s}^{-1}$, $\Omega_e = 10^7\ \text{s}^{-1}$, $v_A \simeq 10^6\ \text{m s}^{-1}$, $\mathscr{L} = 10^8$ m, we obtain $\langle E \rangle \simeq 10^3\ m_e v_{T_e}^2$, which for coronal temperatures ($T_e \simeq 1$ keV) may reach 1 MeV. The number of electrons produced at these energies is restricted by the free energy available, which in this case gives a density of accelerated particles less than $10^{-7} \times$ the ambient density. Lampe and Papadopoulos (1977) conclude from their analysis that electron non-thermal tails may be formed near shock fronts, giving rise to enhanced bremsstrahlung. This would open the possibility of understanding radiation at ω_{pe} and $2\omega_{pe}$ (Tidman, 1965; Tidman *et al.*, 1966) together with the acceleration of some epithermal particles up to fairly high energies. Of course, this theory is still in quite a preliminary state and deserves more detailed attention in the future.

REFERENCES

Abrami, A. *Solar Physics* **11**, 106 (1970).
Achterberg, A. Preprint (1979).
Alisandrakis, C. E. and Kundu, M. R. *Ap. J.* **222**, 342 (1978).
Alfvén, H. and Carlqvist, P. *Solar Physics* **1**, 220 (1967).
Altyntsev, A. T., Krasov, V. I., Lebedev, V. N. and Paperny, V. Y. *Supplt Journal de Physique* **67**, 723 (1979).
Anglin, J. D., Dietrich, W. F. and Simpson, J. A. 15*th Int. Conf. Cosmic Rays* **5**, 43 (1977).
Antiochos, S. and Sturrock, P. A. *Solar Physics* **49**, 359 (1976).
Bai, T. and Ramaty, R. *Solar Physics* **49**, 343 (1976).
Baum, P. J. and Bratenahl, A., *Solar Physics* **47**, 331 (1976).
Baum, P. J. and Bratenahl, A. *J. Plasma Physics* **18**, 257 (1977).
Baum, P. J., Bratenahl, A. and Kamin, G. *Ap. J.* **226**, 286 (1978).
Bekefi, G., *Radiation Processes in Plasmas*, John Wiley, New York (1966).
Bell, A. R. *Month. Not. Roy. Ast. Soc.* **182**, 147 (1978a).
Bell, A. R. *Month. Not. Roy. Ast. Soc.* **182**, 443 (1978b).
Benz, A. O. *Ap. J.* **211**, 270 (1977).
Benz, A. O. and Tarnstrom, G. L. *Ap. J.* **204**, 597 (1976).
Berger, J. M., Newcomb, W. A., Dawson, J. M., Frieman, E. A., Kulsrud, R. M. and Lenard, A. *Physics of Fluids* **1**, 301 (1958).
Berger, J. M. and Seltzer, S. M., *Tables of Energy Losses* NASA S.P. 3012 (1964).
Berk, H. L. and Roberts, K. V. *Physics of fluids* **10**, 1595 (1967).
Bernstein, I. B., Green, J. M. and Kruskal, M. P. *Phys. Rev.* **108**, 546 (1957).
Bertsch, D. L. and Reames, D. V. *Solar Physics* **55**, 491 (1977).
Biskamp, D. and Chodura, R. *Phys. Rev. Letters* **27**, 1553 (1971).
Biskamp, D. and Schindler, K. *Plasma Physics* **13**, 1013 (1971).
Blandford, R. D. and Ostriker, J. P. *Ap. J. Letters* **221**, L29 (1978).
Block, L. P. *Astrophysics and Space Science* **55**, 59 (1978).
Böhme, A., Furstenberg, F., Hildebrandt, J., Saal, O., Kruger, A., Hoyng, P. and Stevens, G. A. *Solar Physics* **53**, 139 (1977).
Boris, J. P., Dawson, J. M., Roberts, K. V. and Orens, J. *Phys. Rev. Letters* **25**, 706 (1970).
Brown, J. C. *Solar Physics* **26**, 441 (1972).
Brown, J. C., *Solar Physics* **31**, 145 (1973).
Brown, J. C., *Solar Physics* **36**, 371 (1974).
Brown, J. C. and Melrose, D. B., *Solar Physics* **52**, 117 (1977).
Brown, J. C., Canfield, R. C. and Robertson, M. N., *Solar Physics* **57**, 399 (1978).
Brown, J. C., Melrose, D. N. and Spicer, D. S., *Ap. J.* **228**, 592 (1978).
Bulanov, S. V. and Sasorov, P. V., *Astronomietzkii Journal* **52**, 763 (1975).
Buneman, O., *Phys. Rev.* **115**, 503 (1959).
Canfield, R. C. *et al.*, *Solar Flares* (ed. P. Sturrock), Colorado Associated University Press, p. 451 (1980).
Carlqvist, P., *Cosmic Electrodynamics* **3**, 377 (1972).
Crannell, C. J., Frost, C. Mätzler, K., Ohki, J. L., Saba, J. L., *Ap. J.* **223**, 620 (1978).
Cartwright, D. G. and Kellog, P. J., *Nature Phys. Sci.* **231**, no 18, 11 (1971).
Cassé, M., Thesis University of Paris VII (1976).
Cassé, M., Preprint (1979).
Cassé, M., Goret, P., Cesarskii, C. J., *Proceedings 14th International Cosmic Rays Conference Munich* **2**, 646 (1975).
Cassé, M. and Goret, P., *Ap. J.* **221**, 703 (1978).

Chambon, G., Hurley, K., Niel, R., Talon, R., Vedrenne, G., Likine, O. B., Kuznitzov, A. V. and Estuline, I. V., in *Gamma Rays Spectroscopy in Astrophysics*, Cline, T. L. and Ramaty, R., (eds) (1978).

Chew, G. F., Goldberger, M. L. and Low, F. E. *Proc. Roy. Soc.* **A236**, 112 (1956).

Choi, D. and Horton, W., *Physics of Fluids* **17**, 2048 (1974).

Choi, D. and Horton, W., *Physics of Fluids* **18**, 858 (1975).

Chupp, E. L., Forrest, D. J., Hegbie, P. R. and Suri, A. N., in Kane, S. (ed.) *Proceedings of the IAU Cospar Symposium*, Reidel (1975).

Colgate, S.., *Ap. J.* **221**, 1068 (1978).

Colgate, S. A., Audouze, J. and Fowler, W. A., *Ap. J.* **213**, 843 (1977).

Coppi, B., *Ap. J.* **195**, 545 (1974).

Coppi, B., Galvao, R., Rosenbluth, M. N. and Rutherford, P. H., *Fiz. Plasmy* **2**, 961–966 (1976).

Craig, I. J. D. and McClymont, A. N., *Solar Physics* **50**, 133 (1976).

Crannel, J. C., Ramaty, R., Joyce, G. and Werntz, C., *Ap. J.* **210**, 582 (1976).

Datlowe, D. W., O'Dell, S. L., Peterson, L. E. and Elcan, M. J., *Ap. J.* **212**, 561 (1977).

Davis, L., *Phys. Rev.* **101**, 351 (1957).

Davis, J. and Rogerson, J. E., *Solar Physics* **51**, 185 (1977).

Dawson, J., *Advances in Plasma Physics* **1** (1968).

Dawson, J. and Oberman, C., *Physics of Fluids* **6**, 390 (1963).

Dietrich, W. F., *Ap. J.* **180**, 955 (1973).

Dietrich, W. F. and Simpson, J. A., *Ap. J. Letters* **225**, L41 (1978).

Dobrowolny, M., Mangeney, A. and Veltri, P. L., *Astron & Astrophys.* **83**, 26 (1980).

Donnelly, R. F., in *High Energy Phenomena on the Sun*, Ramaty, R. and Stone, R. (eds.), NASA S.P. 342, 242 (1973).

Donnelly, R. F. and Kane, S. R., *Ap. J.* **222**, 1043 (1978).

Drake, J. F. and Lee, Y. C., *Physics of Fluids* **20**, 1341 (1977).

Dum, C. T. and Dupree, T. D., *Physics of Fluids* **13**, 2064 (1970).

Dum, C. T., Chodura, R. and Biskamp, D., *Phys. Rev. Letters* **32**, 1231 (1974).

Dupree, T. H., *Physics of Fluids* **11**, 2680 (1968).

Elwert, G. and Haug, E., *Solar Physics* **15**, 234 (1970).

Emslie, G., *Ap. J.* **224**, 241 (1978).

Emslie, G., Preprint (1979).

Emslie, G., Brown, J. C. and Donnelly, R. F., *Solar Physics* **57**, 175 (1978).

Feldman, W. C., *Solar Wind III* Russell, C. T. (ed.) (1974).

Fermi, E., *Phys. Rev.* **75**, 1169 (1949).

Field, G. B., *Ap. J.* **142**, 531 (1965).

Fisk, L. A., *J.G.R.* **76**, 1662 (1971).

Fisk, L. A., *Ap. J.* **224**, 1048 (1978).

Fried, B. D. and Comte, S. D., *The Plasma Dispersion Function*, Academic Press, New York (1961).

Frost, K., Dennis, B. R. and Lencho, R. J., *New Techniques in Space Astronomy*, Labuhn, N. R. and Lüst, R. (eds.) 185–191 (1971).

Frost, K. J. and Dennis, B., *Ap. J.* **165**, 655 (1971).

Galeev, A. A. and Sagdeev, R. Z., *Non-Linear Plasma Theory*, Benjamin Inc., New York (1969).

Galeev, A. A., Sagdeev, R. Z., Sorov, Y. Z., Shapiro, V. D. and Shevchenko, V. I. *Soviet Journal of Plasma Physics* **1** (1975).

Garrard, J. L., Stone, E. C. and Vogt, R. E., *High Energy Phenomena on the Sun*; Ramaty, R. and Stone, R. G., (eds), NASA SP 342, 341 (1973).

Gary, S. P. and Sanderson, J., *J. Plasma Physics* **4**, 739 (1970).

Ginzburg, V. L. and Syrovatsky, S. I., *The Origin of Cosmic Rays*, McMillan, New York (1964).

Gladd, N. T., *Plasma Physics* **18**, 27 (1976).

Goertz, C. K. and Joyce, G., *Astrophys and Space Science* **32**, 16 (1975).

Gubchenko, V. M. and Zaitsev, V. V., *Solar Phys.* **63**, 337 (1979).

Hall, D. E. and Sturrock, P. A., *Physics of Fluids* **10**, 1593 (1967).

Hammer, D. A. and Rostoker, N., *Physics of Fluids* **13**, 1831 (1970).

Hasselmann, K. and Wibberenz, G., *Journal of Geophysics* **34**, 354 (1968).

Heitler, W., *The Quantum Theory of Radiation*, Oxford University Press (1936).

Henoux, J. C., *Solar Physics* **42**, 219 (1975).

Heyvaerts, J., *Ann. d'Astrophys.* **31**, 129 (1968).
Heyvaerts, J., *Astron. Astrophys.* **37**, 65 (1974).
Heyvaerts, J., *Highlights in Solar Physics*, p. 193, CNRS, Paris (1978).
Heyvaerts, J. and Priest, E. R., *Solar Physics* **47**, 223 (1976).
Heyvaerts, J., Priest, E. R. and Rust, D., *Ap. J.* **216**, 123 (1977).
Heyvaerts, J. and Kuperus, M., *Astron. Astrophys.* **64**, 219 (1978).
Horton, W., Choi, D. I. and Koch, R. A., *Phys. Rev.* **A14**, 424 (1976).
Hoyng, P., *Astron. Astrophys.* **55**, 23 (1977a).
Hoyng, P., *Astron. Astrophys.* **55**, 31 (1977b).
Hoyng, P., Brown, J. C. and Van Beek, H. F., *Solar Phys.* **48**, 197 (1976).
Hoyng, P. and Melrose, D. B., *Ap. J.* **218**, 866 (1977).
Hoyng, P., Knight, J. W. and Spicer, D. S., *Solar Physics* **58**, 139 (1978).
Hsieh, K. C. and Simpson, S. A., *Ap. J. Letters* **162**, L191 (1970).
Hudson, P. D., *Month. Not. Roy. Ast. Soc.* **131**, 23 (1965).
Hudson, H. S., *Ap. J.* **224**, 235 (1978).
Hudson, H. S., Bai, T., Gruber, D. E., Matteson, J. L., Nolan, P. L. and Petterson, L. E., *Astrophys. J.* (1978).
Hurley, K. and Duprat, G., *Solar Physics* **52**, 107 (1977).
Jokipii, J. R., *Ap. J.* **141**, 480 (1966).
Jokipii, J. R., *Ap. J.* **143**, 961 (1966).
Kadomtsev, B. B., *Plasma Turbulence*, Academic Press, New York (1965).
Kadomstev, B. B., *Reviews of Plasma Physics* **2**, 153 (1966).
Kahler, S., *IAU Symposium* **68**, 221 (1975).
Kakimov and Tsytovitch, T. N., *Soviet Physics JETP* **41**, 47 (1975).
Kane, S. R., Anderson, K. A., Evans, W. D., Klebesadel, R. W. and Laros, J., *Astrophys. J.* **233**, L151 (1979).
Kennel, C. and Kindel, J. M., *J.G.R.* **76**, 3055 (1971).
Kingsep, A. S., Rudakov, L. I. and Sudan, R. N., *Phys. Rev. Letters* **31**, 1482 (1973).
Knight, J. W. and Sturrock, P., *Ap. J.* **218**, 306 (1977).
Korchak, A. A., *Doklady AN U.S.S.R.* **173**, 291 (1967).
Korchak, A. A., *Comments Astrophys. Space Physics* **6**, 57 (1974).
Korchak, A. A., *Solar Physics* **56**, 223 (1978).
Krall, N. A. and Liewer, P. C., *Phys. Rev.* **A4**, 2094 (1971).
Krall, N. A. and Smith, D. F., *Ap. J.* **199**, 500 (1975).
Kuijpers J. M., *Astron. Astrophys.* **69**, L9 (1978).
Kuijpers, J. M., Preprint (1979).
Kulsrud, R. M. and Pearce, W. P., *Ap. J.* **156**, 303 (1969).
Kulsrud, R. M. and Ferrari, A., *Astrophysics and Space Science* **12**, 302 (1971).
Kundu, M. R., Velusamy, T. and Becker, R. M., *Solar Physics* **34**, 217 (1974).
Lacombe, C., *Astron. Astroph.* **54**, 1 (1977).
Lacombe, C., Thesis, University of Paris VII (1979).
Lacombe, C. and Mangeney, A., *Astron. Astrophys.* **1**, 325 (1969).
Lampe, M., Mannheimer, W. M., McBride, J. B. and Orens, J. H., *Physics of Fluids* **15**, 2356 (1972).
Lampe, M. and Papadopoulos, K., *Ap. J.* **212**, 886 (1977).
Langmuir, I., *Phys. Rev.* **33**, 954 (1929).
Lee, M. A., Unpublished work: see Ramaty (1979) (1978).
Lifschitz, M. A. and Tsytovitch, V. N., *Nuclear Fusion* **10**, 241 (1970).
Lin, R. P., *Solar Physics* **15**, 453 (1971).
Lin, R. P., *Space Science Rev.* **16**, 189 (1974).
Lin, R. P. and Hudson, H. S., *Solar Phys.* **50**, 153 (1976).
Lin, R. P., Mewaldt, R. A., Stone, E. C., Vogt, R., Paizis, C. and Van Hollebeke, M. A. I., *Bull. Amer. Physical Soc.* **23**, 4-AJ7 (1978).
Lingenfelter, R. E. and Ramaty, R., *High Energy Nuclear Reactions in Astrophysics*, B. S. P. Shen (ed.) W. A. Benjamin, New York, p. 99 (1967).
McBride, J. B., Ott, E., Boris, J. P. and Orens, J. H., *Physics of Fluids* **15**, 2367 (1972).
McGuire, R. E., von Rosenvinge, T. T. and McDonald, F. B., *Topical Conference on Cosmic Ray Astrophysics*, Durham (1978).

McGuire, R. E., Von Rosenvinge, T. T. and McDonald, F. B., Preprint (1979).
Mannheimer, W. M., *Physics of Fluids* **20**, 265 (1977).
Mätzler, C., in *Pleins feux sur la physique solaire* (ed.) C.N.R.S., 251 (1978a).
Mätzler, C., *Astron. Astrophys.* **70**, 181 (1978b).
Mätzler, C., Bai, T., Crannell, C. J. and Frost, K. J., *Ap. J.* **223**, 1058 (1978).
Magelssen, G. R. and Smith, D. F., *Solar Physics* **55**, 211 (1977).
Melrose, D. B., *Astrophysics and Space Science* **2**, 171 (1968).
Melrose, D. B., *Astrophysics and Space Science* **4**, 143 (1969).
Melrose, D. B., *Plasma Astrophysics*, Gordon and Breach, New York (1980).
Melrose, D. B., *Solar Physics* **37**, 353 (1974).
Meyer, J. P., Cassé, M. and Reeves, H., 16*th Int. Cosmic Ray Conf.* **12**, 108 (1979).
Meyer, J. P., comm. at 22*nd Liege astrophysical symposium*, 20–22 June 1978 (1978).
Moreton, G. and Severny, A., *Solar Physics* **3**, 282 (1968).
Morrison, P., Olbert, S. and Rossi, B., *Phys. Rev.* **94**, 440 (1954).
Norman, C. and Smith, R., *Astron. Astrophys.* **68**, 145 (1978).
Northcliff, L. C. and Schilling, R. F., *Nuclear Data Tables*, **A7**, 223 (1970).
Novick, R. and Wolff, R. S., *Proc. IAU Symposium* **41** (1971).
Novick, R., Weisskopf, M. C., Berthelsdorf, R., Linke, R. and Wolff, R. S., *Ap. J.* **174**, L1 (1972).
Orens, J. H. and Dawson, J. M., NRL Memorandum, 2850 (1974).
Orrall, F. Q. and Zirker, J. B., *Ap. J.* **208**, 618 (1976).
Papadopoulos, K., *Rev. Geophysics & Space Physics* **15**, 113 (1977).
Papadopoulos, K. and Coffey, T., *J.G.R.* **79**, 674 (1974a).
Papadopoulos, K. and Coffey, T., *J.G.R.* **79**, 1558 (1974b).
Papadopoulos, K., Goldstein, M. L. and Smith, R. A., *Ap. J.* **190**, 175 (1974).
Papadopoulos, K., Hui, B. and Winsor, N., *Nuclear Fusion Letters* **175**, 1087 (1977).
Parker, E. N. and Tidman, D. A., *Phys. Rev.* **111**, 1206 (1958).
Pikel'ner, S. B., *Soviet Astronomy* **18**, 136 (1974).
Pikel'ner, S. B. and Tsytovitch, V. N., *Soviet Astronomy* **19**, 450 (1975).
Priest, E. R. and Heyvaerts, J., *Solar Physics* **36**, 433 (1974).
Priest, E. R. and Sanderson, J. J., *Plasma Physics* **14**, 951 (1972).
Pustilnik, L. A., *Soviet Astronomy* **22**, 350 (1978).
Quon, B. H. and Wong, A. U., *Phys. Rev. Letters* **37**, 1393 (1976).
Raju, P. K., *MNRAS* **139**, 479 (1968).
Ramaty, R., *Ap. J.* **158**, 753 (1969).
Ramaty, R., Invited paper at la Jolla Workshop on *Particle Acceleration Mechanisms in Astrophysics*, January 1979 (1979).
Ramaty, R. and Lingenfelter, R. E., *J.G.R.* **72**, 879 (1967).
Ramaty, R., Koslovsky, B. and Lingenfelter, R., *Space Science Rev.* **18**, 341 (1975).
Ramaty, R. and Crannel, C., *Ap. J.* **203**, 766 (1976).
Ramaty, R., Kozlovski, B. and Suri, A. N., *Ap. J.* **214**, 617 (1977).
Ramaty, R., Colgate, Dulk, Hoyng, Knight, Liu, Melrose, Paizis, Orrall, Shapiro, D. Smith and Van Hollebecke, Chapter 4 of *Solar Flares*, P. Sturrock (ed.), Colo. Ass. Univ. Press (1980).
Rochester, L. S. and Stix, T., *Phys. Rev. Letters* **36**, 11 (1976).
Rosenbluth, M. and Liu, C. S., *Physics of Fluids* **19**, 815 (1976).
Rudakov, L. I. and Korablev, L. V., *Soviet Phys. JETP* **23**, 145 (1966).
Rudakov, L. I., *Soviet Phys. Doklady* **17**, 1166 (1973).
Rust, D. in *Solar System Plasma Physics*, Kennel, C. F., Lanzarotti, L. J. and Parker, E. N. (eds), p. 51 (1979).
Sagdeev, R. Z., *Rev. of Plasma Physics* **4**, 23 (1966).
Sanderson, J. J. and Priest, E. R., *Plasma Physics* **14**, 959 (1972).
Schlaeffer, O. A. and Zahrringer, J., *Phys. Rev. Letters* **8**, 389 (1962).
Schnack, D. and Killeen, J., U.C.L.A. preprint 79, 261 (1977).
Schnack, D. and Killeen, J., *Theoretical and Computational Plasma Physics*. I.A.E.A., Vienna, p. 337 (1978).
Shapiro, M. M., Silberberg, R. and Tsao, G. H., *Proceedings* 14*th International Cosmic Ray Conference*, 2532 (1975).
Shapiro, P. R. and Knight, J. W., *Ap. J.* **224**, 1028 (1978).

Shapiro, P. R. and Moore, R. J., *Ap. J.* **217**, 621 (1977).
Shmeleva, O. P. and Syrovatsky, S., *Solar Physics* **33**, 341 (1973).
Skilling, J., *M.N.R.A.S.* **173**, 245 and **172,** 557 (1975).
Slottje, C., *Nature* **257**, 520 (1978).
Smith, D. F., *Ap. J.* **212**, 891 (1977a).
Smith, D. F., *Ap. J.* **217**, 644 (1977b).
Smith, D. F., *J.G.R.* **82**, 704 (1977c).
Smith, D. F., Preprint (1979).
Smith, D. F. and Lilliequist, C. G., *Ap. J.* (1978).
Smith, D. F. and Priest, E. R., *Ap. J.* **176**, 487 (1972).
Smith, R. A. and Goertz, C. K., *J.G.R.* **83**, 2617 (1978).
Sokolov, V. S. and Kosovichev, A. G., *Solar Physics* **57**, 73 (1978).
Spicer, D., *Solar Physics* **53**, 305 (1977a).
Spicer, D., *Solar Physics* **54**, 379 (1977b).
Spicer, D., Naval Research lab. memo report 3749 (1978).
Stenzel, R. L., *Physics of Fluids* **21**, 93 and **21**, 99 (1978).
Stix, T., *The Theory of Plasma Waves*, McGraw Hill, Inc., New York (1962).
Swann, W. F. G., *Phys. Rev.* **43**, 217 (1933).
Swift, D. W., *J.G.R.* **80**, 2096 (1975).
Syrovatsky. S. I., *JETP* **40**, 1788 (1975).
Syrovatsky, S. I., *Solar Terr. Physics* **1**, Dyer, M., (ed.) p. 119 (1970).
Syrovatsky, S. I., Franck, A. G. and Khodzkaev, A. Z., *Soviet Physics Tech. Physics* **18**, 580 (1973).
Tange, T. and Ichimaru, S., *J. Phys. Soc. Jap.* **36**, 1437 (1974).
Thomson, J. J., Faehl, R. J. and Kruer, W. L., *Phys. Rev. Letters* **31**, 918 (1973).
Tidman, D. A., *J. Planet, Space Science* **13**, 781 (1965).
Tidman, D. A., Birmingham, T. J. and Stainer, H. M., *Ap. J.* **146**, 207 (1966).
Tidman, D. A. and Krall, N. A., *Shock Waves in Collisionless Plasmas*, Wiley Interscience, New York (1971).
Tindo, I. P. and Somov, B. V., Preprint (1977).
Tindo, I. P., Valmicek, B., Livshits, M. A. and Ivanov, Y. D., *Solar Physics* **27**, 2 (1972).
Title, A. M. and Andelin, J. P., *IAU symposium* **43**, 298 (1971).
Trubnikov, B. A., *Plasma Physics* **1**, 105 (1965).
Trulsen, J. and Fejer, J. A., *J. Plasma Physics* **4**, 825 (1970).
Tsytovitch, V. N., *Non-linear Effects in Plasmas*, Plenum Press, New York (1970).
Tsytovitch, V. N., *Plasma Physics* **13**, 741 (1971).
Tsytovitch, V. N., Stenflo, J. and Wilhelmson, H., *Physica Scripta* **11**, 251 (1975).
Tverskoi, B. A., *Soviet Physics JETP* **25**, 317 (1967).
Tverskoi, B. A., *Soviet Physics JETP* **26**, 821 (1968).
Van Beek, H. F., Hoyng, P. and Stevens, G. A., W.D.C.A. report, U.A.G., **28**, 319 (1973).
Van Hollebeke, M. A. I., Ma Sung, L. S. and McDonald, F. B., *Solar Physics* **41**, 189 (1975).
Van Regemorter, H., *Ap. J.* **136**, 906 (1968).
Vedenov, A. A. and Rudakov, L. I., *Soviet Physics Doklady* **9**, 1075 (1965).
Vedenov, A. A. and Ryoutov, D. D., *Reviews of Plasma Physics* **6** (1975).
Vekshtein, G. E., Sagdeev, R. Z. and Ryoutov, D. D., *Soviet Physics JETP Letters* **12**, 291 (1970).
Vlahos, L. and Papadopoulos, K., *Astrophys. J.* **233**, 717 (1979a).
Vlahos, L. and Papadopoulos, K., *Astrophys. J.* **234**, L217 (1979b).
Völk, H. J., *Astrophysics and Space Science* **25**, 471 (1973).
Wang, H. T. and Ramaty, R., *Solar Physics* **107**, 1065 (1974).
Weber, W. R., Damle, S. V. and Kish, J., *Astronomy and Space Science* **15**, 245 (1972).
Weinstock, J., *Physics of Fluids* **12**, 1045 (1969).
Wentzel, D. G., *Ap. J.* **156**, 303 (1969).
Wentzel, D. G., *Astron. Astrophys.* **12**, 71 (1974).
Wild, J. P., Smerd, S. F. and Weiss, A. A., *Annual Review of Astron.* **1**, 291 (1963).
Wong, A. Y. and Quon, B. H., *Phys. Rev. Letters* **34**, 1499 (1975).
Zakharov, V. E., *Soviet Physics JETP* **35**, 908 (1972).
Zhelezniakoff, V. V., *Soviet Astronomy* **1**, 33 (1967).
Zirin, H., *Solar Phys.* **58**, 95 (1978).

Index

Absorbing feature 101
Abundance ratio 117
Acceleration
- electron 121, 472–475
- Fermi 518–537
- first-phase 42–44, 432, 468–516
- particle 10, 13, 94, 210, 212, 271, 429–551
- proton 49, 447
- second-phase 3, 44–45, 83–85, 113, 433–434, 517–551
- stochastic 471–472

Active region 113–114
Active-region (or plage) filament 3, 368
Active-region loop 219–223, 278, 296–302
Adiabatic change 28, 439
Alfvén-Carlqvist model 17, 24, 496
Alfvén Mach number 185–186
Alfvén speed 29, 140
Alfvén wave 18, 182, 526–529
Alfvén-wave model 18–19
Ampère's law 26
Annihilation, magnetic 172–176
Antidiffusion mode 164
Arcade 15, 76–77, 301, 369, 393, 408
Arcade instability 368–372
Arch (see loop)
Arch-filament system 62, 207
Aspect ratio 237
Asymmetry, line 69–70
Atmospheric model (*see also* loop model) 37–38, 292–302

Beam, particle 24, 289, 291, 455–469
Beta 29, 223
Betatron effect 523, 531
Bipolar field (*see also* magnetic dipole) 57
Boundary conditions 179, 230, 288–289, 374–376
- Dirichlet 343, 349, 360
- Neumann 360

Bremsstrahlung 48, 80, 126, 433, 441, 446–447, 454–455
Brightness temperature 445
Buneman instability 474–475, 502, 504–505
Burst
- elementary flare 82, 87
- EUV 454–455
- extended X-ray 433
- hard X-ray 2, 11, 14, 34, 48, 74, 81–85, 87, 100, 125, 289–290, 433–441
- impulsive X-ray 433–437

Burst (*continued*)
- microwave 2, 49, 58, 74, 85–87, 114, 125, 437, 442–446
- radio 95–101
- radio type II 5, 34, 49, 99, 108
- radio type III 5, 18, 34, 49, 100–101, 119
- radio type IV 5, 34, 49, 95–98
- radio type V 101
- soft X-ray 73–75, 280
- travelling 87

Catastrophe (*see also* nonequilibrium) 343–346, 360
Characteristics, method of 398
Chromospheric flare (*see also* Hα flare) 64–73
Colgate's model 18, 273
Collisionless tearing mode 170–171
Compact (or simple-loop) flare 3, 4, 12, 30, 33, 35–39, 52–55, 58, 64–73, 76–77, 93, 194–207, 217–334
Complex variable 153–158
Conduction, thermal 12, 25, 68, 94–95, 305–315, 466
Conduction cooling 309–310
Conduction front 467
Conduction speed 17, 20
Conductive flare model 293–295, 305–315
Conductive time-scale 37, 287, 323, 327, 419, 464–465
Conductivity, electrical 16, 26, 339
Conformal mapping 356
Constant-pitch (or uniform-twist) magnetic field 228, 245–246, 371
Continuity equation 26, 285, 398
Continuum storm 98
Cooling time (*see also* radiative and conductive time-scale) 78–79
Cosmic rays 52, 60
Coronal arcade 15, 76–77, 301, 369, 393, 408
Coronal condensation 393
Coronal density 7, 34, 35
Coronal enhancement 103
Coronal heating 18, 20, 169, 296–297
Coronal mass 416
Coronal transient 5, 11, 34, 104–106, 405
Correlation function 477
Cosmic-ray flare 52, 60, 112–114
Cosmic rays 448–454
Coulomb collisions 459
Current concentration 507–509

Current density 16, 139–140, 160, 189, 194
Current-driven instability 460–461, 475–476, 480–482
Current filament 16, 163, 509
Current-free (or potential) field 15, 62–63, 144, 155–163, 189, 348, 392
Current sheet 14, 16, 19–23, 34–35, 139–213, 347
  curved 157–158, 176
  formation 34–35, 142, 144–171
  width 140, 173, 187–188, 202
Cyclotron resonance 524–526

Density
  coronal 7, 34, 35
  electron 440
  flare 10, 33, 65–67, 77, 79, 284, 322, 391–392. 413–414, 424
  transition-region 7, 90–94
Differential emission measure 281–284, 297–300, 307, 309–316, 333–334
Diffuse pinch 235–237
Diffusion region 165, 175, 180–182, 186–187, 192, 194, 206
Diffusion (or dissipation) time-scale 16, 19, 24, 164
Dimensional analysis 287–288
Dipole, magnetic 155–162
Dirichlet boundary condition 343, 349, 360
Disparition brusque (or filament eruption) 5, 11, 12, 102–107, 115, 196, 211, 371, 381, 405–410
Dissipation spreading 470–471
Dissipation (or diffusion) time-scale 16, 24, 164
Doppler broadening 89
Double-layer 17, 496–506
Double tearing mode 171, 255, 265
Downflow 77, 312, 317, 389, 393
Dreicer (or runaway) field 210, 469, 473–474, 512–515

Eigenvalue problem 363, 367
Ejecta, flare 102–112
Electric current density 16, 139–140, 160, 189, 194
Electric field 66, 268–269, 374, 469–470, 513, 545
Electrical conductivity 16, 26, 339
Electron acceleration (*see also* acceleration) 121, 472–475
Electron beam (*see also* particle beam) 430, 482
Electron collisional heating 289–291, 296
Electron density 440
Electron energy spectrum 122
Electron number 441–442, 454–455, 511
Electron spectrum 447, 520–522
Electrons, non-relativistic 119–121
Electrons, relativistic 118–119
Elementary flare burst 82, 87
Ellerman bomb (or moustache) 111
Emerging flux 7–9, 33, 61–64, 76, 205, 409
Emerging flux model 19, 21, 23, 24, 30, 35, 62, 158–159, 194–212
Emission measure 75, 90–93, 280, 318, 413, 439
  differential 281–284, 297–300, 307, 309–316, 333–334
Energetics, tearing 260–271
Energy
  flare 6, 10, 12, 54, 106, 208, 265
  magnetic 15–16, 62–63, 139, 158, 199, 420–424
  particle 441–442
Energy equation 26, 27, 198–200, 286, 297, 314
Energy (or variational) principle 231–235, 364–370
Energy storage 21
Energy transfer 94–95, 305, 308
Ephemeral active region 57
Equation
  continuity 26, 285, 398
  energy 26, 27, 198–200, 286, 297, 314
  Euler-Lagrange 234, 245, 364, 370
  hydrodynamic 284–292
  induction 25, 27, 172
  Liouville 462
  magnetohydrodynamic 25–29, 338–342
  motion 25, 27, 285, 396
Eruption, filament 5, 11, 12, 102–107, 115, 196, 211, 371, 381, 405–410
Eruptive instability 31–32, 376
Euler-Lagrange equation 234, 245, 364, 370
Euler potential 346, 372–375
EUV burst 454–455
EUV flare 10, 34, 48, 65, 88–94, 278–279
EUV loop 11, 217–219
EUV spectrum 89–90, 279–284
Evaporation 12, 211, 296, 299, 307, 310–312, 321, 323–324, 394–395, 417–418
Evolving flux 197
Evolving magnetic features 61
Existence condition 342–346
Explosive instability 13, 303
Explosive phase 55–56

Extended X-ray burst 433

Fast (or high-energy) particles 72, 112–127, 509–516
Fe-rich flare 448
Fermi acceleration 518–537
Fibrils 63, 208
Filament 3,
    active-region (or plage) 3, 368
    current 16, 163, 509
    quiescent 3, 102–103
    winking 102, 108
Filament activation 63–64, 78, 102–104, 390
Filament eruption (or disparition brusque) 5, 11, 12, 102–107, 115, 196, 211, 371, 381, 405–410
Filamentary structure 69
Fireball 455–468
First-phase acceleration 42–44, 432, 468–516
Flare
    chromospheric (*see also* Hα flare) 64–73
    compact 3, 4, 12, 30, 33, 35–39, 52–55, 58, 64–73, 76–77, 93, 194–207, 217–334
    cosmis-ray 52, 60, 112–114
    EUV 10, 34, 48, 65, 88–94, 278–279
    Hα 2, 8–10, 49–50, 55–56
    homologous 7, 59
    limb 65–68
    proton 34, 112–116
    simple-loop 3, 4, 12, 30, 33, 35–39, 52–55, 58, 64–73, 93, 194–207, 217–334
    spotless 57, 78
    sympathetic 7, 107
    thermal 3, 8, 33, 48, 73–81, 94, 205
    two-ribbon 3–14, 31–33, 39–42, 52–55, 59, 76–80, 91, 196–197, 207–212, 304, 337–425
    umbral 58
    white-light 33, 49, 71–73, 121, 125, 454
    X-ray 10, 75–80, 91, 278–279
Flare area 50
Flare classification 50–52
Flare core 10
Flare density 10, 33, 65–67, 77, 79, 284, 322, 391–392, 413–414, 424
Flare duration (or lifetime) 51, 77–78, 140
Flare ejecta 102–112
Flare energy (*see also* magnetic energy) 6, 10, 12, 54, 106, 208, 265
Flare height 76
Flare importance 50
Flare intensity 2
Flare kernel 59, 68, 315–317
Flare knot 56
Flare magnetic field 56–64, 285, 392, 413, 510
Flare model (or mechanism) 14, 29–32, 193–213
    Alfvén-Carlqvist 17, 24, 496
    Alfvén-wave 18–19
    Colgate 18, 273
    conductive 293–295, 305–315
    emerging flux 19, 21, 23, 24, 30, 35, 62, 158–159, 194–212
    hydrodynamic 317–332
    numerical 38, 317–332
    Piddington 18, 24
    Spicer 18, 272–273
    Sturrock 19, 22–24, 212, 394
    Sweet 20, 22
    Syrovatsky 20–21, 24, 148–150
Flare observations 33–34, 47–127, 278–284, 380–392
Flare occurrence rate 60
Flare spectrum 64–73
Flare temperature 10, 47–48, 67–68, 75, 79, 280, 304, 322–330, 390, 424
Flare thickness 67–69
Flare transition-layer 48, 88–95, 321, 332
Flare velocity 284
Flare volume 76
Flare (or Moreton) wave 5, 107–110
Flash phase 2–5, 22, 55–56, 198, 211, 305
Flux function 342
Flux loop (*see also* loop) 14, 15, 17–18
Footpoint displacement 344–346, 349–360
Force-free field 10, 23, 36, 39–40, 62–63, 151, 208, 222–226, 239–242, 245–247, 346–362, 365–372
    constant-α 224–225, 239–240, 251
Forerunner 11, 34, 106
Frozen-in flux 148, 154, 156, 180, 254, 375

Gamma rays 8, 48–49, 121–127, 446–448, 522
Geomagnetic substorm 213
Global instability 235–237
Gravitational mode (*see also* resistive instability) 167
Green's function 172
Gyrosynchrotron radiation 433, 440

$H_{\parallel}=0$ (or polarity inversion) line 8, 58, 60–61, 63
Hα flare 2, 8–10, 49–50, 55–56

Hα kernel 9
Hα knot 8, 14, 56, 207
Hα ribbon 11, 31, 207, 211, 379, 386, 388–391, 404
Hard X-ray burst 2, 11, 14, 34, 48, 74, 81–85, 87, 100, 125, 289–290, 433–441
Hard X-ray source 434–437, 440
Heating coronal 18, 20, 169, 296–297
Height of flare 76
Helium-rich event 448–449, 537–544
Helmet streamer 22, 394
Homologous flare 7, 59
Hydrodynamic equations 284–292
Hydrodynamic flare model 317–332
Hydrodynamic oscillation 323, 326–327, 329–331
Hydrodynamic time-scale 287–288, 323

Impulsive phase 2, 5, 12, 20, 22, 47–48, 89, 149, 197–199, 210, 289
Impulsive X-ray burst 433–437
Incompressible perturbation 233
Induction equation 25, 27, 172
Instability 13, 163–171,
- arcade 368–372
- Buneman 474–475, 502, 504–505
- collisionless tearing-mode 170–171
- current-driven 460–461, 475–476, 480–482
- double tearing-mode 171, 255, 265
- eruptive 31–32, 376
- explosive 13, 303
- global 235–237
- ideal 163, 248
- interchange 176–177
- ion-acoustic 17, 469, 474, 476, 478, 480
- ion-cyclotron 481–482, 543
- kink 15, 194–196, 238–248, 371–372
- local 235–239
- magnetohydrodynamic 35–37, 40, 196, 217–273, 304, 362–372
- micro- (or microscopic) 24, 198, 205, 475–482, 545
- modulational 485
- neutral-point 144–146
- oscillating two-stream 469, 491–492
- radiative (or thermal) 301–302, 319, 321
- resistive (*see also* tearing, rippling and gravitational modes) 13, 164–171, 248–273
- resistive kink 18, 196, 255–260
- superheating 508
- tearing-mode 30, 32, 37, 144, 165,

Instability (*continued*)
- tearing-mode (*continued*) 169–171, 177–178, 190, 196, 200, 208, 250–271, 509–516
- thermal (or radiative) 163–164, 301–302, 319, 321
- two-stream 546–547

Intensity, flare 2
Interchange instability 176–177
Ion-acoustic instability 17, 469, 474, 476, 478, 480
Ion acoustic (or ion-sound) turbulence 20, 149–150, 188, 289–290, 458, 472–473, 483–485, 506–509, 545
Ion-acoustic wave 466
Ion-cyclotron instability 481–482, 543
Ion-cyclotron wave 466, 502–504, 538
Ionization 67–68
Isotopes 123, 448–454

Joule mode 163–164, 508

Kadomtsev spectrum 479
Kernel
- Hα 9,
- flare 59, 68, 315–317
- X-ray 10, 11

Kink instability 15, 194–196, 238–248, 371–372
Knot
- Hα 8, 14, 56, 207
- X-ray 76

Kruskal-Shafranov condition 238, 244

Langrangian variable 292, 294, 327
Landau damping 478, 483, 549
Langmuir turbulence (*see also* turbulence) 482–485, 494, 545
Langmuir (or plasma) wave 482–485, 493–496
Limb flare 65–68
Line asymmetry 69–70
Line-tying 146, 177, 196, 244–248, 369–372
Liouville equation 462
Local instability 235–239
Loop (or arch — *see also* flux loop) 96
- active-region 219–223, 278, 296–302
- EUV 11, 217–219
- "post"-flare (*see also* loop prominence) 5, 7, 11, 31, 33–34, 41–42, 196, 211, 379–425
- thermally isolated 299–300, 331
- twisted 12, 15, 17, 57, 208, 371
- X-ray 10, 11, 75–76, 78, 219–222

Loop magnetic field 222–223
Loop model 296–302

Loop pressure 221
Loop prominence (*see also* "post"-flare loop) 5, 10, 11, 63, 78, 211
Loop stability 229–248, 301–302
Loop superposition 315–316
Loop velocity 379, 381, 386–389, 413, 424
Lorentz force 27, 145
Lower-hybrid wave 546–551
Lundqvist magnetic field 224, 247

Mach number 403
  Alfvén 185–186
Macrospicule 57
Magnetic annihilation 172–176
Magnetic diffusivity 26, 27, 167, 191
Magnetic dipole 155–162
Magnetic energy (*see also* flare energy) 15–16, 62–63, 139, 158, 199, 420–424
Magnetic field
  constant-pitch (or uniform-twist) 228, 245–246, 371
  flare 56–64, 285, 392, 413, 510
  force-free 10, 23, 36, 39–40, 62–63, 151, 208, 222–226, 239–242, 245–247, 346–362, 365–372
  loop 222–223
  Lundqvist 224, 247
  potential 15, 62–63, 144, 155–163, 189, 348, 392
  variable-twist 229
Magnetic field gradient, 59–60, 78, 113
Magnetic field strength 15, 36
Magnetic induction 25
Magnetic neutral (or polarity inversion) line 8, 58, 60–61, 63
Magnetic permeability 26
Magnetic pressure 25, 27, 145
Magnetic reconnection (see reconnection)
Magnetic Reynolds number 16, 29, 188–193, 218, 250
Magnetic scalar potential 159
Magnetic tension 25, 27, 145
Magnetic vector potential 146–147, 183, 342
Magnetoacoustic (or magnetosonic) wave 18
Magneto-gravitational mode 410
Magnetohydrodynamic (or MHD) equations 25–29, 338–342
Magnetohydrodynamic (or MHD) instability 35–37, 40, 196, 217–273, 304, 362–372
Magnetostatic equilibria 36, 39–41, 337–377
Main phase 2, 12, 149–150, 197–199, 211
Marginal stability 231–232, 235, 410, 471
Mass, post-flare loop 414–415, 425
Mass motions (*see also* flare ejecta) 94–95, 287
Maximum reconnection rate 180, 185–186, 191
Mean atomic weight 27
Mean-free-path 465, 467, 525, 527–529
Mechanism
  flare 14, 29–32, 193–213
  Petschek 19, 35, 182–186, 191–193
Mercier criterion 238
Metastable 13, 23, 199, 347
Method of characteristics 398
Micro- (or microscopic) instability 24, 198, 205, 475–482, 545
Microwave burst 2, 49, 58, 74, 85–87, 114, 125, 437, 442–446
MKS units 27
Mode
  antidiffusion 164
  gravitational 167
  Joule 163–164, 508
  magneto-gravitational 410
  rippling 165, 167–168, 178
  tearing 30, 32, 37, 144, 165, 169–171, 177–178, 190, 196, 200, 208, 250–271, 509–516
Mode coupling 479
Model (*see also* flare model)
  atmospheric 37–38, 292–302
  loop 296–302
  two-temperature 291, 318–319
Modulational instability 485
Momentum equation 25, 27, 285, 396
Moreton (or flare) wave 5, 107–110
Moustache (or Ellerman bomb) 111
Multiple equilibria 342–362

Neumann boundary condition 360
Neutral point 20, 142, 144–150, 153, 156–157, 393–395, 420–422
Neutral sheet 141, 178, 344, 395
Neutron 123–124
Newcomb method 232–235
Non-equilibrium (*see also* catastrophe) 35, 142, 150–155
  thermal 3, 30, 194, 300, 302–305
Nuclear reaction 8, 121–126
Nuclei 116–118, 448–454, 537–544
Nuclei abundance 118
Numerical flare model 38, 317–332
Numerical reconnection experiment 190–193, 195
Numerical scheme 292

Ohm's law 25, 26, 338–339
Ohmic dissipation 16, 25, 420, 459
Oscillating two-stream instability 469, 491–492
Oscillation, hydrodynamic 323, 326–327, 329–331

Particle acceleration (see also acceleration) 10, 13, 94, 210, 212, 271, 429–551
Particle beam (see also electron beam) 24, 289, 291, 455–469
Particle energy 441–442
Particle energy spectrum 115–116, 431, 534–537
Particle, fast (or high-energy) 72, 112–127, 509–516
Particle number 430
Particle propagation 429–430
Passive evolution 372–376
Perfect gas law 26
Periodic variations 86–88
Petschek mechanism 19, 35, 182–186, 191–193
Piddington model 18, 24
Pinch
- diffuse 235–237
- reversed-field 226, 237, 239, 244, 260, 265
- z- 225, 229, 273

Pitch 225
Plage (or active-region) filament 3, 368
Plasma beta 29, 223
Plasma turbulence (*see also* turbulence) 16, 17, 24
Plasma (or Langmuir) wave 482–485, 493–496
Plasmoid 96–97
Polarity inversion (or $H_{\|}=0$) line 8, 58, 60–61, 63
Polarization 84, 86, 95–97, 446, 456–457
Polytropic gas 173
"Post"-flare loop (*see also* loop prominence) 5, 7, 11, 31, 33–34, 41–42, 196, 211, 379–425
"Post"-flare loop mass 414–415, 425
Potential (or current-free) field 15, 62–63, 144, 155–163, 189, 348, 392
Power law 116, 119–120, 290, 431, 441, 512, 534, 536
Preflare phase 2, 5, 12, 197–198, 209–210
Pressure (or sound) wave 321–322
Prominence (*see* filament)
Proton/electron ratio 126–127
Proton stream 115
Pulsation 98
Pure electron event 431

Quasilinear theory 478, 490, 494, 524–531
Quiescent filament 3, 102–103

Radiative (or thermal) instability 163–164, 301–302, 319, 321
Radiative loss function 27, 286
Radiative time-scale 37, 287, 323, 388, 414, 419, 464
Radio burst 95–101
- type II 5, 34, 49, 99, 108
- type III 5, 18, 34, 49, 100–101, 119
- type IV 5, 34, 49, 95–98
- type V 101

Rankine-Hugoniot relations 404
Reconnection 8, 13, 19, 33, 35, 57, 142–143, 177–193, 212, 248–250, 392–405, 411–413
- fast 181–186
- laboratory experiment 188–190, 516
- maximum rate 180, 185–186, 191
- nonlinear 179–188, 263–267
- numerical experiment 190–193, 195

Relativistic electrons 118–119
Resistive instability (*see also* tearing, rippling and gravitational modes) 13, 164–171, 248–273
Resistive kink instability 18, 196, 255–260
Resonance, cyclotron 524–526
Return (or reverse) current 11, 289–291, 457–469
Reversed-field pinch 226, 237, 239, 244, 260, 265
Ribbons, Hα 11, 31, 207, 211, 379, 386, 388–391, 404
Rippling mode (*see also* resistive instability) 165, 167–168, 178
Runaway effect 149, 472–475, 478
Runaway (or Dreicer) field 210, 469, 473–474, 512–515

Safety factor 226
Satellite sunspot 8, 11
Saturation 478–480, 548
Scale-height 340, 342
Scaling law 299
Scattering 524, 530–531, 533
Second-phase acceleration 3, 44–45, 83–85, 113, 433–434, 517–551
Sector boundary 115
Self-similar (or similarity) solution 146, 173–174, 183–186, 306–308, 488
Separable solution 358
Separatrix 162, 188

Sheared magnetic field 33, 59, 62–63, 169, 225
Shell model 69
Schrödinger potential 363, 367
Shock wave 5, 19, 99, 141, 312, 319, 396, 403–405, 420, 532–537, 544–551
 perpendicular 544
 slow magnetoacoustic 143, 182, 189, 191–194, 211, 403
Similarity (or self-similar) solution 146, 173–174, 183–186, 306–308, 488
Simple-loop (or compact) flare 3, 4, 12, 30, 33, 35–39, 52–55, 58, 64–73, 76–77, 93, 194–207, 217–334
 thermal evolution 277–334
Singular layer 234, 236, 249, 251, 259
Skin depth 170
Skylab 10–13, 49, 75–79, 91, 104, 108, 126, 218, 379–392, 424
Slow wave 108–110
Slow magnetoacoustic wave (*see also* shock wave) 18
Soft X-ray burst 73–75, 280
Soft X-rays 2, 11, 48, 73–81
Solar cycle 114
Solar wind 393
Soliton 488–490
Sound speed 28, 398
Sound (or pressure) wave 321–322
Source, hard X-ray 434–437, 440
Specific heats ratio 27
Spectral index 438, 443, 515
Spectral power density 477
Spectrum
 electron 447, 520–522
 EUV 89–90, 279–284
 flare 64–73
 Kadomtsev 479
 particle energy 115–116, 431, 534–537
 X-ray 80–81, 279–284
Speed
 Alfvén 29, 140
 conduction 17, 20
 sound 28, 398
Spicer's model 18, 272–273
Spotless flare 57, 78
Spray 12, 106–107, 111–112
Stability, magnetohydrodynamic 229–248
Static model atmosphere 37–38
Stochastic acceleration 471–472
Streamer, helmet 22, 394
Strong turbulence 485–493
Sturm's theorem 235–236
Sturrock's model 19, 22–24, 212, 394
Subflare 11, 50, 58–59, 283
Sunspot 7, 10, 348
 satellite 8, 111
Superheating instability (*see also* thermal instability) 508
Surge 3, 12, 111–112
Suydam criterion 236, 238
Sweet's model 20, 22
Sympathetic flare 7, 107
Synchrotron radiation 97–98, 443
Syrovatsky's model 20–21, 24, 148–150

Tangential discontinuity 140
Tearing energetics 260–271
Tearing-mode instability (*see also* resistive instability) 30, 32, 37, 144, 165, 169–171, 177–178, 190, 196, 200, 208, 250–271, 509–516
Temperature
 brightness 445
 flare 10, 34, 47–48, 67–68, 75, 79, 280, 304, 322–330, 390, 424
Temperature minimum 7, 71, 95
Thermal conduction 12, 25, 68, 94–95, 305–315, 466
Thermal conduction coefficient 28, 204, 286, 466
Thermal conduction tensor 27, 339
Thermal flare 3, 8, 33, 48, 73–81, 94, 205
Thermal (or radiative) instability 163–164, 301–302, 319, 321
Thermal nonequilibrium 3, 30, 194, 300, 302–305
Thermally isolated loop 299–300, 331
Thick target 83–84, 88, 441–442, 454–455, 523
Thin target 83–84, 441, 522
Time-scale
 conductive 37, 287, 323, 327, 419, 464–465
 cooling 78–79
 diffusion (or dissipation) 16, 19, 24, 164
 hydrodynamic 287–288, 323
 magnetohydrodynamic 229
 radiative 37, 287, 323, 388, 414, 419, 464
Tokamak 237–239
Topological dissipation 151
Toroidal effects 237–239
Transient, coronal 5, 11, 34, 104–106, 405
Transition layer, flare 48, 88–95, 321, 332
Transition-region density 7, 90–94
Travelling burst 87

Trigger 7, 13, 61, 62, 64, 88, 197, 199, 204–205, 303, 377, 482, 507
Turbulence 198, 203–205, 209
 ion-acoustic (or ion-sound) 20, 149–150, 188, 289–290, 458, 472–473, 483–485, 506–509, 545
 Langmuir 482–485, 494, 545
 plasma 16, 17, 24
 strong 485–493
Turbulent velocity 89
Twist 18, 151, 226, 228–229, 246, 369, 371
Twisted loop (or flux tube) 12, 15, 17, 57, 208, 371
Two-ribbon flare 3–14, 31–33, 39–42, 52–55, 59, 76–80, 91, 196–197, 207–212, 304, 337–425
Two-stream instability 546–547
Two-temperature model 291, 318–319

Umbral flare 58
Uniform-twist (or constant-pitch) magnetic field 228, 245–246, 371

Variable-twist magnetic field 229
Variational (or energy) principle 231–235, 364–370
Velocity
 flare 284
 loop 379, 381, 386–389, 413, 424
 photospheric 10, 12, 60–61
 turbulent 89
Velocity shear 9
Viscosity 177
Vlasov theory 364
Volume, flare 76

Wave
 Alfvén 18, 182, 526–529

Wave (*continued*)
 flare (or Moreton) 5, 107–110
 ion-acoustic 466
 ion-cyclotron 466, 502–504, 538
 Langmuir (or plasma) 482–485, 493–496
 lower hybrid 546–551
 magnetoacoustic (or magnetosonic) 18
 Moreton (or flare) 5, 107–110
 plasma 482–485, 493–496
 pressure (or sound) 321–322
 shock 5, 19, 99, 141, 312, 319, 396, 403–405, 420, 532–537, 544–551
 slow 108–110
White-light flare 33, 49, 71–73, 121, 125, 454
Winking filament 102, 108

X-ray arch (or loop) 10, 11, 75–76, 78, 219–222
X-ray bright point 3, 50, 56–57, 196
X-ray burst
 hard 2, 11, 14, 34, 48, 74, 81–85, 87, 100, 125, 289–290, 433–441
 soft 73–75, 280
X-ray flare 10, 75–80, 91, 278–279
X-ray flux 52
X-ray illumination 95
X-ray kernel 10, 11
X-ray knot 76
X-ray loop (or arch) 10, 11, 75–76, 78, 219–222
X-ray spectrum 80–81, 279–284
X-rays, soft 2, 11, 48, 73–81

Z-pinch 225, 229, 273